Embedded Microprocessor System Design using FPGAs

Uwe Meyer-Baese

Embedded Microprocessor System Design using FPGAs

 Springer

Uwe Meyer-Baese
Tallahassee, FL, USA

ISBN 978-3-030-50535-6   ISBN 978-3-030-50533-2   (eBook)
https://doi.org/10.1007/978-3-030-50533-2

This Springer imprint is published by the registered company Springer Nature Switzerland AG
The registered company address is: Gewerbestrasse 11, 6330 Cham, Switzerland

To
Anke, Lisa, and my Parents

Preface

Embedded microprocessor systems are everywhere, just look around you. You will find them in your cellular phones, digital clocks, GPS, video recorder, and internet router as well as in household electronic entertainment devices. A modern car typically uses 50–100 microprocessors. Embedded systems are often resource-limited by price, power dissipation, memory, or storage. A general-purpose computer typically uses up hundreds of watts, while a clock or remote control only consumes in microwatts to be able to run for a year on a single AAA battery. Although many embedded systems require low-power dissipation, the implemented algorithms, like the error turbo correction coding used in UMTS phones, are computational demanding. Nevertheless, today, embedded processors perform sophisticated task and run these complex algorithms. The microprocessors in a car use an estimated 100 million lines of code, the GPS and radio alone account for 20 million lines of code.

FPGAs are the best choice to start with embedded system design space exploration since these are fine grain logic programmable COTS devices with much lower NRE costs than cell-based systems available today. The newest generation of FPGA boards and devices allow to design microprocessor systems using soft, parameterized, or hardcore microprocessor all with the same board. These boards are great starting points for many designs since they also have a substantial number of peripheral components such as Audio CODEC, Video HDMI connector, or SD cards. It would be very time consuming to include these components in your project if using a non-FPGA, standard COTS microprocessor system.

At the beginning of the second decade of the twenty-first century, we find that the two programmable logic device (PLD) market leaders (Altera/Intel and Xilinx) both report revenues greater than US$2 billion. FPGAs have enjoyed steady growth of more than 20% in the last decade, outperforming ASICs and PDSPs by 10%. This comes from the fact that FPGAs have many features in common with ASICs, such as reduction in size, weight, and power dissipation; higher throughput; better security against unauthorized copies; reduced device and inventory cost and reduced board test costs; and claim advantages over ASICs, such as a reduction in development time (rapid prototyping), in-circuit (re)programmability, and lower

NRE costs, resulting in more economical designs for solutions requiring less than 1000 units. Another trend in the hardware design world is the migration from graphical design entries to hardware description language (HDL). It has been found that "code reuse" is much higher with HDL-based entries than with graphical design entries. There is a high demand for HDL design engineers and we already find undergraduate classes teaching logic design with HDLs. Unfortunately, only two HDL languages are popular today. The US West Coast and Asia area prefer Verilog, while US East Coast and Europe more frequently use VHDL. For embedded microprocessor design with FPGAs both languages seem to be well suited, although some VHDL examples are a little easier to read because of the supported fixed and floating-point data type in VHDL-2008. Other constraints may include personal preferences, EDA library and tool availability, readability, capability, and language extensions using PLIs as well as commercial, business, and marketing issues, to name just a few. Tool providers acknowledge today that both languages must be supported, and this book covers examples in both design languages. We are now fortunate that "baseline" FPGA tools are available from different sources at essentially no cost for educational use. We take advantage of this fact in this book. It includes code for Altera/Intel Quartus 15.1 Lite Edition as well as Xilinx Vivado 2016.4 tools, which provides a complete set of design tools, from a content-sensitive editor, microprocessor configurator, compiler, and simulator to a bitstream generator. All examples presented are written in VHDL and Verilog and should be easily adapted to other propriety design-entry systems.

The book is structured as follows. The first chapter starts with a snapshot of today's microprocessors and basic microprocessor principles in general and FPGA-based microprocessors in particular. It also includes an overview on design with IP blocks and a PLL IP core design example. The second chapter discusses devices, boards, and tools used to design state-of-the-art FPGA systems. It discusses a detailed case study of the ultimate RISC (URISC) microprocessor, including model discussion, compilation steps, simulation, performance evaluation, power estimation, and floor planning using Quartus and Vivado. This case study is the basis for many other design examples in subsequent chapters. Chapters 3 and 4 deal with VHDL and Verilog language elements that are used in the microprocessor design. Chapter 5 reviews the ANSI C language and also discusses debug methods and differences to C++. In Chap. 6, software tool development for microprocessors is presented. We will cover lexical analysis using GNU Flex and parser implementation with GNU Bison. We will design an assembler for PicoBlaze microprocessor and a basic and full featured C compiler for a 3-address machine. Instruction set simulator and SW debugger are discussed too. In Chap. 7, the softcore PicoBlaze is developed in a step-by-step fashion adding more and more architecture features. Loop control and data memory design are studied and implemented in HDL. The whole instruction set of the most popular 8-bit FPGA-based microprocessor is then fully discussed in Chap. 8. The Chaps. 9 and 10 discuss the two most popular parameterized core for Altera/Intel and Xilinx devices called Nios II and MicroBlaze, respectively. We will develop a top-down and a bottom-up system design approach. Adding custom IP to the microprocessor will be demonstrated with a floating-point co-processor for Nios and an HDMI decoder for the MICROBLAZE. We will also build a Tiny RISC version of the processors called TRISC3N and TRISC3MB that includes the reduced instruction set and can run basic program generated with the vendor GCC tools. In Chap. 11, the most popular 32-bit COTS ARM Cortex-A9 hard IP processor core is discussed, which is part of Altera/Intel and Xilinx's newest devices. Again top-down and bottom-up design are presented and a Tiny RISC version TRISC3A is developed that shows some ideas why ARM Cortex-A9 is a superior architecture. A CIP to accelerate FFT address computation is designed and speed-up is measured. The Appendix A contains for all five (tiny) processor models (URISC, TRIC2, TRISC3N, TRISC3MB, and TRISC3A) the Verilog source code for Quartus and Vivado and xsim simulations. Appendix B contains the acronym list. Additional files, HDL language reference cards, and utilities are include in the CD-ROM and will be posted online at GitHub and the author's personal web page.

Acknowledgments

This book is based on an embedded microprocessor system design class I am teaching every year since 2003 at the FAMU-FSU College of Engineering in Tallahassee. I have supervised more than 60 Master's and Ph.D. thesis projects in the last 17 years. I wish to thank all my colleagues who helped me with critical discussions in the lab and at conferences and my students that inspired me with their design ideas, some of which are discussed in the game designs in Chaps. 9, 10, and 11.

I am particularly thankful to my friends and colleagues from the University of Granada with whom I have had the pleasure of working since 1998 and have also visited them in Spain in the summer of 2014. They recently helped in getting my lab started with the research on FECG analysis. Special thanks to A. Garcia, E. Castillo, L. Parrilla, A. Lloris, and J. Ramírez.

I would like to thank my colleagues from the University of Kassel summer research stay in Germany in 2015 for their effort to help me with the Vivado Xilinx designs and ZyBo boards. Special thanks go to P. Zipf, M. Kumm, and K. Möller.

Special thank also to Francesco Poderico for his confidence in me to continue the support of his PCCOMP compiler for PICOBLAZE.

I would like to thank my colleagues from the ISS at RHTH Aachen for their time and efforts in teaching me LISA and C-compiler design during my Humboldt award– sponsored summer research stay in Germany in 2006 and 2008. Special thanks go to H. Meyr, G. Ascheid, R. Leupers, D. Kammler, and M. Witte.

I am particular thankful to Guillermo Botella and Diego Gonzalez from the University of Madrid for help with the image and video processing applications.

I would like to thank Rebecca Nevin, B. Esposito, M. Phipps, A. Vera, M. Pattichis, and C. Dick at Altera/Intel and Xilinx for software and hardware support and the permission to include datasheets on the CD of this book. From my publisher (Springer-Verlag) I would like to thank Charles Glaser and Dr. Merkle for their continuous support and help over recent years.

    If you find any errata or have any suggestions to improve this book, please contact me at Uwe.Meyer-Baese@ieee.org or through my publisher.

Tallahassee, FL, USA   Uwe Meyer-Baese

April 2020

Contents

Chapter 1
Embedded Microprocessor Systems Basics

Abstract This chapter gives an overview of the algorithms and technology we will use in the book. It starts with an overview of embedded microprocessor system design, and we will then discuss FPGA-based systems in particular.

Keywords Embedded microprocessor · Instruction set · Field programmable gate array · FPGA technology · Intellectual property core · FPGA benchmarks · Memory architecture · Central processing unit (CPU) · Ken Chapman programmable state machine (KCPSM) · Addressing modes · Data flow architectures · Soft core · Parameterized core · Hard core · Phase-locked loop (PLL) · ModelSim · Complex instruction set computer (CISC) · Reduced instruction set computer (RISC) · Application-specific integrated circuit (ASIC) · Cell-based integrated circuit (CBIC)

1.1 Introduction

Embedded systems are usually characterized that they include a microprocessor (μP) but do not have the typical components of a computer like keyboard, monitor, or mouse [VG02, HHF08, Wol08].

Today the majority of microprocessors are employed in embedded systems. Embedded systems are everywhere, just look around you. You will find them in your cellular phones, digital clocks, GPS, video recorder, www router to household end electronic entertainment devices [A08]. A modern car typically uses 50–100 microprocessors [C09]. A few embedded systems may also have real-time constrains such that a certain computation must be done at a certain deadline, like in the ABS system of your car, but most do not. Embedded systems are often resource-limited by price, power dissipation, memory, or storage. A general-purpose computer typically uses hundreds of watts, while a clock or remote control only consumes μW to be able to run a year on a single AAA battery. Although many embedded systems require low-power dissipation, the implemented algorithms, like the error turbo correction coding used in UMTS phones, are computational demanding. Nevertheless, today embedded processors perform sophisticated task and run these complex algorithms. The microprocessors in a car use an estimated 100 million lines of code, the

© Springer Nature Switzerland AG 2021   1
U. Meyer-Baese, *Embedded Microprocessor System Design using FPGAs*,
https://doi.org/10.1007/978-3-030-50533-2_1

GPS and radio alone account for 20 million lines of code [C09]. Typical embedded microprocessor system design goals can be summarized as follows:

- Optimize for power, performance, memory, cost
- Compatibility
- JTAG debug support
- Architecture style: Microcontroller, PDSP, SIMD, MIMD
- Superscalar: Instruction level parallelism, # of function units, coprocessors
- Memory hierarchy
- Instruction set (CISC, RISC, controller)

Given the wide range of embedded systems applications, it's not a surprise that not a single microprocessor can cover all these requirements in hardware (HW) and software (SW) and customization has to be done. This is exactly the job description of a modern embedded system Engineer: To have an intimate knowledge of the hardware (i.e., microprocessor and its peripherals) and the software (i.e., algorithms and coding in computer language such as assembler or C/C++). To master such a complex system design different skills are needed, ranging from digital logic, microprocessors, computer architecture, compiler design, programming languages (VHDL/Verilog/C/C++), DSP/image processing/math algorithms to knowledge of the background on the task to be solved; see Fig. 1.1.

Several milestones in the microprocessor history have contributed to the success of todays embedded systems. The invention of the integrated circuits (ICs) in 1960 along with the early Intel microprocessors and the successively observed true to the Moore law doubling the transistor count every 18 month allows us today to build millions of gates at low cost. The invention of the RISC processor in the 1980 allows to build high performance μP at a fraction of the power consumption of general-purpose processors (GPP). Last but not least, the FPGAs allow us today to design custom μPs with a commercial off-the-shelf (COTS) device without involving a ASIC foundry in the design process. But before we go into details of FPGA-based μP systems, let briefly review important architecture choice of microprocessor.

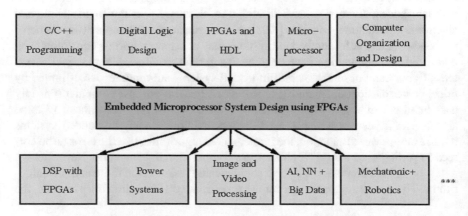

Fig. 1.1 Overview on Embedded Microprocessor prerequisite and possible application areas

Fundamental Microprocessors Aspects

Usually microprocessors are classified into three major classes: the general-purpose or CISC processor, reduced instruction set processors (RISC), and microcontrollers. Let us now have a brief look how these classes of microprocessor have developed and their major differences.

By 1968 the typical general-purpose minicomputers in use were 16-bit architectures using about 200 MSI chips on a single circuit board. The MSI had about 100 transistors each per chip. A popular question [Maz95] then was: can we also build a single CPU with (only) 150, 80, or 25 chips?

At about the same time, Robert Noyce and Gordon Moore, formerly with Fairchild Corp., started a new company first called NM Electronics and later renamed Intel, whose main product was memory chips. In 1969 Busicom, a Japanese calculator manufacture asked Intel to design a set of chips for their new family of programmable calculators. Intel did not have the manpower to build the 12 different custom chips requested by Busicom, since good IC designers at that time were hard to find. Instead Intel's engineer Ted Hoff suggested building a more-general 4-chip set that would access its instructions from a memory chip. A programmable state machine (PSM) with memory was born, which is what we today call a *microprocessor*. After 9 months with the help of F. Faggin the group of Hoff delivered the Intel 4004, a 4-bit CPU that could be used for the BCD arithmetic used in the Busicom calculators. The 4004 used 12-bit program addresses and 8-bit instructions, and it took five clock cycles for the execution of one instruction. A minimum working system (with I/O chips) could be build out of two chips only: the 4004 CPU and a program ROM. The 1 MHz clock allowed additions of multi digit BCD numbers at a rate of 80 ns per digit [FHM96].

Hoff's vision was now to extend the use of the 4004 beyond calculators to digital scales, taxi meters, gas pumps, elevator control, medical instruments, vending machines, etc. Therefore, he convinced the Intel management to obtain the rights from Busicom to sell the chips to others too. By May 1971, Intel gave a price concession to Busicom and obtained in exchange the rights to sell the 4004 chips for applications other than calculators. One of the concerns was that the performance of the 4004 could not compete with state-of-the-art minicomputers at the time. But another invention, the EPROM, also from Intel by Dov Frohamn-Bentchkovsky helped to market a 4004-development system. Now programs did not need an IC factory to generate the ROM, with the associated long delays in the development. EPROMs could be programmed and reprogrammed by the developer many times if necessary.

Some of Intel's customer asked for a more-powerful CPU, and an 8-bit CPU was designed that could handle the 4-bit BCD arithmetic of the 4004. Intel decided to build the 8008, and it also supported standard RAM and ROM devices; custom memory was no longer required as for the 4004 design. Some of the shortcoming of the 8008 design was fixed by the 8080 design in 1974 that used now about 4500 transistors. In 1978, the first 16-bit µP was introduced, the 8086. In 1982 the 80286 followed: a 16-bit µP but with about six times the performance of the 8086. The 80386, introduced in 1985, was the first µP to support multitasking. In the 80387, a mathematics coprocessor was added to speed up floating-point operations. Then in 1989 the 80486 was introduced with an instruction cache and instruction pipelining as well as a mathematics coprocessor for floating-point operations. In 1993, the Pentium family was introduced, with now two pipelines for execution, i.e., a super-scalar architecture. The next generation of Pentium II introduced in 1997 added multimedia extension (MMX) instructions that could perform some parallel vector-like MAC operations of up to four operands. The Pentium 3 and 4 followed with even more advanced features, like hyper threading and SSE instructions to speed up audio and video processing. The largest processor in 2006 was the Intel Itanium with two processor cores and a whopping 592 million transistors. It includes 9 MB of L3 cache alone. The whole family of Intel processors is shown in Table 1.1.

Looking at the revenue of semiconductor companies, it is still impressive that Intel has maintained its lead over many years mainly with just one product, the microprocessor. Other microprocessor-dominated companies like Texas Instruments, Motorola/Freescale, or AMD have much lower revenue. Other top companies such as Samsung or Toshiba are dominated by memory technology, but still do not have Intel's revenue, which has been in the lead for many years now; see Fig. 1.2.

The Intel architecture just discussed is sometimes called a complex instruction set computer (CISC). Starting from the early CPUs, subsequent designs tried to be compatible, i.e., being able to run the same programs. As the bit width of data and programs expanded, you can imagine that this compatibility came at a price: perfor-mance. Moore's law allowed this CISC architecture to be quite successful by adding new components and features like numeric coprocessors, data and program caches, MMX, and SSE instructions and the appropriate instructions that support these additions to improve performance. Intel µPs are characterized by having many

Table 1.1 The family of Intel microprocessors

Name	Year introduced	MHz	IA/bits	Process technology/µm	#Transistors
4004	1971	0.108	4	10	2300
8008	1972	0.2	8	10	3500
8080	1974	2	8	6	4500
8086	1978	5–10	16	3	29K
80286	1982	6–12.5	16	1.5	134K
80386	1985	16–33	32	1	275K
80486	1989	25–50	32	0.8	1.2M
Pentium	1993	60–66	32	0.8	3.1M
Pentium II	1997	200–300	32	0.25	7.5M
Pentium 3	1999	650–1400	32	0.25	9.5M
Pentium 4	2000	1300–3800	32	0.18	42M
Xeon	2003	1400–3600	64	0.09	178M
Itanium 2	2004	1000–1600	64	0.13	592M

IA instruction set architecture

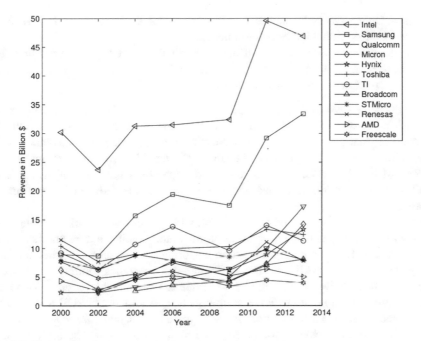

Fig. 1.2 Top semiconductor company revenue

instructions and supporting many addressing modes. The µP manuals are usually 700 or more pages thick.

Around 1980, research from CA University of Berkeley (Prof. Patterson), IBM (later called PowerPC) and at Stanford University (Prof. Henessy, on what later becomes the microprocessor without interlocked pipeline stages, i.e., MIPS µP family) analyzed µPs and came to the conclusion that, from a performance standpoint, CISC machines had several problems. In Berkeley, this new generation of µPs was called RISC-1 and RISC-2 since one of their most important feature was a limited number of instructions and addressing modes, leading to the name reduced instruction set computer (RISC). Let us now briefly review some of the most important features by which (at least early) RISC and CISC machines could be characterized.

- The *instruction set* of a CISC machines is rich, while a RISC machine typically supports fewer than 100 instructions.
- The *word length* of RISC instructions is fixed and typically 32 bits long. In a CISC machine, the word length is variable. In Intel machines we find instructions from 1 to 15 bytes long.
- The *addressing modes* supported by a CISC machine are rich, while in a RISC machine only very few modes are supported. A typically RISC machine supports just immediate and register base addressing.
- The *operands* for ALU operations in a CISC machine can come from instruction words, registers, or memory. In a RISC machine, no direct memory operands are

used. Only memory load/store to or from registers is allowed and RISC machine are therefore called load/store architectures.

*   In *subroutines*, parameters and data are usually linked via a stack in CISC machines. A RISC machine has a substantial number of registers, which are used to link parameter and data to subroutines.

In the early 1990s, it becomes apparent that neither RISC nor CISC architectures in their purest form were better for all applications. CISC machines, although still supporting many instructions and addressing modes, today take advantage of a larger number of CPU registers and deep pipelining. RISC machines today like the ARM, MIPS, PowerPC, SUN Sparc or DEC alpha have hundreds of instructions, and some need multiple cycles and hardly fit the title reduced instruction set computer.

While modern 32-bit or 64-bit RISC and CISC processor perform complex task at high speed, we find specially in automotive and appliance area another set of requirements. Here we do not need high performance, i.e., an 8-bit processor running at a kHz or a few MHz is often sufficient, that consumes a minimum of power (i.e., long battery live), but should be a "complete system" with program ROM on-chip, multi I/O, minimum external components, and sometimes ADC and/or DAC integrated on-chip. Such a microprocessor typically is called *microcontroller*, and these tiny work horses are produced by the billions each year. The 8-bitters have become the favorite microcontrollers. Sales are about 3 billion controllers per year, compared with 1 billion 4- or 16/32-bit controllers. The 4-bit processors usually do not have the required performance, while 16- or 32-bit controllers are usually too expensive or power hungry for many applications. In cars, for instance, only a few high-performance microcontrollers are needed for audio or engine control; the other more than 50 microcontrollers are used in such functions as electric mirrors, air bags, speedometer, and door locking, to name just a few can be 8-bitter.

Sometimes the distinction between RISC and microprocessor becomes a little blurry, since microcontroller also have a small aka "reduced" instruction set. However, a microcontroller usually does not have the high pipelining delays, larger register files, but have very low power consumption and have PROM and peripheral component such as ADC or DAC on-chip, which is usually not found on RISC machines.

1.2 Embedded Microprocessor on FPGAs

When you think of current state-of-the-art microprocessors, the Intel Itanium processor with 592 million transistors may come to mind. Designing this kind of microprocessor with an FPGA, you may think that is too much to ask of today's FPGAs. Clearly today's FPGAs will not be able to implement such a top-of-the-range microprocessor with the programmable resource, although some FPGA families nowadays have high performance RISC Hardcore processor such as ARM

Fig. 1.3 A typical FPGA-based Microprocessor: The Xilinx KCPSM a.k.a. `PicoBlaze`

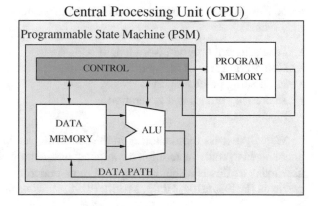

Cortex-A9 but that is still an embedded processor with substantial lower footprint. But there are many applications where a less-powerful microprocessor can be quite helpful. Remember a microprocessor trades performance of the hardwired solution with gate efficiency. In software, or when designed as an FSM, an algorithm like a convolution may ran slower, but usually needs much less resources. So, the microprocessors we build with FPGAs are more of the microcontroller type than a fully featured modern Intel Pentium or TI VLIW PDSP. A typical application, we discuss later would be a microcontroller design for simple game. Now you may argue that this can be done with an FSM. And yes, that is true, and we basically consider our FPGA microprocessor design nothing else then an FSM augmented by a program memory that includes the operation the FSM will perform; see Fig. 1.3. In fact, the early versions of the Xilinx PICOBLAZE processor were called Ken Chapman programmable state machine (KCPSM) [Xil05a]. Important embedded FPGA-based microprocessor constrains (that we will discuss in more details in later chapters) can be summarized as follows:

- $100 \times$ more embedded processor produced than powerful GPP
- Embedded processor designed by small teams with hard time-to-market constraints
- Hard performance/cost/time-to-market issue
- Tool consistence: architecture $\rightarrow$ compiler $\rightarrow$ FPGA recommended to use high level tools
- Embedded processor often more programmable FSM $\rightarrow$ No Itanium (580M transistors)
- Instruction set design based on application
- µP using FPGA's (small) on-chip block memory $\rightarrow$ program often in assembler code
- Use flat memory for high performance
- Use C $\rightarrow$ HDL compiler or custom instruction for improved performance.

A complete microprocessor design usually involves several steps, such as the architecture exploration phase, the instruction set design, and the development tools. Design decision for instruction set and μP features details for microprocessor may include:

- Data path and program word length
- Number and size of registers
- Fix/floating point support
- Addressing modes and generation
- Pipeline stages

We will discuss these choices in the following section in more details. You are encouraged to study in addition a computer architecture book; there are many available today as this is a standard topic in most undergraduate computer engineering curricula [BH03, HJ04, HP03, MH00, PH98, Row04, Sta02].

1.3   Microprocessor Instruction Set Design

The instruction set of a microprocessor (μP) describes the collection of actions a μP can perform. The designer usually asks first, which kind of arithmetic operation I need in the applications for which I will use my μP. For digital signal processing (DSP) applications, for instance, special concern would be applied to support for (fast) add and multiply. A multiply done by a series of smaller shift-adds is probably not a good choice in heavy DSP processing. Besides these ALU operations we also need some data move instructions and some program flow-type instructions such as branch or goto.

The design of an instruction set depends mainly on the underlying μP architecture. Without considering the hardware elements, we cannot fully define our instruction set. Because instruction set design is a complex task, it is a good idea to break up the development into several steps. We will proceed by answering the following questions:

1. What are the addressing modes the μP supports?
2. What is the underlying data flow architecture, i.e., how many operands are involved in an instruction?
3. Where can we find each of these operands (e.g., register, memory, ports)?
4. What types of operations are supported?
5. Where can the next instruction be found?

Addressing Modes

Addressing modes describe how the operands for an operation are located. A CISC machine may support many different modes, while a design for performance like in RISC or PDSPs requires a limitation to the most often used addressing modes. Let us now have a look at the most frequently supported modes.

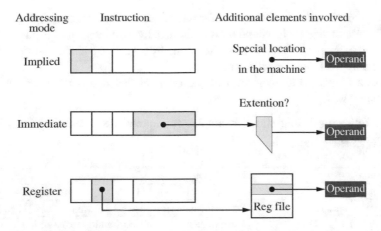

Fig. 1.4 Implied, immediate, and register addressing

Implied addressing   In implied addressing, operands come from or go to a location that is implicitly and not explicitly defined by the instruction; see Fig. 1.4. An example would be the ADD operation (no operands listed) used in a stack machine. All arithmetic operations in a stack machine are performed using the two top elements of the stack. Another example is the ZAC operation of a PDSP, which clears the accumulator and product register in a TMS320 PDSP [TI83]. The following listing shows some examples of implied addressing for different microprocessors:

Instruction	Description	µP
RETURN	The pc is loaded with the value of the stack register	PicoBlaze
bret	Copies the b status into the status register and loads the pc with the ba register value	Nios II
imm 0	Set the upper 16 bit of the immediate operand register for the next instruction to zero	MicroBlaze
eret	Exception return, i.e., load pc with lr value	ARM

Immediate addressing   In the immediate addressing mode, the operand (i.e., constant) is included in the instruction itself. This is shown in Fig. 1.4. A small problem arises due to the fact that the constant provided within one instruction word is usually shorter in terms of number of bits than the (full) data words used in the µP. There are several approaches to solve this problem:

(a) Use a sign extension since most constants (like increments or loop counters) used in programs are small and do not require full length anyway. We should use sign extension (i.e., MSB is copied to the high word) rather than zero extension so that negative values such as −1 are extended correctly.

(b) Use two or more separate instructions to load the low and high parts of the constant one after the other and concatenate the two parts to a full-length word (e.g., MOV and MOVT in ARM).

(c) Use a double-length instruction format to access long constants. If we extend
    the default size by a second word, this usually provides enough bits to load a
    full-precision constant. This is often done by so-called pseudo instructions that
    are mapped to low/upper instructions.
(d) Use a (barrel) shift alignment of the operand that aligns the constant in the
    desired way.

The following listing shows examples for immediate addressing for different
microprocessors.

Instruction	Description	µP
LOAD s1, 05	Load data 5 in register 1	PicoBlaze
roli r6, r6, 1	Rotate the register 6 bits by 1 position	Nios II
lwi r3, r19, 4	Store value in r3 at memory address r19+4	MicroBlaze
subs r3, r3, #1	Subtract 1 from register 3	ARM
ldr r2,[r6,r8,lsl #4]!	Load register 2 from memory location at r6+r8<<4	ARM

To avoid having always two memory accesses, we can also combine method (a)
sign extension and (b) high/low addressing. We then only need to make sure that the
load high is done after the sign extension of the low word.

Register addressing In the register addressing mode, operands are accessed from
registers within the CPU, and no external memory access takes place; see Fig. 1.4.
The following listing shows some examples of register addressing for different
microprocessors:

Instruction	Description	µP
AND sA, sB	Logic bit-by-bit sA AND sB computation result is placed in register sA	PicoBlaze
addk r19, r1, r0	Place the sum of r1 and r0; keep the carry flag	MicroBlaze
xor r6, r7, r8	XOR of the register 7 and register 8 and store result in register 6	Nios II
mul r2, r4, r5	Multiply register 4 and register 5 and put result in register 2	ARM

Since in most machines register access is much faster and consumes less power
than regular memory access, this is a frequently used mode in RISC machines. In
fact all arithmetic operations in a RISC machine are usually done with CPU registers only; memory access is only allowed via separate load/store operations.

Memory addressing To access external memory direct, indirect, and combination
modes are typically used. In the direct addressing mode, part of the instruction word
specifies the memory address to be accessed; see Fig. 1.5. Here the same problem as
for immediate addressing occurs: the bits provided in the instruction word to access

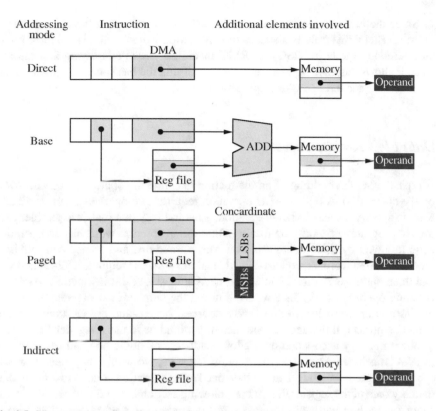

Fig. 1.5 Memory addressing: direct, based, paged, and indirect

the memory operand is too small to specify the full memory address. The full address length can be constructed by using an auxiliary register that may be explicitly or implicitly specified. If the auxiliary register is added to the direct memory address, this is called based addressing; see the LDBU example below. If the auxiliary register is just used to provide the missing MSBs, this is called page-wise addressing, since the auxiliary register allows us to specify a page within which we can access our data. If we need to access data outside the page, we need to update the page pointer first. Since the register in the based addressing mode represents a full-length address, we can use the register without a direct memory address. This is then called indirect addressing; see the STR example below. The following listing shows examples for typical memory addressing modes for different microprocessors:

Instruction	Description	µP
`FETCH s4, 3F`	Read scratch pad RAM location 3F into register 4	PicoBlaze
`ldbu r6,100(r5)`	Compute sum register 5 and 100 and load data from that address into register 6	Nios II
`str r1, [r0]`	Store register value 1 to memory using address specified in register 0	ARM

Since the indirect address mode can usually only point to a limited number of index registers, this usually shortens the instruction words, and it is the most popular addressing mode in RISCs. In RISC machines, based addressing is preferred, since it allows easy access to an array by specifying the base and the array element via an offset; see the LDBU example above.

Data Flow Architectures

A typical assembler coding of an instruction lists first the operation code followed by the operand(s). A typical ALU operation requires two operands, and, if we also want to specify a separate result location, a natural way that makes assembler easy for the programmer would be to allow that the instruction word has an operation code followed by three operands. However, a three-operand choice can require a long instruction word. A modern CPU that addresses 4 GB requires a 32-bit address, and three operands in direct addressing need at least a 96-bit instruction words not counting operation code. As a result, limiting the number of operands will reduce the instruction word length and save resources. A zero-address or stack machine would be perfect in this regard. Another way would be to use a register file instead of direct memory access and only allow load/store of single operands as is typical in RISC machines. For a CPU with eight registers, we would only need 9 bits to specify the three operands. But we then need extra operation code to load/store data memory data in the register file. In the following sections, we will discuss the implications regarding hardware resources and instruction sets when we allow zero to three operands in the instruction word.

Stack machine: a zero-address CPU   A zero-address machine, you may ask, how can this work? We need to recall from the addressing modes, that operands can be specified implicitly in the instruction. For instance, in a TI PDSP all products are stored in the product register P, and this does not need to be specified in the multiply instruction since all multiply results will go to P. Similarly, in a zero-address or stack machine, all two-operand arithmetic operations are performed with the two top elements of a stack [Koo89]. A stack by the way can be considered as a last-in first-out (LIFO). The element we put on the stack with the instruction PUSH will be the first that comes out when we use a POP operation. Let us briefly analyze how an expression like

$$e = a - b + c \times d \tag{1.1}$$

is computed by a stack machine. The left side shows the instruction and the right side the contents of the stack with four entries. The top of the stack is to the left.

Instruction	Stack			
	Top	2	3	4
push a	a	—	—	—
push b	b	a	—	—
sub	$a - b$	—	—	—
push c	c	$a - b$	—	—
push d	d	c	$a - b$	—
mul	$c \times d$	$a - b$	—	—
add	$c \times d + a - b$	—	—	—
pop e	—	—	—	—

It can be seen that all arithmetic operations (ADD, SUB, MUL) use the implicitly specified operands top-of-stack and second-of stack and are in fact zero-address operations. The memory operations PUSH and POP however require one operand.

The code for the stack machine is called postfix (or reverse Polish) operation, since first the operands are specified and then the operations. The standard arithmetic as in (1.1) is called infix notation, e.g., we have the two congruent representations:

$$\text{Infix} : a - b + c \times d \leftrightarrow \text{Postfix} : ab - cd \times +. \tag{1.2}$$

In the Exercises (1.45–1.47) some more examples of these two different arithmetic modes are shown. Some may recall that the postfix notation is exactly the same coding the HP41C pocket calculator requires. The HP41C too used a stack with four values. Figure 1.6a shows the machine architecture for the stack machine.

Accumulator machine: a one-address CPU Let us now add a single accumulator to the CPU and use this accumulator both as the source for one operand and as the destination for the result. The arithmetic operations are of the form

$$accu \leftarrow accu \square op1 \tag{1.3}$$

where $\square$ describes an ALU operation like ADD, MUL, or AND. The underlying architecture of the TI TMS320 [TI83] family of PDSPs is of this type and is shown in Fig. 1.6b. In ADD or SUB operations, for instance, a single operand is specified. The example Eq. (1.1) from the last section would be coded in the TMS320C50 [TI95] assembler code as follows:

Instruction	Description
ZAP	;Clear accu and product register
ADD DAT1	;Add DAT1 to the accu
SUB DAT2	;Subtract DAT2 from the accu
LT DAT3	;Load DAT3 in the T register
MPY DAT4	;Multiply T and DAT4 and store in P register
APAC	;Add the P register to the accu
SACL DAT5	;Store (lower bits) accu at the address DAT4

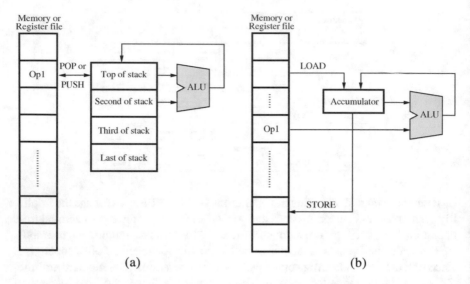

Fig. 1.6 (**a**) Stack CPU architecture. (**b**) Accumulator machine architecture

The example assumes that the variables a-e are mapped to data memory words DAT1-DAT5. Comparing the stack machine with the accumulator machine, we can make the following conclusions:

- The size of the instruction word has not changed, since the stack machine also requires POP and PUSH operations that include an operand.
- The number of instructions to code an algebraic expression is not essentially reduced (seven for an accumulator machine; eight for a stack machine).

A more-substantial reduction in the number of instructions required to code an algebraic expression is expected when we use a two-operand machine, as discussed next.

The two-address CPU In a two-address machine, we have arithmetic operations that allows us to specify the two operands independently, and the destination operand is equal to the first operand, i.e., the operations are of the form

$$op1 \leftarrow op1 \square op2 \tag{1.4}$$

where $\square$ describes an ALU operation like SUB, DIV, or AND. The PICOBLAZE from Xilinx [Xil02a, Xil05a] and the first generation Nios processor [Alt03a] from Altera use this kind of data flow. The basic data flow architecture is shown in Fig. 1.7a. The limitation to two operands allows in these cases the use of a 16-bit instruction word

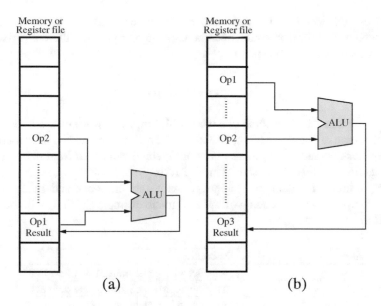

Fig. 1.7 (**a**) Two address CPU architecture. (**b**) Three address machine architecture

format. The coding of our algebraic example Eq. (1.1) would be coded in assembler for the PICOBLAZE[1] style microprocessor as follows:

Instruction	Description
LOAD sE, sC	; Store register C in register E
MUL sE, sD	; Multiply register D with register E
ADD sE, sA	; Add register A to register E
SUB sE, sB	; Subtract register B from E

In order to avoid an intermediate result for the product a rearrangement of the operation was necessary. Note that PICOBLAZE does not have a separate MUL operation and the code is therefore for demonstration of the two-operand principle only. The PICOBLAZE v6 uses 2×16 registers, each 8 bits wide. With two operands and 8-bit constant values, this allows us to fit the operation code and operands or constant in one 16- or 18-bit data word. We can also see that two-operand coding reduces the number of operations essentially compared with stack or accumulator machines.

---

[1] The PICOBLAZE nowadays uses 18-bit instruction words and has still no multiply operation in the ISA.

The three-address CPU The three-address machine is the most flexible of all. The two operands and the destination operand can come or go into different registers or memory locations, i.e., the operations are of the form

$$op1 \leftarrow op2 \square op3 \tag{1.5}$$

Most modern RISC machine like the ARM, PowerPC, MICROBLAZE, or Nios II favor this type of coding [Alt03b, Xil02b, Xil05b]. The operands however are usually register operands or no more than one operand can come from data memory. The data flow architecture is shown in Fig. 1.7b.

Programming in assembler language with the three-operand machine is a straightforward task. The coding of our arithmetic example Eq. (1.1) will look for a Nios II machine as follows:

Instruction	Description
sub r5, r1, r2	; Subtract r2 register from r1 and store in r5
mul r6, r3, r4	; Multiply registers r3 and r4 and store in r6
add r5, r5, r6	; Add r5 and r6 and store in r5

assuming that the registers r1–r5 hold the values for the variables a through e. This is the shortest code of all four machines we have discussed so far. The price to pay is the larger instruction word. In terms of hardware implementation, we will not see much difference between two- and three-operand machines, since the register files need separate multiplexer and de-multiplexer anyway.

Comparison of Zero-, One-, Two-, and Three-Address CPUs

To see the difference of the four CPU architectures, let us first do a brief side-by-side comparison, to compute Eq. (1.1), i.e., $e = a - b + c \times d$. We assume the following instruction sets:

- 0-AC (Stack): PUSH Op1, POP Op1, ADD, SUB, MUL, DIV
- 1-AC (Accumulator): LA Op1, STA Op1, ADD Op1, SUB Op1, MUL Op1, DIV Op1
- 2-AC: LD Op1, M; ST Op1, M; ADD Op1, Op2; SUB Op1, Op2; MUL Op1, Op2; DIV Op1, Op2
- 3-AC: LD Op1, M; ST Op1, M; ADD Op1, Op2, Op3; SUB Op1, Op2, Op3; MUL Op1, Op2, Op3; DIV Op1, Op2, Op3

We assume for 2-AC and 3-AC that operands "a" through "d" are already loaded into registers, but the final result should be stored to memory and the original register values for the 2- and 3-AC machines should be preserved. Here is the side-by-side comparison of the four machines:

0-AC: Stack	1-AC: Accumulator	2-AC: (a = sA;b = sB; …)	3-AC: (r1 = a; r2 = b; …)
PUSH a	LA a	LD sE,sC	SUB r5,r1,r2
PUSH b	SUB b	MUL sE,sD	MUL r6,r3,r4
SUB	STA t	ADD sE,sA	ADD r5,r5,r6
PUSH c	LA c	SUB sE,sB	ST r5, e
PUSH d	MUL d	ST sE, e	
MUL	ADD t		
ADD	STA e		
POP e			

Let us summarize our findings:

- The stack machine has the longest program and the shortest individual instructions.
- Even a stack machine needs a one-address instruction to access memory.
- The three-address machine has the shortest code but requires the largest number of bits per instruction.
- A register file can reduce the size of the instruction words. Typically, in three-address machines, two registers and one memory operand are allowed.
- A load/store machine only allows data moves between memory and registers. Any ALU operation is done with the register file.
- Most designs assume that register access is faster than memory access. While this is true in cell-based ICs (CBICs) or FPGAs that use external memory, inside the FPGA register file access and embedded memory access times are in the same range, providing the option to realize the register file with embedded (three-port) memories.

The above finding seems unsatisfying. There seems no best choice and as a result each style has been used in practice as our coding examples show. The question then is: why hasn't one particular data flow type emerged as optimal? An answer to this question is not trivial since many factors, like ease of programming, size of code, speed of processing, and hardware requirements need to be considered. Let us compare the different designs based on this different design goals. The summary is shown in Table 1.2.

The ease of assembler coding is proportional to the complexity of the instruction. A three-address assembler code is much easier to read and code than the many PUSH and POP operations we find in stack machine assembler coding. The design of a simple C/C++ compiler on the other hand is much simpler for the stack machine since it easily employs the postfix operation that can be much more simply analyzed by a parser. Managing a register file in an efficient way is a very hard task for a compiler. Pipelining further complicates efficient register file use. The number of code words in arithmetic operation is much shorter for two- and three-address operation, since intermediate results can easily be computed. The instruction length is directly proportional to the number of operands. This can be simplified by using registers instead of direct memory access, but the instruction length still is much

Table 1.2 Comparison of different design goal in zero- to three-operand CPUs

Goal	Number of operands			
	0	1	2	3
Ease of assembler	Worst	...	...	Best
Simple C/C++ compiler	Best	...	...	Worst
# of code words	Worst	...	...	Best
Instruction length	Best	...	...	Worst
Range of immediate	Worst	...	...	Best
Fast operand fetch and decode	Best	...	...	Worst
Hardware size	Best	...	...	Worst

shorter with less operands. The size of the immediate operand that can be stored depends on the instruction length. With shorter instruction words the constants that can be embedded in the instructions are shorter and we may need multiple load or double word length instructions, see section memory addressing above. The operand fetch and decode is faster if fewer operands are involved. As a stack machine always uses the two top elements of a stack, no long MUX or DEMUX delays from register files occur. The hardware size mainly depends on the register file. A three-operand CPU has the highest requirements, a stack machine the smallest; ALU and control unit are similar in size.

In conclusion we can say that each particular architecture has its strengths and weaknesses and must also match the designer tools, skills, design goal in terms of size/speed/power, and development tools like the assembler, instruction set simulator, or C/C++ compiler.

Register File and Memory Architecture

In the early days of computers when memory was expensive, *von Neuman* suggested a new highly celebrated innovation: to place the data and program in the same memory; see Fig. 1.8a. At that time computer programs were often hardwired in an FSM and only data memory used RAM. Nowadays the technology constrains are different: memory is cheap, but access speed is still for a typical RISC machine much slower than for CPU registers. In a three-address machine, we therefore need to think about where the three operands should come from. Should all operands be allowed to come from main memory, or only two or one, or should we implement a load/store architecture that only allows single transfer between register and memory, but require that all ALU operations are done with CPU registers? The VAX PDP-11 is always quoted as the champion in this regard and allows multiple memory as well as multiple register operations. For an FPGA design, we have the additional limitation that the number of instruction words is typically in the kilo range, and the von Neuman approach is not a good choice. All the requirements for multiplexing data and program words would waste time and can be avoided if we

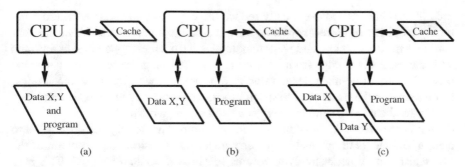

Fig. 1.8 Memory architectures. (**a**) von Neuman machine (GPP). (**b**) Harvard architecture with separate program and data bus. (**c**) Super Harvard architecture with two data busses

use separate program and data memory. This is what is called *Harvard architecture*; see Fig. 1.8b. For PDSPs designs, it would be even better (think of a vector product) if we can use three different memory ports: the coefficient and data come from two separate data memory locations x and y, while the accumulated results are held in CPU registers. The third memory is required for the program memory. Since many algorithms use short loops, some processors try to save the third bus by implementing a small cache. After the first run through the loop the instructions are in the cache and the program memory can be used as a second data memory. This three-bus architecture is shown in Fig. 1.8c and is usually called a *super Harvard architecture*.

A GPP machine like Intel's Pentium or RISC machines usually use a memory hierarchy to provide the CPU with a continuous data stream but also allow one to use cheaper memory for the major data and programs. Such a memory hierarchy starts with very fast CPU registers, followed by level-1, level-2 data and/or program caches to main DRAM memory, and external media like CD-ROMs or tapes. The design of such a memory system is much more sophisticated then what we can design inside our FPGA.

From a hardware implementation standpoint, the design of a CPU can be split into three main parts:

- Control path, i.e., a finite state machine
- ALU
- Register file

of the three items, although not difficult to design, the register file often seems to be the block with the highest cost when implemented with standard logic resources. From these high implementation costs, it appears that we need to compromise between more registers that make a µP easy to use and the high implementation costs of a larger file, such as 32 registers. When designing a RISC register file, we usually have a larger number of registers to implement. One option to save logic resources for the register file is the idea to use two embedded dual-port memory blocks as the register file. We would write the same data in both memories and can read the two sources from the other port of the memory. This principle has been used in the Nios μP and can greatly reduce the required logic resources. From the timing requirement however we now have the problem that Block RAMs are synchronous memory blocks, and we cannot load and store memory addresses and data with the same clock edge from both ports, i.e., replacing the same register value using the current de-multiplexer value cannot be done with the same clock edge. But we can use the rising edge to specify the operand address to be loaded and then use the falling edge to store the new value and set the write enable.

In order to avoid additional instructions for indirect addressing (offset zero) or to clear a register (both operands zero) or register move instructions, sometimes the first register is set permanently to zero as in MICROBLAZE and Nios II. This may appear to be a large waste for a machine with few registers, but simplifies the assembler coding essential, as the following examples show.

Instruction	Description
add r0, r0, r0	; NOP, i.e., do nothing
add r3, r0, r0	; Set register r3 to zero
add r4, r2, r0	; Move register r2 to register r4
ldbu r5, 100(r0)	; Load data from address 100 into r5

Note that the pseudo-instruction above only work under the assumption that the first register r0 is zero.

Operation Support

Most machines have at least one instruction out of the three categories: arithmetic/ logic unit (ALU), data move, and program control. Let us in the following briefly review some typical examples from each category. The underlying data type is usually a multiple of bytes, i.e., 8, 16, or 32 bits of integer data type; some more-sophisticated processors use 32- or 64-bit IEEE floating-point data type.

ALU instructions   ALU instructions include arithmetic, logic, and shift operations. Typical supported arithmetic instructions for two operands are addition (ADD), subtraction (SUB), multiply (MUL), or multiply-and-accumulate (MAC). For a single operand, absolute (ABS) and sign inversion (NEG) are part of a minimum set. Division operation is typically done by a series of shift- subtract-compare instructions since an array divider can be quite large.

The shift operation is useful since in b-bit integer arithmetic a bit growth to $2b$ occurs after each multiplication. The shifter may be implicit as in the TMS320 PDSP from TI or provided as separate instructions. Logical and arithmetic (i.e., correct sign extension) as well as rotations are typical supported. In a block floating-point data format exponent detection (i.e., determining the number of sign bits) is also a required operation.

The following listing shows arithmetic and shift operations for different microprocessors:

Instruction	Description	µP
XOR s3, FF	Bit wise XOR of s3 with FF, i.e. invert s3	PicoBlaze
div r3, r2, r1	Divide r2 by r1 and store quotient in register r3	Nios II
mul r2, r4, r5	Multiply r4 and r5; place 32 LSBs of product in r2	MicroBlaze
lsl r2, r5, #3	Store in r2 the shift left r5 by 3 bits and insert zeros, i.e. multiply by 8	ARM

For µP with small word length such as PICOBLAZE, it's also important to understand how carry-in, carry-out and zero-flags are used and updated by the instructions. These flags are used to build long word arithmetic larger than the original word length. Although logic operations are less often used as arithmetic operation, longer IF conditions evaluations and some more-complex systems that use cryptography or error correction algorithms need basic logic operations such as AND, OR, NOT, or XOR. If the instruction number is critical, we can also use a single NAND or NOR operation, and all other Boolean operations can be derived from these universal functions.

Data move instructions Due to the large address space and performance concerns most machines are closer to the typical RISC load/store architecture than the universal approach of the VAX PDP-11 that allows all operands of an instruction to come from memory. In the load/store philosophy, we only allow data move instructions between memory and CPU registers, or different registers – a memory location cannot be part of an ALU operation. Often register indirect memory access is used, that maybe have additional pre or post de/increment too.

The following listing shows data move instructions for different microprocessors:

Instruction	Description	µP
FETCH s2, 03	Load from scratch pad location 3 data into register s2	PicoBlaze
ldw r4, 4(r6)	Load into register r4 from memory location r6+4	Nios II
str r2, [r4]	Store r2 value in memory at address specified in register r4	ARM

Program flow instructions Under control flow we group instructions that allow us to implement loops, call subroutines, or jump to a specific program location. We may also set the µP to idle, waiting for an interrupt to occur, which indicates new data arrival that need to be processed.

Newer µPs also allow longer loops and nested loops of several levels. In most RISC machine applications, the loops are usually not as short as for PDSPs and the loop overhead is not so critical. In addition, RISC machines use delay branch slots to avoid NOPs in pipeline machines. Function and Procedure call usually require a data stack to save register, return PC values and parameter values. The level of nested call should be known to the programmer before starting the SW development.

The following listing shows program flow instructions for different microprocessors:

Instruction	Description	μP
JUMP NZ, loop	Jump to label loop in case the zero flag is not set	PicoBlaze
CALL func	Call the function named func	PicoBlaze
braid 4b4	Branch without condition to hex address 4b4; next operation is executed since it is in the delay slot	MicroBlaze
beq r5, r0, no_button	Compare register r5 and r0 and if equal jump to label no_button	Nios II
bne wait	Jump to label wait if the equal (i.e., zero) flag from previous operation is zero	ARM

Next Operation Location

In theory, we can simplify the next operation computation by providing a fourth operand that includes the address of the next instruction word. But since almost all instructions are executed one after the other (except for jump-type instructions), this is mainly redundant information and we find no commercial microprocessor today that uses this concept.

Only if we design a one instruction type processor aka an ultimate RISC (URISC) machine, we do need to include the next address or (better) the offset compared to the current instruction in the instruction word [Par05].

1.4   FPGA Technology

VLSI circuits can be classified as shown in Fig. 1.9. FPGAs are a member of a class of devices called field-programmable logic devices (FPLD). FPLDs are defined as programmable devices containing repeated fields of small logic blocks and elements[2]; see Fig. 1.10. It can be argued that an FPGA is an ASIC technology since FPGAs are application-specific ICs. It is, however, generally assumed that the design of a classic ASIC required additional semiconductor processing steps beyond those required for an FPLD. The additional steps provide higher-order ASICs with

---

[2]Called slice or configurable logic block (CLB) by Xilinx, logic cell (LC), logic element (LE), or adaptive logic module (ALM) by Altera.

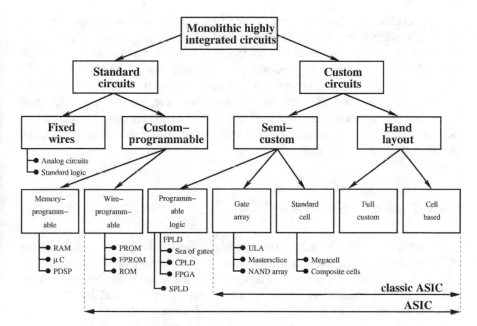

Fig. 1.9 Classification of VLSI circuits

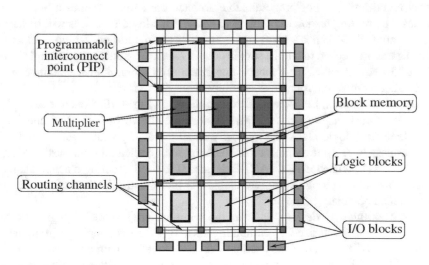

Fig. 1.10 FPGA architecture

their performance and power consumption advantage, but also with high nonrecurring engineering (NRE) costs. At 40 nm, the NRE costs are about $4 million; see [BB08]. Gate arrays, on the other hand, typically consist of a "sea of NAND gates" whose functions are customer provided in a "wire list." The wire list is used during the fabrication process to achieve the distinct definition of the final metal layer. The designer of a *programmable* gate array solution, however, has full control over the actual design implementation without the need (and delay) for any physical IC fabrication facility. A more detailed FPGA/ASIC comparison can be found in section on FPGA Competing Technology.

Classification by Granularity

Logic block size correlates to the *granularity* of a device that, in turn, relates to the effort required to complete the wiring between the blocks (routing channels). In general, three different granularity classes can be found. *Fine-grain devices* were first licensed by Plessey and later by Motorola, being supplied by Pilkington Semiconductors. The basic logic cell consisted of a single NAND gate and a latch. Because it is possible to realize any binary logic function using NAND gates (see Exercise 1.39), NAND gates are called *universal* functions. This technique is still in use for gate array designs along with approved logic synthesis tools, such as ESPRESSO. Wiring between gate-array NAND gates is accomplished by using additional metal layer(s). For programmable architectures, this becomes a bottleneck because the routing resources required to connect the logic functions will be large. A two-input NAND is not a good choice for a PLD. In addition, a high number of NAND gates are needed to build a simple circuit. A fast 4-bit adder, for example, uses about 130 NAND gates. This makes fine-granularity technologies unattractive in implementing most microprocessors components.

The most common FPGA architecture is shown in Fig. 1.10. Concrete examples of contemporary *medium-grain FPGA devices* are shown in Figs. 2.16 and 2.19. The elementary logic blocks are typically small tables, (typically with 4- to 8-bit input tables, 1- or 2-bit output), or are realized with dedicated multiplexer (MPX) logic such as that used in Actel ACT-2 devices [GHB93]. Routing channel choices range from short to long. A programmable I/O block with flip-flops is attached to the physical boundary of the device.

Large granularity devices, such as the complex programmable logic devices (CPLDs), are characterized by combining several so-called simple programmable logic devices (SPLDs), like the classic GAL16V8 together with a fast bus between these SPLDs. SPLDs consist of a programmable logic array (PLA) implemented as an AND/OR array and a universal I/O logic block that includes a memory element such as latch or flip-flop. The SPLDs used in CPLDs typically have 8–10 inputs, 3–4 outputs, and support around 20 product terms. Between these SPLD blocks wide busses (called programmable interconnect arrays (PIAs) by Altera) with short delays are available. By combining the bus and the fixed SPLD timing, it is possible to provide predictable and short pin-to-pin delays with CPLDs. Decoder in SPLD typical are small and fast, arithmetic circuits however even with moderate bit width will require many product terms of a CPLD making SPLD unattractive in implementing most microprocessors.

Classification by Technology

FPLDs are available in virtually all memory technologies: SRAM, EPROM, E^{2}PROM, and antifuse [RGS93]. The specific technology defines whether the device is *reprogrammable* or *one-time programmable*. Most SRAM devices can be programmed by a single-bit stream that reduces the wiring requirements but also increases programming time (typically in the millisecond range) and may be a concern for IP theft. SRAM devices, the dominant technology for FPGAs, are based on static CMOS memory technology and are in-system reprogrammable. They require, however, an external "boot" device for configuration. Electrically programmable read-only memory (EPROM) devices are usually used in a one-time CMOS programmable mode because of the need to use ultraviolet light for erasure. CMOS electrically erasable programmable read-only memory (E^{2}PROM) can be used in-system reprogrammable. EPROM and EEPROM have the advantage of a short setup time. Because the programming information is not "downloaded" to the device, it is better protected against unauthorized use. A recent innovation, based on an EPROM technology, is called "flash" memory. These devices are usually viewed as "page wise" in-system reprogrammable systems with physically smaller cells, equivalent to an E^{2}PROM device. Finally, the important advantages and disadvantages of different device technologies are summarized in Table 1.3.

Table 1.3 Properties of FPLD technology

Technology	SRAM	EPROM	E^{2}PROM	Antifuse	Flash
Reprogrammable	Yes	Yes	Yes	No	Yes
In-system reprogrammable	Yes	No	Yes	No	Yes
Volatile	Yes	No	No	No	No
Copy protected	No[a]	Yes	Yes	Yes	Yes
Examples	Xilinx	Altera	AMD	Actel	Xilinx
	Zynq	MAX5K	MACH	ACT	XC9500
	Altera	Xilinx	Altera		Cypress
	Cyclone	XC7K	MAX 7K		Ultra 37K

[a]Some modern SRAM are protected with DES/AES configuration file encryption

Benchmark for FPLDs

Providing objective benchmarks for FPLDs is a nontrivial task. Performance often depends on the experience and skills of the designer, along with design tool features. To establish valid benchmarks, the Programmable Electronic Performance Cooperative (PREP) was founded by Xilinx [Xil93], Altera [Alt93], and Actel [Act93] and later expanded to more than ten members. PREP has developed nine different benchmarks for FPLDs that are summarized in Table 1.4. The central idea underlining the benchmarks is that each vendor uses its own devices and software tools to implement the basic blocks as many times as possible in the specified device, while attempting to maximize speed. The number of instantiations of the same logic block within one device is called the *repetition rate* and is the basis for all benchmarks. For embedded microprocessor comparisons, benchmarks three and four of a small and large FSM design of Table 1.4 are most relevant.

In Fig. 1.11, repetition rates are reported over frequency, for typical Altera (A_k) and Xilinx (X_k) FPGA and CPLD devices that are currently used on the university development boards. These are not always the largest devices available, but all devices are supported by the web-based version of the design tools. Xilinx seemed to achieve the higher speed, while the Altera FPGAs have a larger number of repetitions. Compared with the CPLDs, it can be concluded that modern FPGA families provide the best embedded microprocessor resources and maximum speed. This is attributed to the fact that modern devices provide fast-carry logic with delays (less than 0.1 ns per bit) that allow fast adders with large bit width, without the need for expensive "carry look-ahead" decoders. Although PREP benchmarks are useful to

Table 1.4 The PREP benchmarks for FPLDs

Number	Benchmark name	Description
1	Data path	Eight 4-to-1 multiplexers drive a parallel-load 8-bit shift register
2	Timer/counter	Two 8-bit values are clocked through 8-bit value registers and compared
3	Small state machine	An 8-state machine with 8 inputs and 8 outputs (see Fig. 3.7)
4	Large state machine	A 16-state machine with 40 transitions, 8 inputs, and 8 outputs
5	Arithmetic circuit	A 4-by-4 unsigned multiplier and 8-bit accumulator
6	16-bit accumulator	A 16-bit accumulator
7	16-bit counter	Loadable binary up counter (see Fig. 3.8)
8	16-bit synchronous prescaled counter	Loadable binary counter with asynchronous reset (see Fig. 3.8)
9	Memory mapper	The map decodes a 16-bit address space into 8 ranges (see Fig. 3.9)

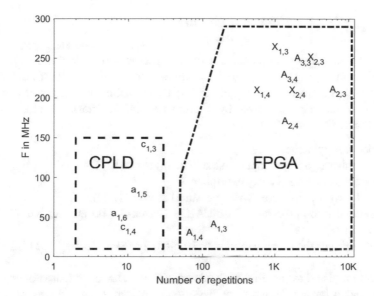

Fig. 1.11 PREP benchmark 3 and 4 (i.e., second subscript) average for FPLDs from the Digilent and TERASIC development boards: A_1 = FLEX10K from UP2; A_2 = Cyclone 2 from DE2; A_3 = Cyclone IV from DE2–115; a_1 = EPM7128 from UP2; X_1 = Spartan 3 from Nexys; X_2 = Spartan 6 from Nexys III; X_3 = Spartan 6 LX45 from Atlys; and c_1 = CoolRunner II CPLD

compare equivalent gate counts and maximum speeds, for concrete applications additional attributes are also important. They include:

- On-chip large block RAM or ROM
- External memory support for ZBT, DDR, QDR, SDRAM
- Embedded hardwired processor system (HPS), e.g., 32-bit ARM Cortex-A9
- Array multiplier (e.g., 18 × 18 bits, 18 × 25 bits)
- Package such as BGA, TQFP, PGA
- Configuration data stream encryption via DES or AES
- On-chip fast analog-to-digital converter (ADC)
- Pin-to-pin delay
- Internal tristate bus
- Readback- or boundary-scan decoder
- Programmable slew rate or voltage of I/O
- Power dissipation
- Hard IP block for ×1, ×2, or ×4 PCIe

Power dissipation of FPLD is another important characteristic, in particular in mobile applications. It has been found that CPLDs usually have higher "standby" power consumption. For higher-frequency applications, FPGAs can be expected to have higher power dissipation. A detailed power analysis example can be found in [MB14, Sect. 1.4.2].

Recent FPGA Families and Features

Some of these features are (depending on the specific application) more relevant to embedded microprocessor design than others. We summarize the availability of some of these key features in Tables 1.5 and 1.6 for Xilinx and Altera, respectively. Device family name is followed by relevant features for most embedded systems applications in columns 2–9:

1. Device family name.
2. The number of address inputs (a.k.a. Fan-in) to the LUT.
3. Size of the embedded array multiplier.
4. Size of the on-chip block RAM measured as kilo (1024) bits.
5. Embedded microprocessor: 32-bit ARM Cortex-A9 on current Xilinx ZYNQ and Altera devices.
6. Xilinx devices the on-chip (Virtex 6: 10 bit, 0.2 MSPS; Series 7: 12 bit 1 MSPS) fast ADC.
7. Target price and availability of the device family. Device that are no longer recommended for new designs are classified as mature with m. Low-cost devices have a single $ and high price range devices have two $$.
8. Year the device family was introduced.
9. Process technology used measured in nanometer.

At the time of writing Xilinx supports four device families: the Virtex family for leading performance and capacity, the Kintex family for DSP intensive application and low cost, and the Artix family of lowest cost, replacing the Spartan family of devices. In addition, an embedded microprocessor centric family called ZYNQ has been introduced. The Virtex-II, Virtex-4-FX, or Virtex-5-FXT families that included one or more IBM PowerPC RISC processor are no longer recommended for new

Table 1.5 Recent Xilinx FPGA family microprocessor system features

Family	Feature							
	LUT Fan-in	Emb. mult.	BRAM size Kbits	Fast A/D	Emb. µP	Cost/mature	Year	Process nm
Spartan 3	4	18 × 18	18	No	–	m	2003	90
Virtex 4	4	18 × 18	36	No	PPC	m	2004	90
Virtex 5	6	18 × 18	36	No	PPC	m	2006	65
Spartan 6	6	18 × 18	18	No	–	$	2009	45
Virtex 6	6	18 × 25	36	Yes	–	$$	2009	40
Artix 7	6	18 × 25	36	Yes	–	$	2010	28
Kintex 7	6	18 × 25	36	Yes	–	$$	2010	28
Virtex 7	6	18 × 25	36	Yes	–	$$	2010	28
Zynq 7Ks	6	18 × 25	36	Yes	1xARM	$$	2011	28
Zynq 7K	6	18 × 25	36	Yes	2xARM	$$	2011	28
Zynq CG	6	18 × 25	36	Yes	2+2ARM	$$$	2015	20
Zynq EV	6	18 × 25	36	Yes	2+4ARM	$$$	2015	16

Table 1.6 Altera FPGA family microprocessor system features

Family	Feature							
	LUT Fan-in	Emb. mult. size	BRAM size Kbits	FP A/D	Emb. μP	Cost/mature	Year	Process nm
FLEX10K	4	No	4	No	–	m	1995	420
Cyclone	4	No	4	No	–	$	2002	130
Cyclone II	4	18 × 18	4	No	–	$	2004	90
Cyclone III	4	18 × 18	9	No	–	$	2007	65
Cyclone IV	4	18 × 18	9	No	–	$	2009	60
Cyclone V	8	18 × 19	0.640,10	No	2xARM	$	2011	28
Cyclone 10	8	18 × 19	0.64,20	No	–	$	2017	20
Arria	8	18 × 18	576	No	–	$	2007	90
Arria II	8	18 × 18	9	No	–	$	2009	40
Arria V	8	27 × 27	10	No	2xARM	$$	2011	28
Stratix	4	18 × 18	0.5,4,512	No	–	$$	2002	130
Stratix II	8	18 × 18	0.5,4,512	No	–	$$	2004	90
Stratix III	8	18 × 18	9,144	No	–	$$	2006	65
Stratix IV	8	18 × 18	9,144	No	–	$$	2008	40
Stratix V	8	27 × 27	20	No	–	$$	2010	28
Stratix 10	8	18 × 19	4500,20,0.64	No	4xARM	$$$	2013	14

designs. The Xilinx devices have 18 × 18-bit or 18 × 25-bit embedded multipliers. Most current devices provide an 18 or 36 Kbit memory. A 0.2 MSPS 10-bit fast ADC on-chip has been added for the sixth Virtex generation. The seventh generation includes a 12-bit 1 MSPS dual channel ADC with additional sensors for power supply and on-chip temperature with possibly 17 sensors sources for the ADCs; see Fig. 1.12b. Keep in mind that a larger number of only the Spartan families are available in the web edition of the development software; most other devices need a subscription edition of the Xilinx ISE software.

Altera offers three main classes of FPGA devices: the Stratix family includes high performance devices, the Arria family has the midrange devices, and the Cyclone devices have lowest cost, lowest power, lowest density, and lowest performance of all three. Logic block size of recent devices has been increased from 4 input LUT to a maximum 8 different inputs, that allow for instance to build 3 input adders at almost the same speed as two input adders. Physically the ALM has two flip-flops, two full adders, two 4-input LUTs and four 3-input LUTs, and many multiplexers that allow a general 6-input function to be implemented; see Fig. 1.12a. The embedded multiplier size in Altera devices ranges from 9 × 9-bit, 18 × 18-bit, to 27 × 27-bit. Larger multipliers can be built by grouping these blocks together at the cost of a reduced speed. Starting with the fifth generation, three 9-bit blocks are combined into one fast 27 × 27-bit multiplier. The memories are available in a wide variety range from 0.5K, M4K, M9K, M10K, M144K, to M512K bit memory. Keep in mind that only a few of the Cyclone devices are available in the Web aka Prime Lite edition of the QUARTUS development software; Arria and Stratix devices need a subscription edition of the software.

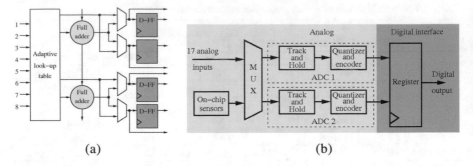

(a)   (b)

Fig. 1.12 New architecture features used in recent FPGA families. (**a**) Altera's ALM block. (**b**) Xilinx series 7 high speed on-chip ADC

FPGAs Competing Technology

The PLD market share, by vendor, is presented in Fig. 1.13. PLDs, since their intro-duction in the early 1980s, have enjoyed in the first two decades steady growth of 20% per annum, outperforming ASIC growth by more than 10%. In 2001 the world-wide recession in microelectronics reduced the ASIC and FPLD growth. Since 2003 we see again a steep increase in revenue (growth about 10% per annum) for the two market leaders. Actel became part of Microsemi Inc. in November 2010. Altera is part of Intel since 2015. The reason that FPLDs outperformed ASICs for many years seems to be related to the fact that FPLDs can offer many of the advantages of ASICs such as:

- Reduction in size, weight, and power dissipation
- Higher throughput
- Better security against unauthorized copies
- Reduced device and inventory costs
- Reduced board test costs

without many of the disadvantages of ASICs such as:

- A reduction in development time (rapid prototyping) by a factor of three to four
- In-circuit re-programmability
- Lower NRE costs resulting in more economical designs for solutions requiring less than 1000 units

CBIC ASICs are used in high-end, high-volume applications (more than ≈ 1000 copies). Compared to FPLDs, CBIC ASICs typically have about ten times more gates for the same die size. An attempt to solve the latter problem is the so-called hard-wired FPGA (Altera named HardCopy ASICs and Xilinx now EasyPath FPGAs), where a gate array is used to implement a verified FPGA design.

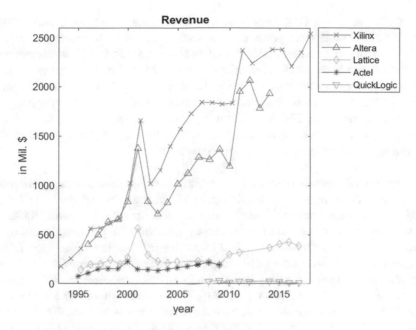

Fig. 1.13 Revenues of the top five vendors in the PLD/FPGA/CPLD market

1.5   Design with Intellectual Property Cores

Although FPGAs are known for their capability to support rapid prototyping, this only applies if the HDL design is already available and sufficiently tested. A complex block like a PCI bus interface, a pipelined FFT, a FIR filter, or a microprocessor may take weeks or even months in development time. One option that allows us essentially to shorten the development time is available with the use of a so-called intellectual property (IP) core. These are predeveloped (larger) blocks, where typical standard blocks like microprocessors, phased-locked loops (PLLs), FIR filters, or FFTs are available from FPGA vendors directly, while more specialized blocks (e.g., AES, DES, or JPEG codec, I2C bus or Ethernet interfaces) are available from third-party vendors. While some blocks are free in the QUARTUS package the larger more sophisticated blocks may have a high price tag. However, as long as the block meets your design requirement, it is most often more cost effective to use one of these predefined IP blocks. Let us now have a quick look at different types of IP blocks and discuss the advantages and disadvantages of each type [H00, VG02, CMG07]. Typically, IP cores are divided into three main forms, as described below.

Soft Core   A *soft core* is a behavioral description of a component that needs to be synthesized with FPGA vendor tools. The block is typically provided in a hardware description language (HDL) like VHDL or Verilog, which allows easy modification by the user, or even new features to be added or deleted before synthesis for a specific vendor or device. On the downside the IP block may also require more work to meet the desired size/speed/power requirements. Very few of the blocks provided by FPGA vendors are available in this form, like the first generation Nios microprocessor from Altera or the PICOBLAZE microprocessor by Xilinx discussed in Chaps. 7 and 8. IP protection for the FPGA vendor is possible [CMG07] but difficult to achieve since the block is provided as synthesizable HDL and can quite easily be used with a competing FPGA tool/device set or a cell-based ASIC. The price of third-party FPGA blocks provided in HDL is usually much higher than the moderate pricing of the parameterized core discussed next.

Parameterized Core A *parameterized* or firm core is a structural description of a component. The parameters of the design can be changed before synthesis, but the HDL is usually not available. The majority of cores provided by Altera and Xilinx come as this type of core. They allow certain flexibility but prohibit the use of the core with other FPGA vendors or ASIC foundries and therefore offer better IP protection for the FPGA vendors than soft cores. Examples of parameterized cores available from Altera and Xilinx include an PLL, FIR filter compiler, FFT (parallel and serial), and embedded processors, e.g., Nios II from Altera and MICROBLAZE from Xilinx. Another advantage of parameterized cores is that usually a resource estimation (LE, multiplier, block RAMs) is available that is expected to be correct within a few percent, which allows a fast design space exploration in terms of size/speed/power requirements even before synthesis. Test benches in HDL (for MODELSIM simulator) that allow cycle-accurate modeling as well as C/C++ models or MATLAB scripts that allow behavior-accurate simulation are also standard for parameterized cores. Code generation usually takes only a few seconds. At the end of this chapter we will study a small parameterized PLL core, and later in Chaps. 9 and 10 we will study the Altera Nios II and Xilinx MICROBLAZE as example for large-scale parameterized core.

Hard Core A *hard core* (fixed netlist core) is a physical description or a hardwired core within the FPGA. The cores are usually optimized and available for a specific device (family), when low power, high speed, or hard real-time constraints are required, like, for instance, a PCI bus interface. The parameters of the design are fixed, like a 16-bit 256-point FFT. For some core behavior, HDL description is available that allows simulation and integration in a larger project. Most third-party IP cores from FPGA vendors, several free FFT cores from Xilinx and the ARM Cortex-A9 embedded processors fall within this type of core class. Since the layout is fixed, the timing and resource data provided are precise and do not depend on synthesis results. However, the downside is that a parameter change is not possible, so if the FFT should have 24-bit input data, the 16-bit 256-point FFT hard core IP block cannot be used. Later in Chaps. 11 and 12, we will study the ARM cortex A9 included in Altera's SoC Cyclone V as well as Xilinx Zynq 7K devices.

IP Core Comparison and Challenges

If we now compare the different IP block types, we have to choose between design flexibility (soft core) and fast results and reliability of data (hard core). Soft cores are flexible, e.g., change of system parameters or device/process technology is easy, but may have longer debug time. Hard cores are verified in silicon. Hard cores reduce development, test, and debug time but no VHDL code is available to look at. A parameterized core is most often the best compromise between flexibility and reliability of the generated core. There are however two major challenges with current IP block technology, which are pricing of a block and, closely related, IP protection. Because the cores are reusable, vendor pricing must rely on the number of units of IP blocks the customer will use. This is a problem known for many years in patent rights and most often requires long license agreements and high penalties in case of customer misuse. FPGA-vendor-provided parameterized blocks (as well as the design tool) have very moderate pricing since the vendor will profit if a customer uses the IP block in many devices and then usually must buy the devices from this single source. This is different with third-party IP block providers that do not have this second stream of income. Here the license agreement, especially for a soft core, has to be drafted very carefully.

For the protection of parameterized cores, the FPGA vendor use FlexLM-based keys to enable/disable single IP core generation. Evaluation of the parameterized cores is possible down to hardware verification by using time-limited programming files or requiring a permanent connection between the host PC and board via a JTAG cable, allowing you to program devices and verify your design in hardware before purchasing a license. For instance, Altera's OPENCORE evaluation feature allows you to simulate the behavior of an IP core function within the targeted system, verify the functionality of the design, and evaluate its size and speed quickly and easily. When you are completely satisfied with the IP core function and you would like to take the design into production, you can purchase a license that allows you to generate non-time-limited programming files. The QUARTUS software automatically downloads the latest IP cores from Altera's website. Many third-party IP providers also support the OPENCORE evaluation flow, but you have to contact the IP provider directly in order to enable the OPENCORE feature.

The protection of soft cores is more difficult. Modification of the HDL to make them very hard to read aka obfuscation [MCB11] or embedding watermarks in the high-level design by minimizing the extra hardware have been suggested [CMG07]. The watermark should be robust, i.e., a single bit change in the watermark should not be possible without corrupting the authentication of the owner.

A comparison of the most important features of the three types of IP cores is shown in Table 1.7.

Table 1.7 IP core comparisons

Feature	IP core type		
	Softcore	Parameterized core	Hardcore
Flexibility	++	++	−
Performance	+	+	++
IP cost	−	+	+
Development time	−	+	++
Vendor independence	++	−	−
Power efficiency	+	+	++
Examples	PICOBLAZE, Nios	FIR Compiler, FFT Compiler, Nios II, MICROBLAZE	PCI, IEEE FP, ARM Cortex-A9

Legend: ++ excellent, + = good, − = poor

IP Core-Based PLL Design Example

Finally, we evaluate the design process of an IP block generation and integration into a project in an example using a design typical with development boards: The board has a fixed high precision 50 MHz oscillator; however, our embedded microprocessor system may require a 2.5 factor higher clock of 125 MHz. Such an increased clock is basically impossible to generate with synchronous digital logic. Most modern FPGAs, however, use phased-locked loops (PLLs) to generate different clock or phase delays. An overall architecture for a typical PLL used in FPGAs is shown in Fig. 1.14. A PLL is a mixed analog/digital circuit and has in its core a phase/frequency detector (PFD), a voltage-controlled oscillator (VCO), and a feedback counter. For steady state we find that both PFD signals (F_{REF} and F_{FB}) are in phase (identical). The loop is stable aka locked and the feedback frequency $F_{FB} = F_{VCO}/M$ is in phase to the incoming reference signal with frequency $F_{REF} = F_{IN}/N$ considering that the reference frequency come from an input divider N. Rearranging such that the output frequency is F_{VCO} is on the left, we choose $M = 5$ and $N = 2$ and the desired output frequency becomes

$$F_{VCO} = N \times F_{IN} / M = 5 \times 50\,\text{MHz} / 2 = 125\,\text{MHz}. \qquad (1.6)$$

The additional post scale counter C can be used to generate additional fraction frequency of F_{VCO} or can be used to generate different phase delays.

You will find in the Altera QUARTUS IP Catalog a variety of PLL architectures, including zero output delay and direct options. For a simple evaluation we name the IP core as our project, i.e., `PLL`. Make a new directory and open a new project in QUARTUS and name it `PLL` and select `Tools → IP Catalog`. In the first step find the `PLL` block in the IP Catalog under `Basic Function → Clock PLL and Resets → PLLs → Altera PLL` window; see Fig. 1.15a. We then select the desired output format (AHDL, VHDL, or Verilog) and specify our working directory. Then the IP GUI tool pops up (see Fig. 1.15b) and we have access to the

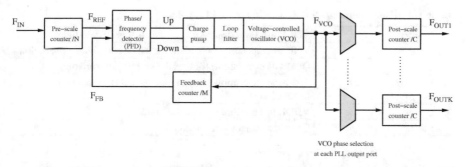

Fig. 1.14 The architecture of a FPGA PLL

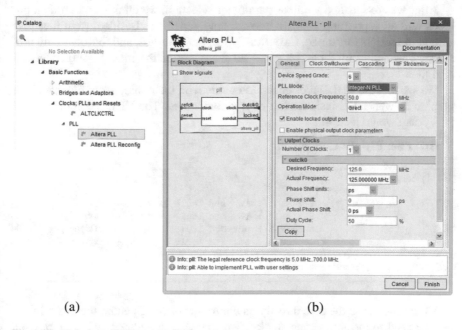

(a)   (b)

Fig. 1.15 IP design of PLL. (**a**) IP Catalog element selection. (**b**) IP parameterization of PLL core according to the 2.5 factor increase reference clock input

PLL documentation and can start with the parameterization of the block. The PLL can be configured with a couple of specialized modes such as zero delay (between reference input and output need for phase delays only), but we used the basic configuration called `direct` operation mode. Since we want to generate the 125 MHz output signal using a 50 MHz reference from the example above, we enter these data in the GUI for `Desired Frequency` and `Reference Clock Frequency`, respectively. We keep the 50% duty cycle and do not require any phase shift. The info window immediately gives us feedback if our frequency choice is realizable with the FPGA PLL. Since a PLL is a mixed analog/digital circuit FPGAs usually

Table 1.8 IP most important files generated for the PLL core

`pll.vhd`	VHDL top-level description of the custom IP core function
`pll.cmp`	VHDL component declaration for the IP core function variation
`pll.bsf`	QUARTUS symbol file for the IP core function variation
`pll_sim/mentor/` `msim_setup.tcl`	Test bench to compile library models and IP component for MODELSIM
`pll.vho`	VHDL IP functional simulation model
`pll.qip`	QUARTUS project information file

have limits on the lower and upper bounds of the VCO signals as can be seen from the `Info` in the parameter window of the PLL. Our 125 MHz are well within the possible range of 5–700 MHz.

After we are satisfied with our parameter selection, we click `Finish` then our block and all supporting files are immediately generated. Beside design files also simulation scripts for ALDEC, CADENCE, MENTORGRAPHICS, and SYNOPSYS are provided. The listing in Table 1.8 gives an overview of the most important generated files.

We see that the VHDL files generated along with their component file that allow graphic as well HDL instantiation of the component. The VHDL component file `pll.cmp` would look like:

VHDL File: PLL IP Component

```
1    component pll is
2            port (
3                    refclk   : in  std_logic := 'X'; -- clk
4                    rst      : in  std_logic := 'X'; -- reset
5                    outclk_0 : out std_logic;        -- clk
6                    locked   : out std_logic         -- export
7                 );
8            end component pll;
```

We decide to use the core directly as our top-level design entry, therefore avoiding the need to instantiate our block in another design and connect the input and outputs. By inspecting the top level VHDL file `pll.vhd` or the component file, we notice that the `ENTITY` has the expected block input `refclk` and output `outclk_0` signals but has some additional useful control signal, i.e., `rst` and `locked`, whose function is self-explanatory. We then start QUARTUS to run a full compilation to generate the file `pll.vho` file for the timing simulation that can be found in `/simulation/modelsim`, and we replace the functional file in the directory `pll_sim` with this generated file.

To simulate the design, we use the generated TCL script to compile the design and model library. Start the MODELSIM simulator and change to the directory `pll/pll_sim/mentor` and then type in the command window `do msim_setup.tcl`. After starting the script, it shows a menu with possible step. We select compilation followed by elaboration. Several libraries and the design files are compiled but no simulation is performed. To run a simulation, we use the following steps:

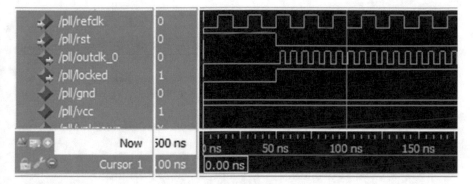

Fig. 1.16 Test bench for PLL IP design. Verification via timing simulation

ModelSim: Command Line Prompt Input Sequence

```
1    type> com
2    type> elab
3    type> add wave *
4    type> force rst 1 0ns, 0 50ns
5    type> force refclk 0 0ns,1 10ns -r 20ns
6    type> run 500ns
```

We zoom in to the first couple of clock cycles to see the PLL locked (after ca. 2 clock cycles) and the high clock signal, as shown in Fig. 1.16. For 2 clock periods of refclk, we have 5 output clock periods, i.e., the desired 2.5 factor higher frequency. We may notice a small problem from the simulation of the IP block that cannot be changed in the GUI settings. Often FPGA board use active low resets because most button go to zero when pressed. In a soft core we would be able to change the HDL code of the design to implement an active low reset, but in the parameterized core we do not have this option. However, we can solve this problem by attaching an inverter to the input of the PLL reset. This is a typical experience with the parameterized cores: the cores provide a 90% or more reduction in design time, but sometimes small extra design effort is necessary to meet the exact project requirements.

Review Questions and Exercises
Short Answer

1.1. Why was Plessy FPLD devices not a commercial success?
1.2. Should we use a Pentium processor to design embedded systems?
1.3. What happens during microprocessor instruction decoding?
1.4. Name and explain five µP addressing modes.
1.5. Draw the data flow architectures for 0–3 address CPUs.

1.6. Convert the following decimal numbers: 123, 1000, 255, 4681 to binary, octal, and hex.
1.7. What is the decimal equivalent of the following hex numbers: A, 123, ABCD, FFF?
1.8. For memories how many bits precise are 1 Kilo Byte (KB), 1 MB, or 1 GB?
1.9. How many bits are needed to store an image with 640 rows, 480 columns and each pixel requires 16 bits?
1.10. Explain the difference between von Neuman, Harvard, and super Harvard architectures.
1.11. Order the 0–3 address CPUs regarding the following properties: Instruction length, ease of assembler, # code words, HW size (best first).
1.12. When would you choose a FPGA and when a cell-based ASIC for your design?
1.13. How does FPLD benchmarks work?
1.14. What is FPLD granularity?
1.15. Order the following technologies by the circuit speed: cell-based ASICs, FPGAs, Full custom and PLDs (slowest speed device first)
1.16. Which FPGA hardware resources are most important for µP designs? (most important first)
1.17. How does a PLL work? When do we need PLLs?

Fill in the Blank

1.18. The Cyclone V devices are a product of _____
1.19. The Zynq 7K devices are a product of _____
1.20. Desired FPLD technology properties are _____
1.21. Undesired FPLD technology properties are _____
1.22. ASIC is an acronym for _____
1.23. CPU is an acronym for _____
1.24. RISC is an acronym for _____
1.25. KCSPM is an acronym for _____
1.26. NRE is an acronym for _____
1.27. PLA is an acronym for _____

True or False

1.28. _____ EPROM FPLDs provide "copy protection," but E^2PROM FPLDs do not.
1.29. _____ SRAM FPLDs are reprogrammable and volatile.

1.30. _____ Power consumption in FPGAs is proportional to the product of frequency and number of logic cells running at this frequency.
1.31. _____ The PREP benchmarks 3 and 4 are FSM type designs.
1.32. _____ The repetition rate in in the PREP benchmark is higher for CPLD than for FPGAs.
1.33. _____ The Cyclone V devices have embedded multipliers, BRAM, and a PPC microprocessors.
1.34. _____ The Zynq devices include an ARM microprocessor.
1.35. _____ The FPGA Revenue Market leader in the last 10 years were Lattice and Actel.
1.36. _____ The FPGA growth was 20% versus ASIC 10% per year in the last decade.
1.37. _____ The BRAM memories in modern FPGAs are asynchronous memory.
1.38. _____ Modern FPGAs often have PLLs.

Projects and Challenges

1.39. Show that the two-input NAND is *universal* by implementing NOT, AND, and OR with NAND gates only.
1.40. Use only two input NAND gates to implement a full adder:
s = a XOR b XOR c_{in}; c_{out} = (a AND b) OR (c_{in} AND (a OR b))
1.41. Show that the two input NOR gate is *universal* by implementing NOT, AND, and OR with NOR gates only.
1.42. Use only two input NOR gates to implement a full adder:
s = a XOR b XOR c_{in}; c_{out} = (a AND b) OR (c_{in} AND (a OR b))
1.43. Show that two input multiplexer f(x,y,s) = x AND (NOT s) OR (y AND s) is *universal* by implementing NOT, AND, and OR with NOR gates only.
1.44. Use only two input multiplexer f(x,y,s) = x AND (NOT s) OR (y AND s) to implement a full adder: s = a XOR b XOR c_{in}; c_{out} = (a AND b) OR (c_{in} AND (a OR b)).
1.45. Convert the following infix arithmetic expressions to postfix expressions (see Eq. 1.2):

(a) a + b − c − d + e
(b) a + b × c
(c) (a − b) × (c + d) + e
(d) (a − b) × (((c − d × e) × f)/g) × h

1.46. Convert the following postfix arithmetic expressions to infix expressions (see Eq. 1.2):

(a) ab + cd /
(b) ab/cd × −
(c) ab + c ^ (with ^ power-of symbol)

1.47. Which of the following pairs of postfix expression are equivalent (see Eq. 1.2)?

   (a) ab + c + and abc + +
   (b) ab − c − and abc − −
   (c) ab × c + and cab × +
   (d) abc + × and ab × bc × +

1.48. Compare 0–3 address coding (AC) machines by writing the shortest programs possible to compute

$$h = (a-b)/\big((c+d)\times(f-g)\big)$$

   for the following instruction sets:

(a) 0-AC (Stack): PUSH Op1, POP Op1, ADD, SUB, MUL, DIV
(b) 1-AC (Accumulator): LA Op1, STA Op1, ADD Op1, SUB Op1, MUL Op1, DIV Op1
(c) 2-AC: LD Op1, M; ST Op1, M; ADD Op1, Op2; SUB Op1, Op2; MUL Op1, Op2; DIV Op1, Op2
(d) 3-AC: LD Op1, M; ST Op1, M; ADD Op1, Op2, Op3; SUB Op1, Op2, Op3; MUL Op1, Op2, Op3; DIV Op1, Op2, Op3

   Assume for 2-AC and 3-AC that all operands are already loaded into registers, but the final result should be stored to memory. Also, for 2AC and 3AC register values need not to be preserved, i.e., you can overwrite the initial register values!

1.49. Repeat Exercise 1.48 for the following arithmetic expression:

$$f = (a-b)/(c+d\times e)$$

   You may rearrange the expression if necessary.

1.50. Repeat Exercise 1.48 for the following arithmetic expression:

$$f = (a-b/c)/(b+d\times e)$$

   You may rearrange the expression if necessary.

1.51. Repeat Exercise 1.48 for the following arithmetic expression:

$$f = (a+b\times c)/(d-e)$$

   You may rearrange the expression if necessary.

1.52. Repeat Exercise 1.48 for the following arithmetic expression:

$$g = (a\times b-c)/(d+e/f)$$

   You may rearrange the expression if necessary.

1.53. Repeat Exercise 1.48 for the following arithmetic expression:

$$g = (a - b \times c) / (d / e + f)$$

You may rearrange the expression if necessary.
1.54. Visit the FPLD vendor webpage to find most recent data for Xilinx and Altera/ Intel FPGA revenue.
1.55. Explain the difference between soft, hard, and parameterized core. What are advantages and disadvantages of each type?
1.56. Check your tool library for IP cores. Write a one-line summary of the function. What is the cost of the cores? Which IP core from your list maybe useful in a μP design?

References

[Act93] Actel: PREP benchmarks confirm cost effectiveness of field programmable gate arrays, in *Actel-Seminar* (1993)
[Alt93] Altera: PREP Benchmarks Reveal FLEX 8000 is Biggest, MAX 7000 is Fastest, in *Altera News & Views San Jose* (1993)
[Alt03a] Altera: Nios-32 Bit Programmer's Reference Manual, Nios Embedded Processor, Ver. 3.1 (2003)
[Alt03b] Altera: Nios II Processor Reference Handbook, NII5V-1-5.0 (2003)
[A08] M. Anderson: Help wanted: embedded engineers: why the United States is losing its edge in embedded systems, IEEE USA Today's Engineer Digest, Mar. (2008)
[BB08] V. Betz, S. Brown. FPGA challenges and opportunities at 40 nm and beyond, in *International Conference on Field Programmable Logic and Applications Prague* (2009), p. 4
[BH03] R. Bryant, D. O'Hallaron, *Computer Systems: A Programmer's Perspective*, 1st edn. (Prentice Hall, Upper Saddle River, 2003)
[C09] R. Charette, This car runs on code. IEEE Spectr. Online **46**(3), 3 (2009)
[CMG07] E. Castillo, U. Meyer-Baese, A. García, L. Parrilla, A. Lloris, IPP@HDL: Efficient intellectual property protection scheme for IP cores. IEEE Trans. Very Large Scale Integr. VLSI Syst. **15**(5), 578–591 (2007)
[FHM96] H. Faggin, M. Hoff, S. Mazor, M. Shima, The history of the 4004. IEEE Microw Mag **16**, 10–20 (1996)
[GHB93] J. Greene, E. Hamdy, S. Beal, Antifuse field programmable gate arrays. Proc. IEEE, 1042–1056 (1993)
[H00] J. Hakewill. Gaining control over silicon IP, Communication Design, online (2000)
[HHF08] J. Hamblen, T. Hall, M. Furman, *Rapid Prototyping of Digital Systems: SOPC Edition* (Springer, New York, 2008)
[HJ04] V. Heuring, H. Jordan, *Computer Systems Design and Architecture*, 2nd edn. (Prentice Hall, Upper Saddle River, 2004)., contribution by M. Murdocca
[HP03] J. Hennessy, D. Patterson, *Computer Architecture: A Quantitative Approach*, 3rd edn. (Morgan Kaufman Publishers, Inc., San Mateo, 2003)
[Koo89] P. Koopman, *Stack Computers: The New Wave*, 1st edn. (Mountain View Press, La Honda, 1989)
[Maz95] S. Mazor, The history of the microcomputer – invention and evolution. Proc. IEEE **83**(12), 1601–1608 (1995)

[MCB11] U. Meyer-Bäse, E. Castillo, G. Botella, L. Parrilla, A. García. Intellectual property protection (IPP) using obfuscation in C, VHDL, and verilog coding, in *Proc. SPIE Int. Soc. Opt. Eng., Independent Component Analyses, Wavelets, Neural Networks, Biosystems, and Nanoengineering IX*, vol 8058, April 2011, pp. 80581F1-12

[MH00] M. Murdocca, V. Heuring, *Principles of Computer Architecture*, 1st edn. (Prentice Hall, Upper Saddle River, 2000)

[MB14] U. Meyer-Baese, *Digital Signal Processing with Field Programmable Gate Arrays*, 4th edn. (Springer, Heidelberg, 2014)

[PH98] D. Patterson, J. Hennessy, *Computer Organization & Design: The Hardware/Software Interface*, 2nd edn. (Morgan Kaufman Publishers, Inc., San Mateo, 1998)

[Par05] B. Parhami, *Computer Architecture: From Microprocessor to Supercomputers*, 1st edn. (Oxford University Press, New York, 2005)

[RGS93] J. Rose, A. Gamal, A. Sangiovanni-Vincentelli, Architecture of field-programmable gate arrays. Proc. IEEE, 1013–1029 (1993)

[Row04] C. Rowen, *Engineering the Complex SOC*, 1st edn. (Prentice Hall, Upper Saddle River, 2004)

[Sta02] W. Stallings, *Computer Organization & Architecture*, 6th edn. (Prentice Hall, Upper Saddle River, 2002)

[TI83] TI: TMS3210 Assembly Language Programmer's Guide, digital signal processor products (1983)

[TI95] TI: TMS320C1x/C2x/C5x Assembly Language Tools user's Guide, digital signal processor products (1995)

[VG02] F. Vahid, T. Givargis, *Embedded System Design* (Wiley, Hoboken, 2002)

[Wol08] W. Wolf, *Computers as Components* (Morgan Kaufmann Publishers, Inc., San Mateo, 2008)

[Xil93] Xilinx: PREP benchmark observations, in *Xilinx-Seminar San Jose* (1993)

[Xil02a] Xilinx: Creating Embedded Microcontrollers, www.xilinx.com, Part 1–5 (2002)

[Xil02b] Xilinx: Virtex-II Pro, documentation (2002)

[Xil05a] Xilinx: PicoBlaze 8-bit Embedded Microcontroller User Guide, www.xilinx.com (2005)

[Xil05b] Xilinx: MicroBlaze – The Low-Cost and Flexible Processing Solution, www.xilinx.com (2005)

Chapter 2
FPGA Devices, Boards, and Design Tools

Abstract In this chapter the specific details of SoC-type FPGAs, prototyping boards, and design tool that are particularly beneficial for embedded system design are discussed in more details. Embedded SW designer may skip this chapter at first reading.

We will take a closer look at two special devices, the Altera EP5CSEMA5F31C6 and the Xilinx Zynq-Z-7010, that will be used in our larger design examples. We will study chip synthesis, timing analysis, floorplan, and power consumption. Finally, in a larger case study, the ultimate reduced instruction set computer (URISC) is designed in HDL, synthesized, and programmed.

Keywords Field-programmable gate array (FPGA) · Field-programmable logic device (FPLD) · Quartus · Vivado · Altera FPGAs · Xilinx FPGAs · TerASIC · Video graphics array (VGA) · High definition multimedia interface (HDMI) · Floor plan · URISC · ZyBo · DE1-SoC · TimeQuest · I2C bus · Universal asynchronous receiver/transmitter (UART) · 7-segment display · Analog-to-digital converter (ADC) · Audio CODEC · CAD design circle

2.1 Introduction

The levels of detail commonly used in VLSI designs range from a geometrical layout of full custom ASICs to system design using the so-called set-top boxes. Table 2.1 gives a survey. Layout and circuit-level activities are absent from FPGA design efforts because their physical structure is programmable but fixed. The best utilization of a device is typically achieved at the gate level using register-transfer level design languages. Time-to-market requirements, combined with the rapidly increasing complexity of FPGAs, are forcing a methodology shift toward the use of intellectual property (IP) macrocells or mega-core cells. Macrocells provide the designer with a collection of predefined functions, such as microprocessors or UARTs. The designer, therefore, need only to specify selected features and attributes

© Springer Nature Switzerland AG 2021   43
U. Meyer-Baese, *Embedded Microprocessor System Design using FPGAs*,
https://doi.org/10.1007/978-3-030-50533-2_2

Table 2.1 VLSI design levels

Object	Objectives	Example
System	Performance specifications	Computer, disk unit, radar
Chip	Algorithm	μP, RAM, ROM, UART, parallel port
Register	Data flow	Register, ALU, COUNTER, MUX
Gate	Boolean equations	AND, OR, XOR, FF
Circuit	Differential equations	Transistor, R, L, C
Layout	None	Geometrical shapes

(e.g., accuracy), and a synthesizer will generate a hardware description code or schematic for the resulting solution.

A key point in FPGA technology is, therefore, powerful design tools to:

- Shorten the design cycle.
- Provide good utilization of the device.
- Provide synthesizer options, i.e., choose between optimization speed vs. size of the design.

A CAE tool taxonomy, as it applies to embedded systems FPGA design flow, is presented in Fig. 2.1. System requirements and hierarchy organization should be well-documented before entering the FPGA design flow.

The design entry can be graphical, text-based, or through a high-level system tool for the softcore and HDL components. A formal check that eliminates syntax errors or graphic design rule errors (e.g., open-ended wires) should be performed before proceeding to the next step. Allocation of predefined block including embedded hardwired processor system (HPS) through Qsys[1] (Altera) is also done within the first design step. A wide variety of test programs for the microprocessor should also be developed (not shown) in parallel to the μP design. In the function extraction, the basic design information is extracted from the design and written in a functional netlist. The netlist allows a first functional simulation (aka RTL level simulation) of the circuit and enables the construction of an example data set called a test bench for later testing of the design with timing information. The functional netlist also allows an RTL view of the circuit that gives a quick overview of the circuit described in HDL. If the RTL view verification or functional simulation is not passed, we start with the design entry again. If the functional test is satisfactory, we proceed with the design implementation, which usually takes several steps and requires much more compile time than the function extraction if the component is soft or parameterized cores and not hard cores. At the end of the design implementation, the microprocessor circuit is completely routed within our FPGA, which provides precise resource data and allows us to perform a simulation with all timing delay information (aka gate-level simulation) as well as performance measurements. Some synthesis tools also offer a technology map view of the circuit that shows how the HDL elements are mapped to LUTs, memory, and embedded multi-

---

[1] Qsys has recently been renamed by Intel to PLATFORM DESIGNER

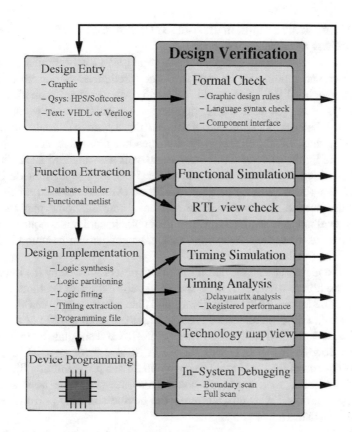

Fig. 2.1 Embedded µP FPGA-based CAD design circle

pliers. If all these implementation data are as expected, we can proceed with the programming of the actual FPGA; if not we have to start with the design entry again and make appropriate changes in our design. Using the JTAG interface of modern FPGAs, we can also directly monitor data processing on the FPGA: we may readout just the I/O cells (which is called a boundary scan), or we can read back all internal flip-flops (which is called a full scan). Modern software tools (e.g., monitor program) may also allow us to go through the microprocessor program step by step and inspect memory and register values. If the in-system debugging fails, we need to return to the design entry.

In general, the decision of whether to work within a fixed, parameterized, or soft microprocessor core is a matter of personal taste, prior experience, and system requirements. A graphical presentation of a system can emphasize the possibility of component-based hierarchy allowing a modular design approach and test. The HDL textual environment is often preferred if the design does not need many large components regarding algorithm control design and allows a wider range of design styles, as demonstrated in the case study later. Specifically, for HDL-based designs, it seemed that with text design, more special attributes and more precise behavior can be assigned in the designs. In a text-based HDL design, if it is VHDL or Verilog, we may use any of the three design strategies:

The *structural* style (via component instantiation) is very similar to graphical netlist design and should only be used for large (predefined or reused) component, never on a gate level since this would make the code very hard to read, verify, and maintain. The *data flow* style (i.e., concurrent statements) is used when a cloud of gate or multiplexer-type circuits are used, but these codings should not have any memory elements. Synthesis tool typically does not recognize a latch or flip-flop when designed with gates or multiplexers and will use standard LUT but not the physical provided flip-flop or latch within the FPGA. A *sequential* design using PROCESS templates (VHDL) or always block (Verilog) is used when memory elements are needed. Of course, these can have combinational part too, just be aware that evaluation is sequential within these blocks and not concurrent as in the data flow part of the code. Details on the HDL languages can be found in Chaps. 3 (VHDL) and 4 (Verilog).

All HDL code starts with the definition of I/O ports followed by the internal nets. In VHDL the library used needs to be specified before the port declarations. VHDL has a larger and more sophisticated library and data type concept than Verilog. Then the actual circuit description starts. That can have one coding style or all three discussed above. A detailed study of how to synthesize and simulate a circuit (according to our flow from Fig. 2.1) will follow in Sect. 2.4 for an URISC processor model. At the end of the CAD tool flow, we will have a programming file ready that can be downloaded to a hardware board (like the prototype boards shown in Fig. 2.2). We proceed with programming the device and may perform additional hardware tests using the read-back methods.

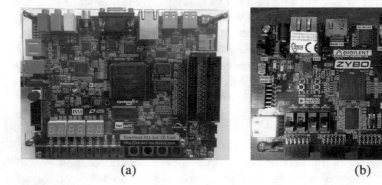

(a)　　　　　　　　　　　　　　　　　　　(b)

Fig. 2.2 Low-cost prototype boards. (**a**) Cyclone V DE1-SoC Altera/TerASIC board [Ter14]. (**b**) Xilinx Zynq Board (ZyBo) [Dig14] with onboard ADC and DAC CODEC

2.2 Selection of Prototype Board

If we like to select an appropriate platform that allows soft, parameterized, and hardcore processor systems, then the number of possible prototype boards is reduced to a few that have a HPS embedded. We also need to provide enough resources (LE, embedded multipliers, block RAMs, PLLs, and number of pins) to host the largest design we plan. On the other side, we may want to select a (low-cost) board that is available through Altera or Xilinx University Program.

Xilinx has very limited direct board support; all boards, for instance, available in the university program are from third parties. However, some of these boards are priced so low that it seems that they are not-for-profit designs. For Xilinx boards the primary provider is Digilent Inc. with the ZedBoard, ZyBo, Nexys 4 DDR, or Basys 3 at time of writing. These are low-cost even for nonuniversity customers. Another goal may be to use boards that are supported by the VIVADO web edition software. Currently the Zynq Evaluation and Development Board (ZedBoard) and Zynq Boards (ZyBo) are supported in the VIVADO web edition. Nexys and Basys will require the use of ISE tool. Table 2.2 gives an overview of some popular Xilinx boards. A good low-cost board for embedded system that has a HPS (with on-chip BRAM, multipliers, audio CODEC, VGA, switches, buttons, and LEDS) is the Zynq Board (ZyBo) offered by Digilent Inc. for only $189–$299; see Fig. 2.2b. The ZyBo has a Zynq XC7Z010-1CLG400C FPGA, 128-Mb flash, 512-MByte DDR, five LEDs, four switches, and six push buttons [Xil18, Dig14]. For DSP and video

Table 2.2 Overview of popular Xilinx boards, their FPGAs, and board components

Board name	ZedBoard	ZyBo[1] /ZyBo-Z7-10	ZyBo-Z7-20
FPGA device	Zynq-7K SoC XC7Z020-CLG484	Zynq-7K XC7Z010-1CLG400C	Zynq-7K XC7Z020-1CLG400C
I/O/LUT/on-chip RAMs	200/53K/612KB	100/17K/270 KB	100/53K/630 KB
Microprocessor	2x ARM Cortex-A9	2x ARM Cortex-A9	2x ARM Cortex-A9
Switches/buttons/ LEDs	8/5/8	4/6/5	4/6/5
Off-chip flash/ DRAM	256 MB/512 MB	–/512 MB	–/1024 MB
Display option	OLED-VGA-HDMI	VGA-HDMI-0/1RGB LEDs	HDMI-Pcam
A/D option	CODEC and XADC	CODEC and 6/5 Pmod	CODEC and 6 Pmod
Price	$495	$189/$199	$299

[1]Recently the ZyBo has been replaced by the ZyBo Z7 boards. Major updates are an additional (unidirectional) HDMI port instead of the VGA and a second version ZyBo-Z7-20 with a larger FPGA [Dig19]

Table 2.3 Overview of popular Altera boards, their FPGAs, and board components

Board name	DE2-115	DE10 standard	DE1 SoC
FPGA device	Cyclone IV EP4CE115F29C7	Cyclone V SoC 5CSXFC6D6F31C6N	Cyclone V SoC 5CSEMA5F31
I/O/LUT/on-chip RAM	529/115K/497 KB	499/111K/707 kB	457/85K/487 KB
Microprocessor	no HPS	2x ARM Cortex-A9	2x ARM Cortex-A9
Switches/buttons/LEDs	18/4/27	10/4/10	10/4/10
Off-chip SRAM/DRAM	2 MB/128 MB	–/64 MB+1 GB	–/64 MB+1 GB
Display options	8 × 7-seg/LCD/VGA/TV decoder	6 × 7-seg./LCD/VGA/TV decoder	6 × 7-seg./VGA/TV decoder
A/D option	24-bit 2 CH. CODEC	8 × 12-bit 500KSPS	24-bit 2 CH. CODEC 8 × 12-bit 500KSPS
Price (std)	$595	$350	$249

experiments, we can take advantage of the on-chip A/D, audio CODEC, and the VGA port.

Altera supports several development boards with a large set of useful prototype components including fast A/D, D/A, audio CODEC, DIP switches, single and seven-segment LEDs, and push buttons (Table 2.3). These development boards are available from Altera directly (university customer) or from TerASIC (standard customer). Altera offers Stratix and Cyclone boards, in the $199–$24,995 price range, which differ not only in FPGA size but also in terms of the extra features, like number, precision, and speed of A/D channels, and memory blocks. For universities a good choice will be the low-cost DE1-SoC Cyclone V board, which is still more expensive than the UP2 or UP3 boards used in many digital logic labs but has a two-channel CODEC, large memory bank outside the FPGA, and many other useful ports (USB, VGA, PS/2, Ethernet, seven segment LEDs, switches, bush buttons, etc.); see Fig. 2.2a.

Let us now have a brief look at common peripherals we have on the boards and will use in later projects.

Memory

FPGA embedded system prototyping boards typically have several different types of memory outside of the FPGA. Let us start with the FPGA configuration memory. Since SRAM-based FPGAs are volatile, they require a configuration file to be downloaded to the device onetime initially that has the FPGA circuit programming information as well as initial on-chip register and block RAM values. During development we download these files via a serial port such as the USB cable. Typically, this communication is done using a JTAG protocol such that partial reconfiguration as well as read-back is enabled. After development, we may wish to use the board without a host computer for configuration, and then often boards provide a flash E2PROM to store the FPGA programming information. These flash configurations ROM are used for programming only and typically can't be part of an overall microprocessor system design to load or store data from.

The second type of memory is SRAM or DRAM memory banks with several MB up to GB in size much larger than the on-chip block RAM resources. SRAM typically have a short access time, while DRAMs typically provide larger memory size. If we use the RAMs as microprocessor memory, we need to be careful with the organization of the memory bank. Other than the block RAM in the FPGA, these external memory banks are fixed in the configuration. For instance, if we build a 32-bit microprocessor system but have a 16-bit memory interface, we need two clock cycles to load one word. A fast 16-bit-wide SRAM may then be slower than a "slower" 32-bit-wide DRAM since DRAM data would available after only one clock cycle.

The third type of external FPGA memory on embedded system boards is often provided through a micro SD card slot. While access time is typically slower than S(D)RAM, the amount of memory is only limited by the size of SD card you buy (typically not provided). This would allow you to compile and configure, for instance, an operating system with a standard PC or copy large amount of data (images, video, etc.) on the SD card and then access these files with your FPGA board through the micro SD card.

Basic I/O Components

There are typically four basic I/O elements on FPGA board: LEDs, seven-segment display, buttons, and switches; see Fig. 2.3a.

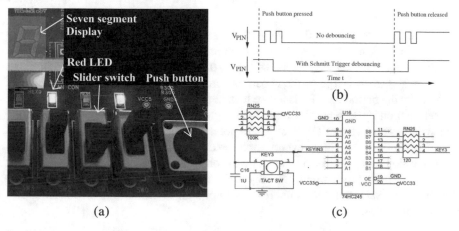

Fig. 2.3 Basic I/O. (**a**) LEDs, buttons, switches, and seven-segment display. (**b**) Timing behavior of switches with and without debouncing. (**c**) Circuit details of the DE1 SoC debouncing logic

As *output* devices all boards have several single LEDs in red or green color or both. They light up when set to logic one and are off when set to logic zero. FPGA board with more space and FPGA I/O pins may add one or more seven-segment displays. Other than single LEDs, the seven-segment displays are often active low, i.e., light up when the input is set to zero since FPGAs or CMOS in general can draw more current at GND than through *Vcc*. Since several seven-segment displays on a board will require a substantial number of FPGA pins, we see sometimes (e.g., Digilent Nexys) that the seven-segment display share the same data line and use a select signal to light up in multiplex mode. The small ZyBo has no seven-segment display, while the DE1 SoC has six non-multiplexed digits where the decimal point is unused.

As basic *input* peripherals to the FPGA, we find switches and buttons. Switch "south" is often used as zero-input logic, while "north" inputs a logic one to the FPGA. ZyBo has four slider switches, while DE1 SoC has ten. The buttons allow us to produce shorter input enable signals to the FPGA than with a slider switch. When using switches or buttons during the switch phase, these devices have the tendency to toggle between zero and one within 2–3 ms of the switching process, i.e., the switch bounces around as shown in Fig. 2.3b. On some boards these buttons are therefore "debounced," i.e., allow a safe single rising or falling edge transition when the button is pressed or released. Debouncing can be done in different ways using external components or with internal FPGA resources. We may use a line diver with hysteresis, the so-called Schmitt trigger, such as the IC 74HC245. These will have a higher threshold for a $0{\rightarrow}1$ transition than for a $1{\rightarrow}0$ transition. In addition, for debouncing the DE board uses a resistor (100 KΩ) and a relatively large capacitor (1 µF). The capacitor is placed parallel to the button to ensure a slower more stable behavior. After the button is released, the capacitor will be charged using the RC time constant according to

$$U_C = \left(1 - e^{-t/(RC)}\right)Vcc \tag{2.1}$$

and will reach 50% when $0.5 = (1 - \mathrm{e}^{-T/(RC)})$ and for our $R = 100$ KΩ and $C = 1$ µF gives $T = R\,C\,\ln(2) \approx 70$ ms, i.e., the minimum time between toggling a button, which is definitely on the safe side. The $1{\rightarrow}0$ will be much shorter since this is just limited by the discharge of the capacitor through the button. As we see also from Fig. 2.3c, the button on DE1 is active low, i.e., will input a logic zero when pressed, other than on ZyBo where its logic is one when a button is pressed. ZyBo buttons, or switches on ZyBo or DE boards, are not debounced and should therefore not be used when counting single events such as stepping through a microprocessor program in single steps. If that is needed, we should add a FPGA debounced circuit that "samples" the input (using a flip-flop and a counter with output enable) every ca. 10 ms or so such that the bouncing does not lead to multiple detection in subsequence modules.

Display Options

FPGA development boards (see Fig. 2.3a) usually come with one or more basic display options such as LCDs, LEDs, and seven-segment displays. In addition, boards often support graphic output such as VGA or HDMI for images and video. The Xilinx ZyBo, for instance, comes with a video graphics array (VGA) interface and with dual-role (Source/Sink) ports for the high-definition multimedia interface (HDMI) that can be used for video and audio. The Altera DE2 SoC has a TV decoder for PAL, SECAM, and NTSC and support for VGA. The VGA interface is one of the classic standards, still supported by most PCs as a default start-up interface to display system information on a CRT monitor. The basic resolution is 640 columns and 480 rows with a refresh rate of 60 Hz. We will use this basic VGA mode since it is supported by most monitors; even LCD monitors usually have a VGA input. This will allow us to use our ZyBo and DE1 SoC directly with a CRT monitor without buying any new display HW.

The basic VGA system setup is shown in Fig. 2.4 used on the DE1 SoC board. The FPGA provides the necessary control and RGB data signals. The VGA chip used on the DE1 is the Analog Devices ADV7123KSTZ140 that includes triple high-speed 10-bit DACs and can run with up to 140 MHz pixel rate [AD98]. We can therefore use VGA (640 × 480), SVGA (800 × 600), XGA (1024 × 768), or SXGA (1280 × 1024) display modes. Note that the DE1 board uses the most significant 8-bit RGB data rather than the 10-bit data of the ADV7123 triple DAC IC since dealing with 10-bit data was too cumbersome and most images have 8-bit RGB resolution anyway. The ZyBo board on the other side has no dedicated VGA chip. Instead it used a R-2R ladder array to transform the RGB digital data directly into analog signal. It used 6 bits for green and 5 bits each for red and blue, for a total of 16 bits or 65K different colors. HSYNC and VSYNC signals are generated with the FPGAs.

Table 2.4 shows the modes and the required pixel clocks (total rows × columns × frame rate), number of pixels (measured in M10K blocks), and grayscale image memory requirements, i.e., rows × columns × 8. Color images need three times as much memory. Note that even a black-white picture with 640 × 480 resolution will

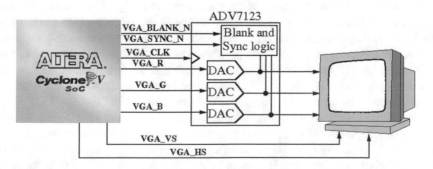

Fig. 2.4 The overall VGA configuration on the DE1 SoC board

Table 2.4 VGA configuration data for 60 Hz refresh rate

Mode	Resolution columns × rows	Total C × R	Pixel clock (MHz)	Pixel (10K)	Size, gray 8 bits (MBits)	Size, color 3 × 8 bit (MBits)
VGA	640 × 480	800 × 525	25.175	30.72	2.4	7.37
SVGA	800 × 600	1056 × 628	40.0	48	3.84	11.52
XGA	1024 × 768	1344 × 806	65.0	78.64	6.29	18.87
SXGA	1280 × 1024	1688 × 1066	108.0	131.07	10.49	31.46

already require over 30 embedded memory blocks of size M10K to store the image on chip. Color or larger images most likely will require additional off-chip memory. The overall timing of a VGA signals consists of the video data and additional synchronization time since we need the CRT beam to be allowed to return to the beginning of the line or the first row. We assume that external synchronization is used and not embedded synchronization using the green video channel. The line in the VGA then starts with an active low horizontal synchronization impulse. Let us have a brief look at the timing of the VGA signals at 60 Hz. The synchronization is 3.8 μs long, and the pixel clock of 25 MHz turns out to be 96 clock cycles. Next comes the horizontal back porch of 1.9 μs or 48 cycles. Then for 640 clock cycles or 25.4 μs, the video signal is present, followed by the front porch signal of 0.6 μs or 16 clock cycles that allows the CRT beam to return to the beginning of the line. Overall it takes 96 + 48 + 640 + 16 = 800 clock cycles for a single line. For the vertical timing, we have a similar specification. Again, let us look at the default VGA signals with 480 lines. The vertical synchronization impulse takes 2 lines, followed by the vertical back porch of 33 lines. Then the next 480 lines contain the video signal, and finally 10 lines of the vertical front porch allow the CRT beam to return to the first line. Overall it takes 2 + 33 + 480 + 10 = 525-line periods to complete the image.

Analog Interface

Embedded systems are often used to monitor, control, or process analog data, e.g., video or audio signals, voltage, temperature, light intensity, speed, accelerometer, etc. FPGA boards therefore often have analog-to-digital converter (ADC) and digital-to-analog converter (DAC). Some of the news sixth- and seventh-generation Xilinx FPGAs even have fast ADC on chip. The Xilinx Zynq-7K device has two ADCs on chip (12 bits, 1 MSPS) that has additional sensors for temperature and power supply. The ADCs have a 17-channel input analog multiplexer that allows to monitor many analog signals at the same time; see Fig. 1.12b.

Before we start with the technical discussion of ADC and DAC, let us briefly review two important design considerations when dealing with analog signals conversions: the *sampling rate* and the *number of bits* to be used. The task to determine an appropriate sampling rate is thanks to the Shannon theorem straight forward.

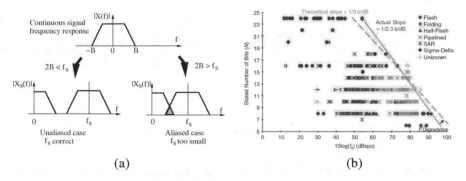

Fig. 2.5 (**a**) Shannon sampling theory graphically explained. (**b**) Overview on ADC principles ©
2005 IEEE [LRR05]

We know that when sampling an analog signal, the Fourier spectrum becomes periodic with the sampling frequency f_S, and as a result, we should use a sampling rate that is twice as high as the highest frequency component of the input signal; otherwise the spectrum will overlap, and the so-called aliasing occurs; see Fig. 2.5a.

So if we like to monitor a power line signal of 60 Hz, a 120 Hz sampling frequency should be minimum according to Shannon. We typically will use a few Hz higher sampling frequency since the antialiasing filter that has a 0–60 Hz passband and a stopband starting at 60+ Hz cannot be built with zero transition band. A realizable filter will have a transition band of a few Hz at least, so our sampling rate should be at least twice as high as the stopband frequency when enough suppression is achieved. The suppression required can be determined from our second major ADC parameter, the number of bits. Now the number of bits unfortunately cannot be determined straightforward mathematically as the sampling rate since this may depend on physiology. Our ear, for instance, are very sensitive to noise, and a high signal-to-noise (S/N) ratio are required so that we find audio signals acceptable. Typical audio signals are therefore processed with 16 bits or more; our eye, on the other side, are less sensitive to quantization noise, and ca. 8 bits are often enough, for calculations color even less. Each bit now contributes about 6 dB to the overall S/N quality (see Exercise 2.46). A good audio anti-aliasing filter therefore should have $16 \times 6 = 96$ dB suppression, while an image processing anti-aliasing filter may need only $8 \times 6 = 48$ dB. We should therefore use a sampling frequency that is at least twice as high as the frequency when the aliasing filter reaches these desired suppressions of the input spectrum.

Now with the two design parameters: sampling rate and number of bits specified for our system, we should start the search for the appropriate ADC and DAC; see Fig. 2.5b for an overview on frequently used ADC architectures. Since DAC are often easier to understand, let us first have a look at typical designs. This will require a little bit of knowledge in electrical circuit that should be part of a modern physics high school curriculum, if not you may just look up the basic of *R/L/C* networks and operational amplifier (OP) from an introduction to EE textbook.

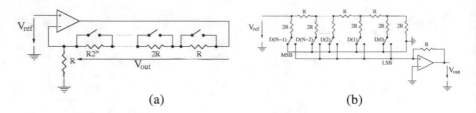

Fig. 2.6 DAC architectures (**a**) $R2^N$ and (**b**) the $R/2R$ network architectures

The $R2^N$ DAC design implements a network with increasing resistor values; see Fig. 2.6a. The OP with its zero-input current will guaranty that the current is always constant at $I = V_{ref}/R$. So, the closing or opening of the switch will add an equivalent voltage values to the output voltage V_{out} since the OP ensures that the current through the $R2^N$ network remains constant. However, from a VLSI standpoint, this is not a very efficient design for larger bit width since the VLSI area needed to build the resistors network is proportional to the size, so a resistor $R2^N$ will require 2^N times the area as the resistor R. That is why the $R/2R$ network shown in Fig. 2.6b that may look more complicated usually gives a smaller design. Independent of the switch position, all outputs (MSB... LSB) are connected to the virtual ground from the OP. The current is divided by 2 in each stage since the parallel circuit of $2R||2R = R$ and $R + R = 2R$. The output voltage is therefore proportional to the sum of the currents, which reflect the coding of the input data word $D(N - 1...0)$. Speed of the converter can be very high and depends only on the speed of the switches. ADC typically requires substantial more hardware resource than an OP and a few registers as shown next.

From the technology overview given in Fig. 2.5b, we conclude that only a few different types of ADC are in use today. The figure vertical axis shows the number of bits and the horizontal the maximum sampling rate. As can be seen from these technologies overview, typically three different ADC types are in use: SAR, *flash converter*, and $\sum\Delta$ ADCs.

Somewhere in the middle in terms of speed and resolution is the successive-approximation register (SAR) analog-to-digital converter. The converter used the just discussed DAC to successively reduce the difference between the input that is typically sampled with a sample and hold (S&H) circuit or track and hold (T&H) and the output of the DAC. This is an iterative process that may take a couple of clock cycles to find the best match between the sampled vales and the DAC output, but on the plus side, the required resources are low. Such a converter from linear technology (LTC2308) is used on the DE1 SoC to provide a 500 kSPS 8-channel 12-bit converter; see Fig. 2.7.

The high-speed, low-resolution converter with 8–10 bits precision and sampling rate over 1 GSPS are only achieved with *flash converter* or half-flash converter that is typically used in video signal processing or high-speed digital oscilloscopes. These consist of a series of resistor/operation amplifiers that "measure" the input using the OP-AMPs in comparator mode, i.e., with an open feedback loop. If the

Fig. 2.7 The successive approximation register (SAR) analog-to-digital converter used on DE1 SoC

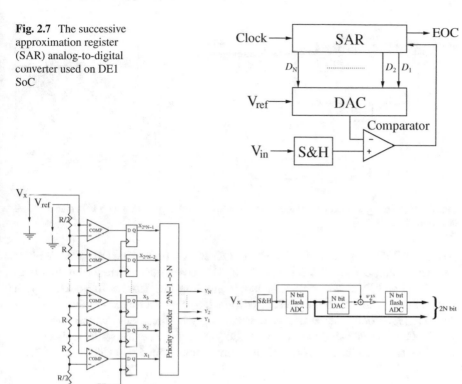

(a) (b)

Fig. 2.8 (**a**) High-speed, low-resolution flash converter. (**b**) Two-flash converter combined in the half-flash converter

input voltage is larger than the specified R array divider value, the output of the OP-AMP will switch, same as all below this OP with lower reference voltage too. Looking at all outputs, we need to find the OP with the highest priority that will determine the input voltage. For 8-bit data, we will need 256 OP-AMPs which limit this method to small bit width. A half-flash will use two smaller flash converters and subtract the first coarse quantization with a second fine-tuned conversion. So, for 8-bit converter, we would need $2 \times 16 = 32$ instead of the 256 converters; see Fig. 2.8b.

The other extreme are the high-resolution, low-speed converter of the $\Sigma\Delta$-type. Here the input is sampled at a very high rate, and these samples are accumulated in an integrator in a feedback loop such that the integrator follows the input and the number of $+1$ vs. -1 impulse indicates a large or small input analog signal. This 1-bit high-speed value sequence is then filtered with a digital filter with downsampler producing the high bit width output data. The integrator loop will also "shape" the noise spectrum moving the quantization noise to higher frequency so that very high resolution can be achieved; see Fig. 2.9b. Such converters are part of the

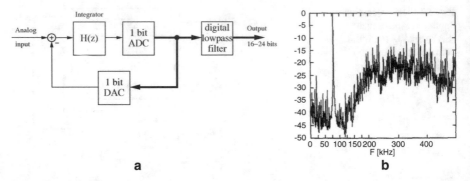

Fig. 2.9 (a) The $\sum\Delta$ converter principle. (b) Spectrum of the $\sum\Delta$ for a 77.5 kHz signal before downsampling [Hut92]

Wolfson WM8731 audio coder/decoder (CODEC) we find on the DE1 SoC [Wol04] and the Analog Device SSM2603 on the ZyBo. These devices are 24-bit dual (stereo) ADC and DAC with programmable sampling rate from 8 to 96 kHz. Both boards have standard 3.5 mm stereo audio jack to connect directly audio signals to the CODEC. Configuration and data streaming with these devices are done via a serial I²C communication protocol that allows to set the device status registers; for sampling rate, data format (16/24/32 bit; stereo/mono), and input gain, see Fig. 2.12.

Communication

A large variety of low- and high-speed communication protocol are in use today in embedded systems, and our boards provide support for the most popular formats. For low bandwidth such as UART, CAN, I²C, PS2, SPI, and IR emitter/receiver, most of the protocol processing can be done using FPGA on-chip resources since the data rate is not too high and the protocols are not too complicated. For high bandwidth communication such as USB 1/2/3 or 10/100/1000 Mbits/s Ethernet, we need beside special connector electrical or optical now also more complex (frame) processing that typically is done with external physical ICs (PHY). The ARM cortex A9 also support several standards with custom IP on chip: 2×SPI, 2× I²C, 2×CAN, 2×UART, 2×USB, and 2×Ethernet. However, a 56-pin multiplex I/O (MIO) is used for these IP blocks such that often only two at a time are connected to I/O pins in one system configuration. Let us briefly review the most important protocols use on both boards.

The *Ethernet* communication is an extreme fast communication widely used in local area network (LAN) technology that is supported on the two boards with 10 Mbit/s (called 10Base-T), 100 Mbit/s (100 Base-TX), and 1 Gbits/s (1000 Base-T) transmission rates that have been standardized by IEEE 802.3. Initially proposed by DEC, Intel, and Xerox as carrier-sense multiple access collision detection (CSMA/CD) protocol, it is now used today in over 80% of LANs. Transmission is synchro-

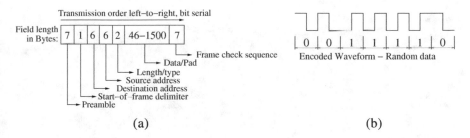

Fig. 2.10   The Ethernet standard frame: (a) frame, (b) signal coding

nous serial over standard CAT-5 unshielded twisted pair cable used for PCs. The two-cable differential lines TRD+ and TRD- are augmented by a pair of LEDs that indicate the link speed (10/100/1000) and activity/no activity on the wires. To avoid baseline wandering, signals are coded with Manchester code; each bit has ether a halve clock length 0 or 1 value, i.e., a transmitted 0 has a falling edge and a 1 a rising edge transition in the middle of the clock; see Fig. 2.10.

On our boards we find that for Ethernet, a Micrel KSZ9021RN PHY chip is used on the DE1, and the Realtek RTL8211E-VL PHY implements a 10/100/1000 Ethernet option on the ZyBo. The physical signal used in Ethernet cable for a typical Ethernet frame is shown in Fig. 2.10. The preamble (PRE) has 7 bytes and consists of alternating 1/0 for synchronization. The start of frame delimiter (SFD) has 1 byte alternation 0/1 and ending with 2 ones. The destination address (DA) has 6 bytes, or 46 bits, same as the source address (SA). The length of the data block uses 2 bytes, but only 1500 bytes data block are allowed, which follow next. A number larger than 1536 will indicate an optional type frame. The frame check sequence (FCS) is a 32-bit cyclic redundancy check word that checks DA, SA, length, and data.

The *Universal Serial Bus* (USB) is a widely used PC standard; millions of flash drives, as well as our FPGA programming cable, uses USB. Other than many other standards, the USB also provides power supply, sometimes enough to power a whole FPGA board as in Xilinx Nexys, and in order to avoid damages to the USB components, it uses different connectors for master and slave. The original USB 1 allowed data rate of full speed at 12 Mbits/s, the USB 2 standard supports 480 Mbits/s, and the newest USB 3 up to 10 Gbits/s.

The DE1 SoC used for USB a SMSC USB3300 controller, ZyBo uses a FTDI FT2232HQ USB bridge, and a Microchip USB3320 transceiver chip supports the USB 2.0 standard, and we will therefore discuss briefly USB 2.0 in the following. The USB 2.0 has four pins: V_{CC} at 5 V, Ground, and the differential signals Data- and Data+, which show opposite voltages most of the time. The signal coding in USB is done with a non-return-to-zero inhibit (NRZI) code with a transition for each zero and no transition for a binary one; see Fig. 2.11 for a coding example. Only at the end packet an all-zero signal is used, called single-ended zero.

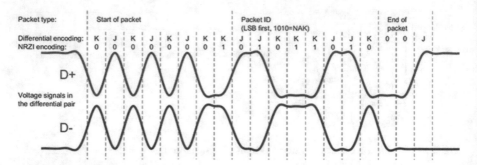

Fig. 2.11 Example USB signals, i.e., a handshake packet, sync sequence; data sequence not accepted PID coding, end of packet

The single bits are grouped into packages. USB uses handshake, token, and data packages class to a total of 16 packages types, which are identified by an 8-bit transmitted identifier. Before we can send a data block, we need to set up the transfer using an address token (IN/OUT) that specify the direction (to-device or to-host) and device number (7 bits and 11-bit address and 5-bit CRC). A data packages start with the sync packet 00000001_2, or, KJKJKJKK, see Fig. 2.11, followed by 8-bit PID. Next are the 0–1024-byte data, followed by a 16-bit CRC check and the EOP signal; see [Axe14] for complete details.

Communication with audio CODEC (Wolfson WM8731 CODEC on the DE1 SoC and the Analog Device SSM2603 on the ZyBo) is done via Inter-Integrated Circuit (I^2C) protocol on both boards. We therefore should look at this standard in more detail. The DE1 has three I^2C masters (HPS1, HPS2, and FPGA) and uses I^2C communication besides the CODEC for two more slaves, the TV decoder (w/r address 0×40 and 0×41) and G-sensor (w/r address 0xA6and 0xA7). I^2C is a serial protocol that mainly uses two wires: a unidirectional clock called SCLK and a bidirectional data signal called SDIN (Fig. 2.12). Both data and clock are wired OR or open drain signal that have a pull-up register, with $R = 2.2$ K for DE1 SoC and $R = 1.5$ K on ZyBo. The master uses GND for logic zero and high impedance for logic one that results to logic one through the pullup registers. High impedance values are displayed in MODELSIM by a blue line at 50% level between zero and logic one levels see Fig. 2.13 below.

Multiple masters are allowed, where the first master lowers the data signal, SDIN will own the bus. The DE1 however has an extra multiplexer chip (TI: TS5A23157) for the three masters to avoid any collisions especially on the clock line. The I^2C is typically used for CODEC system configuration only, for the whole data communication with the ADC and DAC on the CODEC additional clock, data, and left/right channel, select signal is provided.

Figure 2.13 shows the communication example of the DE1 with the CODEC for the first register. Communication is done in groups of nine clock cycles with 7-bit data/address, a R/$\overline{W}$ bit, and an acknowledge bit from the slave. The DE1 CODEC is written only so R/$\overline{W}$ is always low. In HDL first we divide the incoming 50 MHz

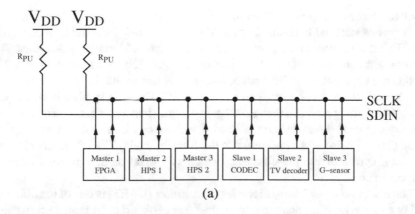

(a)

REGISTER	B8	B7	B6	B5	B4	B3	B2	B1	B0	DEFAULT
R0 Left line in	LRIN BOTH	LIN MUTE	0	0	LINVOL[4:0]					0_1001_0111
R1 Right line in	RLIN BOTH	RIN MUTE	0	0	RINVOL[4:0]					0_1001_0111
R2 Left headphone out	LRHP BOTH	LZCEN	LHPVOL[6:0]							0_0111_1001
R3 Right headphone out	RLHP BOTH	RZCEN	RHPVOL[6:0]							0_0111_1001
R4 Analog audio path control	0	SIDEATT		SIDE TONE	DAC SEL	BY PASS	INSEL	MUTE MIC	MIC BOOST	0_0000_1010
R5 Digital Audio path control	0	0	0	0	HPOR	DAC MU	DEEMPH		ADC HPD	0_0000_1000
R6 Power down control	0	PWR OFF	CLK OUT PD	OSC PD	OUT PD	DAC PD	ADC PD	MIC PD	LINE INPD	0_1001_1111
R7 Digital audio interface format	0	BCLK INV	MS	LR SWAP	LRP	IWL[1:0]		FORMAT[1:0]		0_1001_1111
R8 Sampling control	0	CLKO DIV2	CLKI DIV2	SR[3:0]				BOSR	USB/ NORM	0_0000_0000
R9 Active control	0	0	0	0	0	0	0	0	Active	0_0000_0000
R15 Reset	Reset[8:0]									Not reset

(b)

Fig. 2.12 (a) I2C bus architecture, (b) the DE1 SoC CODEC register assignment

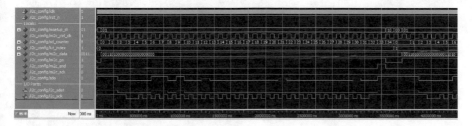

Fig. 2.13 The programming of the first CODEC register to zero

clock to a more appropriate rate for the wired OR processing, say 25 kHz. The transfer sequence starts with the master first driving SDIN low followed by SCLK going low. To set a register values, we would need 1-byte address (0 × 34), then 7-bit register number, and 9-bit register values as in Fig. 2.12b, or a total of 24 bits. This 24 bits are augmented by the 3 acknowledge bits (all should be driven low by the slave); see sd_counter clocks 12, 21, and 30. The stop symbol is coded with clock SCLK going high and then SDIN going high. Then a combination of the three-acknowledgment signal decide the FSM if we can continue with the next register value (lut_index+1) or need to run the transfer again. The whole transfer took 35 clock cycles of the low-frequency clock (25 kHz) as can be seen from the sd_counter.

The *universal asynchronous receiver/transmitter* (UART) is one of the oldest PC and one of the asynchronous standards that was assumed to be more secure due to the use of +/−12 V signal levels. On an old PC, this UART would communicate over the RS232 serial COM port. Many modern PC do no longer have RS232 connectors or ±12 V power supply. Today often cable are connected via USB type and in order to support these legacy standard FPGA embedded system boards provide a UART→USB bridge that use a USB mini-B type connector. On the downside when using USB protocol instead of RS232 voltages, the long cables and high S/N protection are lost. The ZyBo uses a Future Technology Device International Ltd. (FTDI) FT2232HQ bridge and the DE1 a FTDI FT232R. The more sophisticated ZyBo bridge (64 vs. 32 pins) uses a 4-pin JTAG to implement the USB-JTAG circuit. A LED for TXD and RXD show activity on the two data lines.

The UART uses two-wire serial port with a transmitter bit TXD and a receiver bit RXD, while one of the internal, the ARM Cortex-A9 UART cores, will take care of the protocol when the MIO is configured in the right way. Typical default settings are symbol rate f_s = 115,200 baud (unit named after the French communication engineer Emile Baudot), 1 stop bit, no parity, and 8-bit character length. The bit rate is therefore $Rate = f_s \times N$ = 1,152,000 bits/s. The transmission is asynchronous, so the receiver waits for a falling edge (start bit), followed by the data bits (LSB first) and the stop bit. Since no clock is used, a receiver usually monitors the data line at 8–16 oversampling for a falling edge and samples the data in the middle of the bit period or when using an "integrate-and-dump" method at the end to minimize interference. Figure 2.14 shows the 10-bit clock cycle sequence used to transmit the ASCII character S.

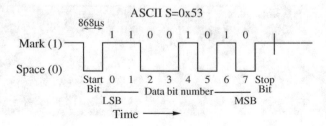

Fig. 2.14 A UART byte transfer

2.3 FPGA Structure

At the beginning of the twenty-first century, FPGA device families now have several attractive features for implementing embedded microprocessors. These devices provide fast-carry logic, which allows implementations of 32-bit (non-pipelined) adders at speeds exceeding 300 MHz [Dil00, Xil93, Alt96], embedded 18 × 18-bit multipliers, large memory blocks, PLLs, and low skew clock networks.

Xilinx FPGA Architecture Overview

Xilinx FPGAs are based on the elementary logic block of the early XC4000 family, and the newest derivatives are called Kintex (low cost) and Virtex (high performance) and Zynq (embedded SoC) (Fig. 2.15). The Xilinx devices have the wide range of routing levels typical of FPGAs.

Since the Zynq-7K device XC7Z010-1CLG400C is part of a popular ZyBo embedded systems board offered by Digilent Inc. (see Fig. 2.2b), we will have a closer look at this FPGA family. The overall device floorplan of the XC7Z010 is shown in Fig. 2.16.

The basic logic elements of the Xilinx Zynq-7K (called slices) are based on the Artix-7 FPGAs and come in two different versions: M and L (no X as in Spartan-6 family), and two slices are combined in a configurable logic blocks (CLB), having a total of eight 6-input one-output LUTs (or sixteen 5-input LUTs), and 16 flip-flops, 256 bit distributed RAM, or 128 bit shift register. Of all slices, 33% are type M and have all these features; 66% slices are of type L and do not have the memory shift register function. Other than for Spartan-6 now, all slices have fast carry logic. Figure 2.15 shows one quarter of a slice and the features for both types. Each LUT in the M type slice can be used as a 64 × 1 RAM or ROM. The Xilinx device has multiple levels of routing, ranging from CLB to CLB to long lines spanning the entire chip.

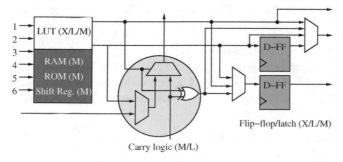

Fig. 2.15 Quarter portion of a Zynq-7K, i.e., Artix-7 slice. The L slice has LUT, two flip-flops, and the fast carry logic. The M slice has all features

Fig. 2.16 The overall
architecture of Zynq-7K
device

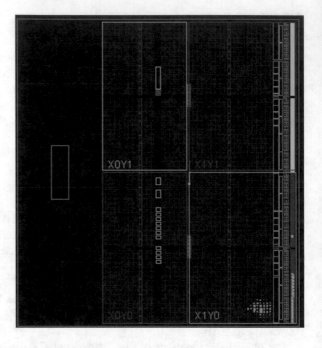

The Zynq-7K device also includes large memory blocks (2 × 18,432 bits or 36,864 bits of data) and can be configured as either two independent 18 Kb RAMs or one 36 Kb used as single- or dual-port RAM or ROM. Each 36 Kb block RAM can be configured as $2^{15} \times 1$, $2^{14} \times 2$, $2^{13} \times 4$, $2^{12} \times 9$, $2^{11} \times 18$, $2^{10} \times 36$, or 512×72, i.e., each additional address bit reduces the data bit width by a factor of two.

Another interesting feature is the embedded multipliers that can save large number of LUTs when multiplier is needed in the microprocessor design. These are fast 25 × 18-bit signed array multipliers. If unsigned multiplication is required, 24 × 17-bit multipliers can be implemented with this embedded multiplier. This device family also includes 2 PLLs and 32 global low skew clock networks that allow one to implement several designs that run at different clock frequencies (or phases) in the same FPGA with low clock skew. The Zynq-7K also includes single- or dual-core ARM Cortex-A9 processor [CEE14] with the following major features:

- 800 MHz single-/dual-core processor
- Dual issue superscalar pipeline with 2.5 DMIPS per MHz
- 32 KB instruction and 32 KB data L1 four-way set-associative cache
- Shared 512 KB, eight-way associative L2 cache for both processors
- 32-bit timer and watchdog

Table 2.5 shows a few more members of the Xilinx Zynq-7K family.

Table 2.5 The Xilinx Zynq 7000 family

Device	LUTs	BRAMs 36 Kbit each	Total Mbits	PLLs	Emb. Mult. 18 × 25-bit	HPS
Z-7007S	14,400	50	1.8	2	66	1xARM
Z-7012S	34,400	72	2.1	3	120	1xARM
Z-7014S	40,600	107	2.5	4	170	1xARM
Z-7010	**17,600**	**60**	**3.8**	**2**	**80**	**2x ARM**
Z-7015	46,200	95	3.3	3	160	2x ARM
Z-7020	53,200	140	4.9	4	220	2x ARM
Z-7030	78,600	265	9.3	5	400	2x ARM
Z-7035	171,900	500	17.6	8	900	2xARM
Z-7045	218,600	545	19.1	8	900	2x ARM
Z-7100	277,400	755	26.5	8	2020	2x ARM

```
XC7Z010S-2CLG484C
  |    | | | |  |
  |    | | | |  |-> Temperature Range(C:0-85;E:0-100;I:-40-100)
  |    | | | |----> Number of Pins
  |    | | |------> Package Type
  |    | |--------> Speed grade (3,2,1 big is fast)
  |    |----------> Single Core Indicator
  |---------------> Part Number: Zynq 7K
```

Fig. 2.17 Device nomenclature for Xilinx Zynq device on the ZyBo board

The ZyBo Xilinx Z-7010 Device

The Xilinx Z-7010 placed on the university program boards are used throughout this book. The device nomenclature is shown in Fig. 2.17 for the FPGA device on the board.

Specific design examples will, wherever possible, target Z-7010 using Xilinx-supplied software. The web-based VIVADO are fully integrated systems with VHDL and Verilog editor, synthesizer, timing evaluations, and bit-stream generator that can be downloaded free of charge from www.xilinx.com. The only limitation in the web version typically is that not all package types and speed grades of every device are supported. The devices we find on prototyping boards are usually supported by these free web license-based tools. Because all examples are available in VHDL and Verilog, any older software version or simulator may also be used. For instance, the Xilinx ISE compiler and the ISIM simulator have successfully been used to compile the examples.

Logic Resources XC7Z010-1CLG400C

The XC7Z010-1CLG400C is a member of the Xilinx Zynq-7K SoC family. The device has four major clock regions each. Each region has 50 rows and 22 column slices and a total of 4400 slices. Each slice has four LUT (total 17600) and eight

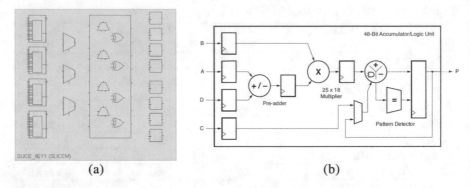

Fig. 2.18 The Zynq (**a**) SLICEM and (**b**) multiplier block architecture

flip-flops, and slices share some control signals such as clock or reset; see Fig. 2.18a. But embedded multiplier, RAM18B, JTEG decoder, and HPS occupy substantial silicon. The HPS alone takes about area of two clock regions. The device also includes three full columns of 36-Kbit memory block or two 18 Kbit blocks (called RAMB36B and RAM18B) that have the height of 5 slices and the total number of RAM36B is therefore $3 \times 2 \times 10 = 60$ or 120 if we use RAM18B. The RAM18Bs can be configured as $2^{15} \times 1$, $2^{14} \times 2$, $2^{13} \times 4$, $2^{12} \times 9$, $2^{11} \times 18$, $2^{10} \times 36$, or 512×72 RAM or ROM. The XC7Z010-1CLG400C also has two full-length columns of 25×18-bit fast array multipliers; see Fig. 2.18b. The highest of the multiplier is 5 slices, and the total number of multipliers is therefore $4 \times 20 = 80$. The lower center section of the XC7Z010-1CLG400C is shown in Fig. 2.16 that shows slices, memory blocks, and I/O resource of the URISC design device floorplan.

Altera FPGA Architecture Overview

Altera devices are based on Flex 10K logic blocks, and the newest derivatives are called Stratix (high performance and SoC) and Cyclone (low cost and SoC). The Altera devices are based on an architecture with the wide busses used in Altera's CPLDs. However, the basic blocks of the Cyclone and Stratix devices are no longer large PLAs as in CPLD. Instead the devices now have medium granularity, i.e., small lookup tables (LUTs), as is typical for FPGAs. Several of these LUTs, called logic elements (LEs) by Altera, are grouped together in a logic array block (LAB) and share some control signals such as clock and reset lines. The number of LEs in a LAB depends on the device family, where newer families in general have more LEs per LAB: Flex10K utilizes 8 LEs per LAB, APEX20K uses 10 LEs per LAB, and Cyclone II-IV has 16 LEs per LAB.

As an example of an Altera device, let us have a look at the Cyclone V SE SoC EP5CSEMA5F31C6 used in the low-cost SoC prototyping board DE1SoC by Altera; see Figs. 2.2a and 2.19. The basic block of the Altera Cyclone V device

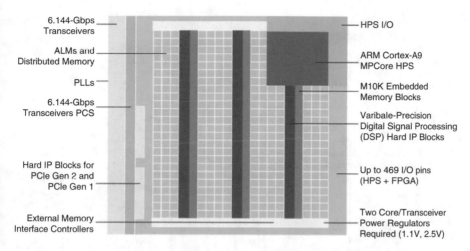

Fig. 2.19 The overall architecture of Cyclone V SoC devices [Alt11]

achieves a medium granularity using small LUTs. The Cyclone V device is similar to the Altera 10K device used in the mature UP2 and UP3 boards, with increased RAM blocks memory size to 10 Kbits, which are no longer called EAB as in Flex 10K or ESB as in the APEX family but rather M10K memory blocks, which better reflects their memory size. The basic logic element in Altera FPGAs is called adaptive logic module (ALM) on the large scale. The ALM consist of several smaller logic element (LE)[2] which includes flip-flops, 2 four-input and 4 three-input LUTs, 32 multiplexer, and 2 dedicated full adders, as shown in Fig. 2.20. Each ALM can be used in normal mode, extended LUT, arithmetic, or shared arithmetic mode. Sixteen LEs are combined in a logic array block (LAB) in Cyclone IV devices. In Cyclone V each LAB has 10 ALM which are equivalent to 20 LEs of Cyclone IV type. Each device contains at least one column with embedded 27×27-bit multipliers and one column M10K memory blocks. One 27×27-bit multiplier can also be used as three-signed 9×9-bit multipliers. The M10K memory can be configured as $2^8 \times 32$, $2^9 \times 16$, 1024×8, 2048×4, 4096×2, or 8192×1 RAM or ROM. In addition, one parity bit per byte is available (e.g., 256×36 configuration), which can be used for data integrity. These M10Ks and LABs are connected through wide high-speed busses as shown in Fig. 2.22. Several PLLs are in use to produce multiple clock domains with low clock skew in the same device. At least 6.5 MB configuration files size is required to program the EP5CSEMA5F31C6. Table 2.6 shows the members of the Altera Cyclone V SE SoC family and Fig. 2.19 the most important component location on the silicon die.

If we compare the two routing strategies from Altera and Xilinx, we find that both approaches have value: the Xilinx approach with more local and less global routing resources is synergistic to logic synthesis use because in many logic opera-

---

[2] Sometimes also called logic cells (LCs) in a design report file

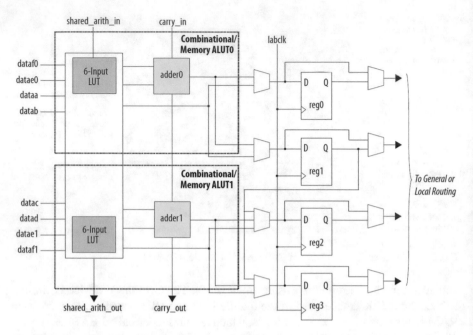

Fig. 2.20 Functional overview of Cyclone V adaptive logic module (ALM) [Alt11]

tions, the data are local. The Altera approach, with wide busses, also has value, because typically not only are single bits processed in bit slice operations but also normally wide data vectors with 16 to 32 bits must be moved to the next processing block.

The Altera EP5CSEMA5F31C6

The EP5CSEMA5F31C6 device for Altera projects placed on the university program boards is used throughout this book. The device nomenclature is shown in Fig. 2.21 for this device.

Specific design examples will, wherever possible, target the Cyclone V device EP5CSEMA5F31C6 using Altera-supplied software. The web-based QUARTUS software are fully integrated systems with VHDL and Verilog editor, synthesizer, timing evaluations, and bit-stream generator that can be downloaded free of charge from www.altera.com. The only limitation in the web version typically is that not all package types and speed grades of every device are supported. The devices we find on prototyping boards are usually supported by these free web license-based tools. Because all examples are available in VHDL and Verilog, any older software version or simulator may also be used.

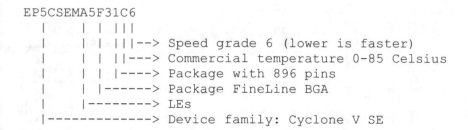

Fig. 2.21 Device nomenclature for Altera Cyclone 5 device on DE1 SoC board

Logic Resources EP5CSEMA5F31C6

The EP5CSEMA5F31C6 is a member of the Altera Cyclone V family. The device
has 80 rows and 88 columns of LABs, and 10 ALM are always grouped together to
build a logic array block (LAB) that shares some control signals such as clock or
reset; but not all LABs are used for ALMs since embedded multiplier, M10K, JTEG
decoder, and HPS occupy substantial silicon. Therefore, the total number of LABs
used for ALMs is not just the product of rows × column. The HPS takes about ¼
area of the device. From Table 2.6 it can be seen that the EP5CSEMA5F31C6
device has 32K ALMs. Since each ALM has two full adders, we have as maximum
number of implementable full adders 64K. The number of LABs that consists of
ALMs is 32,070/10 = 3207. In the left area of the device, two JTAG interfaces are
placed and use the area of about $2 \times 17 \times 9 = 306$ LABs. The device also includes
5 full and 4 half columns of 10-Kbit memory block (called M10K memory blocks)
that have the height of one LAB, and the total number of M10Ks is 397. The M10Ks
can be configured as $2^8 \times 32$, $2^9 \times 16$, 1024×8, 2048×4, 4096×2, or 8192×1
RAM or ROM, where for each byte one parity bit is available. The
EP5CSEMA5F31C6 also has two full-length and one-half columns of 27×27-bit
fast array multipliers that can also be configured as 18×18-bit or 9×9-bit multipliers. A total of 87 multipliers of size 27×27 bits, 174 multipliers of size 18×18 bits,
or 261 of size 9×9 bit are available. The lower center section of the
EP5CSEMA5F31C6 is shown in Fig. 2.31 that shows ALMs, LAB, M10K, and
multiplier from the overall device floorplan.

Additional Resources and Routing EP5CSEMA5F31C6

While the exact number of horizontal and vertical bus lines for the
EP5CSEMA5F31C6 is no longer specified in the data sheet as in the older (e.g.,
Flex 10K) families [MB01], from the `Compiler Report`, we can determine the
overall routing resources of the device; see `Fitter → Resource Section →`
`Logic and Routing Section`. Local connections are provided by block
interconnects and direct links. Each ALM drives all types of interconnections –

Table 2.6 The Cyclone V SE SoC device family

Device	LUTs	ALM	BRAMs 10 Kbit each	Total Mbits	PLLs/Clk networks	Emb. mult. 18 × 18 bit	HPS
EP5CEA2	25K	9430	140	1.4	5/16	72	2x ARM
EP5CEA4	40K	15880	270	2.7	5/16	168	2x ARM
EP5CEA5	**85K**	**32070**	**397**	**3.9**	**6/16**	**174**	2x ARM
EP5CEA6	149K	41910	557	5.5	6/16	224	2x ARM

local, row, column, carry chain, shared arithmetic chain, and direct link interconnects. Each LAB can drive 30 ALMs through fast local and direct link interconnects. Ten ALMs are in any given LAB, and ten ALMs are in each of the adjacent LABs. The next level of routing are the C2 and R3 fast column and row local connections that allow wires to reach LABs at a distance of two-column LABs or three-row LABs, respectively. The longest connections for logic available are R14 and C12 wires that allow connection over 14 rows or 12 column LABs, respectively. It is also possible to use any combination of row and column connections in case the source and destination LAB are not only in different rows but also in different columns. Table 2.7 gives an overview of the available routing resources in the EP5CSEMA5F31C6.

The data sheets provide more details about the available internal LAB control signals and global clocks. The EP5CSEMA5F31C6 has 6 PLLs and a total of 16 clock networks spanning the entire chip to guaranty short clock skew. Each PLL has nine output counters that can be used to generate different frequencies or phase offsets. PLL can be used, for instance, to generate a clock with 0 and −3 ns offset as needed in the DE2-115 board or to increase the clock rate as we did in Sect. 1.5 from the oscillator provided 50 MHz to required 125 MHz by the system.

All 10 ALM in a LAB share the same synchronous clear and load signals (Fig. 2.22). Two asynchronous clear, two clocks, and three enable signals are shared by the ALMs in the LAB. This will limit the freedom of the control signals within one LAB, but since microprocessor signals are usually processed in wide bus signals, this should not be a problem for our designs.

The LAB local interconnect is driven by row or column bus signals or ALMs in the LAB. Neighboring LABs, PLLs, BRAM, and embedded multipliers from left or right can also drive the local LAB. As we can see from the data sheet of the device, the delay in these connections varies widely, and the synthesis tool always tries to place logic as close together as possible to minimize the interconnection delay; see floorplan in Fig. 2.31. A 32-bit adder, for instance, would be best placed in two LABs in two rows one above the other.

The following case study should be used as a guide for the examples and self-study problems in subsequent chapters.

Table 2.7 Routing resources for the Altera's Cyclone IV E device EP5CSEMA5F31C6

Block	C2	R3	C4	R6	C12	R14
289,320	119,108	130,992	56,300	266,960	13,420	12,676

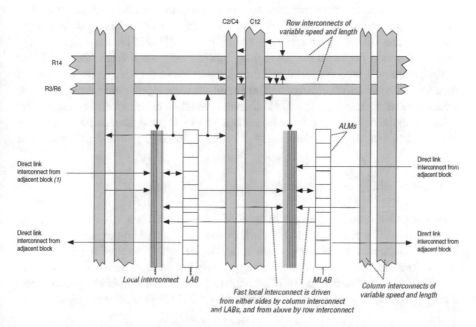

Fig. 2.22 Cyclone V logic array block resources [Alt11]

2.4   Case Study: A PSM Called URISC

The programmable state machine from Fig. 1.3 should be considered as the underlying architecture for this case study. One of the challenges of the design is to have behavioral HDL code that is synthesized by Altera QUARTUS and Xilinx VIVADO software and works well with the recommended simulators (MODELSIM for Altera and XSIM for Xilinx) for both RTL and timing simulation. In the following we walk through all steps that are usually performed when implementing a design using QUARTUS:

- Initial design specification and development
- Compilation of the design
- Design results and floorplan
- Simulation of the design
- Performance evaluation

The URISC Processor Model

Before we start to discuss the use of the design tools, let us have a brief look at this simple PSM model that nevertheless is a complete RISC processor that can basically run any assembler program from simple register clear or move to sophisticated program such as computation of Fibonacci sequence or factorials as shown in the exercises. The URISC machine suggested by Parhami [MP88] has a single instruction only, which performs the following: subtract operand 2 from operand 1 and replace operand 1 with the result and then jump to the target address (or to an offset location from the current instruction) if the result was negative otherwise continue with the next instruction. The URISC therefore takes the reduced instruction set principle to the extreme: It uses only *one* instruction. If using one instruction only is beneficial or not should be investigated in the following. The instructions are of the form.

```
urisc op1, op2, offset
```

where offset specifies the next instruction offset, which is 1 most of the time. At the time URISC was introduced [MP88], memory was small and expensive, and the original design tried to use only one memory device, which resulted in many micro-instructions per operation. Today, memory is large and inexpensive compared with the original multi micro-step URSIC design [MP88, P05]. For this reason, we make a couple of small modifications to reflect today's RISC/FPGA design principles:

1. Use 16 registers and not memory to implement the dst -= src instruction.
2. Initialize all registers to −1 at reset.
3. Use r[0] as input port (in_port); r[1] keep at −1; r[15] as out_port.
4. Use only one register for a program counter (pc) and instruction register (ir).
5. Use single program memory ($127 = 2^7$ words × 16 bits width), no data memory.
6. Allow a relative and absolute PC update through a mode flag.

Here is a small test program that shows how the PSM works. It read from the in_port and outputs the data to the out_port.[3]

---

[3] HDL coding tutorial is provided in Chaps. 3 and 4. The Verilog code for the URISC can be found in Appendix A.

ASM Program 2.1: LED test program using the URISC

```
 1   *================================================================
 2   * FILE:          io.asm
 3   * DESCRIPTION:   Short test program for the URISC
 4   *               processor model.
 5   *               Load word from in_port and output to out_port
 6   * REGISTER USE:  r[0]  = in_port
 7   *               r[1]  = -1
 8   *               r[2]  = temp
 9   *               r[15] = out_port
10   *================================================================
11           URISC r[2],  r[2],   +1    * r[2]=0
12           URISC r[2],  r[0],   +1    * r[2]=-iport
13           URISC r[15], r[15],  +1    * oport=r[15]=0
14           URISC r[15], r[2],   +1    * oport=r[15]=-(-iport)=iport
15           URISC r[2],  r[2],   +1    * r[2]=0
16           URISC r[1],  r[2],   @0    * -1-0=-1<0 always branch
```

Since we have 16 registers, we need a 2 4-bit register index (source and destination). We use 7 bits for address with an extra mode bit to distinguish between relative and absolute address. We only have a single instruction URISC and do not need any op code to use. The 16-bit instruction word then looks as follows:

URISC r[<dst>], r[<src>], @<value: unsigned decimal value>

dst=xxxx			src=xxxx			mode	value=xxxxxxx						
15		12	11		8	7	6						0

We have now enough information to compute the hex code for this test program that we store in an HDL table aka program ROM for convenient use later.

VHDL Program 2.2: LED test program for URSIC translated to VHDL

```
1    -- Test program for the URISC processors
2    library ieee;
3    use ieee.std_logic_1164.all; use ieee.std_logic_unsigned.all;
4    -- =========================================================
5    ENTITY rom128x16 IS
6     PORT(clk      : IN STD_LOGIC;
7           address : IN STD_LOGIC_VECTOR(6 DOWNTO 0);
8           data    : OUT STD_LOGIC_VECTOR(15 DOWNTO 0));
9    END;
10   -- =========================================================
11   ARCHITECTURE fpga OF rom128x16 IS
12
13     TYPE rom_type IS ARRAY (0 TO 127) OF
14                                 STD_LOGIC_VECTOR(15 DOWNTO 0);
15     CONSTANT rom : rom_type := (
16     X"2201", X"2001", X"ff01", X"f201", X"2201", X"2180",
17     OTHERS => X"0000");
18
19   BEGIN
20     data <= rom(conv_integer(address));
21
22   END fpga;
```

The CONSTANT values code in hex presents the steps the FSM will go through. The first two hex digits indicate the two registers involve followed by the mode bit, and finally the last 7 bits show the target address or the relative addressing for the next pc value. We can now develop our FSM to process this data. The HDL coding is shown next.

VHDL Code 2.3: The URISC processor

```
1     LIBRARY ieee;
2     USE ieee.std_logic_1164.ALL;
3     USE ieee.std_logic_arith.ALL;
4     USE ieee.std_logic_unsigned.ALL;
5     -- ==============================================================
6     ENTITY urisc IS                        ------> Interface
7       PORT(clk     : IN  STD_LOGIC; -- System clock
8            reset   : IN  STD_LOGIC; -- Active low asynchronous reset
9            in_port : IN  STD_LOGIC_VECTOR(7 DOWNTO 0); -- Input port
10           out_port : OUT STD_LOGIC_VECTOR(7 DOWNTO 0)); -- Output port
11    END;
12    -- ==============================================================
13    ARCHITECTURE fpga OF urisc IS
14
15      COMPONENT rom128x16 IS
16      PORT(clk      : IN STD_LOGIC;
17           address  : IN STD_LOGIC_VECTOR(6 DOWNTO 0);
18           data     : OUT STD_LOGIC_VECTOR(15 DOWNTO 0));
19      END COMPONENT;
20
21      TYPE STATE_TYPE IS (FE, DC, EX);
22      SIGNAL state      : STATE_TYPE;
23
24    -- Register array definition
25      TYPE RTYPE IS ARRAY(0 TO 15) OF STD_LOGIC_VECTOR(7 DOWNTO 0);
26      SIGNAL r : RTYPE;
27
28    -- Local signals
29      SIGNAL data, ir : STD_LOGIC_VECTOR(15 DOWNTO 0);
30      SIGNAL mode, jump : STD_LOGIC;
31      SIGNAL rd, rs : INTEGER RANGE 0 TO 15;
32      SIGNAL pc, address : STD_LOGIC_VECTOR(6 DOWNTO 0);
33      SIGNAL dif        : STD_LOGIC_VECTOR(7 DOWNTO 0);
34
35    BEGIN
36
37      prog_rom: rom128x16           -- Instantiate the LUT
38      PORT MAP (clk      => clk,    -- System clock
39                address => pc,      -- Program memory address
40                data    => data);   -- Program memory data
41
42      FSM: PROCESS (reset, clk)     ------> FSM with ROM behavioral style
43        VARIABLE result : STD_LOGIC_VECTOR(8 DOWNTO 0);-- Temporary reg.
44      BEGIN
45        IF reset = '0' THEN                 -- asynchronous reset
```

```
46              FOR k IN 1 TO 15 LOOP
47                 r(k) <= CONV_STD_LOGIC_VECTOR( -1, 8); -- all set to -1
48              END LOOP;
49              pc <= (OTHERS => '0');
50              state <= FE;
51           ELSIF rising_edge(clk) THEN
52           CASE state IS
53             WHEN FE =>              -- Fetch instruction
54               ir <= data;          -- Get the 16-bit instruction
55               state <= DC;
56             WHEN DC =>             -- Decode instruction; split ir
57               rd <= CONV_INTEGER('0' & ir(15 DOWNTO 12));  -- MSB has des.
58               rs <= CONV_INTEGER('0' & ir(11 DOWNTO 8));   -- second source
59               mode <= ir(7);          -- flag for address mode
60               address <= ir(6 DOWNTO 0); -- next PC value
61               state <= EX;
62             WHEN EX =>             -- Process URISC instruction
63               result := (r(rd)(7) & r(rd)) - (r(rs)(7) & r(rs));
64               IF rd > 0 THEN
65                 r(rd) <= result(7 DOWNTO 0); -- do not write input port
66               END IF;
67               IF result(8) = '0' THEN  -- test is false PC++
68                 pc <= pc + 1;
69               ELSIF mode = '1' THEN  -- result was negative
70                 pc <= address;  -- absolute addressing mode
71               ELSE
72                 pc <= pc + address; -- relative addressing mode
73               END IF;
74               r(0) <= in_port;
75               out_port <= r(15);
76               state <= FE;
77           END CASE;
78             jump <= result(8);
79             dif  <= result(7 DOWNTO 0);
80           END IF;
81         END PROCESS;
82
83      END ARCHITECTURE fpga;
```

We see the three major parts of the HDL code separated by horizontal (comment) lines: LIBRARY definition, ENTITY, and ARCHITECTURE specification. We see in the coding first the standard logic package including unsigned arithmetic use specification (lines 1–4). The ENTITY (lines 6–11) currently only includes the four ports for input and output. We can monitor additional internal signals too in a functional simulation, but not all signal may be available during timing simulation when physical implementation in LUT sometimes makes local signal non-observable. The ARCHITECTURE part starts with the PROM component definition, followed by general-purpose signals (lines 13–34). Our FSM has three states called fetch (FE), decode (DC), and execute (EX). The URISC uses a 16-register array that is defined next as array definition (lines 24–26). The ARCHITECTURE body has two major blocks: the PROM (lines 37–40) instantiation (that works like a procedure call in C/C++) and the FSM specification (lines 42–81) embedded in a PROCESS that allows a sequential analysis of the code similar to a sequential analysis in any programming language. When our active low reset is activated, all registers are set to -1, program counter is cleared, and next state will be fetch (FE); see (lines 45–50). The FSM is controlled by a rising edge of the clock, i.e., any registers inferred will only change values when a rising edge of clk occurs. After reset become deactivated, our FSM will run through three states:

- In the fetch (FE) state (lines 53–55), the instruction register ir will store the output of the PROM, i.e., data and next state coding is DC.
- In the decode (DC) state (lines 56–61), the instruction word is decomposed in the four components: index of the destination register rd, index for the two-source register rs, the mode bit that distinguishes between relative and absolute addressing, and the lower 7 bit of the address code. The next state is now EX.
- In the execute (EX) state (lines 62–76), we compute the subtract operation. r(rd)-r(rs) and store the results in a register if index is not zero. To avoid overflow problems, we first make a 1-bit sign extension, though. Based on the MSB, we continue with the next instruction and perform a "jump," i.e., use address to compute the next pc value. If the mode flag is set, we use the absolute address value; otherwise we use a pc-relative mode. In the same state, we also store in_port in register zero and r(15) in the out_port register. This completes a full instruction, and we continue with a next state fetch FE.

To better monitor the progress of the PSM, we monitor the "jump" condition and the difference values computed with the two additional SIGNALs jump and dif. That completes the PROCESS as well as the ARCHITECTURE of the URISC.

Design Compilation in Altera QUARTUS

To check and compile the file, start the QUARTUS software and select File → Open Project or launch File → New Project Wizard if you do not have a project file yet. In the project wizard, specify the project directory you would like to use and the project name and top-level design as urisc. Then press Next and specify the HDL file you would like to add, in our case urisc.vhd or urisc.v. Press Next again and then select the device 5CSEMA5F31C6 from the Cyclone V SE SoC family. Click Next and the select MODELSIM-Altera as simulation tool and press Finish. If you use the project file from the CD, the file urisc.qsf will already have the correct file and device specification. Now select File→ Open to load the HDL file. The VHDL has been presented earlier (see VHD Code 2.1 or Appendix A for the Verilog code).

The object LIBRARY, found early in the code, contains predefined modules and definitions. The ENTITY block specifies the I/O ports of the device and generic variables. Next in the coding comes the component in use and additional SIGNAL definitions. The HDL coding starts after the keyword BEGIN. The first PROCESS includes the processor FSM. The URISC has an asynchronous active low reset. The program ROM table is instantiated as a component, and the port signals of the component are connected to the local signal within our design. You may want to have a look at the ROM table design file `rom128x16.vhd`. You can load the file, or you can double-click it in the `Project Navigator` window (top left). In general, a synthesizable ROM or RAM design with initial data loaded in the memory is not a trivial task. A good starting point is either to have a look in the VHDL 1076.6-2004 subset of VHDL for synthesizable code or a little easier is to have a look at the language templates suggest by the tool vendors (Altera: `Edit →` `Insert Template → VHDL → Full Designs → RAMs and ROMs →` `Dual-Port ROM.` or Xilinx: `Edit → Language Template → VHDL →` `Synthesis Constructs → Coding Examples → ROM → Example` `Code`). Altera recommend use of a function call for the initialization and Xilinx a `CONSTANT` array initial definition. It turns out that the latter approach works well with both tools and with the simulators for functional and timing simulators and will be the preferred method for the rest of the book. The synthesis attributes as defined in VHDL 1076.6-2004 are currently not supported by either vendor. In VERILOG RAM and ROM, initialization is already specified in the language reference manual (LRM), and the most reliable method is to use the `$readmemh()` function in combination with an `initial` statement.

To optimize the design for speed, go to `Assignment → Settings →` `Compiler Settings`. Under `Optimization` mode click on `Performance` (`High effort` or `Aggressive`). Optimizing for `Area` can be selected in the same `Compiler Setting` menu. For a good compromise, you can also try the `Balanced` mode which is also the default setting we will use. Timing requirement can be set using the Synopsys Design Constraints (SDC) file. The default speed is set to 1 ns in QUARTUS that usually works well for a first run.

Now start the compiler tool (its the thick right arrow shortcut symbol) that can be found under the `Processing` menu. A window left to our HDL window as shown in Fig. 2.23b will show the progress in the compilation. You can see all the steps involved in the compilation, namely, `Analysis & Synthesis`, `Fitter`, `Assembler`, `Timing Analysis`, `Netlist Writer`, and `Program Device`. Alternatively, you can just start the `Analysis & Synthesis` by clicking on `Processing → Start → Start Analysis & Synthesis` or with `<Ctrl K>`. The compiler checks for basic syntax errors and produces a

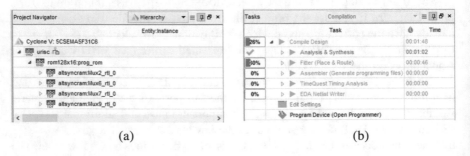

(a)   (b)

Fig. 2.23 (a) Project navigator. (b) Compilation steps in QUARTUS

report file that lists resource estimation for the design. After the syntax check is successful, compilation can be started by pressing the large `Start` shortcut or by pressing `<Ctrl L>`. If all compiler steps were successfully completed, the design is fully implemented. Press the `Compilation Report` button (IC symbol with a paper) from the top menu buttons, and the flow summary report should show 79 ALMs and 384 memory bits used. Check the memory file `rom128x16.vhd` for VHDL and `urisc256x8.hex` for Verilog initialization file. Figure 2.23b summarizes all the processing steps of the compilation, as shown in the QUARTUS compiler window.

For a *graphical* verification that the HDL design describes the desired circuit, using Altera's QUARTUS software, we can use the RTL viewer. The result for the `urisc.vhd` circuit is shown in Fig. 2.24. The RTL view bird's-eye view reveals that the URSIC has already many components that make it hard to verify, but the RTL state machine viewer verifies that we indeed have built a three-state machine. To start the RTL viewer, click on `Tools→ Netlist Viewer→ RTL Viewer`. The other netlist viewer called `Technology Map Viewer` gives a precise picture of how the circuit is mapped onto the FPGA resources. However, to verify the HDL code even for a small design as the function generator, the technology map provides too much detail to be helpful, and we will not use it in our design studies.

Design Compilation in Xilinx VIVADO

The VIVADO software in the `Project Manager` view will have an excellent overview of the implementation steps and the associate design files. It will not only show the files and library used and resources in bar graph or table form, also power dissipation and timing information are shown in the `Project Summary` window, a

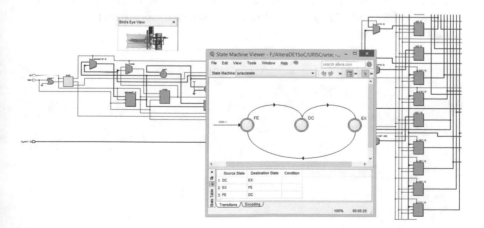

Fig. 2.24 RTL view of the URISC processor including bird's-eye view and state machine viewer

substantial improvement to the ISE software, that has all this information too just more buried in the report files.

A full set of synthesis data in general will require a device specification and a full compile that can take up substantial CPU time for a big device. The map report will show the desired values such as the number of flip-flop, LUT, I/O, Bufg, block RAMs, and embedded multipliers used. Since device families have different types of logic cells, LUT, and block RAM sizes, this data may vary for different devices.

To get the timing data, however, VIVADO has no longer the "speed" or "area" options as ISE, instead we need to constrain the design. The idea comes from the Synopsys ASIC constrain files, where you specify a desired clock frequency, and if the synthesis reaches the clock goal, the rest of the compile effort can be directed to reducing the area of the design. At a minimum we need to specify a constrain file `*.xdc` and set the clock as follows:

```
create_clock -add -name sys_clk_pin -period 8.00 -wave-
form {0 4} [get_ports {clk}];
```

where `clk` is assumed the name of the 125 MHz clock signal. This will set the desired clock period to 8 ns or 125 MHz. If that frequency is too high for the device chosen, then a negative clock skew is reported, and we need to increase the period. Finding the maximum speed is therefore an iterative process, more labor intensive than with ISE. Finding an "area" optimum is simple; we just relax the timing requirement, to say `-period  1000` (i.e., 1 MHz clock), and all efforts of the compiler will be used to minimize the area.

The processing step to set up a project VIVADO is similar to ISE. We start which `Create New Project` or `Open Project`. We will then add sources file and need to specify if these are `Design Sources`, (`*.vhd` or `*.v`), `Constraints` files (`*.xdc`), or `Simulation  Sources` (`*_tb.vhd` or `*_tb.v`); see Fig. 2.25a. We can then `Run  Synthesis` to get a resource estimation or `Run Implementation` to get precise resource data. You also need first to specify your clock goal in the constraints file that can also include the pin assignments for your development board. You may download a "gold" template from the board vendor that you not by accident assign wrong pin and damage the board and then modify it to match your design I/O port names, e.g.:

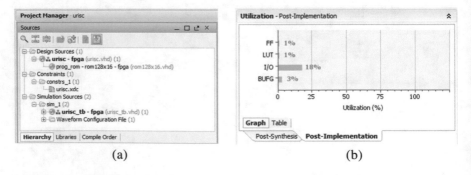

(a)   (b)

Fig. 2.25 VIVADO (**a**) project manager file structure, (**b**) graphic display of synthesis results

```
##Clock signal
set_property -dict { PACKAGE_PIN L16   IOSTANDARD LVCMOS33 } [get_
ports { clk }];
```

is the pin assignment for the clock signals clk on the ZyBo board.

After a full compile with Area goal, we find that the VIVADO URISC uses 63 from 17,600 LUTs, 18 I/Os, and 93 flip-flops. The utilization report shows 27 SLICEL and 9 SLICEM in use. The synthesis tool did not use any multiplier, distributed RAM, or block RAM since the design was too small.

Design Tool Considerations for Simulation

While the design flows for Altera and Xilinx tools are similar, there has been some notable difference over the years. Most noticeable is the handling of the simulation of the designs. We have seen the two FPGA market leaders take opposite directions in recent years.

Xilinx since ISE version 12.3 (end of 2010) no longer provides a free MODELSIM simulator and instead provides a free embedded ISIM simulator that is integrated within the ISE tool, and the VIVADO tool has a similar internal simulator called XSIM tool. The two most important differences are that the ISIM simulator has the option to do a simulation via TCL script, while the XSIM simulator has an analog (aka waveform) display option.

The Altera QUARTUS software on the other hand comes with two free simulator options. In the past Altera favored the internal VWF waveform simulator (up to QUARTUS version 9.1) and now recommends the external MODELSIM-Altera or QSIM. The MODELSIM-Altera allows us to use the professional tool from Mentor Graphic Inc. The second alterative is the Altera QSIM tool that may have a few less feature than MODELSIM (e.g., no analog waveform) but is also a little easier to handle since it does not require one to write HDL test benches or DO file scripts to assign I/O signals. However, at the time of writing, MODELSIM seemed to be the better option since it has been around for many years, is used more often in industry, and supports the Cyclone V devices for functional and timing simulation. Therefore, the MODELSIM-Altera was selected as default simulator. Moving between VHDL and Verilog stimuli file and Altera and Xilinx is also simplified by using MODELSIM-Altera DO files and not HDL test benches.

The typical next step in a project would be a verification via a functional simulation of the software and processor. For a HPS this would typically be an instruction set simulator (ISS) or the GDB debugger for C/C++ programs. ISS and debugger are discussed in Chap. 6.

For a softcore processor as our URISC, we would use an HDL-based functional simulator first. In the following we will first discuss how to simulate the URISC using the MODELSIM-Altera tool followed by Xilinx VIVADO-based simulation consideration.

Altera-Specific Simulation Consideration

When simulating a design with the MODELSIM tools, we can apply directly our input stimuli to our design, and we do not need to write a test bench in HDL. We will discuss test benches later when we talk about the VIVADO simulation. Typically, we will put the MODELSIM instructions in a script, which follows typical TCL aka DO file scripting rules. Also, if we do not use a test bench and simulate our circuit directly using the TCL stimuli script, then we can use the same script for RTL and timing simulation. Furthermore, for VHDL and Verilog, almost the same stimuli file can be used if we replace vlog by vcom or vice versa.

Before we discuss a simulation example, let us first have a look at the file names used, where * stands for the project name. For the RTL simulation, we use the *.vhd and *.v files. For timing simulation four different filenames are used: *.vho and *.vo for VHDL and Verilog using Altera tools and *_tb.vhd and *_tb.v for Xilinx VIVADO. For both RTL and timing simulation, we use as stimuli files *.do (Altera) and *_tb.(v)hd for Xilinx VIVADO.

To simulate, open the MODELSIM-ALTERA tool. You should see that many predefined libraries are already loaded. Use File → Change Directory to move to the directory that includes the HDL files and the simulation scripts. For timing simulation, you need first to compile the design in QUARTUS or VIVADO, where you can simulate the HDL code with timing information. Before we run the script, we need to make sure that HDL and *.do files are in the same directory. Use dir *.do and dir *.vhd to verify that the files are indeed in the current path. Then type do urisc.do in the MODELSIM-ALTERA transcript window. The script below compiles the files, opens a waveform, and simulates the circuit. The do script for this example looks as follows:

MODELSIM TCL/DO File script 2.4: The URISC processor

```
 1   ########## Compile design
 2   vlib  work
 3   vcom -93 rom128x16.vhd
 4   vcom -93 urisc.vhd
 5   vsim work.urisc(fpga)
 6
 7   ########## Add I/O signals to wave window
 8   add wave clk reset
 9   add wave -divider  "Locals:"
10   add wave -hex state pc data
11   add wave -dec rd rs address
12   add wave -dec r(0) r(1) r(2) r(15)
13   add wave jump mode dif
14   add wave -divider  "I/O Ports:"
15   add wave -dec in_port out_port
16   radix -hex
17
18   ######### Add stimuli data
19   force clk 0 0ns, 1 10ns -r 20ns
20   force reset 0 0ns, 1 105ns
21   force in_port 5 0ns
22
23   ########## Run the simulation
24   run 750ns
25   wave zoomfull
26   configure wave -gridperiod 40ns
27   configure wave -timelineunits ns
```

The simulation script has four parts:

1. First the work library (i.e., directory) is created via `vlib`. Then the components in the project are compiled via `vcom` or `vlog` for VHDL and Verilog, respectively. Then the top-level project is compiled, and the simulator is called via `vsim` specifying the top-level ENTITY as well as the ARCHITECTURE name.
2. Next signals are added to the wave window in the order they should appear in the wave window. We start with the input control signals such as clock and reset, followed by local signal that is not part of the ports and then outputs. Dividers are used to help with differentiations or to add extra information such as author names.
3. Next follows (in any order) the stimuli data to the wave form signals we have just defined. We can define periodic signals such as `clk` using the `-r` option and non-periodic as, for instance, `reset` and `in_port`. Value followed by the time of the value should appear. A comma-separated list is used for other values/time pairs. Make sure you specify the radix first if you want the values to be represented in.
4. Finally, we run the simulation for a specific time and zoom to display the full time frame. A grid and time unit can also be defined for the wave window.

As can be seen from the script or the wave window, we have used the following data: as period 20 ns = 1/50 MHz was selected, we use a long active `reset` time larger than 100 ns to avoid any mismatch between functional and timing simulation when using Xilinx global reset feature. The functional simulation aka RTL simulation of the URISC processor is shown in Fig. 2.26.

Figure 2.26 shows after initial active low `reset` the processor completing every three clock cycles one instruction from our program ROM by running through the three-state machine stages FE, DE, and EX. We only display relevant registers 0, 1, 2, and 15 and see that the value 5 from the `in_port` appear as −5 after (inverted) loading in `r(2)` after 210 ns. This values is then forwarded to register `r(15)` after 330 ns. Then the two instruction long jump sequence start, and the URISC starts at `pc=0` again after 450 ns after completion of the 6-instruction loop, as shown in to ASM program 2.1.

Xilinx-Specific Simulation Consideration

Xilinx has announced that in the future, the VIVADO tool set will be supported, and the ISE tool set will be retired. Unfortunately, VIVADO seemed not to support any FPGA family before the seven generation, and many designers if using Virtex-6 devices, for instance, will have to continue to use the ISE tools still. Since introducing the VIVADO software in 2013, no effort has been seen to support older FPGA devices in VIVADO. But the good news is that all Zynq-7K devices are supported in the free VIVADO web version and no license is needed to compile or simulate these devices.

When simulating a design with the ISE simulator, we had the option to use a stimuli file from a `TCL` script like MODELSIM DO files, or we could write a test bench in HDL. Now with VIVADO we need a test bench to simulate our processor. A

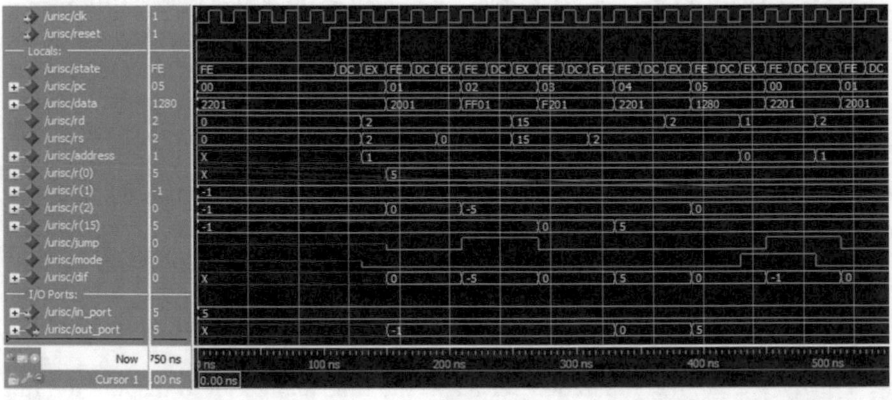

Fig. 2.26 MODELSIM RTL simulation of URISC processor running the I/O program

test bench is a short HDL file, where we instantiate the circuit to be tested and then generate and apply our test signals with a statement like

```
clk <= NOT clk AFTER 5 ns;
```

to generate a clock with a 2×5 ns = 10 ns clock period. However, one difficulty with the XSIM test bench comes from the fact that the circuit with timing information is synthesized directly from a Verilog netlist. In general, in VIVADO a Verilog netlist on LUT-based level is used to have accurate timing simulation. Since a Verilog netlist is used even for VHDL designs, a match with VHDL source code is only possible for STD_LOGIC or STD_LOGIC_VECTOR and Boolean type. We cannot use INTEGER, SIGNED, or FLOAT data types, and even BUFFER or GENERIC parameters are not permitted in the entity. This however applies only to the I/O interface. Within the design we can indeed use INTEGERs, and design reuse with GENERIC parameters can be done in VHDL via CONSTANT definitions and as PARAMETER in Verilog within the designs without interference the coding requirements for the I/O ports of the designs.

Another important design consideration for the Xilinx tools is the handling of the global set/reset (GSR) in the simulator. The idea behind the GSR is that all flip-flops in all FPGAs are set to predefined values after reset within the first 100 ns of the simulation. Only after the first 100 ns any flip-flop operation can occur. Generated timing netlist will always ensure this functionality; however, the behavior simulation does not necessary follows this by default. This is demonstrated by the URISC simulation shown in Figs. 2.27 and 2.28. As can be seen in the behavior simulation (Fig. 2.27), the processor operation starts earlier than in the timing simulation due to the 100 ns GSR in the timing simulation; see Fig. 2.28. As a result, the out_port has the final value 5 after 300 ns in behavior simulation and at 380 ns in timing simulation. To avoid such a mismatch in the behavior/timing simulation, it is therefore highly recommended to hold flip-flop activity via an ENABLE or RESET signal for the first 100 ns. This timing simulation can then be matched with a behavior simulation with an initial 100 ns reset.

Fig. 2.27 Behavior simulation with no GSR consideration. Note that the reset is inactive after 25 ns; the URISC processing, however, start immediately. This results in a behavior/timing simulation mismatch

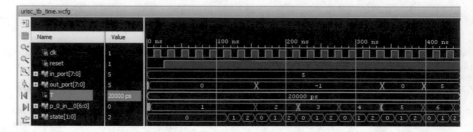

Fig. 2.28 Timing simulation with no GSR consideration. Note that the URISC does not start processing until after 100 ns and now will not match the behavior simulation

Writing Test Benches

With today's high complex designs, a substantial design effort is directed toward the verification of the circuit. As the Pentium bug from year 1995 in the FP divider hardware has shown us, such an insufficient testing can have a large financial impact (over $100 M for Intel) besides the image damage such a recall may have.

Verification can take many different forms. For a small design, we can use the RTL viewer aka RTL Analysis Schematic to inspect the synthesized circuit. For a more complicated system, we may use input test stimuli generated on the fly or via a test vector lookup table generated in MATLAB or with a C/C++ program. The correct output behavior can be text-based, i.e., report "mismatch" of actual and expected results or graphical such as an oscilloscope. Figure 2.29 shows a verification by graphical inspection.

Writing a text-based HDL test bench (TB) is not too complicated but nevertheless can be labor intensive. Until ISE 11 Xilinx offered a tool called "HDL bencher." You had to define waveform for input signals, and you could specify or generate the desired results based on a behavior simulation [VSS00, BE00] (Fig. 2.30).

On the other side, you may have special requirements how a TB should look like, and then in general, it is preferred to use a template as starting point [Ham10]. A template typically will have the following elements:

1. Libraries in use such as IEEE
2. An "empty" ENTITY without any ports

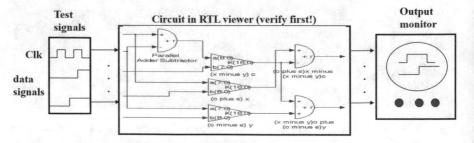

Fig. 2.29 Test bench design for a complex multiplier

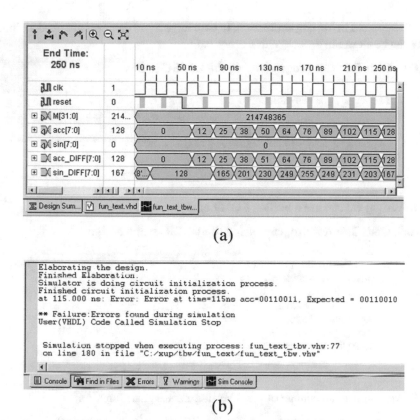

Fig. 2.30 The HDL bencher provided by ISE until version 11. (**a**) Test bench desired values and simulation results. (**b**) Error message for mismatch using MODELSIM simulation at time 115 ns. Value desired was specified as decimal 50, but the simulation shows a value of decimal 51

3. The `Unit Under Test` (UUT) component definition
4. The signals/wires/registers in use
5. The UUT component instantiations
6. Definition of period signals, e.g., `clk`
7. Definition of a-period data signals, e.g., `reset, data input`, etc.

   The Verilog TB will not have items 1 and 3. It is also important to remember that the XSIM simulator orders the displayed signals by default as specified item under 4, in precisely the shown order. That is, in order to avoid rearranging the signals in the simulator window, it is recommended to sort the signals/register/wires in the order we like to see them in the waveform window. The simulator does not care about the ordering of the components port or the order how you assign the ports in the component instantiation. VIVADO Verilog and VHDL simulation will look in general very similar. Only in case we have used VHDL FSM state coding with literal names this will in Verilog be displayed as integer numbers since a literal display is not supported in Verilog simulation. In case large input data sets are needed, input data can be stored in a CONSTANT array, e.g.:

VHDL Code 2.5: test program using the URISC in hex

```
1   TYPE rom_type IS ARRAY (0 TO 127) OF
2                                STD_LOGIC_VECTOR(15 DOWNTO 0);
3     CONSTANT rom : rom_type := (
4     X"2201", X"2001", X"ff01", X"f201", X"2201", X"2180",
5     OTHERS => X"0000");
```

In Verilog we can use a Verilog ROM definition, e.g.:

Verilog Code 2.6: test program using the URISC using assign

```
1   reg [15:0] ROM [127:0];
2   assign data = ROM[addr];
3   initial begin
4      ROM[0]=16'h2201;
5      ROM[1]=16'h2001;
6      ROM[2]=16'hff01;
7      ROM[3]=16'hf201;    ...
```

Or we can take advantage of them Verilog readmemh function

Verilog Code 2.7: test program using the URISC using $readmemh

```
1   reg [15:0] ROM [127:0];
2   assign data = ROM[addr];
3   initial begin
4      ROM[0]=16'h2201;
5      ROM[1]=16'h2001;
6      ROM[2]=16'hff01;
7      ROM[3]=16'hf201;    ...
```

Of course, we have to add the divider and the text by hand after we have started the VIVADO bchavior or timing simulation.

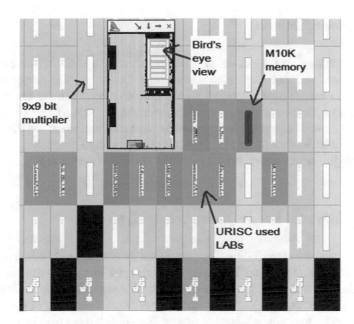

Fig. 2.31 Altera QUARTUS floorplan with bird's-eye view of the URISC processor model

QUARTUS *Floor Planning*

The design results in QUARTUS can be verified by clicking on the fourth shortcut button (i.e., Chip Planner or opening the Tool → Chip Planner) to get a more detailed view of the chip layout. The Chip Planner view is shown in Fig. 2.31.

Use the Zoom in button (i.e., the ± magnifying glass) to produce the screen shown in Fig. 2.31. Zoom in to the area where the LAB and an M10K are highlighted in a different color. Unused ALM LAB are light blue, unused memory is light green, and DSP block are magenta. Used blocks have dark color: You should then see the ALM LAB used by URISC in dark blue and the M10K block highlighted in dark green. If you click on the Bird's Eye View button[4] on the left menu buttons, an additional window will pop up. Now select, for instance, the M10K block and then press the button Generate Fan-In Connections or Generate Fan-Out Connections several times, and more and more connections will be displayed.

---

[4] Note, as with all MS Windows programs, just move your mouse over a button (no need to click on the button), and its name/function will be displayed.

Vivado *Floor Planning*

In the Vivado floorplan, let us verify used resources such as CLB, block RAM, or multiplier used and their particular placement in the design. The device floorplan will typically automatically open up after a full compile, if not we can start it by selecting `Implementation` → `Implemented Design` → `Report Utilization`. The large-scale floorplan has been shown earlier; see Fig. 2.16. To get a more detailed view of the chip layout, we can use the ± magnifying glass, and we may get the device floorplan details for the URISC as shown in Fig. 2.32 below.

Timing Estimates and Performance Analysis

Altera's Quartus and the Xilinx ISE software in the past had only two timing optimization goals: area or speed. However, with increased device density and system-on-chip designs with multiple clock domains, different I/O clock modes, clocks with multiplexers, and clock dividers, this strategy optimizing area or speed for the complete device may not be a good approach anymore. Altera Quartus and Xilinx Vivado now offer a more sophisticated timing specification that looks more like a cell-based ASIC design style and is based on the Synopsys Design Constrain (SDC) specifications. The idea here is that a synthesis tool may have for the same circuit different library elements such as ripple carry, carry save, or fast look-ahead styles for the adder. In order to display timing data, a full compile of the design has to be done first for both Quartus and Vivado. The tool then first optimizes the design to meet the specified timing constraints and in a second step then optimizes area. Vivado will in fact does not produce any timing information unless at a minimum you provide a constrain file with a single line constraining the clock information such as

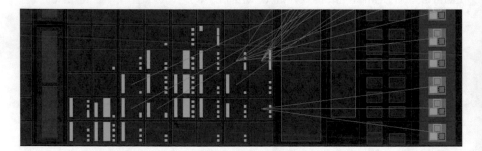

Fig. 2.32 Xilinx Vivado floorplan of the URISC processor model

```
create_clock -add -name sys_clk_pin -period 1.00 [get_ports { clk
}];
```

If you like similar synthesis results as with previous tool such as minimum area or maximum speed, you can achieve this by over- or under-constraining the design. For instance, if specified in the SDC file a target clock delay of 1 ns (1 GHz frequency), then this is most likely over-constraining your design and device. The slack will be negative, and if you add the negative slack let's say 7 ns to your initial 1 ns clock period, then with a clock of 1 ns + 7 ns = 8 ns, you should be pretty close the fastest clock rate you can run your circuit with. You may need a few more iterations (adding or subtraction small ns amounts from your clock period) to have precise maximum clock period until the slack is just a little positive; see Fig. 2.33b below.

If you like only to optimize the area, then you should use a very low target clock rate, since then all efforts are put in the area optimization.

For optimization speed with the Altera tool, we use the default setting with period of 1 ns equivalent to 1 GHz clock frequency that will appear as FMAX_REQUIREMENT "1 ns" in QUARTUS QSF files. Since often our design goal is optimization for speed, when using Altera tools, we can take advantage of Altera default setting (i.e., without an SDC file) that over-constrains the design to run at the desired 1 GHz clock rate. We should expect to see a warning message in the compilation report that timing has not been met but that it is intended and should not be a concern. Since our design most likely runs slower, this will ensure the QUARTUS compiler is synthesizing the fastest design possible. Beside the design constrain of particular signal, Altera QUARTUS also offer optimization mode for the whole device, ranging from Balanced, Performance, Power, to Area. You can select this under Assignment → Settings → Compiler Settings. Under Optimization   mode click on Performance   (High   effort or Aggressive). Optimizing for Area can be selected in the same Compiler Setting menu. For a good compromise, you can also try the Balanced mode which is also the default setting we will use. On the downside we will always get a compiler warning that Synopsys Design Constraints File urisc.

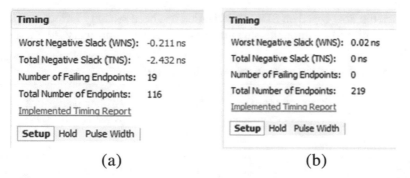

(a)   (b)

Fig. 2.33 Xilinx VIVADO timing result. (**a**) Negative slack shows up in red, (**b**) positive slack is shown in black color

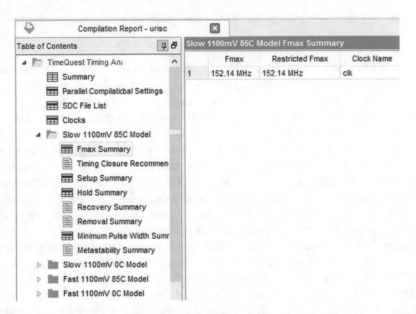

Fig. 2.34 Registered performance of URSIC design from the `TimeQuest Timing Analyzer`

`sdc` was not found and a `Critical Warning: Timing requirements not met`, but we can ignore these messages. If we use `DE1SoC_SystemBuilder.exe` to generate the board support files, a `*.sdc` file is generated with the onboard clock generator of 50 MHz or 20 ns clock period.

For Altera we run the `TimeQuest` analysis, and such a `Compilation Report` is shown in Fig. 2.34. It includes the four sets of timing data for the slow 85C/0C and fast 85C/0C models. We use the most pessimistic, i.e., the slow 85C model. The `Fmax Summary` frequency is the maximum frequency our circuit can run with. Sometime the performance is further restricted due to the maximum I/O pin speed of 250 MHz. If there is no register-to-register path in the circuit as in a pure combinational design, this `Fmax` will come up empty with the message `No paths to report`.

Since the flow based on SDC specification has been recently introduced, it may be a good idea to go through some tutorials. The Altera University Program offers

Table 2.8 The four URISC synthesis data for all tools and HDLs

	Altera QUARTUS 15.1 EP5CSEMA5F31C6 Balanced optimization (20 ns SDC)		Xilinx VIVADO 2016.4 XC7Z010S-2CLG484C	
	VHDL	Verilog	VHDL	Verilog
ALM/LUT	79	146	63	71
BRAM	1 (M10K)	1 (M10K)	0	0
Multiplier	0	0	0	0
Fmax/MHz	152.14	186.25	250	217.39

an introduction in the "Using TIMEQUEST Timing Analyzer" tutorial. More details are available in the "QUARTUS Prime TIMEQUEST Timing Analyzer Cookbook" with many different multi-clock domain examples.

The performance analysis with the Xilinx VIVADO software is very similar. We also need a SDC file to specify the clock design goal. One run is required for area goal using relaxed clock specification. Several iterations are required to find the best timing. The following table shows a VIVADO example timing iteration (error within 2.5%) for the URISC:

Period	2 ns	3.6 ns	3.8 ns	3.9	4 ns
Slack WNS	−2.3	−0.507	−0.342	−0.261	+0.016

Starting with 500 MHz as overambitious timing goal, the final best performance comes out to be clock period of 4 ns or 250 MHz. For our URISC model, the overall synthesis results are shown in Table 2.8.

This concludes the case study of the URISC processor model.

Review Questions and Exercises

Short Answers

2.1. Why was Plessy FPLD devices not a commercial success?
2.2. Why do push button have Schmitt trigger (threshold) circuits at the FPGA inputs?
2.3. Why are seven-segment displays being often active low, i.e., light up when assigned GND signal?
2.4. What method would you use if your FPGA has not enough I/O pins to drive all your seven-segment displays?
2.5. Why is the output register in URISC not 100% of the time of the same value?

Fill in the Blank

2.6. The Cyclone V devices are a product of _____.
2.7. The Zynq-7K devices are a product of _____.
2.8. The EP5CSEMA5F31C6 on the DE1 SoC has _____ 18 × 18-bit embedded multipliers.
2.9. The URISC ISA has _____ different instructions.
2.10. The DE1 SoC board has _____ switches.

2.11. The EP5CSEMA5F31C6 on the DE1 SoC has _____ M10K embedded memory blocks.
2.12. VLSI is an acronym for _____.
2.13. RTL is an acronym for _____.
2.14. VHDL is an acronym for _____.
2.15. SRAM is an acronym for _____.
2.16. VGA is an acronym for _____.
2.17. ADC is an acronym for _____.
2.18. USB is an acronym for _____.
2.19. UART is an acronym for _____.
2.20. LAB is an acronym for _____.

True or False

2.21. _____ Altera's EP5CSEMA5F31C6 has about 85K LEs.
2.22. _____ An Altera M10K can be configured as $8K \times 1$, $4K \times 2$, $2K \times 4$, $1K \times 8$, 512×16, or 256×32-bit memory RAM or ROM.
2.23. _____ To implement an 8-bit adder, Altera's Cyclone family uses 16 LEs.
2.24. _____ The Altera's EP5CSEMA5F31C6 device has 32070 M10K memory blocks.
2.25. _____ The Cyclone V LAB has 10 ALMs.
2.26. _____ The JTAG block on the EP5CSEMA5F31C6 device uses the area of about 306 LABs.
2.27. _____ The Cyclone V DE1-SoC board is available for ca. $300 directly from Xilinx.
2.28. _____ The 18×18-bit multiplier can be used as four 9×9-bit multipliers also.
2.29. _____ The DE1 SoC has a 25 MHZ and 50 MHz oscillators on board.
2.30. _____ The DE1 SoC has a EP5C6 FPGA from the Cyclone VI family.
2.31. _____ The DE1 SoC has red LEDs and six seven-segment displays.
2.32. _____ The ZyBo has six switches and four buttons.
2.33. _____ The ZyBo has a 32-bit audio CODEC.
2.34. _____ The ZyBo is programmed via a parallel printer cable.
2.35. _____ The DE2-115 board has an LCD display, but the DE1 SoC has no LCD.
2.36. _____ A mouse cannot be connected to the DE1 SoC board because a PS/2 connector does not exist.
2.37. _____ The DE1 SoC has a VGA but no HDMI, while the ZyBo had VGA and HDMI ports.
2.38. _____ The user USB can be used to connect a USB stick to the DE1 SoC board.

2.39. _____ The color of the ZyBo user LEDs is green.
2.40. _____ The push buttons on the DE1 SoC board have an extra debounce logic.
2.41. _____ The ZyBo board has a 50 MHz oscillator on board.
2.42. _____ The only difference of the DE2-115 and the DE1 SoC board is the size of the FPGA.
2.43. _____ The Cyclone V DE1 SoC board is available for ca. $300 directly from Xilinx.

Projects and Challenges

2.44. Think of other "one instruction set computers" by using other arithmetic operations and conditions.
2.45. Compare the URISC and standard 32-bit RISC in terms of opcode length, FPGA resource (LE, memory, DSP blocks) and speed, program length, and program latency.
2.46. Determine the S/N of a uniform ADC.
2.47. Develop URISC code for the following expressions:

```
a. clear (src)
b. dest = (src1)+(src2)
c. exchange (src1) and (src2)
d. goto label
e. if (src1>=src2), goto label
f. if (src1-src2), goto label
```

Use r[2] and r[3] as auxiliary registers and assume r[0] is in_port, r[15] out_port, and r[1] = −1.

2.48. For each instruction of the URISC in the previous exercise, develop PROM data and test via HDL simulation.
2.49. Develop a URISC program to implement:

    (a) A squaring operation
    (b) A multiplication
    (c) The Fibonacci sequence
    (d) A factorial

2.50. For each of the URISC programs in the previous exercise, develop PROM data and test via HDL simulation.

References

[AD98] Analog Device Inc., *ADV7123: Triple 10-Bit High Speed Video DAC* (Norwood, 1998)

[Axe14] J. Axelson, *USB Complete: Everything you Need*, 5th edn. (Lakeview Research, Chicago, IL, 2014)

[Alt96] Altera, "Data sheet," FLEX 10K CPLD Family (1996)

[Alt11] Altera, *Cyclone V Device Handbook*, vol 1-4 (2011)

[BE00] A. Bloom, R. Escoto, HDL bencher XE for fast behavior FPGA verification. Xcell J **38**, 26–27 (2000)

[CEE14] L. Crockett, R. Elliot, M. Enderwitz, R. Stewart, *The Zynq Book* (Strathclyde Academic Media, Glasgow, 2014)

[Dig14] Digilent, *ZyBo Reference Manual* (Pullman, WA, 2014)

[Dig19] Digilent, *ZyBo Z7 Migration Guide* (Pullman, 2019). https://reference.digilentinc.com/reference/programmable-logic/zybo-z7/migration-guide

[Dil00] B. Dipert, EDN's first annual PLD directory, EDN pp. 54–84 (2000)

[Ham10] M. Hamid, Writing Efficient Testbenches, Xilinx XAAP199 (May 2010)

[Hut92] R. Huthmann, Design of Bandpass 1 Bit $\sum\Delta$ ADC, Master Thesis, TU Darmstadt (1992)

[LRR05] B. Le, T. Rondeau, J. Reed, C. Bostian, Analog-to-digital converters. IEEE Signal Process. Mag. **22**, 69–77 (2005)

[MP88] F. Mavaddat, B. Parhami, URSIC the ultimate reduced instruction set computer. Int. J. Eng. Educ **25**, 327–334 (1988)

[MB01] U. Meyer-Baese, *Digital Signal Processing with Field Programmable Gate Arrays*, 1st edn. (Springer-Verlag, Berlin, 2001)

[P05] B. Parhami, *Computer Architecture: From Microprocessor to Supercomputers* (Oxford University Press, New York, 2005)

[Ter14] TerASIC, *DE1-SoC User Manual* (Altera University Program, 2014)

[VSS00] Visual Software Solutions, Inc., *HDL Bencher User's Guide* 60 pages (2000)

[Wol04] Wolfson Microelectronics plc, "WM8731," Portable Internet Audio CODEC with Headphone Driver and Programmable Sample Rates (2004)

[Xil93] Xilinx, *Data Book*, XC2000, XC3000 and XC4000 (1993)

[Xil18] Xilinx, Zynq-7000 SoC Data Sheet: Overview, Product Specification DS190 (2018)

Chapter 3
Microprocessor Component Design in VHDL

Abstract This chapter gives an overview of the part of the very high-speed integrated circuit hardware description language (VHDL) used in the book. As with most languages, we start with lexical preliminary and then discuss data types, operations, and statements of the VHDL language. Coding tips, additional help, and further reading complete the chapter. If your preferred HDL is Verilog, you may skip this chapter and continue with Chap. 4.

Keywords Very high-speed integrated circuit (VHSIC) · VHSIC hardware description language (VHDL) · Design waves · PAL assembler · RTL view · Keywords · Reserved words · VHDL operators · VHDL coding style · VHDL data types · VHDL data objects · Design recommendations · Lexical elements · Attributes · Finite state machine (FSM) · Memory initialization · Clock · Sequential statements · Concurrent statements · Language reference manual (LRM)

3.1 Introduction

Looking back to the time when VHDL was introduced, we had the following situation in PLD design (see Fig. 3.1a): The data flow part of a design typically was designed with a graphic user interface using vendor-specific graphic library elements. For the control part such as FSM, often PAL assembler was used. Since there was not a single standard for PAL assembler for each device, a vendor-specific language such as ABEL, PALASM, or CUPL was used. This was a very inflexible design flow and initially the DOD in the USA proposed to combine PAL assembler and data flow design to be integrated into a single nonvendor-specific HDL; see Fig. 3.1a.

This initial effort by the DOD was then finalized by the IEEE into the standard 1076. This initial standard from 1987 underwent minor revision (e.g., XNOR and shift operations were added) in 1993. Substantial addition to the standard such as fixed- and floating-point numbers and programming language interface (PLI) were

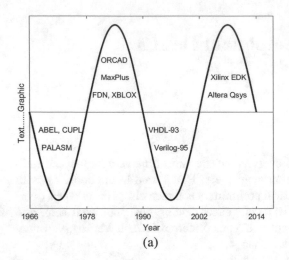

LRM 1087 year	Pages
1987	218
1993	249
2002	309
2008	640

(b)

(a)

Fig. 3.1 (**a**) Design wave in FPGA design. (**b**) LRM pages table

added in 2002 and 2008. The magnitude of each update is revealed by the number of additional pages in the language reference manual (LRM) shown in Fig. 3.1b [I87, I93, I02, I08]. However, tool provider has been slow adopting VHDL 1076-2002 and 2008 features, and it is often a wise approach to restrict the coding to the VHDL 1076-1993 standard that is supported by the majority of tools which also has more than enough features to design even the most complicated microprocessor. We will therefore in the following reference often the VHDL 1087-1993 standard. The associated quick reference cards developed by COMIT Systems and QUALIS can be found on the book CD.

3.2   Lexical Elements

If we plan to learn a new programming language, the first important question to ask is if the language is line sensitive (e.g., MATLAB) and/or column sensitive (e.g., PYTHON). This can often be answered by checking if a semicolon ";" is used to terminate all statements. In VHDL the one-line statement

```
a <= b OR C;
```

will be synthesized to the same circuit if we had written

```
a <=
        b OR
c;
```

Since it is required that each statement is completed with a semicolon, we can conclude that additional space, tab, or returns really do not matter, and therefore VHDL is *not* a line or column-sensitive language (with the exception of comments). Comments in VHDL start with a double minus -- and continue to the end of the line. No multiline comments are supported in VHDL 1993 (VHDL 2008 supported multiline comments with /* … */).

The second lexical question of importance is if a language is *case* sensitive. VHDL is not case sensitive such that INPUT, Input, or input are all considered as the same identifier.

The only parenthesis type supported in VHDL is the round one's (), both to index arrays and to group terms in statements; the square [] or curly { } are prohibited.

Operators used in expressions are one, two, or three character long, need to be written without space, and use more often alphabet character than special symbols (e.g., OR and not ||). Section 3.3 discusses all operators in more details.

Constant and initial values for bit or standard logic vectors use binary, octal, or hexadecimal base code with B, O, and X, respectively. Single bit constants use single quotes '…', while vectors use double quotes "…". INTEGER values are written without quotes with base 10 being default base. Another base can be specified using base values followed by the constant embedded in pound symbols base#const#. A few constant examples are listed in the table below.

Constant	Base	Dec. value	Minimum number of bits
101	10	101	7
B"101"	2	$101_2 = 5$	3
5#101#	5	$1 * 5^2 + 0 * 5^1 + 5^0 = 26$	5
X"101"	16	$1 * 16^2 + 0 * 16^1 + 16^0 = 257$	12

Common data types used in later chapters are BIT, BOOLEAN, STD_LOGIC, STD_LOGIC_VECTOR, and INTEGER that are defined in the library standard (linked by default) and STD_LOGIC package (requires the IEEE library). User data types we often use are INTEGER with range, enumeration type, and multidimensional arrays and vectors. There are plenty more types (NATURAL, POSITIVE, STD_ULOGIC, SIGNED, UNSIGNED, REAL, STRING, CHARACTER, etc.) we do not use in our microprocessor designs. VHDL is a very strictly typed language, and conversion between types requires always a conversion function. There are no implicit conversions such as INTEGER <= BOOLEAN assignment (allowed in C/C++); a VHDL complier will issue an error message. More details on data type can be found in Sect. 3.4.

A *basic user identifier* in VHDL used for project name, label, user type, signal, variable, constant name, entity, architecture, packages name, etc. has to start with a

```
ABS ACCESS AFTER ALIAS ALL always And ARCHITECTURE ARRAY ASSERT assign ASSUME
ASSUME_GUARANTEE ATTRIBUTE automatic Begin BLOCK BODY buf BUFFER bufif0
bufif1 BUS Case casex casez cell cmos config COMPONENT CONFIGURATION CONSTANT
CONTEXT COVER deassign Default defparam design disable DISCONNECT DOWNTO edge
Else   ELSIF   End   endcase   endconfig   endfunction   endgenerate   endmodule
endprimitive endspecify endtable endtask ENTITY event EXIT FAIRNESS FILE For
Force  forever  fork  Function  Generate  GENERIC  genvar  GROUP  GUARDED  highz0
highz1  If  ifnone  IMPURE  IN  incdir  include  INERTIAL  initial  Inout  input
instance  integer  IS  join  LABEL  large  liblist  Library  LINKAGE  LITERAL  LOOP
localparam macromodule MAP medium MOD module Nand negedge NEW NEXT nmos Nor
noshowcancelled  Not  notif0  notif1  NULL  OF  ON  OPEN  Or  OTHERS  OUT  output
PACKAGE  Parameter  pmos  PORT  posedge  POSTPONED  primitive  PROCEDURE  PROCESS
PROPERTY   PROTECTED    pull0    pull1    pulldown    pullup    pulsestyle_onevent
pulsestyle_ondetect PURE RANGE rcmos real realtime RECORD reg REGISTER REJECT
Release  REM  repeat  REPORT  RESTRICT  RESTRICT_GUARANTEE  RETURN  rnmos  ROL  ROR
rpmos  rtran  rtranif0  rtranif1  scalared  SELECT  SEQUENCE  SEVERITY  SHARED
showcancelled SIGNAL signed OF SLA SLL small specify specparam SRA SRL STRONG
strong0 strong1 SUBTYPE supply0 supply1 table task THEN time TO tran tranif0
tranif1 TRANSPORT tri tri0 tri1 triand trior trireg TYPE UNAFFECTED UNITS
unsigned UNTIL Use VARIABLE VMODE VPROP VUNIT vectored Wait wand weak0 weak1
WHEN While wire WITH wor Xnor Xor
```

Fig. 3.2 Verilog 1364-2001 and VHDL 1076-2008 keyword aka reserved words. VHDL only in capital letters, Verilog only in small letters, both first capital

letter and then have (single) underline and alphanumeric characters, in BNF we would write

```
letter{[_]alphanumeric}*
```

We are not allowed to use any VHDL keyword, and since we like to provide also Verilog version of our designs, it is in general recommended to avoid Verilog keyword too. Figure 3.2 lists all 215 VHDL, Verilog, and common keyword of the two languages using the newest standards.

Here are a couple of examples of valid and invalid user identifiers:

Legal: x x1 x_y Cin clock_enable
Not legal: entity (keyword) x_ _ y (two underline) CA$H (special character)
    4you (number first)

Extended identifiers use a \ID\ coding, are case sensitive, and allow easy transition from another language to VHDL. We do *not* use extended identifiers in this book.

Finally let us have a brief look at the overall coding organization and coding styles in VHDL. At maximum a VHDL program has five blocks: library, library body, entity, architecture, and component binding. At minimum we only have ENTITY and ARCHITECTURE. Our code typically starts with the library and associate packages including user-defined component. The ENTITY then contains port definition and their direction (IN, OUT, or BUFFER if the output is needed again within the design) and data type. Finally, the ARCHITECTURE has the details on the implementation. As coding styles, we distinguish three: In the *structural style*, predefined components are instantiated and connected via a netlist. This can be considered a one-to-one map of the graphic design used in early PLD design days.

As behavior (i.e., the gate implementation not immediate apparent) coding can be done using concurrent statements (called *data flow* style that emphasizes the parallel nature of hardware) or with *sequential coding* within a PROCESS statement that uses a sequential analysis. Here is a short three gate examples demonstrating all three design styles.

VHDL Example 3.1: The Three Coding Styles

```
 1   LIBRARY ieee;                     -- Using predefined packages
 2   USE ieee.std_logic_1164.ALL;
 3   USE work.lib74xx.ALL;
 4   -- ------------------------------------------------------------
 5   ENTITY example IS
 6     PORT (a, b, c, d : IN  STD_LOGIC;
 7              f, g, h     : OUT STD_LOGIC);
 8   END ENTITY example;
 9   -- ------------------------------------------------------------
10   ARCHITECTURE fpga OF example IS
11     SIGNAL n1, n2, t1, t2 : STD_LOGIC;
12   BEGIN
13   -- Structural style
14     C1: lib7432 PORT MAP(a, b, n1);
15     C2: lib7486 PORT MAP(n1, c, n2);
16     C3: lib7408 PORT MAP(n2, d, f);
17   -- Concurrent style
18     t1 <= a OR b;
19     t2 <= c XOR t1;
20     g <= d AND t2;
21   -- Sequential style
22     PROCESS (a, b, c, d)
23       VARIABLE t : STD_LOGIC;
24     BEGIN
25       t := a OR b;
26       t := c XOR t;
27       h <= d AND t;
28     END PROCESS;
29   END ARCHITECTURE;
```

All three code segments will be synthesized to the same circuit, as shown in Fig. 3.3.

VHDL 1993 does not have any direct compiler directive within the language definition; however, VHDL has a couple of internal flags such as the synthesis ATTRIBUTE to tell the compiler desired feature [P10]. Additional compiler switches maybe embedded in the comment section via --pragram(on) and --pragma(off) used in obfuscation [MCB11]. VHDL 2008 add flag via the `protect directive for encryption, but all these are tool/vendor-specific or 2008 VHDL and will not be used in later chapters [I08].

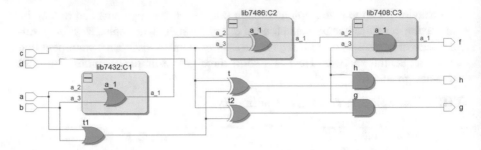

Fig. 3.3 The RTL view for `example.vhd` with the components expanded

3.3　Operators and Assignments

The VHDL operators with increasing precedence are listed in Table 3.1.

There are a couple of observations that can be drawn from Table 3.1. VHDL has "only" seven priority levels. Logical operators all have the same priority, so mixing AND and OR expressions will require grouping with parentheses (). No parentheses are needed for the NOT operator since it has the highest priority. On the plus side, we do not have to remember special symbols such as || for OR as logic operation in VHDL plain text is used. Another simplification is the fact that logical and bitwise operations use the same operator. Here is a short example for logic operations assuming `a='1'; b=c='0'` and `v="0011"` and `u="1010"`:

Number of values type	Scalar	Vector
Boolean expression	$a * b' + c$	$v' + u$
VHDL code	`(a AND NOT b) OR c`	`NOT v OR u`
Values	$1 * 0' + 0 = 1$	$(0011)' + 1010 = 1110$

Two of the relation operations in VHDL are different from C/C++: the equal (single =) and the not equal (/=). Relation operation most often are used as IF condition since the argument must be a BOOLEAN type, so in case `reset` is type BIT or STD_LOGIC, the statement `IF reset THEN` ... is not correct, and we need to write `IF reset='1' THEN` ... to make the argument a BOOLEAN type.

Table 3.1 Operators in VHDL in increasing precedence

Logical operators	AND	OR	NAND	NOR	XOR	XNOR
Relation operators	=	/=	<	<=	>	>=
Shift operators	SLL	SRL	SLA	SRA	ROL	ROR
Adding operators	+	−	&			
Sign operators	+	−				
Multiply operators	*	/	MOD	REM		
Miscellaneous	**	ABS	NOT			

Shift operations that have been added to the VHDL 1993 standard compared to VHDL 1987 and are available in both directions (left/right) as logic (zero extension) or arithmetic (sign extension) shifts. Also, rotations operations are defined. The shift amount can be positive or negative basically reversing the direction (left/right) of the shift operation. Shift operation is also usually "cheaper" than a multiple operations, so we may, for instance, consider replacing the expression $x*4$ with x SLL 2. The same works for division and right shifts; however, we should use the arithmetic shift in case we have sign number, e.g., $y/8$ can be replaced by y SRA 3. The expression y SRL 3 should be used for unsigned numbers. Unfortunately, the shift operations are not supported for STD_LOGIC_VECTOR in VHDL 1993 [I93], and custom function must be used [MB14] for the VHDL 2008 standard [I08, R11].

Evaluation is left associative, e.g., for a=2; b=3; c=4, we get:

```
left     <= (a-b) - c;   -- force left associative gives -5
right    <= a - (b-c);   -- forcing right associative gives 3
standard <= a - b - c;   -- VHDL default is left associative -5
```

Besides the usual arithmetic adding operations, we often use the concatenation "&" that combines two short vectors to one longer. Since VHDL is a strictly typed language, we often need to match precise the size of the left and right operand. The "&" can help here to make, for instance, a zero extension: x_zxt <= "0000" & x; to extend x by 4 bits. Similarly we can slice a vector into smaller pieces by using the index or range within the vector, e.g.:

```
op5   <= ir(17 DOWNTO 13);   -- Op code for ALU ops
kflag <= ir(12);             -- Immediate flag
imm12 <= ir(11 DOWNTO 0);    -- 12 bit immediate operand
```

The other arithmetic operations have the usual mathematical meaning, where ** is the potentiation or power-of operator, e.g., $2**4=16$. MOD and REM are modulo and remainder operations and are hardware intense, and we usually try to avoid these. Both produce the same result if both operands are of the same sign. In general, for $r = x$ REM y, the remainder r will have the same sign as x. For $r = x$ MOD y, the remainder r will have the same sign as y.

VHDL uses three types of *assignments*: for a SIGNAL the assignment symbol is <=, for a VARIABLE the assignment symbol is :=, and for the assignment arrow => is used for CONSTANT or PORT MAP. All three have lowest priority in a statement to make sure all expressions are computed first before making an assignment. All three have the same priority, so if we mix two in one statement such as <= and => in one statement, we need to use parentheses, e.g., we may define:

```
uPbus <= (OTHERS => 'Z'); -- a bus signal high impedance signal
x_sxt(15 DOWNTO 8) <= (OTHERS=>x(7)); -- sign extension
```

3.4   Data Types, Data Objects, and Attributes

VHDL Data Types

Modern programming languages typically have data types such as BOOLEAN, INTEGER, floating point, and enumeration types. However, if we look through the keywords in Fig. 3.2, none of these data types are listed for VHDL. The reason is that in VHDL, data types (and the associated operations) are defined in separate library files. The packages STANDARD is linked by default, and we often use types BOOLEAN, BIT, and INTEGER; we do not use CHARACTER, REAL, TIME, STRING, or BIT_VECTOR. Our reference cards list brief definitions of these, and the LRM 1993 [I93] section 14.2 lists all operations defined for these types. Let us in the following briefly look at the data types we use in later chapters. The enumeration type BOOLEAN can have values TRUE or FALSE. Here are two examples of BOOLEAN type go_eat and jc:

```
go_eat <= hungry AND NOT bankrupt;
jc <= (op6=jumpz AND z='1') OR (op6=jumpnz AND z='0') OR (op6=jump);
```

Logic operations and comparison are predefined operations for the type BOOLEAN.

The type BIT can have values '0' or '1' and is used for single bit data. Logic operations including NOT and comparisons are ok, but no arithmetic operations are defined for BIT, e.g.:

```
    two <= sw(0) AND NOT sw(1); -- ok for BIT sw
    sum <= sw(0) + sw(1); -- not ok
```

INTEGER type includes all values from $-2147483647\ldots2147483647$ or $-2^{31}-1\ldots2^{31}-1$, and 2 s complement arithmetic is assumed for signed numbers. For INTEGER we can use arithmetic including power-of ** as well as comparisons. However, we cannot select a single bit or use logic operations with type INTEGER.

Another library-defined data type we (and all ASIC industry) use often is the STD_LOGIC and STD_LOGIC_VECTOR data types. The BIT type with only two values is expanded to nine values, and the additions are high impedance 'Z', don't care '-', unknown 'X', uninitialized 'U', and not so often use values weak zero 'L', weak one 'H', and weak unknown 'W'. The STD_LOGIC is a "resolved"

type in contrast to STD_ULOGIC, meaning that the result of two STD_LOGIC
signals driving the same bus is defined [A08]. A microprocessor bus behavior is so
much easier described. The STD_LOGIC_VECTOR is an array of STD_LOGIC
bits, and with the additional arithmetic and (un)signed packages all arithmetic of
INTEGERs, logic and relation operations are supported. VHDL 2008 now also
includes shift operations that were not part of VHDL 1993 or a package. The STD_
LOGIC was important enough that the IEEE defined a special standard 1164. Here
are some STD_LOGIC and STD_LOGIC_VECTOR example operations:

```
opand   <= x AND y; -- Bitwise or logic operation
sum     <= x + y;   -- Sum of two vectors
prod    <= x * y;   -- The product of double length result
shift   <= x & "00"; -- shift left by two bit, i.e., multiply by 4
gt      <= x >= y;  -- comparison of two vectors or bits
```

As user-defined types we often use array types and enumeration types. Array
types are a collection of the scalar types, and we use 1D (e.g., register) or 2D (memory). The syntax to define a new type is as follows:

```
-- Memory definition
TYPE MTYPE IS ARRAY(0 TO 255) OF STD_LOGIC_VECTOR(7 DOWNTO 0);
```

We now can use the new type with

```
SIGNAL pram, dram : MTYPE;
```

The enumeration type is an abstract definition of item list that is particularly useful in FSM design, e.g.:

```
TYPE STATE_TYPE IS (FE, DC, EX);
SIGNAL state       : STATE_TYPE;
```

Now transition through the FSM states can be done with a more intuitive state
name than just integer numbers.

Since INTEGER are by default 32-bit long, this will often result in larger than
necessary circuits. If we like to continue to use the arithmetic, comparison, and
conversion function defined for INTGER type, we can build a derived type via
RANGE and preserve the properties. Here is an example:

```
SIGNAL rs : INTEGER RANGE 0 TO 15;
```

Now just 4 bits will be needed by the synthesis tool to implement this reduced
range INTEGER; no new library is needed.

Conversion Functions

Our VHDL library include all standard conversion functions between predefined data type if appropriate. The popular conversions in later chapters are between STD_LOGIC_VECTOR and INTEGER types. Assuming i is an INTEGER and s a STD_LOGIC_VECTOR, we have:

```
i <= CONV_INTEGER(s); -- conversion to an INTEGER
s <= CONV_STD_LOGIC_VECTOR(i, 8); -- use 8 LSBs for conversion
```

The conversion to STD_LOGIC_VECTOR has a second parameter indicating the number of bits to convert. If we have specified initially to use the STD_LOGIC_SIGNED package and like to force an unsigned conversion, we can zero extend the input, e.g.:

```
rs <= CONV_INTEGER('0' & ir(11 DOWNTO 8));
```

Attributes

The LRM section 14.1 and QUALIS reference card item 6 list an impressive number of *attributes* associated with our value type, range function, or signal. FPGA tool vendor provides additional synthesis attributes (we do not use these to stay tool/vendor independent) such as enum_encoding, chip_pin, keep, preserve, or noprune [P10]. Let us have a brief look at the popular attributes we use in later chapters:

The S'EVENT attribute let us design clock edge sensitive circuits, e.g.:

```
IF clk='1' AND clk'EVENT THEN
   Q <= d;
END IF;
```

Within a PROCESS will generate a positive-edge-triggered flip-flop, and the clk='1' AND clk'EVENT is equivalent to the STD_LOGIC function RISING_EDGE(clk).

The T'RANGE attribute allows a more generic coding than an explicit coding of the bound such as FOR k=start TO end LOOP ... as shown in the following example:

```
FOR  k=x'RANGE LOOP
  IF  x(k)= '0' THEN sign_bits <= k;
    EXIT;
  END IF;
END FOR;
```

to detect the number of zero's (i.e., the guard bits) in a vector x.

The attribute T'LEFT rather than T'HIGH should be used for the sign bit since it gives both times the correct values if the TO or DOWNTO indexing in the vector was used.

Data Objects

The four data object classes in VHDL 1993 are CONSTANT, SIGNAL, VARIABLE, and FILE. CONSTANTs are defined at the beginning and cannot be overwritten. A typical use of a CONSTANT would be operation code of a microprocessor. A SIGNAL is defined between ARCHITECTURE and BEGIN of the code and is considered "global" accessible. They represent a unique net and are updated at the end of the simulation cycle. An I/O port is also considered a SIGNAL. SIGNAL should be preferred in concurrent coding or needed multiple times within the design. A VARIABLE is defined local within a PROCESS, is iterative value holder updated immediately, not accessible outside the PROCESS, and may represent multiple nets. More on SIGNAL vs. VARIABLE can be found in Sect. 3.5. The FILE objects are not used in later chapters.

3.5   VHDL Statements and Design Coding Recommendations

VHDL has two types of statements: *sequential* (see LRM section 8 or item 4 QUALIS reference card) and the *concurrent* aka parallel statements (see LRM section 9 or reference card item 5). Both have special rules we need to comply with. If we try to link the statement type with our coding styles described in Example 3.1, we have that behavior coding within a PROCESS uses sequential statements and data flow and structural style are based on concurrent statements. This is the first time we see a major difference to a typical programming language, and that is why some authors insist that we refer to VHDL *coding* implying a parallel evaluation of the statements rather than *programming* associated typically with a sequential program evaluation [P10]. Anyway, the two most important rules for concurrent coding are:

- SIGNALs represent unique nets
- The order of the statement sequence does not matter

Applying these rules to Example 3.1, we see why two different auxiliary signals t1 and t2 were used (lines 18–20). We could *not* use the typical sequential coding style. The concurrent code

```
t <= a OR b;
t <= c XOR t;
g <= d AND t;
```

would not compile. However, based on the second rule, we can rearrange the statements as

```
t1 <= a OR b;
g <= d AND t2;
t2 <= c XOR t1;
```

and it would compile fine and produce precise the same circuit as shown in Fig. 3.3. Evaluation of the equation (or a component list) is done concurrent/parallel, and the ordering of the statement does not matter. This is a strict contrast to a usual (sequential) programming language: you cannot use the result t2 in the second statement before it is computed in the third statement.

The evaluation within a PROCESS is more in line with a typical sequential program evaluation. The two most important rules for sequential coding are:

- VARIABLEs do *not* represent unique nets.
- The order of the statement within a PROCESS does matter.

The first rule is confirmed by our Example 3.1. The VARIABLE t represents both the output of the first OR gate and the output of the XOR gate. These nets both represented by t do not need to be the same during simulation as required for a SIGNAL.

The second rule for the sequential statement can be verified by switching the last two statements within the PROCESS environment:

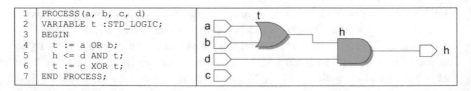

The VHDL compiler will synthesize these statements to an OR/AND circuit shown above on the right, leaving out the XOR gate from line 6. The compiler may issue a warning that input c is not used. So, the order of the statement does indeed matter within a PROCESS environment.

The use of a SIGNAL as local net within a PROCESS is possible but usually not recommended and may result in unexpected synthesis/simulation behavior. If we had coded in the PROCESS using a SIGNAL t, i.e.:

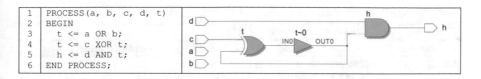

1	PROCESS(a, b, c, d, t)
2	BEGIN
3	t <= a OR b;
4	t <= c XOR t;
5	h <= d AND t;
6	END PROCESS;

we would have then somehow produced a contradiction of the sequential evaluation within a PROCESS and the requirement that a SIGNAL represents a unique net. VHDL solves this by "using the last assignment to a SIGNAL," and as a consequence, the first assignment from line 3 is ignored. The RTL view circuit shown on the right above only has the XOR/AND circuit leaving out the OR gate. The compiler may notify us via a warning that "inputs a and b are not in use."

Multistatements with similar content are usually coded using loops in programming languages. Loops in concurrent VHDL coding can also be used to GENERATE multiple copies of a component, e.g.:

```
U1: FOR i IN 0 TO REP-1 GENERATE
U2: BM1 -- Instantiate bm1 REP times
PORT MAP (clk => clk, rst => rst, s0 => s0, s1 => s1,
          s1 => s1, id => id, ipd => x(i), q => y(i));
END GENERATE;
```

Note that from a syntax standpoint, these are the only two cases in VHDL where labels are required: component instantiation and the GENERATE.

For sequential coding we can use a WHILE loop or FOR loops:

```
PROCESS(r)
BEGIN
    FOR k IN r'RANGE LOOP
        r(k) <= CONV_STD_LOGIC_VECTOR( -1, 8); -- all set to -1
    END LOOP;
END PROCESS;
```

Combinational Coding Recommendations

After this initial discussion of sequential vs. concurrent coding, you may wonder what the general coding recommendations of typical microprocessor components would be. An important goal of your code should be that your code should be short/compact and the behavior coding should be easily understood. Table 3.2 shows the matching general recommendations.

Table 3.2 VHDL circuit design recommendations

Circuit	VHDL coding recommendations
Combinational	Use Boolean equation for logic cloud (e.g., full adder equations)
Multiplexer	Use concurrent WHEN or complete IF for multiplexer with two inputs; use CASE for more than two inputs to multiplexer within a PROCESS
LUT	Use a CASE statement within a PROCESS for LUT-based designs (e.g., seven-segment decoder)
Latch	Use incomplete IF in a PROCESS
Register/flip-flop	Use RISING_EDGE in a PROCESS
Counter or accumulator	Use coding like accu <= accu + 1; within a PROCESS
FSM	Use enumeration type for states and a CASE statement for the next state assignments. Use two PROCESSs to separate next state and output decoder
Components	Use (large) predefined components whenever possible; do not instantiate gate-level circuits

Small combinational circuit such as the go_eat or jc code (see Sec. 3.4) will be more compact using concurrent code. For a 2:1 multiplexer, we can skip the PROCESS syntax by using the WHEN statement:

```
Y <= imm8 WHEN kflag='1'
     ELSE s(rs);
```

The behavior of the IF statement within a PROCESS maybe a little easier to digest but requires a PROCESS with sensitivity list, and this overhead can be avoided. The other danger always using the IF within a PROCESS is to accidentally infer a latch; see Fig. 3.4. The syntax of the VHDL IF statement differ a little from the C/C++ style: an END IF is required, and the condition located between IF ... THEN does not require parentheses (), although many authors [BV99, P03, P10, S96] and the Altera VHDL language template suggest otherwise.

A multiplexer with more than two inputs or a lookup table (LUT) as needed in ALU or BCD to seven-segment converter should be designed using a CASE statement. Here is an ALU example:

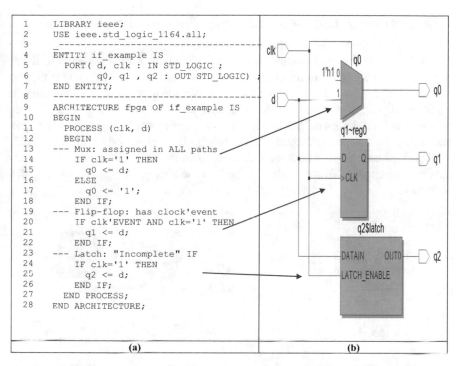

```
1      LIBRARY ieee;
2      USE ieee.std_logic_1164.all;
3      ------------------------------------
4      ENTITY if_example IS
5        PORT( d, clk : IN STD_LOGIC ;
6              q0, q1 , q2 : OUT STD_LOGIC) ;
7      END ENTITY;
8      ------------------------------------
9      ARCHITECTURE fpga OF if_example IS
10     BEGIN
11       PROCESS (clk, d)
12       BEGIN
13     --- Mux: assigned in ALL paths
14         IF clk='1' THEN
15           q0 <= d;
16         ELSE
17           q0 <= '1';
18         END IF;
19     --- Flip-flop: has clock'event
20         IF clk'EVENT AND clk='1' THEN
21           q1 <= d;
22         END IF;
23     --- Latch: "Incomplete" IF
24         IF clk='1' THEN
25           q2 <= d;
26         END IF;
27       END PROCESS;
28     END ARCHITECTURE;
```

(a)   (b)

Fig. 3.4 (a) Behavior code for 2:1 mux, latch, and flip-flop all done with the IF statement and (b) the synthesized circuit

```
ALU: PROCESS (reset, clk, x0, y0, c, op5, op6, in_port)
    CASE op5 IS
        WHEN add     =>   res   <= x0 + y0;
        WHEN addcy   =>   res   <= x0 + y0 + c;
        WHEN sub     =>   res   <= x0 - y0;
        WHEN subcy   =>   res   <= x0 - y0 - c;
        WHEN opxor   =>   res   <= x0 XOR y0;
        WHEN load    =>   res   <= y0;
        WHEN opinput =>   res   <= '0' & in_port;
        WHEN OTHERS  =>   res   <= x0; -- keep old
    END CASE;
  END PROCESS ALU;
```

Basic Sequential Circuit Coding: Flip-Flop and Latches

Figure 3.4 shows the recommended behavior coding style for latches and flip-flops. We seldom need latches in μP components; however, we should be aware that an "incomplete" IF, i.e., an IF with no assignment in the false path, will infer a latch. To design flip-flops, we need a S'EVENT attribute or the RISING_EDGE() function. A gate-level design of latches and flip-flops is often not recognized by the tools and will not be mapped to the embedded latch/flip-flop resources; see Exercise 3.53.

Flip-flops may require additional features such as (a)synchronous set or reset, data load, or enable. We can modify the code of the VHDL IF statement in Fig. 3.4 as follows:

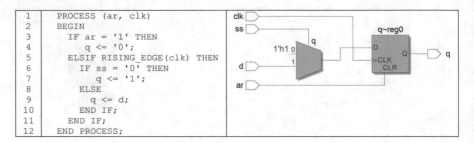

```
1    PROCESS (ar, clk)
2    BEGIN
3        IF ar = '1' THEN
4            q <= '0';
5        ELSIF RISING_EDGE(clk) THEN
6            IF ss = '0' THEN
7                q <= '1';
8            ELSE
9                q <= d;
10           END IF;
11       END IF;
12   END PROCESS;
```

This is an example for a positive-edge-triggered flip-flop with active high asynchronous reset (ar) and an active low synchronous set (ss).

A close relative to flip-flops is the counter which can be designed asynchronously using toggle flip-flop (T-FFs). But we prefer synchronous counters which uses an increment of a flip-flop array aka *register* that only need a single clock network. As with flip-flops additional features of a counter maybe desired such as enable, set, reset, data load, modulo, or an up/down switch. The program counter (pc) in a μP is often build as synchronous counter with asynchronous reset and immediate data load input (used for jump operations):

```
PROCESS (clk, reset)
BEGIN
  IF reset = '0' THEN
     pc <= (OTHERS => '0'); -- reset pc
  ELSIF FALLING_EDGE(clk) THEN
    IF jc THEN
       pc <= imm12; -- load 12 bit immediate
    ELSE
       pc <= pc + X"001"; -- increment counter
    END IF;
  END IF;
END PROCESS;
```

Memory

The memory of a μP typical has both RAM for data and ROM for program storage. We need to carefully read the device-tool vendor recommendation for HDL coding that indeed the FPGAs embedded memory blocks are used and our code is not mapped to LE which contain a single flip-flop but will require thousands of LEs compared to a single 4–10 Kbit embedded memory block. The memory blocks in modern FPGAs are often synchronous memory (unlike the "old" Altera Flex 10K devices), and we therefore should only try to build synchronous memory with a register for data and/or address. We may consider vendor-specific IP code blocks such as lpm_rom, but that would make transitions to other devices or tools hard. We should also try to enable MODELSIM for verification. Table 3.3 shows possible options, and the most flexible method seemed at time of writing the ROM with := initializations. A 128-word ROM with 16 bits can be coded as follows:

```
TYPE rom_type IS ARRAY (0 TO 127) OF STD_LOGIC_VECTOR(15 DOWNTO 0);
CONSTANT rom : rom_type := (
X"2201", X"2001", X"ff01", X"f201", X"2201", X"2180",
OTHERS => X"0000");
```

Finite State Machine

Finite state machines (FSMs) are the core of every microprocessor and come in two flavors: Moore and Mealy as shown in Fig. 3.5. A Moore machine is often preferred when we like to avoid glitches at the output (e.g., keyboard), while a Mealy machine allows immediate attention visible at the output port (e.g., brakes in a car). In HDL we must code three blocks: the next state logic, the machine state aka state register, and the output logic. We may use one, two, or three PROCESS blocks. We should

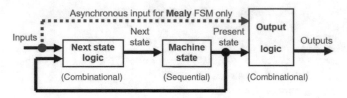

Fig. 3.5 Moore and Mealy FSM

also add a `reset` to start the FSM in a define initial state. A three-state machine just showing the state transition (without associate URISC actions) is shown next.

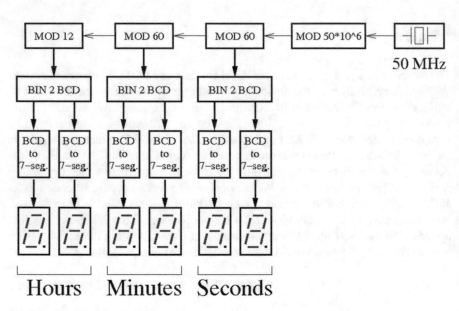

Fig. 3.6 Clock top-level overview with hours, minutes, and seconds display

Table 3.3 Memory design with data initialization

VHDL coding style	Altera Quartus	Xilinx Vivado/ ISE	ModelSim
MIF files + `lpm_rom`	X		
COE files + core gen		X	
Function init_ram (Altera HDL templates)	X		X
ATTRIBUTE `ram_INIT_FILE` (vendor attributes)	X		
Initial array values	X	X	X

VHDL coding goal: works for all three tools and is synthesized to block memory and not LE/CLB

```
TYPE STATE_TYPE IS (FE, DC, EX);
SIGNAL state      : STATE_TYPE;
...
 PROCESS (clk, reset)
BEGIN
  IF reset = '0' THEN -- asynchronous reset
     state <= FE; -- reset pc
  ELSIF RISING_EDGE(clk) THEN
    CASE state IS
      WHEN FE =>              -- Fetch instruction
                  state <= DC;
      WHEN DC =>              -- Decode instruction; split ir
                  state <= EX;
      WHEN EX =>              -- Process URISC instruction
                  state <= FE;

    END CASE;
   END IF;
 END PROCESS;
```

After reset is active, the FSM runs through the three states FE, DC, and EX and
then start again with FE.

Design Hierarchy and Components

Modular designs are used to split a large design into smaller pieces that can be build,
reused, tested, and combined later by one or a whole team of designers more efficient. Components are the preferred method in VHDL to build a modular design.
Building a design with elementary gate components as we did in Example 3.1 is in
general not recommended. This would make code hard to read and is even worse
than the graphic design we try to substitute with VHDL. As an example, think of a
clock design with hours, minutes, and seconds displays. We will need two digits
each and a binary to BCD converter and modulo 50×10^6, modulo 12, and modulo
60 counters; see Fig. 3.6.

We see that we will need four similar modulo counters, three binary-to-BCD
converters, and six seven-segment display decoders. In a modular design, we would
design just 3 vs. 13 in a single large design without components. There are different
methods to use component in VHDL. We may develop a separate library and instantiate this library as in Example 3.1. A more direct approach to use components is
used in later chapter by:

• Placing the VHDL file of the component in the same directory as the top-level
  design.

- Define the component declaration in the `SIGNAL` section of the top-level code.
- Use positional or the named (preferred) signal connections to local nets.

For one seven-segment display, the partial coding would look as follows:

```
ARCHITECTURE fpga OF DE2_lab1 IS
  COMPONENT seg7_lut IS
  PORT(idig : IN STD_LOGIC_VECTOR(3 DOWNTO 0);
       oseg : OUT STD_LOGIC_VECTOR(6 DOWNTO 0));
  END COMPONENT;
  .....
BEGIN
C4: seg7_lut PORT MAP (idig => accu(31 DOWNTO 28), oseg => hex1);
```

VHDL Coding Style, Resources, and Common Errors

There are plenty of coding guidelines you may need to follow based on your company or project rules. If you like to develop your own set of rules, you may want to start with the OpenMORE design recommendation given out by EDA leaders Synopsys Inc. and Mentor Graphics Inc. (over 250 rules with score values see `openmore.xls` from book CD) or look through the coding recommendation that comes with your FPGA design tools. Here is a small set of recommendation (we use in later design examples) you may consider at a minimum:

- For each file use a header with title, project name, author, tools, date, and short description with resource and timing data/estimations.
- Use `IN/OUT` avoid `BUFFER` mode.
- Use one I/O identifier per line and briefly explain the function of the I/O port.
- Use capital letters for keywords, small for user identifiers, or vice versa.
- Use indentations 2–4, e.g., align `IF`, `ELSE`, and `END IF`.
- Use full length identifiers with under_line or CamelCase to combined words in one identifier.
- Use white space (don't use Tab) consistent, e.g., no space before;,")().
- Use component for design hierarchy; avoid `FUNCTION` or `PROCEDURE`.
- Use `GENERIC` for bit width, `RISING_EDGE` for flip-flops.
- Use `INTEGER` (with `RANGE`) or `STD_LOGIC` data types with "`DOWNTO`" indexing.

When you start your VHDL coding, you may want to shorten your design time by remembering the following rules, aka frequently encounter errors in VHDL coding:

- Names used in `ENTITY` and `ARCHITECTURE` must be the same.
- Every VHDL statement must end with (single) semicolon (;).
- Quotes: single '...' for bits; "..." for vectors, no quotes for integers

Examples: $10 \rightarrow$ integer ten, but $"10" \rightarrow 10_2 = 2_{10}$.

- Labels: GENERATE and components must have label; starting label and closing label must match.
- PROCESS sensitivity list: WAIT $\rightarrow$ no list, all SIGNALs on the right and in IF/CASE condition go into the list.
- Only () parentheses are used (for vector elements or grouping terms) $\rightarrow$ [ ] { } $, or % are illegal character.
- Logical AND | OR | XOR | NAND | NOR | XNOR (all have same priority).
- Relation = | /= | < | <= | > | >= ({not} equal different to C/C++ programs)
  CASE statements: combine same pattern with symbol |.
- IF condition: type is Boolean! Example: x:bit requires IF x='1' THEN ....
- There are three applications for IF statements:

  Latch: design using incomplete IF statement (i.e., no assignment in false path)
  Flip-flop: RISING_EDGE(clk) or clk'EVENT AND clk='1'
  MUX: IF assignments in all branches

3.6   Further Readings

At first as with any new programming language, there is a lot of information to digest. Here are a couple of tips where you may find additional useful information to get you started coding in VHDL:

- Help:

  - Language reference manual (LRM) free from IEEE explore (PDF) or print version ca. $50 from IEEE
  - VHDL and STD_LOGIC reference cards by COMIT systems or QUALIS (PDF on book CD)
  - Altera's and Xilinx's VHDL help under the Help menu
  - Altera's and Xilinx's VHDL templates under the Template menu

- Tip: Collect as many VHDL examples as you can find:

  - VHDL books, e.g.:

    V. Pedroni [P03] Good introduction of *all* important VHDL statements, cheap
    V. Pedroni [P10] Introduction of *all* important VHDL statements incl. VHDL 2008
    J. Bhasker [B98] Many examples of mapping VHDL statements to gates
    R. Jasinski [J16] Many tips on excellent coding style incl. 2008 language improvements, not an introduction to VHDL

M. Smith [Ch. 10, S96] Good introduction; free book in the EDACafé; and ASIC data dated

Free online tutorials such as "The VHDL Cookbook" by P. Ashenden

- Put post-it pointers to important examples in your favorite textbook.
- Analyze these examples and make sure you have them handy when you code.
- Altera/Xilinx online tutorials, examples, and videos, see university program resources
- DSP with FPGA examples [MB14] under /vhdl.

Review Questions and Exercises

Short Answer

3.1. Why was VHDL developed by the DOD?

3.2. What are advantage and disadvantages of graphic vs. text-based CAD tool?

3.3. Why does VHDL keyword do not include data types?

3.4. Which data types are included in the VHDL STANDARD library?

3.5. Why are parentheses required in logic expression x AND y OR z?

3.6. What are the differences between sequential and concurrent coding? What are language requirements for both?

3.7. Can we modify a loop variable within the loop body? If not explain why.

3.8. Determine the number of bits in the following VHDL SIGNALs:

    (a) STD_LOGIC_VECTOR of length 8 with index 7...0 named vec

    (b) Array of 16 words of 8-bit unsigned integers name regs

    (c) Memory array with 256 elements of 16-bit STD_LOGIC_VECTOR with index 1...16 named dmem

    (d) 320 × 240 array of 16-bit unsigned INTEGER type named image

3.9. Write SIGNAL declaration for:

    (a) STD_LOGIC_VECTOR of length 8 with index 7...0 named vec

    (b) Array of 16 words of 8-bit unsigned integers name regs

    (c) Memory array with 256 elements of 16-bit STD_LOGIC_VECTOR with index 1...16 named dmem

    (d) 320 × 240 array of 16-bit unsigned INTEGERs named image

3.10. Given the following declarations:

```
CONSTANT u : BIT := '1';
CONSTANT v : BIT_VECTOR(5 DOWNTO 0) := "111000";
CONSTANT w : BIT_VECTOR(5 DOWNTO 0) := "000011";
```

Determine the length of the result x_k for $1 \leq k \leq 9$ and the binary value for the VHDL statements below:

    (a) x1 <= u & w;

(b) x2 <= w & v;
(c) x3 <= w SRL -1;
(d) x4 <= u NAND v(1);
(e) x5 <= NOT v;
(f) x6 <= v SRA 1;
(g) x7 <= w SLA 3;
(h) x8 <= v XOR w;
(i) x9 <= u AND NOT v(2) AND NOT w(2);

3.11. Develop VHDL code to implement the circuits given in the table. For $(0 \leq k \leq 3)$ PROCESS Pk uses input d(k) and output q(k) and STD_LOGIC data type. The synthesized circuit types are latch, D-flip-flop, or T-flip-flop, and the function of a, b, and c, i.e., can be clock, a-synchronous set (as) or reset (ar), or synchronous set (ss) or reset (sr). All flip-flops are rising edge triggered, and all control signals are actively high.

Process	Circuit type	clk	as	ar	ss	sr
P0	Latch	a				
P1	Latch	a	b	c		
P2	T-FF	b		a		c
P3	D-FF	b	a		c	

3.12. Determine for the following three process statements on the left the synthesized circuit. Also label I/O ports. Briefly describe the three circuits including the control signals:

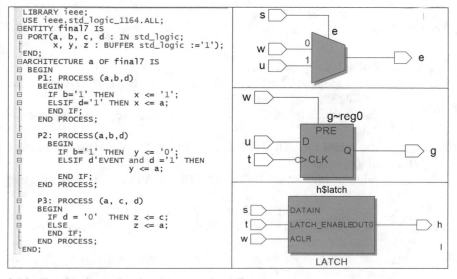

```
LIBRARY ieee;
USE ieee.std_logic_1164.ALL;
ENTITY final7 IS
  PORT(a, b, c, d : IN std_logic;
       x, y, z : BUFFER std_logic :='1');
END;
ARCHITECTURE a OF final7 IS
BEGIN
  P1: PROCESS (a,b,d)
  BEGIN
    IF b='1' THEN    x <= '1';
    ELSIF d='1' THEN x <= a;
    END IF;
  END PROCESS;

  P2: PROCESS(a,b,d)
  BEGIN
    IF b='1' THEN y <= '0';
    ELSIF d'EVENT and d ='1' THEN
                   y <= a;
    END IF;
  END PROCESS;

  P3: PROCESS (a, c, d)
  BEGIN
    IF d = '0'  THEN z <= c;
    ELSE             z <= a;
    END IF;
  END PROCESS;
END;
```

3.13. For the three circuits above on the right:

(a) Briefly describe the circuits including the control signals.
(b) Write VHDL code using STD_LOGIC data type.
(c) Complete the simulation below for e, g, and h.

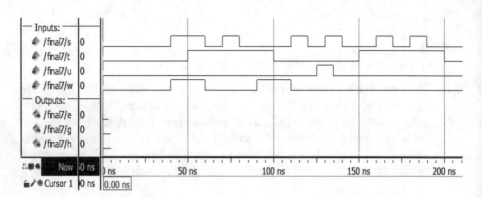

Fill in the Blank

3.14. The type INTEGER RANGE 10 TO 20 requires _____ bits.
3.15. The type INTEGER RANGE -2**6 TO 2**4-1 requires _____ bits.
3.16. The type INTEGER RANGE -10 TO -5 requires _____ bits.
3.17. The type INTEGER RANGE -7 TO 8 requires _____ bits.

True or False

3.18. _____ VHDL is case sensitive.
3.19. _____ At minimum VHDL code has two blocks: ENTITY and ARCHITECTURE.
3.20. _____ To use the STD_LOGIC_VECTOR data type, we need to include the STANDARD library.
3.21. _____ The values for STD_LOGIC and BIT_VECTOR can use base binary (B), octal (O), or hex (H).
3.22. _____ Signed arithmetic for INTEGER in VHDL is done in two complements.
3.23. _____ The default size for INTEGER is 32 bits.
3.24. _____ Shift operations in VHDL 1993 only defined to BIT_VECTOR and not the STD_LOGIC_VECTOR data type.
3.25. _____ The most popular attribute is S'EVENT used to design flip-flop and latches.

3.26. _____ To convert a STD_LOGIC_VECTOR to INTEGER, the function TO_INT() is used.

Determine if the following VHDL identifiers are valid (true) or invalid (false)

3.27. _____ Ports
3.28. _____ _Y_E_S_
3.29. _____ One-Way
3.30. _____ BEGIN_IF
3.31. _____ 4you
3.32. _____ P!NK
3.33. _____ THIS_IS_a_VeryLOOOONG_IDENTIFIER

Determine if the following VHDL string literals are valid (true) or invalid (false)

3.34. _____ B"0_0_0_0"
3.35. _____ O"678"
3.36. _____ X"678"
3.37. _____ 10#987654321#
3.38. _____ 2#210#
3.39. _____ 5#_4321_#
3.40. _____ 20#ABCD#

3.41. Determine the error lines (Y/N) in the VHDL 1993 code below and explain what is wrong.

VHDL code	Error (Y/N)	Give reason
LIBRARY ieee; // Using predefined packages		
ENTITY error is		
PORTS (x: in BIT; c: in BIT;		
Z1: out INTEGER; z2 : out BIT);		
END error		
ARCHITECTURE error OF has IS		
SIGNAL s ; w : BIT;		
BEGIN		
w := c;		
Z1 <= x;		
P1: PROCESS (x)		
BEGIN		
IF c='1' THEN		
x <= z2;		
END;		
END PROCESS P0;		
END OF has;		

3.42. Determine the error lines (Y/N) in the VHDL 1993 code below and explain what is wrong.

VHDL code	Error (Y/N)	Give reason
`LIBRARY ieee;  // Using predefined packages`		
`USE altera.std_logic_1164.ALL;`		
`ENTITY shiftregs IS`		
`GENERIC (WIDTH : POSITIVE = 4);`		
`PORT(clk, din : IN STD_LOGIC;`		
`    dout       : OUT STD_LOGIC);`		
`END;`		
`ARCHITECTURE a OF shiftreg IS`		
`COMPONENT d_ff`		
`  PORT (clock, d : IN  std_logic;`		
`     q        : OUT std_logic);`		
`END d_ff;`		
`SIGNAL b : logic_vector(0 TO width-1);`		
`BEGIN`		
`d1: d_ff PORT MAP (clk, b(0), din);`		
`g1: FOR j IN 1 TO witdh-1 GENERATE`		
`  d2: d-ff`		
`    PORT MAP(clk => clock,`		
`        din => b(j-1),`		
`        q => b(j));`		
`END GENERATE d2;`		
`dout <= b(width);`		
`END a;`		

Projects and Challenges

3.43. Verify via simulation that the following gates are *universal* by implementing NOT, AND, and OR

    (a) With two inputs NAND gates only

    (b) With two inputs NOR gates only

    (c) With 2:1 multiplexer

3.44. Design a full adder

```
s=a XOR b XOR cin ;
cout  = (a AND  b) OR (cin  AND (a OR b))
```

in VHDL and verify via simulation using only:

(a) AND, OR, and NOT gates
(b) With two inputs NAND gates only
(c) With two inputs NOR gates only
(d) With 2:1 multiplexer

3.45. Design a (a) binary, (b) gray, (c) Johnson, and (d) one-hot eight-state counter with asynchronous reset. Determine size and speed for the counters using the Balanced synthesis option.

3.46. Implement the LS163 counter and match the following simulation:

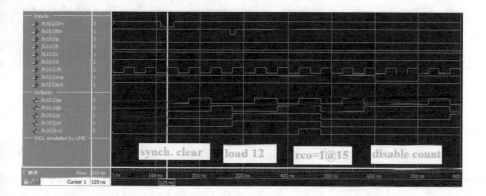

3.47. Implement an 8-bit ALU with the following eight operations: addition, subtraction, negation aka 2's complement, Boolean AND, Boolean OR, Boolean NOT, multiplication output MSBs, and multiplication output LSBs. Verify via simulation shown below.

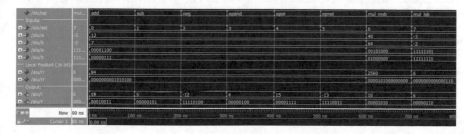

3.48. Implement an 8-bit data shift ALU that has the following operations: SL0, SL1, SR0, SR1, ROR, and ROL. Verify via simulation.

3.49. Implement the LS181 ALU (logic operations only) and match the simulation below:

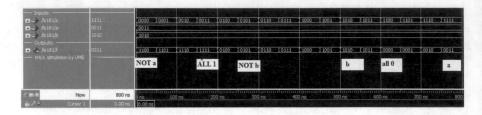

3.50. Briefly explain the FPGA compiler steps: synthesis, fitter, place and route, assembler, and programmer.

3.51. Develop the VHDL code for the one-input one-output FSM described below through the next state table. Use one-hot state encoding. The FSM has an active low asynchronous `reset`, and `reset` state is A. Develop a next state diagram and a test bench with 100% cover of all branches.

| Present state | Next state | | Output z |
	x = 0	x = 1	
A	B	C	0
B	B	A	1
C	B	C	1

3.52. Develop the VHDL code for a FSM that works like the Mustang turn light: 00X, 0XX, and XXX for a left turn where 0 is off and X is LED on. Use X00, XX0, and XXX for right turn signal. For emergency light both slider switch for the left/right selection should be on. The turn signal sequence should be repeated once per second.

3.53. Determine the resources for the following gate-level and behavior level designs:

   (a) A concurrent coded gated D-latch using AND/OR/NOT gates or one 2:1 multiplexer
   (b) A gated D-latch using the incomplete IF statement within a PROCESS
   (c) A concurrent coded flip-flop using a master/slave latch configuration using two multiplexers
   (d) A flip-flop using the S'EVENT within a PROCESS

3.54. Determine the resources for the following gate-level and behavior level designs:

   (a) 8-bit adder using GENERATE of eight full adder equations.
   (b) 8-bit adder using vector operator +.
   (c) A 4 × 4-bit multiplier using full adder only.
   (d) A 4 × 4-bit multiplier using the vector operator *.

3.55. (a) Design the PREP benchmark 3 shown in Fig. 3.7a in VHDL. The design
is a small FSM with eight states, 8-bit data input i, clk, rst, and an 8-bit
data-out signal o. The next state and output logic are controlled by a positive-
edge-triggered clk and an asynchronous reset rst; see the simulation in
Fig. 3.7c for the function test. The following table shows next state and output
assignments.

Current state	Next state	i (Hex)	o (Hex)
start	start	(3c)′	00
start	sa	3c	82
sa	sc	2a	40
sa	sb	1f	20
sa	sa	(2a)′(1f)′	04
sb	se	aa	11
sb	sf	(aa)′	30
sc	sd	–	08
sd	sg	–	80
se	start	–	40
sf	sg	–	02
sg	start	–	01

(b) Design the multiple instantiation for benchmark 3 as shown in Fig. 3.7b.
Verify the correct connections using three instantiations and make a snapshot.
Determine the registered performance Fmax and the used resources (LEs, multipli-
ers, and block RAMs) for the design with the maximum number of instantiations of
PREP benchmark 3 within a 10% error. As device use the one from your develop-
ment board and synthesis option balanced.

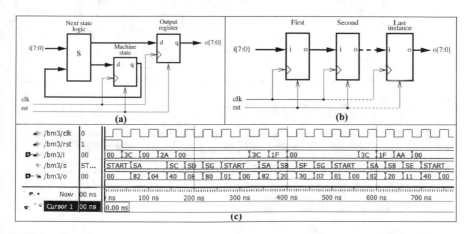

Fig. 3.7 PREP benchmark 3. (**a**) Single design. (**b**) Multiple instantiation. (**c**) Test bench to check
function

3.56. (a) Design the PREP benchmark 7 (which is equivalent to benchmark 8) shown in Fig. 3.8a in VHDL. The design is a 16-bit binary up-counter. It has an asynchronous reset rst, an active-high clock enable ce, an active-high load signal ld, and 16-bit data input d[15...0]. The registers are positive-edge triggered via clk. The simulation in Fig. 3.8c shows first the count operation to 5, followed by a ld (load) test. At 490 ns a test for the asynchronous reset rst is performed. Finally, between 700 and 800 ns, the counter is disabled via ce. The following table summarizes the functions:

clk	rst	ld	ce	q[15...0]
X	0	X	X	0000
↑	1	1	X	D[15...0]
↑	1	0	0	No change
↑	1	0	1	Increment

(b) Design the multiple instantiation for benchmark 7 as shown in Fig. 3.8b. Verify the correct connections using three instantiations and take a snapshot. Determine the registered performance Fmax and the used resources (LEs, multipliers, and block RAMs) for the design with the maximum number of instantiations of PREP benchmark 7 within a 10% error. As device use the one from your development board and synthesis option balanced.

3.57. (a) Design the PREP benchmark 9 shown in Fig. 3.9a using VHDL. The design is a memory decoder common in microprocessor systems. The

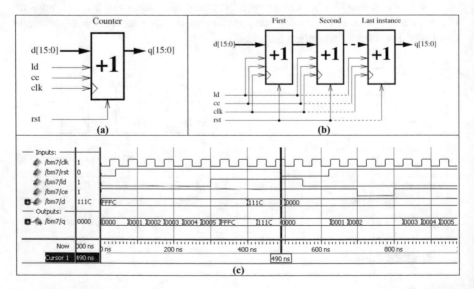

Fig. 3.8 PREP benchmark 7. (**a**) Single design. (**b**) Multiple instantiation. (**c**) Test bench to check function

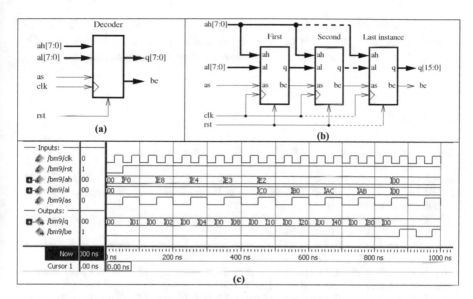

Fig. 3.9 PREP benchmark 9. (**a**) Single design. (**b**) Multiple instantiation. (**c**) Test bench to check function

addresses are decoded only when the address strobe as is active. Addresses that fall outside the decoder activate a bus error be signal. The design has a 16-bit input a[15...0], an asynchronous active-low reset rst, and all flip-flops are positive-edge triggered via clk. The following table summarizes the behavior (X = don't care):

rst	as	clk	a[15...0] (hex)	q[7...0] (binary)	be
0	X	X	X	00000000	0
1	0	↑	X	00000000	0
1	1	0	X	q[7...0]	be
1	1	↑	FFFF to F000	00000001	0
1	1	↑	EFFF to E800	00000010	0
1	1	↑	E7FF to E400	00000100	0
1	1	↑	E3FF to E300	00001000	0
1	1	↑	E2FF to E2C0	00010000	0
1	1	↑	E2BF to E2B0	00100000	0
1	1	↑	E2AF to E2AC	01000000	0
1	1	↑	E2AB	10000000	0
1	1	↑	E2AA to 0000	00000000	1

(b) Design the multiple instantiation for benchmark 9 as shown in Fig. 3.9b and match the simulation shown in Fig. 3.9c. Verify the correct connections using three instantiations and take a snapshot. Determine the registered performance Fmax and the used resources (LEs, multipliers, and block RAMs) for the design with the maximum number of instantiations of PREP benchmark 9 within a 10% error. As device use the one from your development board and synthesis option balanced.

References

[A08] P. Ashenden, *The Designer's Guide to VHDL*, 3rd edn. (Morgan Kaufman Publishers, Inc., San Mateo, 2008)

[B98] J. Bhasker, *AVHDL Synthesis Primer*, 2nd edn. (Star Galaxy Publishing, Allentown, 1998)

[BV99] J.S. Brown, Z. Vranesic, *Fundamentals of Digital Logic with VHDL Design* (McGraw-Hill, New York, 1999)

[I87] IEEE Standard VHDL Language Reference Manual. Institute of Electrical and Electronics Engineers, Inc., USA, VHDL 1076– (1987)

[I93] IEEE Standard VHDL Language Reference Manual. Institute of Electrical and Electronics Engineers, Inc., USA, VHDL 1076– (1993)

[I02] IEEE Standard VHDL Language Reference Manual. Institute of Electrical and Electronics Engineers, Inc., USA, VHDL 1076– (2002)

[I08] IEEE Standard VHDL Language Reference Manual. Institute of Electrical and Electronics Engineers, Inc., USA, VHDL 1076– (2008)

[J16] R. Jasinski, *Effective Coding with VHDL*, 1st edn. (The MIT Press, Cambridge, MA, 2016)

[MCB11] U. Meyer-Bäse, E. Castillo, G. Botella, L. Parrilla, A. García, Intellectual property protection (IPP) using obfuscation in C, VHDL, and Verilog coding. Proc. SPIE Int. Soc. Opt. Eng., Independent Component Analyses, Wavelets, Neural Networks, Biosystems, and Nanoengineering IX **8058**, 80581F1–80581F12 (2011)

[MB14] U. Meyer-Baese, *Digital Signal Processing with Field Programmable Gate Arrays*, 4th edn. (Springer, Heidelberg, 2014)

[P03] V. Pedroni, *Circuit Design with VHDL*, 1st edn. (The MIT Press, Cambridge, MA, 2003)

[P10] V. Pedroni, *Circuit Design and Simulation with VHDL*, 2nd edn. (The MIT Press, Cambridge, MA, 2010)

[R11] A. Rushton, *VHDL for Logic Synthesis*, 3rd edn. (John Wiley & Sons, New York, 2011)

[S96] D. Smith, *HDL Chip Design* (Doone Publications, Madison, AL, 1996)

Chapter 4
Microprocessor Component Design in Verilog

Abstract This chapter gives an overview of the part of the Verilog hardware description language used in later chapters. As with most languages, we start with lexical preliminary and then discuss data types, operations, and statements of the Verilog language. Coding tips, additional help, and further reading complete the chapter. If your preferred HDL is VHDL, you may skip this chapter at first reading.

Keywords Verilog · Operators · Compiler directives · Assignments · Coding style · Verilog data types · Value set · Lexical elements · Attributes · Finite state machine (FSM) · Design recommendations · Coding error · Clock · Design hierarchy · Components · Memory initialization · Procedural statements · Continuous statements · Language reference manual (LRM) · PREP benchmark

4.1 Introduction

The development of Verilog had a little different motivation than VHDL. In VHDL the DOD try to define a unified HDL such that the design can be moved freely over different tool platforms and vendors. ASIC and FPLD developers in California do not so heavily depend on large DOD projects, and the goal of Verilog consequently was a little different than with VHDL. The two key points in the history of Verilog HDL were to have a language that was close in syntax to the most popular computer programming in Engineering which was (and still is) the C/C++ language (see Fig. 4.1 for a comparison) and secondly a standardization of the language seemed not too important. So initially Gateway Design Automation (later acquired by Cadence Design Systems) developed the Verilog language. Cadence is still the registered Verilog trademark owner. Open Verilog International (OVI) was formed to control public domain Verilog [S00]. It took another 10 years until the Verilog Language Reference Manual (LRM) was then finally also put into an IEEE standard 1364-1995; see [I95]. The Verilog 1995 standard comes with comprehensive language manual including a PLI interface and has a total of 644 pages, much more

© Springer Nature Switzerland AG 2021

U. Meyer-Baese, *Embedded Microprocessor System Design using FPGAs*,
https://doi.org/10.1007/978-3-030-50533-2_4

| ```
count:=0;
FOR k IN 0 TO size-1 LOOP
  IF Data(k) = '1' THEN count := count +1;
  END IF;
END LOOP;
``` | (a) VHDL |
| ```
count=0;
for (k=0; k<size; k=k+1) begin
  if (Data[k] == 1) count = count +1;
end
``` | (b) Verilog |
| ```
count=0;
for (k=0; k<size; k=k+1) {
  if (Data[k] == 1) count = count +1;
}
``` | (c) C/C++ |

Fig. 4.1 Comparison of an one-counter program segment

than initial VHDL LRM (218 pages); see Fig. 3.1. The IEEE Verilog update [I01] in 2001 added several useful language elements such as signed data type, ANSI C style I/O, and arithmetic shift and potentiation. The IEEE 1364-2001 is the standard most synthesis tools support today and comes with an 856-page LRM [A04, S03, Xil04].

We will therefore in the following reference often the Verilog 1364-2001 standard. The Verilog quick reference cards developed by COMIT Systems and QUALIS use the 1364-1995 standard and can be found on the book CD.

4.2   Lexical Elements

If we plan to learn a new programming language, the first important question to ask is if the language is line sensitive (e.g., MATLAB) and/or column sensitive (e.g., PYTHON). This can often be answered by checking if a semicolon ";" is used to terminate all statements. In Verilog the one-line statement

```
assign A = B | C;
```

will be synthesized to the same circuit if we had written

```
assign A =
          B |
C;
```

Since it is required that each statement is completed with a semicolon, we can conclude that additional space, tab, or returns really do not matter, and therefore Verilog is *not* a line- or column-sensitive (with the exception of single-line comments) language. Single-line comments in Verilog start with a double slash // and continue to the end of the line. Multiline comments are supported comments within /* ... */; however, nested multiline comments are *not* supported in Verilog.
```

The second lexical question of importance is if a language is case sensitive. Verilog is case sensitive such that REG, Reg, or reg are all considered different identifier.

All three parentheses types are in use in Verilog: round ones () are used to group terms, function, and component parameters, [] to index arrays, and curly {} in concatenations.

Operators used in expressions are 1, 2, or 3 character long, need to be written without space, and use more often special symbols (e.g., || and not OR) than alphabet character. Section 4.3 discusses all operators in more details.

Constant and initial values for vectors use binary, octal, decimal, or hexadecimal base code with B, O, D, and X, respectively (base letter can be small or capital). Constants may start with the number of bits aka size, then one single quote ', (optional) signed number flag, base, and finally the value, i.e.,

```
[size]['[s]base]value
```

Without the initial size number, we have a so-called unsized constant which is 32 bits long, the same as integer numbers that require 32 bits. Boolean true is integer value 1, and false is integer value 0. A few constant coding examples are listed in Table 4.1.

The four values in Verilog are 0, 1, x, and z, where 0/1 is also the false/true condition. x or X is *unknown* and z or Z *high impedance*. Again, we can use small or capital letter x/X or z/Z. Verilog data types used in later chapter are net and reg and do not require a library. The most common net type is the generic type wire that is used as connection between structural elements. reg is used for storage, including signals that need to be stored within a cyclic evaluation, i.e., within an always statement, so not necessary a memory element. More details on data types can be found in Sect. 4.4.

A *basic user identifier* in Verilog used as project name, label, signal, constant name, module, component name, etc. has to start with a letter or underline and then has $, alphanumeric, or underline characters; in BNF, we would write

Table 4.1 Verilog constant examples

Constant	Base	#Bits	Decimal value
3'B101	Binary	3	$101_2 = 5$
9'O101	Octal	9	$1*8^2 + 0*8^1 + 8^0 = 65$
101	Decimal	32	101
12'H101	Hexadecimal	12	$1*16^2 + 0*16^1 + 16^0 = 257$
-8'SH01	Hexadecimal	8	-1 same as 8'SHFF

```
letter|_{alphanumeric|$|_}*
```

We are not allowed to use any Verilog keyword, and since we like to provide also VHDL version of our designs, it is in general recommended to avoid VHDL keywords too. Figure 3.2 in Chap. 3 lists all 215 Verilog, VHDL, and common keyword of the two languages using the newest standards. Here are a couple of examples of valid and invalid basic user identifiers in Verilog:

Legal: x x1 _y_ CA$H clock_enable
Not legal: module (keyword) P!NK (special character) 4you (number first)

Escaped identifier starts with a backslash, i.e., use a **ID** coding, are case sensitive, and allow easy transition from another language to Verilog. We do *not* use escape identifiers in this book.

Finally let us have a brief look at the overall coding organization and coding styles in Verilog. A Verilog program is embedded between module and endmodule. Our code typically starts with the ANSI C style (new in 2001 standard) port definition and their direction (input, output, or inout for bidirectional ports) and data type, and, finally, the details on the implementation. As coding style, we distinguish three: In the *structural style*, predefined components (user defined or using Verilog primitives) are instantiated and connected via a netlist. This can be considered a one-to-one map of the graphic design used in early PLD design days. As behavior (i.e., the gate implementation may not immediate apparent), coding can be done using *continuous* statements (i.e., the *concurrent* style that emphasizes the parallel nature of hardware) or with *procedural statements* within an always block that uses a *sequential* analysis. Here is a short three-gate example demonstrating all three design styles.

Verilog Example 4.1: The Three Coding Styles

```
1    /*******************************************************
2     * IEEE STD 1364-2001 Verilog file: example.v
3     * Author-EMAIL: Uwe.Meyer-Baese@ieee.org
4     *******************************************************/
5    module example   //----> Interface
6     (input   A, B, C ,D, // single bit input
7      output F, G,
8      output reg H);   // output single bits
9    // -----------------------------------------------------------
10     wire  N1, N2, T1, T2;
11   //  reg t;
12
13   // Structural style using user defined components
14      lib7432 C1 (.A_3(A), .A_2(B), .A_1(N1));
15      lib7486 C2 (.A_3(N1), .A_2(C), .A_1(N2));
16      lib7408 C3 (.A_3(N2), .A_2(D), .A_1(F));
17
18   //   Same using the 26 primitives
19   //      or (N1, A, B);
20   //      xor(N2, N1, C);
21   //      and(F, N2, D);
22
23   //-- Concurrent style
24     assign T1 = A | B;
25     assign T2 = C ^ T1;
26     assign G = D & T2;
27
28   //---- Sequential style
29     always @(*)
30     begin : P1
31       reg T;
32       T = A | B;
33       T = C ^ T;
34       H <= D & T;
35     end
36
37   endmodule
```

All three code segments will be synthesized to the same circuit, as shown in Fig. 4.2.

Verilog 2001 has a few compiler directives such as `define, `include, `ifdef, `ifndef, and `endif. These can be used for global control of the synthesis process, and some authors make heavy use of it, such as in the PACOBLAZE design by Pablo Bleyer Kocik that uses one master source file to synthesis many different architecture features using compiler directives. Additional compiler switches maybe embedded in the comment section via //pragram(on) and //pragma(off) used in obfuscation [MCB11]. This is not our main goal, and compiler directive are therefore spare in our Verilog code.

4.3 Operators and Assignments

The Verilog operators with increasing precedence are listed in Table 4.2.

There are a couple of observations that can be drawn from Table 4.2. Verilog has 13 priority levels, almost twice what VHDL uses. Logical operators for AND have a higher priority than OR, so mixing AND and OR expressions will not require grouping terms with parentheses (). No parentheses are needed for the NOT operator ether, since it has the highest priority. Logic scalar and bitwise operators on vectors have different coding. Here is a short example for logic, bitwise, and unary operations assuming A=1; B=C=0 and V=4'B0011 and U=4'B1010:

Number of values types	Logic scalar	Bitwise vector	Unary
Boolean expression	$A \times B' + C$	$V' + U$	$\prod U(k)$
Verilog code	(A && !B) || C	~V | U	& U
Values	$1 \times 0' + 0 = 1$	$(0011)' + 1010 = 1110$	$1 \times 0 \times 1 \times 0 = 0$

Unary operations reduce the input vector based on the operation specified to a single bit.

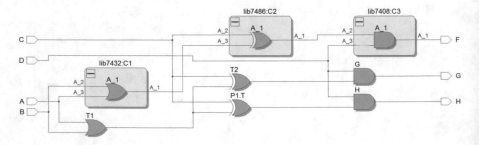

Fig. 4.2 The RTL view for example.v with the components expanded

Table 4.2   Operators in Verilog in increasing precedence

Conditional operator	? :			
Logical OR	||			
Logical AND	&&			
Bitwise OR/NOR	|	~|		
Bitwise XOR/XNOR	^	^~	~^	
Bitwise AND/NAND	&	~&		
Equality ex/including z / x	==	!=	===	!==
Relational	<	<=	>	>=
Shift logic and arithmetic	<<	>>	<<<	>>>
Add, subtract	+	–		
Multiply, divide, modulus	*	/	%	
Exponentiation	**			
Unary	+	–	!	~

All six relation operations form C/C++ have the same coding in Verilog. In addition we have the length-three operators === and !== that also considers the z and x
values; otherwise, comparison would always result in x if one operand contains x
or z. Relation operation most often are used within if condition. Since there is no
Boolean type in Verilog, all single bit scalar can be used as logic or bitwise coding,
e.g., if (reset)... and if (reset==1)... are both correct Verilog syntax.

Verilog 1995 contained just the logic shift. Verilog 2001 standard adds arithmetic
shift operations available in both directions: left <<< as alternative to a power-of-
two multiplications and right >>> as an alternative to power-of-two divide with a
correct sign propagation for negative number, e.g., with START = 4'B1000 we have

```
assign RESULT = (START >>  2); // Gives 0010
assign RESULT = (START >>> 2); // Gives 1110
```

No rotations operators are defined in Verilog. The shift amount should be positive since negative shifts in MODELSIM Verilog do not reverse the direction (left/
right) of the shift operation. Shift operation are usually "cheaper" than a multiply
operations, so we may, for instance, consider replacing the expression T * 4
with T << 2.

The other arithmetic operations have the usual mathematical meaning, where **
is new in Verilog 2001 and used for potentiation aka power-of operator, e.g.,
2**4=16. The modulo operation % is hardware intense, and we usually try to
avoid. For R = X % Y, the remainder R will have the same sign as X.

Evaluation in arithmetic is left associative, e.g., for a=2; b=3; c=4, we get

```
assign left    = (a-b) - c; // force left associative gives -5
assign right   = a - (b-c); // forcing right associative gives 3
assign default = a-b-c; // implicit left associative -5
```

Besides the usual arithmetic operations, we often use the concatenation {,} that combines two short vectors to one longer vector. Concatenation can help here to make, for instance, a zero extension: T_ZXT = {4'B000,T}; to extend T by 4 bits. Similarly we can slice a vector into smaller pieces by using the index or range within the vector, e.g.,

```
assign op5   = ir[17:13];  // Op code for ALU ops
assign kflag = ir[12];     // Immediate flag
assign imm12 = ir[11:0];   // 12 bit immediate operand
```

Verilog uses two types of *assignments*: for non-blocking assignment symbol is <=, and for blocking the assignment symbol = is used. Data flow style will require the use of non-blocking assignments and in procedural statements both assignments are allowed but may synthesize to different circuits. More about that in Sect. 4.5.

The port assignment for components is done using the *named* style via .component_port_name(net_name),... syntax. In the *positional* port assignments in Verilog, local nets are listed in the exact order as defined in the component definition.

4.4 Data Types and Value Set

Verilog Values Set

The Verilog values set consists of four basic values:

0 represents logic zero, or Boolean false
1 represents logic one, or Boolean true
x or X represents an unknown logic value
z or Z represents a high-impedance state

Verilog Data Types

A Verilog has just two groups of data types: the variable data types and the net data types. The net data types are used for physical connections between two circuit elements such as gates. A net cannot be a storage element such as a flip-flop. A general net type we use most often is the generic wire type. There are more precise net types such as `supply0`, `tri`, `wand`, etc. that describe the physical nature of the net more precise, but we do not use these. Variable is an abstraction of a data storage element including combinational output that needs to be stored in a cycle evaluation. Verilog variables include type `reg` and `integer` we use frequently, and three more we do not use (`real`, `realtime`, and `time`).

Verilog is not such a strong-typed language as VHDL. Length of left and right sides in assignments for instance does not need to match; `integer` and `reg` type can be exchanged without conversion functions, and Boolean or single-bit scalar are equivalent.

Let us in the following briefly look how the data types are used in later chapters. In many programming language type, Boolean usually has values `true` or `false`, but in Verilog the standard logic value `0` and `1` are used. So coding single bits aka scalar gives the same if coding with Boolean or bitwise operators

```
assign go_eat = hungry && !bankrupt; // Boolean ops: !,&&,||
assign go_eat = hungry  & ~bankrupt; // bitswise: ~,&,|,…
```

Bitwise operations or Boolean operations will result in the same synthesis result for single bits. For vectors we need to be more careful. Here Boolean logic operations will result in a single bit true/false aka 1/0 result and not a bit-by-bit vector evaluation as in the bitwise operation. If the LHS is a vector this 1-bit result is zero extended. Now for the vector logic AND, we get a true if just 1 bit is a match: so let's assume we have A=4'B1111, B=4'B0000, and C=4'B0100, and then for 4 bit output vectors, we get the following results (assume that LHS are 4 bit vectors):

```
assign logicAB = A && B; // single bit=0; zero extend result= 0000
assign bitwiseAB = A & B;// result bit-by-bit= 0000
assign logicAC = A && C; // single bit=1; zero extend result= 0001
assign bitwiseAC = A & C;// result bit-by-bit= 0100
```

No surprise for the first two results when Boolean evaluation gives 0 and no bit of the two vectors match. But in the second part, we have 1-bit match, i.e., results is true = 1 and extended by three zeros to the left, hence `logicAC=0001`. The bitwise operation reflects that the second bit from the left matches, i.e., `bitwiseAC=0100`.

According to Verilog LRM [I01], the `integer` type is tool dependent but must include a minimum of 32–bit range, i.e., $-2^{31}...2^{31} - 1$, and two's complement arithmetic is assumed for signed numbers. In principle we can use `integer`s to build 1D register, 2D memory, etc. that work as well as a `reg` type. However, since Verilog 2001 `integer`s do not allow a range reduction, always at least 32 bits are used for implementation with the potential of being more hardware demanding than what we need since we often need less in our data path than 32 bits.

As user-defined types, we often use array types and enumeration types. Array types are a collection of the scalar types, and we use 1D (e.g., register) or 2D (memory). Constants in Verilog can be defined using the `parameter` keyword. This

parameter can be used to set global values such as memory size or bitwidth. The
syntax to define a new memory with 4096 word of size 18 bit each is as follows:

```
parameter DATA_WIDTH=18, parameter ADDR_WIDTH=12;
...
// Declare the ROM variable
reg [DATA_WIDTH-1:0] rom[2**ADDR_WIDTH-1:0];
```

The parameter can be defined within the ANSI C style port list that would
also allow to overwrite the parameter value when the module is instantiate as
component; see Sec. 12.2 LRM [I01].

The enumeration type is an abstract definition of item list that are very useful in
FSM design. Verilog has no special enumeration type for that purpose, but using
parameter list is the recommended work around to define states in a FSM to
make the code more readable, e.g.,

```
parameter FE=0, DC=1, EX=2;
reg [1:0] STATE;
```

Now transition through the FSM states can be done with a more intuitive state
name than just integer numbers. A FSM transition would now be coded as STATE
<= DC; while the simulation will still show the integer values, at least the Verilog
code can be easier digested.

Another interesting feature is that Verilog computed the LHS length based on the
maximum length of the right and left operands. So if right and left have 8 bits and
we have input data A=255 and B=1, the 8 bit addition summation gives A+B=0; so
we better use an additional guard bit or alternatively use the left concatenation using
the curly brackets:

```
assign  Sum8 = A + B;     // All 8 bits; no carry bit: Sum8=0
assign  Sum9 = A + B; // left 9 bits; carry bit: Sum9 = 256
assign  {Cout, Sum} =  A + B; // left concat.: Sum=0; Cout=1
```

where Sum8 is are 8-bit word and Sum and Sum9 are 9-bit words.

If we now define all operands as signed type, then even in Sum9 we get a correct sign extension for the input operands and with all three methods will get, for
instance, $100 - 1 = 99$ as simulation result.

4.5   Verilog Statements and Design Coding Recommendations

Verilog has two types of statements: *procedural* aka sequential assigning values to *variables* (see LRM Chap. 6 [I01] or item 5 QUALIS reference card) and the *continuous* aka concurrent statements assigning values to *nets* (QUALIS reference card item 3). Both have special rules we need to comply with. If we try to link the behavior statement types with our coding styles described in Example 4.1, we see that procedural statements are embedded in an `always` block in sequential style. The concurrent style is based on continuous statements. This is the first time we see a major difference to a typical programming language and that is why some authors insist that we refer to Verilog *coding* implying a parallel evaluation of the statements rather than *programming* associated typically with a sequential program evaluation. Anyway, the two most important rules for continuous Verilog coding are the following:

- A `net` represents a unique wire; `reg` type is not permitted.
- The order of the statement sequence does not matter.
- Each statement starts with `assign` keyword, and the blocking assignment (=) must be used.

Applying these rules to Example 4.1, we see why two different auxiliary signals `T1` and `T2` were used. We could *not* use the typical sequential programming style. The continuous code

```
assign T = A | B;
assign T = C ^ T;
assign G = D & T;
```

would *not* compile due to multiple constant drivers for net "T." However, based on the second rule, we can rearrange the statements as

```
assign T1 = A | B;
assign G  = D & T2;
assign T2 = C ^ T1;
```

and it would compile fine and produce precisely the same circuit as shown in Fig. 4.2. Evaluation of the equation (or a component list) is done in parallel, and the ordering of the statement does not matter. This is a strict contrast to a usual (sequential) programming language: you cannot use the result `T2` in the second statement before it is computed in the third statement.

The statement evaluation within an `always` block is more in line with a typical sequential program evaluation. The three most important rules for procedural coding are the following:

- `variables` do *not* represent unique wires.
- The order of the statement within an `always` block does matter if blocking (=) assignments are used.
- A `variable` that needs to be saved during cyclic evaluation requires the `reg` type (even combinational circuit elements).

The first rule is confirmed by Example 4.1. The variable `T` represents both the output of the first `OR` gate (line 32) and the output of the `XOR` gate (line 33). These nets both represented by `T` do not need to be of the same value during simulation as required for a `wire`.

The second rule for the sequential statement can be verified by switching the last two statements within the `always` block:

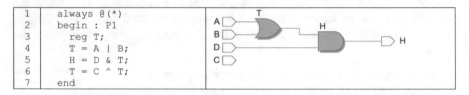

The Verilog compiler will synthesize these statements to an `OR`/`AND` circuit as shown on the right, leaving out the `XOR` gate from line 6. The compiler may issue a warning that input `c` is not used. So, the order of the statement does indeed matter within the `always` environment. Note also that a label (line 2: `P1`) is required if we declare a `variable` inside an `always` block.

The use of a variable `reg` as local net within a parallel aka non-blocking assignment with an `always` block is possible but usually not recommended and may result in unexpected synthesis/simulation behavior. If we had coded in the `always` block using non-blocking assignment for T, i.e.,

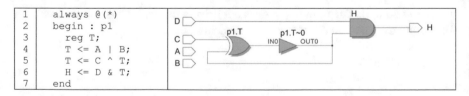

we would have then somehow produced a contradiction of the sequential evaluation within an `always` and the requirement that LHS of a non-blocking statement represents a unique net. Verilog solves this by "using the last assignment to a non-blocking variable," and consequently the first assignment to T from line 4 is ignored. The RTL view circuit shown on the right only has the `XOR`/`AND` circuit. The compiler may notify use via a warning that "input a and b do not drive logic."

Multi-statements with similar content are usually coded using loops in programming languages. For sequential coding, we can use `while`, `repeat`, `forever`, or `for` loops. We often use the `for` loop:

```
always @(negedge RESET)
begin
    if (!RESET) begin     // Set all registers to -1
        for (k=1; k<=15; k=k+1) R[k] = -1;
    end
end
```

Loops in continuous Verilog code can also be used to GENERATE multiple copies of a component, e.g.,

```
genvar i;
generate  // Instantiate bm1 REP times
   for (i=0; i<REP; i=i+1) begin : U1
      bm1 U2 (.clk(clk), .rst(rst), .s0(s0), .s1(s1),
              .s1(s1), .id(id), .ipd(x[i]), .q(y[i]));
   end
endgenerate
```

Note that from a syntax standpoint, there are two more cases in Verilog where labels are required: component instantiation and the GENERATE with `for` loop; see Sec. 12.1 LRM [I01].

Combinational Coding Recommendations

After this initial discussion of continuous vs. procedural assignments, you may wonder what general coding recommendations of typical microprocessor components would look like. An important goal of your code should be that your code should be short/compact and the behavior coding should be easily understood. Table 4.3 shows the matching general recommendations.

Small combinational circuit such as the `go_eat` code (see Sec. 4.4) will be more compact using continuous code. For a 2:1 multiplexer, we can skip the `always` syntax by using the `? :` statement:

```
assign y = (kflag) ? imm8 : s[rs]; // MPX second source ALU
```

The behavior of the if statement within an always block may be a little easier to digest but requires an always with sensitivity list, and this overhead can be avoided. The other danger using the if within an always block is to accidentally infer a latch; see Fig. 4.3. The syntax of the Verilog if statement is a match to the C/C++ style, but a begin...end and not {...} is required for multiple statements; see Fig. 4.1.

A multiplexer with more than two inputs or a look-up table (LUT) as needed in ALU or BCD to seven-segment converter should be designed using a case statement. Here is an ALU example:

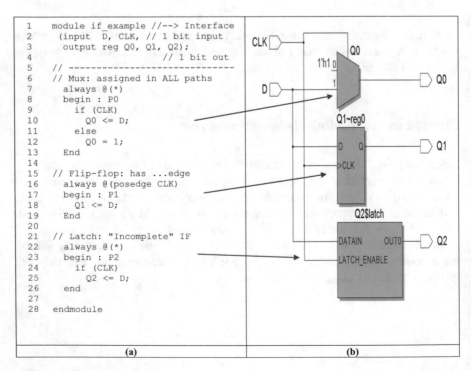

```
1    module if_example //--> Interface
2     (input  D, CLK, // 1 bit input
3      output reg Q0, Q1, Q2);
4                         // 1 bit out
5    // -------------------------------
6    // Mux: assigned in ALL paths
7      always @(*)
8      begin : P0
9        if (CLK)
10         Q0 <= D;
11       else
12         Q0 = 1;
13     End
14
15   // Flip-flop: has ...edge
16     always @(posedge CLK)
17     begin : P1
18       Q1 <= D;
19     End
20
21   // Latch: "Incomplete" IF
22     always @(*)
23     begin : P2
24       if (CLK)
25         Q2 <= D;
26     end
27
28   endmodule
```

(a)   (b)

Fig. 4.3 (a) Behavior code for 2:1 mux, latch, and flip-flop all done with the if statement and (b) the synthesized circuit

Table 4.3 Verilog circuit design recommendations

Circuit	Verilog coding recommendations
Combinational	Use Boolean equation for logic cloud (e.g., full adder equations)
Multiplexer	Use continuous assignment ?: or complete if for multiplexer with two inputs; use case for more than two inputs to multiplexer within an always block
LUT	Use a case statement within an always block for LUT-based designs (e.g., seven-segment decoder).
Latch	Use incomplete (i.e., without else) if in an always block
Register/flip-flop	Use posedge or negedge in the sensitivity list of an always block
Counter or accumulator	Use coding like accu = accu + 1; within an always block
FSM	Define states using parameter and a case statement for the next state assignments. Use two always blocks to separate next state and output decoder
Components	Use (large) predefined components whenever possible; do not instantiate gate primitives circuits

```
always @(*)
begin : P3
  case (op5)
     add    :  res  = x0 + y0;
     addcy  :  res  = x0 + y0 + c;
     sub    :  res  = x0 - y0;
     subcy  :  res  = x0 - y0 - c;
     opand  :  res  = x0 & y0;
     opor   :  res  = x0 | y0;
     opxor  :  res  = x0 ^ y0;
     load   :  res  = y0;
     opinput :  res = {1'b0 , in_port};
     fetch  :  res  = {1'b0 , dmd};
     default :  res  = x0;
  endcase
end
```

Basic Sequential Circuit Coding: Flip-Flop and Latches

Figure 4.3 shows the recommended behavior coding style for latches and flip-flops. We seldom need latches in μP components; however, we should be aware that an "incomplete" if, i.e., an if with no assignment in the false (i.e., else) path, will infer a latch. To design flip-flops, we need a posedge or negedge in the always sensitivity list. A gate level design of latches and flip-flops is often not recognized by the tools and will not be mapped to the embedded latch/flip-flop resources; see Exercise 4.55.

Flip-flops may require additional features such as (a)synchronous set or reset, data load, or enable. We can modify the code of the Verilog if statement in Fig. 4.3 as follows:

```
1   always@(posedge CLK or negedge AR)
2       if (!AR) Q <= 0;
3       else if (SS) Q <= 1;
4           else      Q <= D;
```

This is an example for a positive edge triggered flip-flop with active low asynchronous reset (AR) and an active high synchronous set (SS). Note that asynchronous control signals are listed in the always sensitivity list with an edge control signal while synchronous are not listed.

A close relative to flip-flops is the counter which can be designed as asynchronous counter using toggle flip-flop (T-FFs) we do not use in later chapters. We prefer a synchronous counter which uses an increment of a flip-flop array aka *register*. As with flip-flops, additional features of a counter maybe desired such as enable, set, reset, data load, modulo, or an up/down switch. The program counter (pc) in a μP is often built as synchronous counter with asynchronous reset and immediate data load input (used for jump operations):

```
always @(negedge clk or negedge reset) // use falling edge
  if (~reset) begin
    pc <= 0;           // reset the program counter
  end else begin
    if (jc)
      pc <= imm12;     // new address for jump
    else
      pc <= pc + 1; // increment the program counter
  end
```

Memory

The memory of a μP typical has both RAM for data and ROM for program storage. We need to carefully read the device/tool vendor recommendation for HDL coding that indeed the embedded memory blocks are used and our code is not mapped to LE which contain a single flip-flop but will require thousands of LEs compared to a single 4–10 Kbit-embedded memory block. The memory blocks in modern FPGAs are often synchronous memory (unlike the "old" Altera Flex10K devices), and we therefore should only try to build synchronous memory with a register for data and/ or address. We may consider vendor-specific IP code block such as `lpm_rom`, but that would make transitions to other devices or tools hard. We should also try to enable MODELSIM for verification. We can therefore use the initial style:

```
reg [15:0] ROM [639:0];
assign data = ROM[addr];
initial begin
  ROM[0]=16'h0000;
  ROM[1]=16'h04a9;
...
```

This direct value coding method would require generating a complete Verilog file if the μP program is modified. It is therefore more convenient to take advantage of the readmemh() functions that also works well with all three tools. For VIVADO Xilinx tools, it is recommended to use a *.mif file extension for the ROM data and our typical synchronous ROM block therefore for 4096 words with 18-bit data each can be coded as follows:

```
module rom4096x18
#(parameter DATA_WIDTH=18, parameter ADDR_WIDTH=12)
  (input  clk,                      // System clock
   input  reset,                    // Asynchronous reset
   input [(ADDR_WIDTH-1):0] address, // Address input
   output reg [(DATA_WIDTH-1):0] q); // Data output
// ----------------------------------------------------
// Declare the ROM variable
  reg [DATA_WIDTH-1:0] rom[2**ADDR_WIDTH-1:0];

  initial
  begin
    $readmemh("testdnest.mif", rom);
  end

  always @ (posedge clk or negedge reset)
    if (~reset)
      q <= 0;
    else
      q <= rom[address];
endmodule
```

The *.mif file is a plain ASCII file with one word per line.

Finite-State Machine

Finite-state machines (FSMs) are the core of every microprocessor and come in two flavors: Moore and Mealy as shown in Fig. 4.4. A Moore machine is often preferred when we like to avoid glitches at the output (e.g., keyboard), while a Mealy machine allows immediate attention visible at the output port (e.g., brakes in a car). In HDL we must code three blocks, the next state logic, the machine state aka state register, and the output logic, and we may use 1, 2, or 3 `always` blocks. We should also add a `reset` to start the FSM in a defined initial state. A three-state machine just showing the state transition (without associate URISC actions) is shown next:

```
parameter FE=0, DC=1, EX=2;
reg [1:0] state;
...

always @(posedge clk or negedge reset) //FSM behavioral style
begin : States                       // URISC in behavioral style
  if (~reset) begin    // all set register to -1
    state <= FE;
  end else begin     // use rising edge
    case (state)
    FE: begin            // Fetch instruction
          state <= DC;
        end
    DC: begin          // Decode instruction; split ir
          state <= EX;
        end
    EX: begin      // Process URISC instruction
          state <= FE;
        end
    default : state <= FE;
    endcase
  end
end
```

After `reset` is active, the FSM runs through the three states, FE, DC, and EX, and then starts again with FE.

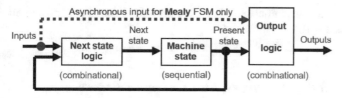

Fig. 4.4 Moore and Mealy FSM

Design Hierarchy and Components

Modular designs are used to split a large design into smaller pieces that can be built, reused, tested, and combined later by one or a whole team of designers more efficiently. Components are the preferred method in Verilog to build a modular design. Building a design with gate primitives as we did in Example 4.1 is in general not recommended. This would make code hard to read and is even worse than the graphic design we try to substitute with Verilog.

As an example, think of a clock design with hours, minutes, and seconds displays. We will need two digits each and a binary to BCD converter, BCD to seven-segment decoder, and modulo 50×10^6, 12, and 60 counters; see Fig. 4.5.

We see that we will need four similar counters, three binary-to-BCD converters, and six seven-segment display decoder. In a modular design, we would design just 3 vs. 13 in a single large design without components. In order to use components, we should

- Place the HDL file of the component in the same directory as the top-level design
- Use positional or the named (preferred) signal connections to local nets

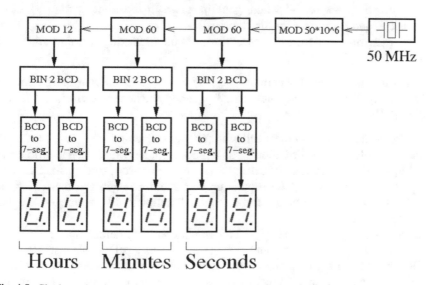

Fig. 4.5 Clock top-level overview with hours, minutes, and seconds display

A coding example for instantiation of user component has been presented in Example 4.1 (lines 14–16).

Verilog Coding Style, Resources, and Common Errors

There are plenty of coding guidelines you may need to follow based on your company or project rules. If you like to develop your own set of rules, you may want to start with the OPENMORE design recommendation given out by EDA leaders Synopsys Inc. and Mentor Graphics Inc. (over 250 rules with score values see `openmore.xls` from book CD) or look through the coding recommendation that comes with your FPGA design tools. Here is a small set of recommendation (we use in later design examples) you may consider at a minimum:

- For each file use a header with title, project name, author, tools, date, and short description with resource and timing data/estimations.
- Use the ANSI C style I/O port definition from Verilog 2001 to avoid the K&R style double specification from Verilog 1995.
- Use one I/O identifier per line, and briefly explain the function of the I/O port.
- Use capital letters for short identifiers and parameter aka constant identifiers.
- Use indentations 2–4, e.g., align `module`/`endmodule`, `if`/`else`, `begin`/ `end`, etc.
- Use full-length identifiers with under_line or CamelCase to combine words in one identifier.
- Use white space (don't use Tab) consistently, e.g., no space before ;,"()[]{}.
- Use (nested) component for design hierarchy; avoid `function` or `task`.
- Use `parameter` for bit width, generic `always` sensitivity list `@{*}` when possible, and `posedge` for flip-flops.
- Use `(signed)` `integer` and "downto" data indexing if doing arithmetic operations.

When you start your Verilog coding, you may want to shorten your design time by remembering the following rules, aka frequently encountered errors in Verilog coding [BV99]:

- Every Verilog statement must end with (single) semicolon ";".
- No semicolons are placed after "end" environment keywords such as `endmodule`, `end`, `endtask`, `endfunction`, `endgenerate`, `endcase`, etc.
- Quotes: use one single ' for separation `size` and `base` in constants; "..." are used for strings; no quotes for integers. Examples: $10 \rightarrow$ integer ten, but $2'B10 \rightarrow 10_2 = 2_{10}$.
- Labels are required for `always` blocks with variable definitions, `generate` with `for` loops, and components (LRM Ch. 12 calls label module instance name [I01]); `defparam` uses component label not component name. A `for` loop within `generate` blocks requires a label too. Syntax: `generate for(...)` `begin: label ... end`.

- `always` sensitivity list: all `variables` on the right and in `if`/`case` condition go into the list; In Verilog 2001, `@ (*)` or `@*` can be used for combinational-only designs to include all variables.
- Nets are used in continuous coding together with starting `assign` keyword and the (=) assignment operator.
- In `always` blocks sequential coding non-blocking (<=) and blocking (=) assignment can be used, but blocking will use sequential analysis only. In the cycle evaluation, even combinational variables need to be of the `reg` type. Combinational and edge-triggered outputs *cannot* be combined in one `always` block.
- All three parenthesis types are used: { } are for concatenation, [] for vector elements, and () for grouping terms, function calls, component ports, etc.
- The loop variable for `generate` requires a definition via `genvar`. When using a `for` loop, a label is required too.
- Logical operators use length 2 and bitwise single-length operator for AND (&) and OR (|). AND has higher priority than OR. Logic NOT (!) and bitwise NOT (~) are different too.
- Relation: `Y=1` assigns `Y` value one; `Y==1` checks if `Y` is 1; `Y===1` checks for `Z` and `X` values too.
- The `if` condition must be included in (), and a single statement can follow; for more than one, the `begin...end` environment is required. Accidental latches can be inferred in `if` or `case` statements when assignments are not done in all paths to the output.
- There are three applications for `if` statements within the `always` block:

    Latch: design using incomplete `if` statement (i.e., no assignment in false path)
    Flip-flop: use `posedge(clk)` for rising edge and `negedge(clk)` for falling edge
    MUX: `if` assignments in all branches

4.6   Further Reading

At first as with any new programming language, it is a lot of information to digest. Here are a couple of tips where you may find additional useful information to get you started coding in Verilog:

- Help:

    - Language reference manual (LRM) free from IEEE explore (PDF) or print version ca. $50 from IEEE [I01].
    - Verilog reference cards by COMIT Systems or QUALIS (PDF on book CD).
    - Altera's and Xilinx's Verilog help under the Help menu
    - Altera's and Xilinx's Verilog Templates under the Template menu

- Tip: Collect as many Verilog examples as you can find:

  - Verilog books, e.g.:

    J. Bhasker [B98]. Many examples of mapping Verilog statements to gates
    M. Ciletti [C04]. Many tips and new 2001 language improvements
    M. Smith [Ch.11, S96]. Good introduction; free book in the EDACafé;
    ASIC data dated
    Free online tutorials such as "Handbook on Verilog HDL" D. Hyde [H97].

  - Put post-it pointers to important examples in your favorite text book.
  - Analyze these examples, and make sure you have them handy when you code.
  - For Altera/Xilinx online tutorials, examples, and videos, see University
    Program resources.
  - DSP with FPGA examples [MB14] under /verilog.

Review Questions and Exercises
Short Answer

4.1. What were the two key principles when Verilog was developed?
4.2. What are the advantages and disadvantages of graphic- vs. text-based CAD
     tool?
4.3. Why does the Verilog data type reg does not imply a memory element?
4.4. What are the four values used in Verilog language?
4.5. Why are no parentheses required in logic expression A && B || C?
4.6. What are the differences between blocking and non-blocking assignments?
     What are the language requirements for both?
4.7. Can we modify a loop variable within the loop body? If not, explain why.
4.8. Determine the number of bits in the following Verilog variable
     definitions:

     (a) Vector of length 8 with index 7...0 named vec
     (b) Array of 16 words of 8-bit signed data named regs
     (c) Memory array with 256 elements of 16 bits with index 1...16 named
         dmem
     (d) 320 × 240 array of 16-bit unsigned data named image

4.9. Write Verilog declaration for:

     (a) Vector type wire of length 8 with index 7...0 named vec
     (b) Array of 16 words type reg of 8-bit signed data name regs
     (c) Memory array type reg with 256 elements of 16-bit data with index
         1...16 named dmem
     (d) 320 × 240 array type reg of 16-bit unsigned data named image

4.10. Given the following declarations:

```
wire U = 1'b1;
wire signed [5:0] V = 6'b111000;
wire signed [5:0] W = 6'b000011;
```

Determine the length of the RHS result X_k for $1 \le k \le 9$ and the binary value for the Verilog statements below:

(a) `assign X1 = {U, W};`
(b) `assign X2 = {W, V};`
(c) `assign X3 = W / 2;`
(d) `assign X4 = ~(U & V[1]);`
(e) `assign X5 = ~V ;`
(f) `assign X6 = V >>> 1;`
(g) `assign X7 = | W;`
(h) `assign X8 = V + W;`
(i) `assign X9 = U & ~V[2] & ~W[2];`

4.11. Develop Verilog code to implement the circuits given in the table. For $(0 \le k \le 3)$ `always` statement, Pk uses input `D[k]` and output `Q[k]` and `reg` data type. The synthesized circuit types are latch, D-flip-flop, or T flip-flop. The function of A, B, and C can be clock, asynchronous set (AS) or reset (AR), or synchronous set (SS) or reset (SR). All flip-flops are rising edge triggered, and all control signal are active high.

always	Circuit Type	CLK	AS	AR	SS	SR
P0	Latch	A				
P1	Latch	A	B	C		
P2	T-FF	B		A		C
P3	D-FF	B	A		C	

4.12. Determine for the following three `always` statements (Fig. 4.6 left) the synthesized circuit. Also label I/O ports. Briefly describe the three circuits including the control signals:

4.13. For the three circuits (Fig. 4.6 right) above on the right:

(a) Briefly describe the circuits including the control signals.
(b) Write Verilog code `final7.v` using `reg` data type.
(c) Complete the simulation below for E, G, and H.

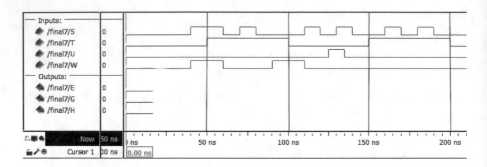

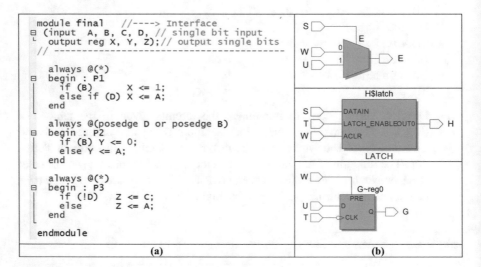

Fig. 4.6 Coding for Exercise 4.12 (**a**) and RTL view for Exercise 4.13 (**b**)

Fill in the Blank

Complete the table data below.

	Constant	Base	#Bits	Decimal value
4.14.	8'B00001111			
4.15.	12'o1234			
4.16.	1234			
4.17.	16'H1234			

True or False

4.18. _____ Verilog is not case sensitive.

4.19. _____ Verilog code is enclosed between `module` and `endmodule`.

4.20. _____ Verilog uses a nine-value system: `0,1,Z,X,U,L,H,-,W`.

4.21. _____ Verilog has no enumeration type, Boolean, or floating-point data type.

4.22. _____ The four bases in Verilog constants are binary (`B`), octal (`O`), decimal (`D`), and hex (`H`).

4.23. _____ Signed arithmetic for `integer` in Verilog is done in two's complement.

4.24. _____ The size for all `integer` is 32 bits.

4.25. _____ Shift right operations in Verilog 2001 can be logic using `>>` or arithmetic using the `>>>` operator.

4.26. _____ There is no semicolon after `end` and `endmodule`.

4.27. _____ Logic value `True` is coded with 1 and `False` with 0 in Verilog.

Determine if the following Verilog user identifiers are valid (True) or invalid (False).

4.28. _____ `inputs`

4.29. _____ `_NOT_VALID_`

4.30. _____ `38Special`

4.31. _____ `one-way`

4.32. _____ `begin_if`

4.33. _____ `3DoorsDown`

4.34. _____ `F!NAL`

4.35. _____ `THIS_IS_a_VeryLOOOONG_IDENTIFIER`

Determine if the following Verilog string literals are valid (True) or invalid (False).

4.36. _____ `4'B0_0_0_0`

4.37. _____ `9'o678`

4.38. _____ `12'H678`

4.39. _____ `'D987654321`

4.40. _____ `3'B210`

4.41. _____ `12'O_4321_`

4.42. _____ `16'XABCD`

4.43. Determine the error lines (Y/N) in the Verilog 2001 code below and explain what is wrong.

Verilog code	Error (Y/N)	Give reason
`module error    ##----> Interface`		
`(input   X, C;`		
`output Z1,`		

Verilog code	Error (Y/N)	Give reason
`output reg Z2);`		
`wire  S, W,`		
`assign W <= C;`		
`assign Z1 = X;`		
`always @{C}`		
`begin : S`		
`if (CLK==1) begin`		
`X <= Z2;`		
`end if;`		
`end`		
`end_module`		

4.44. Determine the error lines (Y/N) in the Verilog 2001 code below and explain what is wrong.

Verilog code	Error (Y/N)	Give reason
`module shiftregs begin`		
`(input CLK, DIN,`		
`output    DOUT);`		
`parameter WIDTH <= 4;`		
`wire [0:WIDTH-1] B;`		
`d_ff D1 (CLK, DIN, B(0));`		
`genvariable j;`		
`generate`		
`for (j=1;j<WITDH;j=j+1) begin`		
`d-ff D3 (.D(B[j-1]),`		
`.CLOCK(CLOCK),`		
`.Q(B[j]));`		
`End`		
`endgenerate`		
`assign DOUT = B[WIDTH];`		
`endmodule`		

Projects and Challenges

4.45. Verify via Verilog code simulation that the following gates are *universal* by implementing NOT, AND, and OR:

(a) With two inputs NAND gates only

   (b) With two inputs NOR gates only

   (c) With 2:1 multiplexer

4.46. Design a full adder

```
        s=a XOR b XOR cin ;
  cout  = (a AND  b) OR (cin  AND (a OR b))
```

in Verilog, and verify via simulation using only:

   (a) AND, OR, NOT gates

   (b) With two inputs NAND gates only

   (c) With two inputs NOR gates only

   (d) With 2:1 multiplexer

4.47. Design a (a) binary, (b) gray, (c) Johnson, and (d) one-hot 8-state counter with asynchronous reset. Determine size and speed for the counters using the Balanced synthesis option.

4.48. Implement the LS163 counter and match the following simulation.

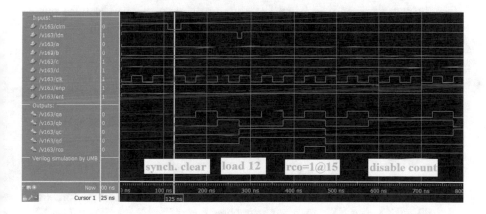

4.49. Implement an 8-bit ALU with the following eight operations: addition, subtraction, negation aka two's complement, Boolean AND, Boolean OR, Boolean NOT, multiplication output MSBs, and multiplication output LSBs. Verify via simulation shown below.

4.50. Implement an 8-bit data shift ALU that has the following operations: SL0, SL1, SR0, SR1, ROR, and ROL. Verify via simulation.

4.51. Implement the LS181 ALU (logic operations only), and match the simulation below.

4.52. Briefly explain the FPGA compiler steps: synthesis, fitter, place & route, assembler, and programmer.

4.53. Develop the Verilog code for the one-input one-output FSM described below through the next state table. Use one-hot state encoding. The FSM has an active low asynchronous `reset`, and `reset` state is A. Develop a next-state diagram and a test bench with 100% cover of all branches.

	Next state		Output z
Present state	x = 0	x = 1	z
A	B	C	0
B	B	A	1
C	B	C	1

4.54. Develop the Verilog code for an FSM that works as the Mustang turn light: 00X, 0XX, and XXX for a left turn where 0 is off and X is LED on. Use X00, XX0, and XXX for right turn signal. For emergency light, both slider switches for the left/right selection should be on. The turn signal sequence should repeat once per second.

4.55. Determine the resources for the following gate level and behavior level designs:

   (a) A continuous coded gated D-latch using AND/OR/NOT gates or one 2:1 multiplexer
   (b) A gated D-latch using the incomplete `if` statement within an `always`
   (c) A continuous coded flip-flop using a master/slave latch configuration using two multiplexers
   (d) A flip-flop using the `posedge` with an `always`

4.56. Determine the resources for the following gate level and behavior level designs:

(a) 8-bit adder using `generate` of eight full adder equations
(b) 8-bit adder using vector operator +
(c) A 4 × 4-bit multiplier using full adder only
(d) A 4 × 4-bit multiplier using the vector operator *

4.57. (a) Design the PREP benchmark 3 shown in Fig. 4.7a in Verilog. The design is a small FSM with eight states, 8-bit data input i, `clk`, `rst`, and an 8-bit data-out signal o. The next state and output logic are controlled by a positive edge triggered `clk` and an asynchronous reset `rst`; see the simulation in Fig. 4.7c for the function test. The following table shows next state and output assignments.

Current state	Next state	i (Hex)	o (Hex)
start	start	(3c)′	00
start	sa	3c	82
sa	sc	2a	40
sa	sb	1f	20
sa	sa	(2a)′(1f)′	04
sb	se	aa	11
sb	sf	(aa)′	30
sc	sd	–	08
sd	sg	–	80
se	start	–	40
sf	sg	–	02
sg	start	–	01

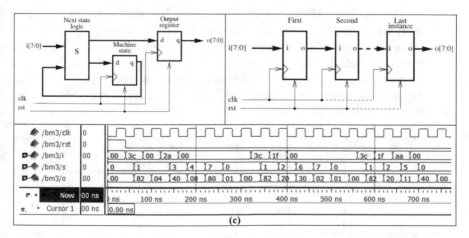

Fig. 4.7 PREP benchmark 3. (**a**) Single design. (**b**) Multiple instantiation. (**c**) Test bench to check function

(b) Design the multiple instantiation for benchmark 3 as shown in Fig. 4.7b. Verify the correct connections using three instantiations, and make a snapshot. Determine the registered performance Fmax and the used resources (LEs, multipliers, and Block RAMs) for the design with the maximum number of instantiations of PREP benchmark 3 within a 10% error. As device, use the one from your development board and synthesis option balanced.

4.58. (a) Design the PREP benchmark 7 (which is equivalent to benchmark 8) shown in Fig. 4.8a in Verilog. The design is a 16-bit binary up-counter. It has an asynchronous reset rst, an active-high clock enable ce, an active-high load signal ld, and 16-bit data input d[15..0]. The registers are positive edge triggered via clk. The simulation in Fig. 4.8c shows first the count operation to 5 and followed by a ld (load) test. At 490 ns a test for the asynchronous reset rst is performed. Finally, between 700 and 800 ns, the counter is disabled via ce. The following table summarizes the functions.

clk	rst	ld	ce	q[15...0]
X	0	X	X	0000
↑	1	1	X	D[15...0]
↑	1	0	0	No change
↑	1	0	1	Increment

(b) Design the multiple instantiation for benchmark 7 as shown in Fig. 4.8b. Verify the correct connections using three instantiations, and take a snapshot. Determine the registered performance Fmax and the used resources (LEs, multipliers, and Block RAMs) for the design with the maximum number of instantiations of PREP benchmark 7 within a 10% error. As device, use the one from your development board and synthesis option balanced.

4.59. (a) Design the PREP benchmark 9 shown in Fig. 4.9a using Verilog. The design is a memory decoder common in microprocessor systems. The addresses are decoded only when the address strobe as is active. Addresses that fall outside the decoder activate a bus error be signal. The design has a 16-bit input a[15..0] and an asynchronous active-low reset rst, and all flip-flops are positive edge triggered via clk. The following table summarizes the behavior (X = don't care).

rst	as	clk	a[15...0] (hex)	q[7...0] (binary)	be
0	X	X	X	00000000	0
1	0	↑	X	00000000	0
1	1	0	X	q[7...0]	be
1	1	↑	FFFF to F000	00000001	0
1	1	↑	EFFF to E800	00000010	0
1	1	↑	E7FF to E400	00000100	0
1	1	↑	E3FF to E300	00001000	0

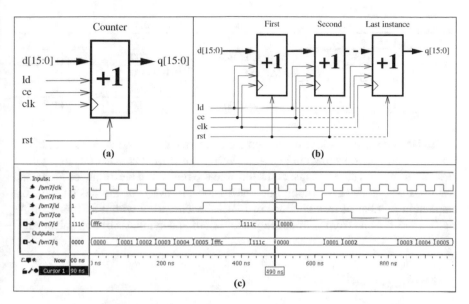

Fig. 4.8 PREP benchmark 7. (**a**) Single design. (**b**) Multiple instantiation. (**c**) Test bench to check function

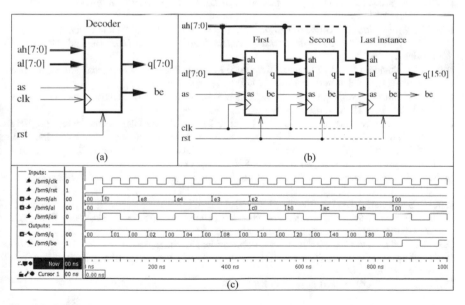

Fig. 4.9 PREP benchmark 9. (**a**) Single design. (**b**) Multiple instantiation. (**c**) Test bench to check function

rst	as	clk	a[15...0] (hex)	q[7...0] (binary)	be
1	1	↑	E2FF to E2C0	00010000	0
1	1	↑	E2BF to E2B0	00100000	0
1	1	↑	E2AF to E2AC	01000000	0
1	1	↑	E2AB	10000000	0
1	1	↑	E2AA to 0000	00000000	1

(b) Design the multiple instantiation for benchmark 9 as shown in Fig. 4.9b, and match the simulation shown in Fig. 4.9c. Verify the correct connections using three instantiations, and take a snapshot. Determine the registered performance Fmax and the used resources (LEs, multipliers, and Block RAMs) for the design with the maximum number of instantiations of PREP benchmark 9 within a 10% error. As device, use the one from your development board and synthesis option balanced.

References

[A04] Altera: "Quartus II support for Verilog 2001," Quartus II version 4.2 help (2004)
[B98] J. Bhasker, *Verilog HDL Synthesis* (Start Galaxy Publishing, Allentown, 1998)
[BV99] J.S. Brown, Z. Vranesic, *Fundamentals of Digital Logic with Verilog Design* (McGraw-Hill, New York, 1999)
[C04] M. Ciletti, *Starter's Guide to Verilog 2001* (Prentice Hall, Upper Saddle River, 2004)
[H97] D. Hyde, *CSCI 320 Computer Architecture Handbook on Verilog HDL* (Bucknell University, Lewisburg, 1997)
[I95] IEEE Standard Verilog Language Reference Manual. Institute of Electrical and Electronics Engineers, Inc., USA, Verilog 1364– (1995)
[I01] IEEE Standard Verilog Language Reference Manual. Institute of Electrical and Electronics Engineers, Inc., USA, Verilog 1364– (2001)
[MCB11] U. Meyer-Bäse, E. Castillo, G. Botella, L. Parrilla, A. García, Intellectual property protection (IPP) using obfuscation in C, VHDL, and Verilog coding. Proc. SPIE Int. Soc. Opt. Eng., Independent Component Analyses, Wavelets, Neural Networks, Biosystems, and Nanoengineering IX **8058**, 80581F1–80581F12 (2011)
[MB14] U. Meyer-Baese, *Digital Signal Processing with Field Programmable Gate Arrays*, 4th edn. (Springer, Heidelberg, 2014)
[S00] S. Sutherland, "The IEEE Verilog 1364-2001 Standard: What's New, and Why You Need It," in *Proceedings 9th Annual International HDL Conference and Exhibition*, Santa Clara, CA (2000), p. 8, http://www.sutherland-hdl.com
[S96] D. Smith, *HDL Chip Design* (Doone Publications, Madison, 1996)
[S03] Synopsys: "Common VCS and HDL compiler (Presto Verilog) 2001 constructs," Solv Net doc id: 002232 (2003)
[Xil04] Xilinx: "Verilog-2001 support in XST," XST version 6.1 help (2004)

Chapter 5
Microprocessor Programming in C/C++

Abstract This chapter gives an overview of the C/C++ language elements used in embedded microprocessor examples in later chapters as well as used in the utility program we will develop. As with most programming languages, we start with lexical preliminary and then discuss data types, operations, and control flow statements of the C/C++ language. Comparison of ANSI C with C++, debugging recommendations, and further reading complete the chapter.

Keywords ANSI C · DEC PDP-7 · Lexical elements · C vs C++ · ASCII table · LED counter · Monitor program · C Data types · Storage class · Header files · C operators · C assignments · Code hierarchy · Function · Flow control · Debugger · SPEC · Dhrystone · DMIPS · VAX MIPS

5.1 Introduction

The language C is one of the early high-level programming language originally developed by Dennis Ritchie as a high-level assembler language for early computer from DEC UNIX system called PDP-7 [KR78]. It was derived from the language B by Ken Thompson which was a typeless language. C contrarily provides a variety of data types: characters, integers, and floating-point numbers of several sizes are used in C. Hierarchy of derived data types are created with pointers, arrays, structures, and unions. Expressions are formed from operators and operands; any expression, including an assignment or a function call, can be a statement. Pointers provide for machine-independent address arithmetic. C also provides the fundamental control-flow constructions required for well-structured programs: statement grouping, decision-making (`if-else`), selecting one of a set of possible cases (`switch`), looping with the termination test at the top (`while`, `for`) or at the bottom (`do`), and early loop exit (`break`) [KR78].

You may have heard about *modern* programming languages everybody talks about today such as C++, C#, Python, or Java, and you may wonder what the favorite programming language today for embedded microprocessor is. It turns out that most of the excitement that came with a new language were not important enough to replace the advantages of the C programs such as efficient and short programs

© Springer Nature Switzerland AG 2021   159
U. Meyer-Baese, *Embedded Microprocessor System Design using FPGAs*,
https://doi.org/10.1007/978-3-030-50533-2_5

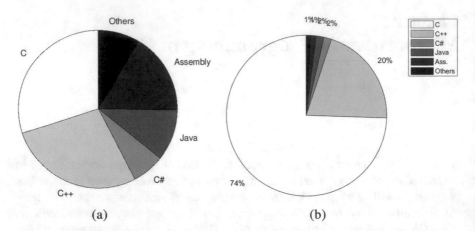

Fig. 5.1 The market share of programming languages used in embedded μP designs. (**a**) In year 2006 [HHF08] (**b**) in year 2016 [GB16]

due to the low-level language elements and complete language definition. In fact about 10 years ago, C++ gained a substantial market share when new embedded microprocessor project started – today we see that all the new features in C++ made the code less compact, and we have seen that C programs have recaptured substantial market share; see Fig. 5.1.

We will therefore for software languages focus our discussion on the C language with a few key facts discussed in Sect. 5.7 about differences to C++. The C language original definition is from year 1989 sometimes called Standard C, C89, ISO C, or ANSI C. Later language updates such as C90, C95, C99, C11, etc. maybe supported by your compiler; however, it is more secure to use ANSI C only, which has more than enough language features for our embedded systems.

5.2   Lexical Elements

If we plan to learn a new programming language, the first important question to ask is if the language is line sensitive (e.g., MATLAB) and/or column sensitive (e.g., PYTHON). This can often be answered by checking if the semicolon ";" is used to terminate all statements. In ANSI C the one-line statement

```
a = b + c;
```

will be compiled to the same program if we had written

```
a =
          b +
c;
```

Since it is required that each statement is completed with a semicolon, we can conclude that additional space, tab, or returns really do not matter, and therefore ANSI C is *not* a line- or column-sensitive language (with the exception of C++ style single-line comments and compiler directives). Comments in C are enclosed in /* ... */ and can run over multiple line. C++ style single-line comments start with a double slash // and continue to the end of the line and are supported by most C compilers too (i.e., part of C99 standard).

The second lexical question of importance is if a language is *case* sensitive. C/C++ is case sensitive such that MAIN, Main, or main are all considered different identifier. All keywords are lower case. It is still highly recommended to use each variable name only one time (small, camel, or capital).

A *basic user identifier* in ANSI C used as variable name, label, constant name, functions name, etc. has to start with a letter or underline and then have alphanumeric or underline characters; in BNF we would write

```
letter|_{alphanumeric|_}*
```

Table 5.1 below lists all ANSI C keywords. Here are a couple of examples of valid and invalid basic user identifiers in ANSI C:

Legal: x x1 _y_ Clock clock_enable
Not legal: main (keyword) P!NK (special character) 4you (number first)

All three parenthesis types are in use in C/C++: round ones () are used for function parameter, to group terms, and type casts, [] to index arrays, and the curly {} in grouping statements.

Operators (including combined assignments) used in expressions are 1, 2, or 3 character long, need to be written without space, and use more often special symbols (e.g., ‖ and not OR) than alphabet character. Section 5.4 discuss all operators in more details.

Constant and initial values of variables use decimal or hexadecimal base code, e.g., decimal 16 is coded as 0x10 in hex. Binary or octal bases are not supported in C. Character and strings use 8 bits ASCII code, where numbers range from decimal code 48 to 57, capital letters decimal code range from 65 to 90, and small letter decimal code from 97 to 122. A complete ASCII table for printable character (see Fig. 5.2b below) is developed as an exercise at the end of the chapter.

Table 5.1 ANSI C keyword aka reserved words sorted into function groups

Data types	char, double, enum, float, int, void
Data attributes	long, short, signed, struct, typedef, union, unsigned
Control flow	break, case, continue, default, do, else, for, goto, if, return, switch, while
Storage class	auto, const, extern, register, static, volatile

Dec	Hex	Char
0	0	Null
1	1	Start of heading
2	2	Start of text
3	3	End of text
4	4	End of transmission
5	5	Enquiry
6	6	Acknowledge
7	7	Bel
8	8	Backspace
9	9	Horizontal tab
10	A	New line
11	B	Vertical tab
12	C	New page
13	D	Carriage return
14	E	Shift out
15	F	Shift in
16	10	Data link escape
17	11	Device control 1
18	12	Device control 2
19	13	Device control 3
20	14	Device control 4
21	15	Negative acknoledge
22	16	Synchronous idle
23	17	End of transmission block
24	18	Cancel
25	19	End of medium
26	1A	Substitute
27	1B	Escape
28	1C	File separator
29	1D	Group separator
30	1E	Record separator
31	1F	Unit separator

(a)

Dec	Hex	Char		Dec	Hex	Char		Dec	Hex	Char
32	20	Space		64	40	@		96	60	`
33	21	!		65	41	A		97	61	a
34	22	"		66	42	B		98	62	b
35	23	#		67	43	C		99	63	c
36	24	$		68	44	D		100	64	d
37	25	%		69	45	E		101	65	e
38	26	&		70	46	F		102	66	f
39	27	'		71	47	G		103	67	g
40	28	(		72	48	H		104	68	h
41	29	)		73	49	I		105	69	i
42	2a	*		74	4a	J		106	6a	j
43	2b	+		75	4b	K		107	6b	k
44	2c	,		76	4c	L		108	6c	l
45	2d	-		77	4d	M		109	6d	m
46	2e	.		78	4e	N		110	6e	n
47	2f	/		79	4f	O		111	6f	o
48	30	0		80	50	P		112	70	p
49	31	1		81	51	Q		113	71	q
50	32	2		82	52	R		114	72	r
51	33	3		83	53	S		115	73	s
52	34	4		84	54	T		116	74	t
53	35	5		85	55	U		117	75	u
54	36	6		86	56	V		118	76	v
55	37	7		87	57	W		119	77	w
56	38	8		88	58	X		120	78	x
57	39	9		89	59	Y		121	79	y
58	3a	:		90	5a	Z		122	7a	z
59	3b	;		91	5b	[		123	7b	{
60	3c	<		92	5c	\		124	7c	|
61	3d	=		93	5d	]		125	7d	}
62	3e	>		94	5e	^		126	7e	~
63	3f	?		95	5f	_		127	7f	DEL

(b)

Fig. 5.2 The ASCII table. (**a**) Control code 0–31. (**b**) Printable character 32–127 (Nios II terminal print)

Much can be learned from the functionality of a compiler by looking at the supported keywords. That indeed can be also the good starting point for the evaluation of data type, attributes, and control flow instructions supported. There are 32 reserved words in ANSI C, and it is usually a good idea to order these by the four major classes: data types, data attributes, control flow, and storage class.

The four basic data types used in later chapter are char, int, float, and double and may be characterized further with attributes such as unsigned, short, long, or long long (C99 only), or for the compiler storage class, we can specify auto, const, extern, register, or static. Additional data types we often use are arrays and pointers and functions; we less often use structures and unions. More details on data type can be found in Sect. 5.3. The control flow operations are discussed in Sect. 5.5.

Finally let us have a brief look at the overall C program organization. Here is a short-running LED examples that has most segments we see in all our programs.

C Example Program 5.1: LED Running Light with Software Monitor

```
1    #include <stdio.h> /* for printf */
2    #include "address_map_nios2.h"
3
4    /* function prototypes */
5    void wait (int s );
6
7    /* This program demonstrates use of Altera Monitor
8     * Program communication in the DE1-SoC board
9     *
10    * It performs the following:
11    *  1. displays the counter values using the red leds
12    *  2. speed of counter is determined by the SW switches
13    *  3. any change in SW value is displayed in the Altera
14    *     Monitor Terminal window
15    */
16   int main(void)
17   {
18   /* Declare volatile pointers to I/O registers. volatile means
19      that the compiler cannot optimize the variable and
20      need to use the verbatim code */
21   /* Constant values on the right are from address_map_nios2.h */
22   volatile int *red_LED_ptr=(int *) LEDR_BASE;/* red LED addr. */
23   volatile int *SW_switch_ptr = (int *) SW_BASE; /* slider SW */
24
25      int SW_value, SW_old; // Holds current and last switch value
26      int k;  // Running variable
27
28      printf("This simple LED counter program\n");
29      printf("running on the Nios II.\n\n");
30      printf("Watch the LEDs ....\n");
31      printf("Use SW to change the speed\n");
32      k=1;
33      while (1) { /* Run forever */
34        SW_value = *SW_switch_ptr; /* Read the switch value */
35        *red_LED_ptr = k;     /* Run one LED at a time */
36        wait(SW_value); /* call the sleep function */
37        if (k<1024) k *= 2;
38        else k=1;
39        if (SW_old != SW_value) { /* Check for new SW value */
40          printf("New SW value = %d\n",SW_value);
41          k=1;
42          SW_old=SW_value;
43        }
44        SW_old=SW_value;
45      }
46      return 0;
47   }
48
49   /****************************************************************
50    * Custom wait if usleep() is not available
51    ****************************************************************/
52   void wait ( int s )
53   { /* volatile so the C compiler doesn't remove the loop */
54      volatile int u, v, sum=0;
55      for (u=1;u<100000;u++)
56        for (v=1;v<s;v++) sum+=v;
57   }
```

A typical program has five blocks: library(s), compiler directives, functions prototypes, main program, and functions definitions. The library functions provide urgently needed routines such as I/O, file operations, or constant definitions (line 1). Local files use double quotes "...", while global library files use the <...>.

Compiler directive such as #define, #ifdef, #ifndef, #else, or #endif statements can be used to define global constant or set switches such as DEGUG display on/off. We may also specify macro operations that will be used as inline substitutions for frequently needed (small) tasks, such as print header information or SWAP two variables. It is common practice in embedded system programming to put this definition in an external file. For Nios II TerASIC board, the file address_ map_nios2.h is used (line 2). The Table 5.2 gives an overview of often used library files.

If we use user functions, we should defines these next (lines 4–5), just specifying the I/O ports followed by a semicolon. This way we can use these within the main program or within other functions and do not need to be concerned about the compilation order. Next follows the main program (lines 16–47). We may add a parameter list main(int argc, char *argv[]) such that input file name or parameter values can be given our main functions when we start the program. Finally we define the functions we like to use in our program (lines 49–57).

The above program will produce a "running" light on the LEDs of our development board. The speed of the run is determined by the slider switches of the board. Any change in the switch values is shown in the monitor program of our development system; see Fig. 5.3 lower left panel called Terminal. A video of the running light can be found on the book CD under led_countQT.MOV and led_counterWMT.MOV for QuickTime and Windows media player video orientations, respectively.

Table 5.2 Common header files in ANSI C [Jon91]

Header file	Content	Examples
limits.h	Integral limits	CHAR_BIT, INT_MAX, INT_MIN
float.h	Floating point limits	FLT_MANT_DIG, DBL_MANT_DIG
stdio.h	I/O facilities	printf(), sprintf(), fopen(), fprintf()
stdlib.h	Utility functions	abs(), atoi(), atof(), malloc(), free(), rand(), srand(), qsort()
unistd.h	Low-level time functions	usleep()
time.h	High-level time functions	time(), difftime(,)
math.h	Math functions	sin(x), sqrt(), pow(,), floor(), log()
string.h	String manipulations	strlen(), strcpy(,), strcat(,), strcmp(), strcat(,)

Fig. 5.3 The LED counter program running on DE board

5.3 Data Types, Data Attributes, and Storage Class

Basic C Data Types

ANSI C has four basic types:

char   represents a single character
int   represents the whole numbers
float   floating-point numbers in single precision
double   floating-point numbers in double precision

The whole number type int can be further quantified with short, long, and long long in C99.

The range of the number is mainly determined by the number of bits used. We can look at the compiler header files (limits.h and float.h) to find out, or we can take advantage of the predefined C function sizeof(). This function will tell us the number of bytes ($\times 8$ gives bits *B*) a type occupies in memory. Then for unsigned we can use a range $0 \ldots 2^{B-1}$ and $-2^{B-1} \ldots 2^{B-1}-1$ for signed numbers. By default we have signed data; to define unsigned numbers, we add the data attribute unsigned in front of the type. Table 5.3 gives an overview of the size of the basic data types.

Table 5.3 Data type sizes and (signed) range as used by Nios II GCC

Data type	Size	Range
char	8 bits	−128...127
short int	16 bits	−32768...32767
int	32 bits	−2147483648...2147483647
long long	64 bits	−9223372036854775808 ...9223372036854775807
float	32 bits: sign bit; 8 bit exponent; 23 bit mantissa	3.402823466e+38 (maximum) $\varepsilon = 1.192092896e-07$
double	64 bits: sign bit; 11 bit exponent; 52 bit mantissa	1.7976931348623158e+308 (maximum) $\varepsilon = 2.2204460492503131e-016$ (smallest)

We may also add a storage class specification such as `auto`, `const`, `extern`, `register`, `static`, or `volatile`. Most often we use the `volatile` specifier to tell the compiler not to optimize our variable processing, e.g.

```
volatile int delay_count;
...
for (delay_count=100000; delay_count!=0; --delay_count);
```

this delay loop (that is used to reduce the speed of our LEDs toggling) would be optimized away almost certainly by every half good C compiler since it is not really doing any computation, just running down to delay counter to zero. The `volatile` keyword tells the compiler still to run down the counter.

ANSI C allows us to build derived or composite data types. Arrays and pointers are heavy in use in embedded processing since we often communicate with peripheral component via the I/O port address; see, for instance, `SW_switch_ptr` in coding Example 5.4. Pointer use `*p` for dereference and `&i` for the address of `i`. Pointer will allocate the same memory space independent of the data type (typical int) since it needs to store the whole address of the microprocessor memory which does not depend on the data type. Array pointers should match the array type such that increment `++` or decrement `−−` points to the correct next array element. Let us demonstrate the use of pointers in the following short code sequence:

```
int i, k;                /* define two integer variables */
int *i_ptr, *k_ptr;      /* define two pointers to int */
i_ptr = &i;              /* i_ptr now points to i */
*i_ptr = 5;              /* i has now value 5 */
k_ptr = i_ptr;           /* k_ptr now points also to i */
k = *k_ptr;              /* k has now also value 5 */
```

Pointer and arrays are close relatives. An 1D or 2D array, etc. can be defined by using `var_name[num]` in the variable definition. The index of an array always starts at 0 and ends at `num-1`. We can also include initial values for the array using {} and this way also implicitly specify the length. The end of a character string is usually indicated with \0. Here are a couple of examples (verbatim copies from our Nios examples):

```
char text_top_row[40] = "Altera DE1-SoC\0";  /* VGA message */
short buffer[512][256];   /* Pixel Buffer */
unsigned char    seven_seg_decode_table[] = {
0x3F, 0x06, 0x5B, 0x4F, 0x66, 0x6D, 0x7C, 0x07,
0x7F, 0x67, 0x77, 0x7C, 0x39, 0x5E, 0x79, 0x71 };
```

We have always two options to access the array; we can use the array index representation or the pointer style; the following two representations are equal:

```
array_name[i] == *(array_name + i);
&array_name[i] == array_name + i;
```

We do not use `struct` or `union` too often, only in some of the utility programs. There you may find a definition to build a linked list such as

```
struct symbol {
    char *symbol_name;
    int symbol_value;
    struct symbol *next;
};
```

That assigns each name a unique value. We may then use a loop to look through our list and return the assigned value if found using the following code:

```
int lookup_symbol(char *symbol)
{     int found = -1;
      struct symbol *wp = symbol_list;
      for(; wp; wp = wp->next) {
              if(strcmp(wp->symbol_name, symbol) == 0)
         { return wp->symbol_value;} /* Found symbol */
         }
      return -1;  /* Symbol not found */
}
```

The function will run through our list and return the symbol values or a −1 if the symbol was not found.

Table 5.4 Operation support and precedence in ANSI C

Precedence	Associativity	Operator	Description
15	L	() [] -> .	Function call, scope, array/member access
14	R	! ~ -* & sizeof (*type cast*) ++x --x x++ x--	(most) unary operations, size of and type casts
13	L	* / %	Multiplication, division, modulo
12	L	+ -	Addition and subtraction
11	L	<< >>	Bitwise shift left and right
10	L	< <= > >=	Comparisons: less-than, ...
9	L	== !=	Comparisons: equal and not equal
8	L	&	Bitwise AND
7	L	^	Bitwise exclusive OR (XOR)
6	L			Bitwise inclusive (normal) OR
5	L	&&	Logical AND
4	L	||	Logical OR
3	R	? :	Conditional expression
2	R	= += -= *= /= %= &= |= ^= <<= >>=	Assignment operators
1	L	,	Comma operator

5.4   C-Operators and Assignments

The ANSI C operators with increasing precedence are listed in Table 5.4.

There are a couple of observations that can be drawn from Table 5.4. ANSI C uses 15 priority levels, more than twice what VHDL uses. Bitwise and logical operators for AND have a higher priority than OR, so mixing AND and OR expressions will not require grouping terms with parenthesis (). No parenthesis are needed for the NOT operator ether, since it has the priority 14. There is also a XOR available for bitwise operations but not for logical. Logical and bitwise operations on integers may produce different results. The logical AND produces one, if both operands are nonzero; similarly, a logic OR is zero only if both operands are zero. We need to be careful with white space since the AND symbol & is used for address arithmetic too; see Exercise 5.7. Here are a few short examples for logic and bitwise operations assuming A = 1 and B = 2:

Logic expression	Logic result	Bitwise expression	Bitwise result
A && B	1	A & B	0
A || B	1	A	B	3
! A	0	~A	0xFF...E

We just witness the working of the unary (i.e., one operand) NOT operations. Other unary operations include plus, minus, the post-increments/post-decrements, and pre-increments/pre-decrements that increase/decrease the variable by one. Assuming we have B = 2 and use one of the following statements, we get

```
R = B++; // Gives R=2
R = ++B; // Gives R=3
R = B--; // Gives R=2
R = --B; // Gives R=1
```

ANSI C has all six relation operations typical for most programming languages. However, the equal and not equal have a lower priority than the other four relation operations.

ANSI C supports shift left and right. Depending on the hardware and compiler, a shift left can be more efficient than a multiply by power-of-two and a shift right more efficient than a divide by power-of-two. You may want to verify that your compiler handles negative numbers in the two's complement sense, i.e., with a correct sign propagation for negative number, e.g., with char START = -4 (i.e., -100_2 binary; 0xFC hex), we have

```
RESULT = (START << 1); // Gives -8
RESULT = (START >> 1); // Gives -2(dec)=0xFE with correct sign
    RESULT = (START >> 1); // Gives 0x7E=126(dec) with incorrect
sign
```

No rotations operators are defined in ANSI C. The shift amount should be positive.

The other arithmetic operations have the usual mathematical meaning. The modulo operation % is time consuming, and we usually try to avoid. For R = X % Y, the remainder R will have the same sign as X.

Evaluation in arithmetic is left associative, e.g., for a=2; b=3; c=4 we get

```
left    = (a-b) - c; // force left associative gives -5
right   = a - (b-c); // forcing right associative gives 3
default = a-b-c; // ANSI C implicit is left associative -5
```

A verification of each operation can be done with a small test sequence such as:

C Program 5.2: Verification of Operations with Test opc.c

```c
#include <stdio.h>
#include <math.h>
#include <errno.h>
#include <stdlib.h>
#include <ctype.h>
// Check the support operation
char i, x, y; // char/integer 8 bits
char a[11]; // array of numbers
void main(){
    x=5;y=14;
  // Arithmetic test first
  a[0]= x + y - 19;
  a[1]= y - x - 8;
  a[2]= x * y - 68;
  a[3]= y / x + 1;
  a[4]= y % x;
// Bitwise test next
  a[5]= (x & y) + 1;
  a[6]= (x | y) - 9;
  a[7]= (x ^ y) - 4;
  a[8] = (~x) + 14;
// Shift test
  a[9]= (x << 2) - 11;
  a[10]= (y >> 1) + 3 ;

  for (i=0; i<=10; i++) // print increased values
        printf("%d) res= %d\n",i, a[i]);
}
```

After compilation and running the C Code, we have an increasing array of values printed. Two more operators, the comma operator and condition operator, are discussed together with control flow constructs in the next section.

Not many arithmetic functions we typically use with our pocket calculators are included in the standard language operations. However, most compilers come with a substantial number of library functions. A few additional math functions such as random number generator rand() and seed srand(), or qsort() are available with stdlib.h. The math.h library contains many useful math operations such as power-of, square-root, logarithm, and trigonometric functions. Many of these are defined for arguments type double, so substantial time and program size maybe needed. For many applications the use of float data type precision is more than enough, and double precision is not really needed. In C99 for all these math functions, float data type versions have been added. This is an appreciated addition for most embedded processors. While on a PC or laptop we may not see substantial runtime changes since these CPU have double precision floating-point units, for embedded processors, often only float type aka 32 bit is supported, and substantial performance increase will be observed. The "reduced" precision function names are the original double precision function names with an added "f", e.g., sqrtf(), powf(,), logf(), sinf(), cosf(), etc.

ANSI C uses two groups of *assignments*: For three-operand assignments, symbol = is used. The same assignment operator is used to specify initial value. For the two-operand processing type, the following two assignments are identical:

```
var = var op expr; // var on left and right side
var op= expr;      // can be simplified this way
```

This works for all five arithmetic operations $(+, -, \times, /, \%)$ and the bitwise operations $(\&, |, \char94, <<, >>)$.

5.5   Control Flow Constructs

Typical control flow constructs in programming languages are single and multiple select statements and loops. Let us first discuss the select statements and then loops. For simple selection in ANSI C, an if statement is used that performs statement based on the condition(s) specified. Here is a coding example how we would determine the maximum of two inputs:

```
if (A > B)
   MAX = A;     // A is larger
else
   MAX = B;     // B is larger
```

In case we have more than one statement for each branch of the if statement, we can group these with the curly parenthesis {...}. For multiple conditions we may consider the switch statement, e.g.

```
KEY_value = *(KEY_ptr); // Read the pushbutton
...
switch (KEY_value)
{  case 2 : k=k/2; break; // Move bar to the right
   case 4 : k=32;  break; // Move bar to center position
   case 8 : k=k*2; break; // Move bar to the left
   default : break;       // No default action
}
...
*(red_LED_ptr) = k;       // Light up the red LEDs
```

All case actions should be terminated with the break statement to avoid a fall through. A default action can be used to cover the most common case.

Most compilers support all three-loop types such as ANSI C: while, for, and do loop, but all three can basically do the same. More interesting is the question what the most efficient way is to access an array in a loop. A small example to build the sum of five array elements shows basic syntax for the three loops and coding options.

C Program 5.3: Loop Coding and Array Access `loops.c`

```
 1    // Check supported loops and array access
 2
 3    int sum;// sum of the array
 4    int i; // char/integer 8 bits
 5    int a[5]; // array of numbers
 6    int *ptr, *ptrlast; // address values
 7
 8    void main(){
 9
10     // Set test values
11       a[0]= 5;  a[1]= 10; a[2]= 15; a[3]=20; a[4]=25;
12       sum=0; // reset sum; should be 3*75=225 at the end
13
14       i=0;
15       while (i<5) {
16          sum += a[i];
17          i++;
18       } // use standard array access
19
20       ptr =(int *) &a; ptrlast = ptr+5;
21       do {        // use array pointer offset
22          sum += *ptr++;
23       } while (ptr < ptrlast);
24
25       ptr =(int *) &a;   //increment array pointer
26       for (i=0; i<5; i++) sum += *(ptr+i);
27    }
```

The first loop shows the use of the `while` loop together with standard array access (C code lines 14–18). The `while` condition is checked first before entering the loop. Next is a `do/while` sequence using post address increment access to the array (C code lines 20–23). In contract to the `while` loop, the `do/while` runs at least one time, and the condition for another run is checked at the end. Last is the `for` loop with address pointer offset access (C code line 25–26). In the `for` loop, we have three-loop control elements: the initial statement, a terminating condition, and a loop update statement all separated by a semicolon. Each of the three control parts can have multiple statements separated by a comma. Compiler directives can be used that only one loop type is active such that we can measure the different program length. It turns out that for most compilers, the `for` loop type with address increment gives the fastest and shortest program code, i.e.

```
for (i=0; i<5; i++) sum += *ptr++;
```

We may also use a "loop unrolling," i.e., do not use a loop at all and write five statements:

```
//Loop unrolling
ptr = (char *) &a;
sum += *ptr++;
```

```
sum += *ptr++;
sum += *ptr++;
sum += *ptr++;
sum += *ptr;
```

The unrolled loop often produces a longer code, but due to the reduced computational overhead, the unrolled loop often runs faster. The only exception to this rule
would be if the `for` loop fits in the cache but the unrolled loop does not as in
TMS320 PDSPs with a small cache.

5.6   Code Hierarchy and I/O

The use of *functions* in your C/C++ project not only builds a hierarchy in your code
but also makes it easier to maintain and reuse your code. The overall code size reduction is often a very important feature with limited program ROM size in embedded
systems. If we allow also recursive function calls, your μP need to have a substantial
size in the pc level stack such that multilevel subroutine calls are not producing an
overflow. We can try a recursive function call with the following small example.

C Program 5.4: Recursive Function Call Verification with `fact.c`

```
 1    #include <stdio.h> /* for printf */
 2    /* Factorial using STD loop and recursive calls
 3     * Author: Uwe Meyer-Baese
 4     */
 5    /* function prototypes */
 6    int fact(int n);
 7
 8    int N, f, r, i;
 9
10    void main()
11    {
12        N= 5; f=1;
13        for (i=1; i<=N; i++) f = f * i; // STD loop
14        r= fact(N); // Recursive calling function
15
16        printf("fact(%d) STD = %d    recursive = %d\n",N,f,r);
17    }
18    int fact(int n)
19    {
20      if (n==0)
21        return 1;
22      else
23        return (n * fact (n-1));
24    }
```

We see the conventional computation of the factorial using the `for` loop in line
13. We would expect for $N = 5$ as result $f = 1 \times 2 \times 3 \times 4 \times 5 = 120$. Now in the second approach, we use a recursive function call using the recursive function shown in

lines 18 to 24. Notice how the function call itself again in line 23 with `fact(n-1)`. Now let us verify our program with a run; the result as expected is as follows:

```
fact(5) STD = 120    recursive = 120
```

The limitation for a function we have discussed so far is that the function only provides a single return value. If we have more than one value to return, we replace the "call by value" we have used so far by a "call by reference." Here is a small example that is used in a nested function call to sort an array:

C Program 5.5: Function Call with Multiple Return Values `swap.c`

```
1    #include <stdio.h> // For printf
2    /* SWAP/SORT example to demonstrate multiple return values
3     * Author: Uwe Meyer-Baese
4     */
5     /* function prototypes */
6    void swap(int *, int *);
7    int sort(int *, int);
8
9    int a, b, c, L=5;
10   int v[] = { 8, 1, 13, 5, 7 }; //5 element vector
11
12   int main(int  argc, char *argv[])
13   {
14    a = 5; b = 7;
15    printf("Before swap: a = %d   b = %d\n",a,b);
16    swap( &a, &b);
17    printf("After  swap: a = %d   b = %d\n",a,b);
18    printf("Before sort: (%d,%d,%d,%d,%d)\n",v[0],v[1],v[2],v[3],v[4]);
19    c = sort(v, L);     // sort data
20    printf("After  sort: (%d,%d,%d,%d,%d)\n",v[0],v[1],v[2],v[3],v[4]);
21    printf("FYI: Used %d swap in sort\n", c);
22    // Program termination message
23    printf("Done with %s.exe -- Good bye\n", argv[0]);
24   }
25   /********** swap 2 elements ********/
26   void swap(int *x, int *y)
27   { int t; // Use temporary
28     t = *x;
29     *x = *y;
30     *y = t;
31   }
32   /********** sort array ********/
33   int sort(int *x, int len)
34   { int k, done, count=0;
35     do { done = 1;
36       for (k = 1; k < len; k++)
37         if (x[k - 1] > x[k])
38         { swap( &x[k - 1], &x[k]); done = 0; count++; }
39     } while (!done);
40     return(count);
41   }
```

To enable the call by reference, we use pointers and address operators when connecting scalars (line 16). For arrays we do not use the address specified to make a call by reference (line 19) since an array name is a pointer. Because we have specified the function header first, we can call `sort()` that now calls `swap()` multiple times, i.e., a nested function call. We first test our swap functions (lines 14–17) and then continue with a sorting (lines 18–21) array `v[]` that has been initialized with random integer values (line 10). The output of a program run will look as follows:

```
Before swap: a = 5    b = 7
After  swap: a = 7    b = 5
Before sort: (8,1,13,5,7)
After  sort: (1,5,7,8,13)
FYI: Used 5 swap in sort
Done with swap.exe -- Good bye
```

Finally let us discuss how to input data to our program and to produce output. Let us discuss the three most popular ways to provide input to a program. We may "hold" program processing and "ask" via a `scanf()` the user to input data to our program. Alternatively, we may also input data directly when starting it in command-line and attach arguments to, such as

```
io.exe 65 hallo 1989.0
```

The third method would be to place our data in a normal text file (e.g., `data.txt`) we may compose with programs such as `notepad` or another program and then specify the file as argument when starting the program in the command-line, i.e.

```
io.exe data.txt
```

If we use `main(int argc, char *argv[])` routine with parameters, then our command-line inputs will be stored in `argv[]` array locations 1, 2, 3, etc. while 0 hold the program name, while `argc` contains the argument counter. Here is a small program that demonstrates input options as well as output to standard output (i.e., screen) and file.

C Program 5.6: Input and Output Demonstrations `io.c`

```
1    #include <stdio.h> // For printf, fopen, fprintf
2    #include <stdlib.h> // For atoi, atof
3    #include <string.h> // For strlen, strcpy
4
5    // Check I/O functions
6    int k, i, ic; // Local variables
7    char s[20];
8    char fname[] ="example.txt"; // File name
9    float f;
10   FILE  *fin, *fout; // File pointer
11
12   int main(int  argc, char *argv[])
13   { // enter: 65 hallo 1989.0
14     // print line inputs
15     for (i = 0; i < argc; i++) printf("%d: %s\n", i, argv[i]);
16     switch (argc) { // Switch based on line inputs
17       case 1: printf("Please enter INT STRING FLOAT\n");
18               ic = scanf("%d %s %4f\n", &i, s, &f);//no inputs
19               break;
20       case 2: fin = fopen(argv[1], "r"); // Filename given
21               if (fin != NULL) /* read from file*/
22                  fscanf(fin, "%d %s %4f\n", &i, s, &f);
23               fclose(fin);
24               printf("Done reading 3 inputs from FILE\n");
25               break;
26       case 4: printf("Done reading 3 inputs from STDIN\n");
27               i = atoi(argv[1]);   // all 3 given
28               f =(float)atof(argv[3]); strcpy(s, argv[2]);
29               break;
30       default: printf("Wrong parameter number\n"); exit(1);
31     }
32     printf("Reading done -- Start Processing\n");
33   // print integer in different formats
34     printf("dec=%d  octal=%o  hex=%04x  char=%c\n", i,i,i,i);
35   // print first and last character and whole string
36     k = strlen(s);
37     printf("first=%c  last=%c  whole=%s ...\n",s[0],s[--k],s);
38   // printf float in 3 formats
39     printf("exp=%e  std=%f  short=%g\n",f,f,f);
40
41   // print to file example.txt
42     fout = fopen(fname, "w");
43     if (fout != NULL) {
44        printf("Writing the file:  %s\n", fname);
45        fprintf(fout,"dec=%d  octal=%o  hex=%04x  char=%c\n",i,i,i,i);
46        k = strlen(s);
47        fprintf(fout,"first=%c  last=%c  whole=%s ...\n",s[0],s[--k],s);
48        fprintf(fout, "exp=%e  std=%f  short=%g\n", f, f, f);
49        fclose(fout);
50     }
51     else { printf("Unable to write the file:  %s\n", fname); }
52   // Program termination message
53       printf("Done with %s.exe -- Good bye\n", argv[0]);
54   }
```

The main variable argc from line 12 holds the number of input parameters and the 2D array *argv[] the parameter stored as strings. argv[0] holds the program name. The for loop (line 15) will print the command-line arguments. The switch statement (lines 16–31) will take care of the three different input modes. Our program will require to enter an integer, a string, and a float number. If no inputs are provided, the program will ask to enter the three data. If one input is given, we assume it is the file name, and we try to open the file with fopen() for reading using the "r" parameter and read the data (lines 20–25). If the file does not exist, the function fopen() will return a NULL pointer. For argc equal to four, all three inputs were provided, and we convert the strings to the appropriate variable type using the atoi() function for conversion to integer and the atof() function for the conversion to float type. We then print out (aka display on screen) the three inputs in different formats.

All format specifier starts with the percent symbol %, then an optional length specifier, and followed by the base letter. For the integer variable, we can use base decimal %d, octal %o, small hex %x, capital hex %X, or character display %c. If we enter 65 as integer, the associated character in the ASCII table (see Fig. 5.2) is the first capital letter A. We may also force the output to a specific length, and if we also like to see leading zeros (as often required in assembly coding), we can use a format such as %04x such that always four digits including leading zeros are displayed. Then for hex value 41 the display will be 0041. A string uses the %s format specifier and displays the string starting an index zero until a \0 (i.e., the string termination symbol) is discovered. For s[20] rather than the full specified string length of 20, only 5 characters are displayed. For strings of length one, we can also use the single character format %c. For float numbers typical format choices are %e that shows a normalized exponential representation, the %f format that shows standard float format with 6 fractional digits, and the %g format that somehow tries to minimize the display in case the fractional bit are all zero. A program run will produce the following outputs when reading from file data.txt that contains the data: 65 hallo 1989.0

```
0: io
1: data.txt
Done reading 3 inputs from FILE
Reading done -- Start Processing
dec=65   octal-101   hex=0041   char=A
first=h   last=o   whole=hallo ...
exp=2.018000e+03   std=2018.000000   short=2018
Writing the file:  example.txt
Done with io.exe -- Good bye
```

Next in coding (line 42) we open an output file example.txt for writing using the specifier "w". Writing to file is pretty much the same as writing to standard output, just replace printf(...) with fprintf(FID,...), i.e., adding the initial f and the file identifier. Finally, we generate a program termination message in case the program completed successfully.

5.7   Additional Considerations and Recommendations

Finally let us discuss three more important topics when developing ANSI C code: The question if we should use a debugger or debug printing and what must be considered moving from C to C++ or vice versa. Finally we like to discuss programs to measure microprocessor performance aka benchmarks.

Debugger Versus Debug Printing

In a perfect world a programmer would write code that is 100% correct. That is highly unrealistic since every programmer makes mistakes. In fact, if you think about today's sophisticated programming task, you may find yourself spending more time verifying code and fixing bugs in your code than with concept and implementation phase of the project. Typical errors in embedded software developing include

- Typing mistakes such as syntax or typos in constant values, variable name, address values, etc.
- Error in communication with I/O devices, specially missing interrupt enable or to clear flags
- Incorrect subroutine linkage including version control
- Error in memory addressing, location, and wait states
- Poorly written programs that does not implement the correct algebra or data flow

Question is how we can keep the time it takes to find these bugs as short as possible. There are two major concepts call *debugger* and *debug printing* (for someone from outside this discussion looks almost like a philosophical fight with no winner ever), and both have its advantages and disadvantages.

Now let us start with the first method called *debug printing*. Often you will have the need that you need to monitor your data, checking for values range, array overflows, divide by zeros, etc. An obvious method that will not require any additional tools or skills we have already acquired is the display of these critical variables using `printf()` on a terminal/screen or `fprintf()` the data to a file. We may use coding like

```
printf("Debug printing (y/n) ?\n"); scanf("%c",&ask);
...
if (ask=='y') printf("New SW value = %d\n", SW_value);
```

However, the problem with this coding style is that your final program will have all your debug code which can take CPU time but more importantly makes a substantial increase in program size. Here are the program download volumes for led_ count (see Altera Nios II Example Program 5.1) with and without the debug code:

Style	No `printf()` or `scanf()`	With `printf()`	With `printf()` and `scanf()`
Size	3 KB	55 KB	91 KB

As can be seen, if you include printf() and scanf() functions in your code, assume you need for each about 40–50 KB code! That is quite substantial in embedded systems, maybe larger what is available. There are reduced size printf() functions provided by our FPGA vendors such as custom xil_printf() or alt_printf() that are much smaller since these do not support floating-point number format prints but still will require substantial additional size. A better approach is therefore to use compiler directives that will remove the debug code in the preprocessing compiler phase in case the switch is off. We may use method like

```
#define DEBUG 0
...
if (DEBUG) printf("New SW value = %d\n", SW_value);
```

That coding looks similar as the one above; the major difference is that DEBUG is now a compiler constant and the compiler maybe takes advantage of this fact. This coding style should work fine with most C/C++ compilers that notice that the printf() is dead code in case the DEBUG flag is set to zero. If the compiler is not that smart, we may use the second option that needs a little more coding:

```
//#define DEBUG
#ifdef DEBUG
     printf("New SW value = %d\n", SW_value);
#endif /* DEBUG */
```

The compiler directive will guarantee a printf() free code if DEGUG is not defined, since the preprocessor will remove the code between #ifdef and #endif during the preprocessing phase. The compiler not even needs to evaluate if the code is dead or not. So, in conclusion for debug printing, you should use the compiler directive mode using #ifdef and #endif that is part of debug printing.

Now let us talk about another major concept to make your code error free: *Debugging*. Most of the professional C/C++ compilers today come with a sophisticated debugger tool too. The GNU C/C++, for instance, uses a tool called GDB. A debugger may have one or more of the following capabilities:

- Set breakpoint
- Run in single steps, skip, and/or dive into subroutines
- Set conditional breakpoints based on register or memory values
- Allow to modify memory and registers
- Allow you to localize where the program crashed

The GDB will produce a core dump file that can be used to localize where the program crashed. If the error is less severe, then we may set breakpoints at key elements of our code, verify variables, register values or memory contend, and even modify these. Besides GDB we can also use the Altera Monitor Program that performs most of these tasks in a GUI environment such that we do not need to learn another (often verbose) debugger language.

To understand the basic concept of the debugger, let us run a short flash.c program that read in the slider switches values and toggle the red LEDs after some time. Using the slider switches position shown in Fig. 5.5a (i.e., SW0 and SW2 = ON), the program should read the value 5. If we run the Altera debugger, we would always use a breakpoint at main. In AMP click on gray area left next to the instruction, and you can turn the breakpoint on and off; see Fig. 5.4. Next, we would try to identify the next useful location for a breakpoint. We used D4 within the infinity loop so that we can modify by hand what the program has read from the input slider switches. Let us assume we like to modify this sensor value to 15 or Hex F and then we need to identify the register that stores the variable named data. From the disassembly code for stwio, we conclude it to register r16, and we double click the value and change it to F. If we now run another iteration until we reach our breakpoint again, then the LED display on the board will change according to our new value, i.e., LED0-4 are ON; see Fig. 5.5b. Here is the source code for flash.c.

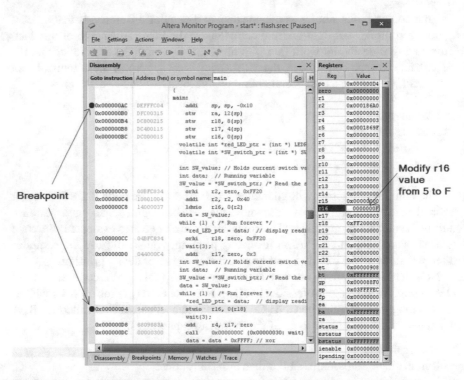

Fig. 5.4 The Altera monitor debugger running flash.c on Nios II

(a)   (b)

Fig. 5.5 The LED (a) original display of five in binary. (b) Modified display setting r16 = F

C Program 5.7: Short Test Program Used in the Debugger flash.c

C Program 5.15: Short test program used in the debugger flash

```
1    #include <stdio.h> /* for printf */
2    #include "address_map_nios2.h"
3
4    /* function prototypes */
5    void wait (int s );
6
7    /* This program demonstrates use of Altera Monitor
8     * Program debugging in the DE1-SoC board
9     *
10    * It performs the following:
11    *   1. Reads SW value and displys on red leds
12    *   2. Toggle On/OFF leds after about 1 second
13    */
14   int main(void)
15   {
16   /* Constant values on the right are from address_map_nios2.h*/
17   volatile int *red_LED_ptr=(int *) LEDR_BASE;/*red LED addr.*/
18   volatile int *SW_switch_ptr = (int *) SW_BASE; /*slider SW*/
19
20     int SW_value; // Holds current switch value
21     int data;   // Running variable
22     SW_value = *SW_switch_ptr; /* Read the switch value */
23     data = SW_value;
24     while (1) { /* Run forever */
25        *red_LED_ptr = data;  // display reading
26        wait(3); // waist some time
27        data = data ^ 0xFFFF; // xor i.e. toggle
28     }
29
30   }
31   /****************************************************************
32    * Custom wait if usleep() is not available
33    ****************************************************************/
34   void wait ( int s )
35   {  /* volatile so the C compiler doesn't remove the loop */
36      volatile int u, v, sum=0;
37      for (u=1;u<100000;u++)
38         for (v=1;v<s;v++) sum+=v;
39   }
```

Let us briefly summarize the pros and cons of the two debug options:

- The debug print method does not require to learn a new tool or language.
- The debug print program runs faster since register and memory are not monitored all the time.
- If moving to another platform (compiler or µP), a single flag allows us to monitor many intermediate results to locate faster at which point our program runs out of sync.

- The debugging method has the advantage that not a single-line extra C/C++ is required.
- The debugged code is 100% already the final product, i.e., much cleaner code without any "dead code."
- The debugging method allows more advanced test and monitoring such as modify interrupt request (register), memory contend, and peripherals.

As you see there is not really a clear winner. It's often a matter of personal choice and experience. This concludes to discussion on debugging versus debug print. More details on Altera debugging method can be found online on Altera/Intel University program webpage [Int17, Alt15].

C Versus C++

Sorry but I usually do not like to spoil the party, but the name C++ is a little misleading. It suggests that C++ is the ANSI C language with additional features as indicated by the ++. However, that is not entirely correct. Yes, C++ can do basically everything (and more) than ANSI C can deliver, but a couple of coding are just done different in C++ and C and you cannot always use your C code in C++. It's sometime necessary to rewrite it. Let us in the following first summarize the similarities and then the differences. Table 5.5 shows the major similarities of C and C++.

The way the identifier for a variable name, label, constant name, function name, etc. is defined is the same. Both are case sensitive such that main, Main, and MAIN are all different identifiers. The four basic data types are the same, and flow control uses the same elements to build loops or conditions.

But we also find a substantial number of coding requirements that are different in C++ and ANSI C. Table 5.6 give an overview of the most important issues.

Usually we can use the same tool to compile C and C++. The file extensions are different such that a compiler can easily distinguish both. Library files are quoted using with extension *.h in ANSI C, none in C++. Many header files are renamed, so, for instance, printf or scanf in ANSI C we find in the header stdio.h. to use cout << in C++, we need the iostream header file and maybe iomanip to set the output precision. Similarly, limits.h becomes climits, float.h becomes cfloat, stdio.h becomes cstdio, stdlib.h becomes cstdlib, time.h becomes ctime, math.h becomes cmath, and string.h becomes cstring. Great additions in C++ toolbox are classes and operations defined for classes. We can, for instance, define a complex type and associate operations like multiply and addition and use the standard operator for the new class; see Exercise 5.37–5.39. In ANSI C the work around (WA) would be to use a struct and define the operations via function call; see Exercise 5.34–5.36.

Table 5.5 Similarities of C++ and ANSI C

Item	ANSI C	C++ code
ID syntax	L|_{L|D|_}	
Case sensitive	Case sensitive	
Data types	char, int, float, double etc.	
Flow control	for, while, if, switch, function, ?:	

Table 5.6 Difference in C++ and ANSI C coding

Item	ANSI C	C++ code
File extension	*.c	*.cpp
Library	#include <stdio.h>	#include <iostream>
Output	printf, fprintf	cout <<
Input	scanf	cin >>
Print to string	sprintf	N/A
Custom types	use struct	use class
Operator overloads	N/A	OK

Performance Evaluation of Microprocessor Systems

Performance evaluation of general-purpose computer systems are typically done with collections of standard programs such as the program sets defined by the System Performance Evaluation Cooperation (SPEC) for integer and floating-point benchmarks. Founded by HP, DEC, MIPS, and SUN microsystems, the runtime of popular programs such as GO, GCC, COMPRESS, LI, PERL, etc. are measured and an average score determines the system performance. This makes sense for a general-purpose computer system that should be universal and used for applications ranging from text processing, compiler, to gaming. In embedded microprocessor systems, we only find that very few microprocessor systems such as your mobile phone processors that run many different apps need to be fast for all kinds of applications. Furthermore, most embedded systems have requirements on power dissipation, real-time requirements, or memory limits and may not even be able to run these "big" SPEC benchmark programs. In addition, the floating-point SPEC requires a FORTRAN compiler not available with all embedded processors. For an embedded system that runs a single application such as MPEG decoder or SSD error correction, it is also not useful if the processor is fast compiling GCC programs or in the Chinese game GO. For embedded systems, therefore, we find more often that so-called synthetic benchmarks such as Whetstone or Drystone are more practical and useful. This single C-programs try to reproduce the statistic of a typical program in terms of operations used, control flow, subprogram calls, and datatype used. These short routines are run million times to produce DMIPS and MWIPS rating of a microprocessor system. Let us have a closer look at the DMIPS benchmark, since this is often found in microprocessor data sheets. The data sheet typically shows a DMIPS/MHz value for the processor such that based on the actual processor speed you can compute the expected DMIPS value by multiplication of DMIPS/MHz value with the maximum processor speed F_{max}. The value ranges from 0.02 DMIPS/MHz for a Nios II without any cache to 2.1 for an ARM Cortex-A9. A high DMIPS/MHz ratio indicates a superior architecture and/or system design. The DMIPS rating will depend not only on the CPU architecture but also on the overall system design such as compiler (options) used, if data and/or instruction cache is available and if on-chip or SDRAM memory is used for programs. Table 5.7 shows some measurements data for different system configurations. Modification in the compiler options can give performance difference up to a factor of 3.87. The modification in the cache architecture and program location brought a maximum factor difference of 13.2 in performance. With a maximum clock speed of 111 MHz, we measure a maximum DMIPS performance of 102 DMIPS for the Nios II/f on our DE1 SoC board. Increasing the cache size further (I/D = 16K) had only minor impact on DMIPS/MHz = 0.99, but F_{max} dropped by 10% to 100.05 MHz, so overall DMIPS rating did not improve.

The DMIPS source code is available online for many different compilers, including the original Ada program by R. Weicker or our preferred C-code [Wei84]. Since this is a pre-ANSI-C code, the benchmark may need some editing such as function prototypes and function return types as required by ANSI-C. A working timer for the system under test is also required. Dhrystone DMIPS source code is available with the Xilinx SDK New Project Templates and was available with Altera Nios 2 EDS in the past (no longer in QUARTUS 15.1 version). For ARM you can use the internal XScuTimer, but for MICROBLAZE you need to add an AXI timer and for Altera a Qsys component timer is needed. Dhrystone uses scalar and arrays of integer, char, enumeration, Boolean, pointer, and string, but no floating-point types. DMIPS benchmark code contains all elements of a typical program such as arithmetic operation (+,−,*,/), comparisons (=,/=,<,>,<=,>=) and logic operations (and, or not), control constructs (if, for, while etc.), and function calls; see Fig. 5.6.

Table 5.7 DMIPS/MHz measurement for Nios II/f

ALM	1625	1456	1920	1683	
Memory	<1%	52%	2%	53%	
Fmax/MHz	106.53	108.87	108.97	111.5	
On-chip program		X		X	
SDRAM program	X		X		
D-cache	0	0	4K	4K	
I-cache	0	0	2K	2K	
Optimization					Max/min
-O1	0.02	0.08	0.19	0.23	9.9
-O2	0.04	0.13	0.25	0.50	12.4
-O3	0.06	0.21	0.31	0.81	12.6
-O4	0.07	0.22	0.64	0.92	13.2
Max/min	2.9	2.87	3.27	3.87	

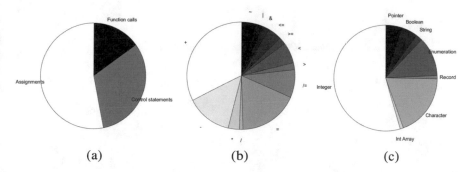

Fig. 5.6 The Dhrystone C language distributions. (**a**) Elements. (**b**) Operators. (**c**) Datatypes

The frequency of the operators in Dhrystone reflects typical program statistic, e.g., add operations is used 13 times more often that the divide operation. The Dhrystone benchmark uses repeated call of eight "procedures" and three functions. The number of Dhrystone runs per second is normalized by 1757.0 according to the VAX MIPS rating, a popular processor at the time the benchmark was defined.

If you look at the majority of examples we have discussed so far, you may think that embedded microprocessors are used only for trivial things such a light control, keyboards, or turning your car mirrors. There are, however, many much more sophisticated applications embedded microprocessor systems are used for today such as JPEG, MPEG encoder and decoder, mechatronic systems, communication protocols, or biomedical signal processing. Assume, for instance, we have measured the multichannel ECG signal of a pregnant woman (with our eight-channel ADC on the DE1 SoC board for instance) and like to extract the fetal ECG that is ten times smaller in amplitude than the mother ECG [MMS16]. A well-working procedure for this type of task is the principle component analysis (PCA). Here we compute the cross-correlation matrix R_{xx} between the channels and then compute the eigenvectors of this R_{xx} matrix and can then synthesize the FECG. The eigenvalue and eigenvector computation is done with a series of matrix operations, and these recursive matrix multiplications are often done in floating-point to avoid arithmetic overflow. The whole FECG algorithm will require a substantial number of floating-point operations that need to be computed in real time. Since DMIPS does not provide any floating-point measurement, we may use the Whetstone benchmark, but in the MWIPS score, only three out of the eight loops will measure floating-point performance. Let us therefore do our own floating-point measurement including time needed to compute `sqrt()` function needed in the Jacobi and the QR eigenvalue algorithms. To reduce the loop overhead, we use always 10 times the same operation and run each loop 10K times, for a total of 100K of each operation.

We measure the time T and can compute the latency of a single operations $\tau = T/100K$. The reciprocal scaled by 10^6 will give the floating-point million instructions per second (FMIPS) score, shown in Fig. 5.7. We will do these measurements for all basic operations $(+,-,*,/)$, and the conversion from/to integers. Finally, we may be interested in some more advanced operations/functions such

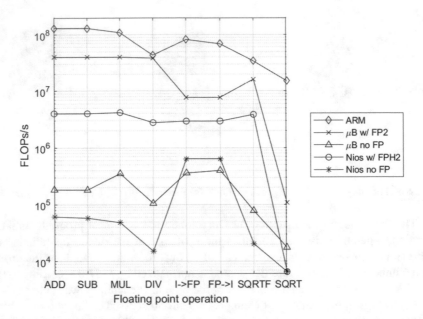

Fig. 5.7 The floating-point performance for basic operations

as trigonometric (sin, cos, atan) or exp and sqrt. Since sqrt() is used
in the EV algorithms discussed earlier, we measure this too. With the advanced
operations, we need to be careful since from the definition in math.h we see that
all are defined for 64-bit double precision. So, if we like to use 32-bit single precision aka float, then we should use the C99 functions instead. A "f" is added to the
function name such that sqrt() becomes now sqrtf() in C99. The results in a
substantial performance difference can be seen from Fig. 5.7 comparing sqrt()
and sqrtf(). Figure 5.7 shows the measured FMIPS performance for three
CPUs: ARM, MicroBlaze, and Nios II/e. This benchmark for the floating-point
performance is reflected in the actual measurement for the FHR monitor. Since
we sampled at 250 Hz, a real-time system will need to provide at least 250 frames
per second (FPS). The ARM allows a real-time performance of 3570 FPS, while
MICROBLAZE 10 FPS and Nios II/e at 2 FPS do not allow a real-time FHR monitor.

5.8   Further Readings

At first as with any new programming language, it is a lot of information to digest.
ANSI C is a pretty mature topic, and we may find excellent old references in the
library and not too often new books, e.g., [Sha05]. C++ is nowadays often part of
computer engineering curriculum, and many modern books with modern learning
features should be available [Bro05, Eck00, Str11, Ski92]. Here are a couple of classic and modern books and tips to get you started coding in C/C++:

- Help:

  - C/C++ frequently asked questions: http://c-faq.com/
  - Formal language definition in BNF online at http://www.lysator.liu.se/c/index.html
  - YACC grammar at http://www.lysator.liu.se/c/ANSI-C-grammar-y.html
  - Most University have C/C++ course such as COP3014 at FSU that comes with lecture notes and project ideas, e.g., http://www.cs.fsu.edu/~vastola/cop3014/
  - Check http://www.cplusplus.com/reference/clibrary/ that contains good working examples for individual C/C++ language elements.
  - Try do download, order, or print a C/C++ reference card such as the one provided by quickstudy.com, cse.msstate.edu/~crumpton/reference, http://web.pa.msu.edu/people/duxbury/courses/phy480/Cpp_refcard.pdf or Gaddis book appendix.

- Tip: Collect as many C/C++ examples as you can find:

  - Put post-it pointers to important examples in your favorite textbook.
  - Analyze these examples and make sure you have them handy when you code.
  - ANSI C books:

    K&R [KR78] classic compact reference with sophisticated project ideas by creator of ANSI C.
    Z. Shaw [Sha05] Detail discussed on important modern issue when coding in C.
    If you know another high-level language, you may try to find a transition book such as [Bro88].

  - Popular C++ books:

    T. Gaddis [Gad17] popular C++ book used at FSU, FAMU, and many other universities
    B. Stroustrup [Str04] Book by the creator of C++
    S. Meyers [Mey14] Popular brief compact 334 pages C++ book
    Free online tutorials on YouTube or WWW such as http://www.cplusplus.com/doc/tutorial/ or https://www.cprogramming.com/tutorial/c++-tutorial.html

Review Questions and Exercises

Short Answer

5.1. What are the four keyword groups in ANSI C?

5.2. Explain the header file syntax using "..." and <...>. Which library has I/O functions, which has trigonometric and which random numbers functions?

5.3. Compare the use of a function swap with a macro definition SWAP, and what are advantages and disadvantages of both?

5.4. Why should we in general avoid goto statements? When would you use the goto statement?

5.5. Why should we avoid statement like a[i] = i++; ?

5.6. When should arrays be created dynamically via malloc() instead of creating them using definitions?

5.7. Briefly explain the expressions below: Is it valid? Is this a logic or bitwise AND, address operations, or both?

   I. Y & Y
  II. Y && Y
  III. Y &Y
  IV. Y & &Y

5.8. Briefly explain the following declarations. Is it a pointer, function, or a variable?

    i. int name;
   ii. int *name;
  iii. int name[];
   iv. int name();
    v. int *name[];
   vi. int (*name)[];
  vii. int *name();
 viii. int (*name)();

5.9. Name three properties that are the same and three properties that are different in C coding and C++ coding.

5.10. Explain briefly the methods debugger versus debug printing. Name two advantages of both methods.

Fill in the Blank

5.11. The function prototype

```
int  alt_up_char_buffer_string(alt_up_char_buffer_dev
*char_buffer, const char *ptr, unsigned int x, unsigned
int y)
```

displays a text message in the VGA character buffer 80×60 array.
Show the display result for the following Nios II code:
alt_up_char_buffer_string (char_buffer_dev, "Altera\0", 0,30)
alt_up_char_buffer_string (char_buffer_dev, "DE2-115\0", 75,0)
alt_up_char_buffer_string (char_buffer_dev, "VGA\0", 75,59)
alt_up_char_buffer_string (char_buffer_dev, "NIOS II\0", 0,59)
Recall (0,0) is at up/left and y↓ as in a beam writing from upper left to lower right.
VGA 80×60 Character Display:

5.12. Repeat previous Exercise 5.11 for the following function calls:

```
alt_up_char_buffer_string (char_buffer_dev, "Test\0", 40, 30);
alt_up_char_buffer_string (char_buffer_dev, "EuPSD\0", 0, 0);
    alt_up_char_buffer_string (char_buffer_dev, "three\0", 75,
59);
    alt_up_char_buffer_string (char_buffer_dev, "FOR\0", 75, 30);
```

5.13. The function prototype

```
    void alt_up_pixel_buffer_dma_draw_box(alt_up_pixel_buffer_
dma_dev *pixel_buffer,
int x0, int y0, int x1, int y1, int color, int backbuffer)
```

draws a filled box in the scaled VGA 320×240 array, and the color code is

15 ··· 11 10 ··· 5 4 ··· 0

For the following Nios II code

```
    alt_up_pixel_buffer_dma_draw_box (pixel_buffer_dev, 80, 60,
120, 90, 0xF800, 0);
```

Determine:
Color of box: _____.
Lower left corner of the box: x = _____   y = _____
Upper right corner of box: x = _____   y = _____
X-width of box: _____ Pixel
Y-height of box:_____ Pixel

5.14. Repeat the previous exercise for the following function call:

    alt_up_pixel_buffer_dma_draw_box (pixel_buffer_dev, 100, 110, 220,
    160, 0x07E0, 0);

5.15. The id syntax in C/C++ in BNF is _____.
5.16. The C-functions printf and scanf are coded in C++ as _____ and
_____.

True or False

5.17. _____ The size of all pointers is the same for one microprocessor.
5.18. _____ The size for char, int, float, and double in Nios GCC are 8,
      32, 32, and 64 bits.
5.19. Which of the following are not keywords in the ANSI C language? (circle all)

        I. goto
       II. int
      III. loop
       IV. while
        V. begin
       VI. register

5.20. Which of the following shows the correct syntax for an ANSI C language if
      statement?

        I. if [expression]
       II. if expression then
      III. if (expression)
       IV. if {expression}

5.21. Which of the following are loops in the C language? (circle all)

        I. while
       II. for

III. `generate`
IV. `repeat until`
  V. `do while`

5.22. Which of the following gives the value stored at the address pointed to by the pointer q?

  I. `val(q)`
 II. `&q`
III. `*q`
IV. `(val) q`
  V. `q->`
VI. `#q`

5.23. Which of the following correctly accesses the fifth element stored in A, an array with 10 elements?

  i. `A[5]`
 ii. `A[4]`
iii. `A(4)`
iv. `A{4}`
  v. `A(5)`
vi. `A*5`

5.24. Which of the following are correct ANSI C variable types? (circle all).

  I. `float`
 II. `real`
III. `int`
IV. `bit`
  V. `natural`
VI. `char`
VII. `boolean`

Projects and Challenges

5.25. Try to give a full description for the following declarations:

  i. `int *(*name())[];`
 ii. `int (*name[4][6])();`
iii. `int *(*name[])();`

5.26. Write an ANSI C program that allocates the maximum amount of memory using `malloc()`. How many images of VGA size $640 \times 480 \times 24$ bits can you store in the measured memory?

5.27. Write a program to print the ASCI table in the range 32–127 (decimal). Print a table like Table 5.2b with decimal, hex, and ASCII entry.
5.28. Write a program to print the factorial numbers from 1 to 10.
5.29. Write a program to print the Fibonacci sequence for the first 20 elements.
5.30. Write a program for conversion of Fahrenheit to Celsius using the formula.

```
C = (F-32)*5.0/9.0
```

Print a table of values $F = 0, 10, 20, \ldots, 100$.

5.31. Write a program to print the odd magic square using the Siamese method aka De la Loubère method of size 3, 5, 7, and 9.
5.32. Develop an ANSI C program that works as a watch. Use the `time.h` library function to compute 1-second clock ticks. Your clock command-line input should be `clock hour minutes seconds`, e.g., `clock 11 45 30`. Print each second the new time.
5.33. Implement a reaction timer using `time.h` functions. Display a number 1–9, and measure the time it takes a user to press the key. Use an average from ten trials for the overall reaction time.
5.34. Develop an ANSI C function library for complex numbers. As type use the custom type.

```
typedef struct fcomplex {float r, i;} fcomplex;
```

The library should have the following functions:

Function prototype	Function details
`fcomplex RCmul(float x, fcomplex a);`	Scale with a constant
`fcomplex Cadd(fcomplex a, fcomplex b);`	Complex addition
`fcomplex Csub(fcomplex a, fcomplex b);`	Complex subtractions
`fcomplex Cmul(fcomplex a, fcomplex b);`	Complex multiply
`fcomplex Complex(float re, float im);`	Build a complex variable
`fcomplex Conjg(fcomplex a);`	Compute conjugate complex
`fcomplex Cdiv(fcomplex a, fcomplex b);`	Complex division
`float Cabs(fcomplex z);`	Absolute |C|
`void pC(fcomplex z);`	Print a complex number

Develop the ANSI C code and test each function separately.

5.35. Use the library designed in the previous Exercise 5.34 to develop a DFT and IDFT for any length.

Function prototype	Function details
`void dftC(fcomplex *a, int n);`	Forward DFT: $X[k] = \sum_{n=0}^{N-1} x[n]e^{-j2\pi kn/N}$
`void idftC(fcomplex *a, int n);`	Inverse $x[n] = \dfrac{1}{N}\sum_{n=0}^{N-1} X[k]e^{j2\pi kn/N}$

Test your program with a length 8; DFT have (a) real-only data x = 1,2,...8 (b) imaginary only data x = j1,j2,...j8, and (c) combination of both x = 1+j1,2+j2,...8+8j. Make sure the IDFT produced the original data.

5.36. Use the library designed in Exercise 5.34 to build an ANSI C complex pocket calculator for addition, multiply, divide, and subtract. The calculator should ask to enter first complex number, operator, and second complex number, and then compute the result.

5.37. Develop a C++ class library for complex numbers. As type use the class

```
complex::complex(float a, float b)
```

The body library should have

Function prototype	Function details
`float real();`	Return real part
`float imag();`	Return imaginary part
`friend complex operator +(complex a, complex b);`	Complex additions
`friend complex operator -(complex a, complex b);`	Complex subtract
`friend complex operator *(complex a, complex b);`	Complex multiply
`friend complex operator /(complex a, complex b);`	Complex division
`friend int operator ==(complex a, complex b);`	Compare equal
`friend istream& operator >>(istream& is, complex& c);`	Read a complex number
`friend ostream& operator <<(ostream& os, complex& c);`	Print a complex number

Develop the ANSI C code and test each function separately. DO not use the predefined C++ `<complex>` class.

5.38. Develop a C++ program to implement the DFT

$$X[k] = \sum_{n=0}^{N-1} x[n] e^{-j2\pi kn/N}$$

using the library from Exercise 5.37. Your program should ask for the DFT length N and then generate the test sets: $x = 1-j1, 2-j2, 3-j3 \ldots (N-1)-j(N-1)$.

5.39. Use the C++ class library designed in Exercise 5.37 to build a C++ complex pocket calculator for addition, multiply, divide, and subtract. The calculator should ask to enter the first number, followed by operator and finally the second complex number, and then compute the result.

5.40. Use the DMIPS source code from the book CD and measure the performance of your PC and and/or an embedded system.

5.41. Compare runtime of programs that use `swap` as function or a compiler `SWAP` macro. As test case, use sorting of two large random array, i.e., length > 1 M elements.

5.42. Develop a program to report the size aka number of bits of data types: `char`, `short`, `int`, `long`, `float`, and `double` using the `sizeof` operator.

5.43. Develop a program to report the size aka number of bits of the mantissa for type `float` and `double`. Hint (see, e.g., [PTV92]): for size $b = 1, 2, \ldots$ compute $x = 1<<b; y = x+1$; then if $x == y$, you found b.

References

[Alt15] Altera, Altera Monitor Program Tutorial for Nios II (Quartus Prime 15.1, October 2015)

[Bro05] G.J. Bronson, *A First Book of C++ From Here to There*, 2nd edn. (Course Technology, Boston, 2005)

[Bro88] T. Brown, *C for Pascal* (Silicon Press, Summit, 1988)

[Eck00] B. Eckel, *Thinking in C++: Introduction to Standard C++*, vol 1 & 2, 2nd edn. (Prentice Hall, Upper Saddle River, 2000)

[Gad17] T. Gaddis, *Starting Out with C++ Brief: From Control Structures Through Objects*, 9th edn. (Pearson, Boston, 2017)

[GB16] A. Girson, M. Barr, *2016 Embedded Systems Safety & Security Survey* (Barr Group, Gaithersburg, MD, 2016)

[HHF08] J. Hamblen, T. Hall, M. Furman, *Rapid Prototyping of Digital Systems: SOPC Edition* (Springer, New York, 2008)

[Int17] Intel Debugging of Application Programs on Intel's DE-Series Boards (Quartus Prime 17.0, June, 2017)

[Jon91] R. Jones, *The C Programmer's Companion* (Prentice Hall, Upper Saddle River, 1991)

[KR78] B. Kernighan, D. Ritchie, *The C Programming Language*, 2nd edn. (Prentice Hall, Upper Saddle River, 1978)

[Mey14] S. Meyers, *Effective Modern C++* (O'Reilly Media, Sebastopol, CA, 2014)

[MMS16] U. Meyer-Baese, H. Muddu, S. Schinhaerl, M. Kumm, P. Zipf, Real-time fetal ECG system design using embedded microprocessors, in *Proc. SPIE Commercial + Scientific Sensing and Imaging*, ed. by L. Dai, Y. Zheng, H. Chu, A. Meyer-Baese, vol. 9871, (April 2016), p. 987106-1-14

[PTV92] W. Press, W. Teukolsky, W. Vetterling, B. Flannery, *Numerical Recipes in C*, 2nd edn. (Cambridge University Press, Cambridge, 1992)

[Sha05] Z. Shaw, *Learn C the Hard Way* (Addison Wesley, New York, 2005)

[Ski92] M. Skinner, *The C++ Primer* (Prentice Hall, Upper Saddle River, 1992)

[Str04] B. Stroustrup, *Programming: Principles and Practice Using C++*, 2nd edn. (Addison Wesley, Upper Saddle River, 2005)

[Str11] S. Prata, *C++ Primer Plus*, 6th edn. (Addison Wesley, Upper Saddle River, 2011)

[Wei84] R. Weicker, DHRYSTONE: A synthetic systems programming benchmark. Commun. ACM **27**(10), 1013–1030 (1984)

Chapter 6
Software Tool for Embedded Microprocessor Systems

Abstract This chapter gives an overview of the software tools used in embedded microprocessor system design. It starts with a motivation followed by fundamental concepts for assembler, C/C++ compiler, and ISS. Further reading recommendations are given along in the text when different topics appear.

Keywords Programming tools · Compiler · Interpreter · Assembler · GNU Bison · GNU Flex · Yet another compiler-compiler (YACC) · GDB · HEX code · psm2hex · 3-address code · Scanner · Parser · Gimple · LISA · URISC · Instruction set simulator (ISS) · SW debugger · Bakus-Naur form · Expression

6.1 Introduction

According to Altera's Nios online net seminar [Alt04], one of the main reasons why Altera's Nios development systems have been a huge success (10,000 systems were sold in the first 4 years after introduction) is based on the fact that, besides a fully functional microprocessor, also all the necessary software tools including a GCC-based C compiler is generated at the same time when the IP block parametrization takes place. You can find many free µP cores on the web on general code share sides such as `GitHub.com` or `sourceforge.net`. or more specialized sites; see for instance:

- `http://www.opencores.org/` open core
- `http://www.fpgacpu.org/` FPGA CPU

The processor overview pages such as `https://en.wikipedia.org/wiki/Soft_microprocessor` shows many of the processors and their features. But most of these processors lack a *full* set of development tools and are therefore less useful. A set of development tools (best case) should include:

- Assembler, linker, and loader/basic terminal program
- C/C++ compiler
- Debugger or Instruction set simulator

© Springer Nature Switzerland AG 2021
U. Meyer-Baese, *Embedded Microprocessor System Design using FPGAs*,
https://doi.org/10.1007/978-3-030-50533-2_6

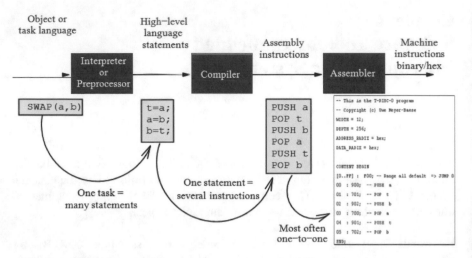

Fig. 6.1 Models and tools in programming

Figure 6.1 explains the different levels of abstraction in the development tools from C/C++ compiler to assembler program. In the following, we will briefly describe the main programs used to develop these tools. You may also consider using the electronic system level tools such as language for instruction set architecture (LISA) originally developed at ISS, RWTH Aachen [HML02], and now a commercial product of Synopsys Inc., which automatically generates an assembler and instruction set simulator, and the C/C++ compiler with a few additional specifications in a semiautomatic way. Writing a compiler can be a time-consuming project. A good C/C++ compiler, for instance, requires up to 50 man-years of work [LM01, ASU88, Leu02]. Developing a good assembler is not quite so demanding as a good C/C++ compiler since it is mainly a line-by-line translation of the text style program to processor instruction in HEX code. Nowadays we can benefit from the programs developed in the GNU project that provides several useful utilities that speed up compiler development:

- The GNU tool FLEX [Pax95] is a scanner or lexical analyzer that recognizes patterns in text, similar to what the UNIX utility grep or the line editor sed can do for single pattern.
- The GNU tool BISON [DS02] a, YACC-compatible parser generator [Joh75] allows us to describe a grammar in Bakus–Naur form (BNF), which can initiate actions if expressions are found in the text.
- For the GNU C/++ compiler gcc, we can take advantage of the tutorial written by R. Stallman [Sta90] to adapt the C/C++ compiler to the actual μP we have or plan to build.

All three tools are freely available under the terms of the GNU public license and for all three tools we have included documentation on the book CD under the SW folder, which also includes many useful examples.

6.2 Assembler Development and Lexical Analysis

A program that can recognize lexical patterns in a text is called a *scanner*. FLEX, compatible with the original AT&T Lex, is a tool that can generate such a scanner [LS75]. FLEX uses an input file (usual extension *.l) and produces C source code that can then be compiled on the same or a different system. A typical scenario is that you generate the parser under UNIX or LINUX with the GNU tools and, since most Altera tools run on a PC, we compile the scanner under MS-DOS so that we can use it together with the QUARTUS software. The default UNIX file that is produced by FLEX uses a lex.yy.c file name. We can change this by using the option -oNAME.C to generate the output NAME.C instead. Note that there is no space between -o and the new name under UNIX. Assume we have a FLEX input file vsanner.l then we use the two steps:

```
flex    -ovscanner.c    vscanner.l
gcc     -o vscanner.exe    vscanner.c
```

to generate a scanner called vscanner.exe under UNIX. Under MS-DOS, we would use the *.com extension that has the higher priority than *.exe. Even a very short input file produces about 1800 lines of C-code and has a size of 46 KB. We can already see from these data the great help this utility can be. We can also FTP or copy the C-code scanner.c to an MS-DOS PC and then compile it with a C/C++ compiler of our choice. The question now is how do we specify the pattern in FLEX for our scanner, i.e., how does a typical FLEX input file look like? Let use first have a look at the formal arrangement in the FLEX input file. The file consists of three parts:

```
%{
C header and defines come here
%}
definitions ...
%%
rules ...
%%
user C code ...
```

The three sections are separated by two %% symbols. Here is a short example of an input file that performs a lexical analysis of a Verilog file and reports the types of number of items he finds. The FLEX file vscanner.l is shown in listing 6.1.

FLEX Input File 6.1: vscanner.l

```
1   /* Scanner for Verilog (key)word, lines and character count */
2   /* Author-EMAIL: Uwe.Meyer-Baese@ieee.org */
3   %{
4     int nchars, nwords,  nlines, kwords = 0;
5   %}
6   DELIMITER        [.;,)(:\"+-/*@]
7   KW  module|input|output|reg|begin|end|posedge|always|endmodule
8   WORDS            [a-zA-Z][a-zA-Z0-9$_]*
9   %%
10  \n               { nlines++; nchars++; }
11  {DELIMITER}      { nchars++; }
12  {KW}             { kwords++, nchars += yyleng; }
13  {WORDS}          { nwords++, nchars += yyleng; }
14  [ \t]+              /* eat up whitespace */
15  .                { nchars++; }
16  %%
17  int yywrap(void) { return 1; }
18
19  int main(void) {
20       yylex();
21        printf("*** Results from Verilog scanner:\n");
22       printf("Verilog keywords=%d  lines=%d  words=%d  chars=%d\n",
23  kwords, nlines, nwords+kwords, nchars);
24       return 0;
```

The most important part is the rule section. There we specify the pattern followed by the actions. The pattern "." is any character except new line, and \n is new line. The bar | stands for the "or" combination. You can see that most coding has a heavy C-code flavor. We count the keyword in variable kwords, other words in nwords, number of characters in nchars, and the number of lines in nlines. So our scanner works similarly to the statistic we get for MS word file through Properties → Details you may have used before. Note that FLEX is column sensitive. Only patterns in the rule section can start in the first column; not even a comment is allowed to start here. Between the pattern and the actions, or between multiple actions combined in parenthesis, a space is needed.

We have already discussed two special symbols used by FLEX: the dot "." that describes any character and the new line symbol \n. Table 6.1 shows the most often used symbols. Note that these are the same kinds of symbols used by the utility grep or the line editor sed to specify regular expressions. Here are some examples of how to specify a pattern:

Table 6.1 Special symbols used by FLEX

.	Any single character except new line
\n	New line
*	Zero or more copies of the preceding expression
+	One or more copies of the preceding expression
?	Zero or one of the preceding expression
^	Begin of line or negated character class
$	End of line symbol
|	Alternate, i.e., or expressions
()	Group of expressions
"+"	Literal use of expression within quotes
[]	Character class
{}	How many times an expression is used
\	Escape sequence to use a special symbol as a character only
.	Any single character except new line

Pattern	Matches
a	The character a.
a{1,3}	One to three a's, i.e., a | aa | aaa.
a|b|c	Any single character from a, b, or c
[a-c]	Any single character from a, b, or c, i.e., a | b | c.
ab*	a and zero or more b's, i.e., a | ab | abb | abbb...
ab+	a and one or more b's, i.e., ab | abb | abbb...
a\+b	string a+b.
[\t\n]+	One or more space, tab or new lines.
^L	Begin of line must be an L
[^a-b]	Any character except a, b, or c

Assume we have the following small Verilog example:

Verilog Code 6.2: Scanner Example d_ff.v

```
1    module D_FF (CLK, D, Q); // Example flip-flop
2    input CLK;
3    input D;
4    output Q;
5    reg Q;
6        always @(posedge CLK) //--> gives rising edge FF
7            Q = D;
8    endmodule
```

then calling our scanner with `vscanner.exe < d_ff.v` will produce the following output:

```
*** Results from Verilog scanner:
Verilog keywords=8  lines=8  words=26  chars=129
```

After the introductory example, we can now take on a more-challenging task. Let us build an `psm2hex` converter that reads in assembler code and outputs an HEX file that can be loaded into the block memory as used by the FPGA software. To keep things simple, let us use the assembler code of 8-bit processor PICOBLAZE aka KCPSM6 discussed in Chaps. 7 and 8 with the following 16 initial operations (sorted by their operation code):

LOAD, XOR, INPUT FETCH, ADD, ADDCY, STAR, SUB, SUBCY, CALL, JUMP, RETURN, OUTPUT, STORE

Since we should also allow forward referencing labels in the assembler code, we need to have a two-pass analysis. In the first pass we make a list of all labels and their code lines. In the second run we can then translate our assembler code line-by-line into HEX code. We store the plain file under the default name `psm.hex`. Each line represents a single operation and will have 5 digits: 2 digits for the op code and then 3 digits for 0–2 operators including 12-bit immediate values. We augment our file at the end to fill the full FPGA block RAM until we have 4K words. During the run a protocol with the detected tokens is displayed on the standard output screen. Here is the FLEX input file for our two-pass scanner:

FLEX Description 6.3: `psm2hex.l`

```
1     /* Scanner for PicoBlaze psm assembler to HEX file converter */
2     %{
3     #include <stdio.h>
4     #include <string.h>
5     #include <math.h>
6     #include <errno.h>
7     #include <stdlib.h>
8     #include <time.h>
9     #include <ctype.h>
10    #define DEBUG 0
11    int    state=0; /* end of line prints out IW */
12    int    icount=0; /* number of instructions */
13    int    lcount=0; /* number of labels */
14    int    pp=1; /** preprocessor flag **/
15    int    vimm, imm=0; /* 2. op is kk flag */
16    int    offset=0; /* offset opc for jump code */
17    char   opis[6],lblis[4],immis[4];
18    FILE   *fid;
19    struct inst {int adr; int opc; int x; int y; int kk; char *txt;} iw;
20    struct init  {char *name; int code;} op_table[20] = {
21    "LOAD"   , 0x00, "STAR"   , 0x16, "FETCH" , 0x0A, "STORE"   , 0x2E,
22    "INPUT"  , 0x08, "OUTPUT" , 0x2C, "XOR"   , 0x06, "ADD"     , 0x10,
23    "ADDCY"  , 0x12, "SUB"    , 0x18, "SUBCY" , 0x1A, "JUMP"    , 0x22,
24    "CALL"   , 0x20, "RETURN" , 0x25, 0,0};
25    FILE       *fid;
26    int add_symbol(int value, char *symbol);
27    int lookup_symbol(char *symbol);
28    void list_symbols();
29    int h2i(char c);
30    int  lookup_opc(char *opc);
31    %}
32    HEX           [a-fA-F0-9]{1,2}
33    REG           [s|S][a-fA-F0-9]
34    NZ            [n|N][z|Z]
35    DELIMITER     [,]
36    COMMENT       ";"[^\n]*
37    LABEL         [a-zA-Z][a-zA-Z0-9]*[:]
38    GOTO          [a-zA-Z][a-zA-Z0-9]*
39    %%
40    \r                /* avoid trouble with FTP files */
41    \n                {if (pp) printf( "end of line \n");
42                      else { if ((state==3) && (pp==0))
43                        /* print out an instruction at end of line */
44                        {
45                          printf("%02X",iw.opc+imm+offset);
46                                   /* First two digits have op code */
47                          fprintf(fid, "%02X",iw.opc+imm+offset);
```

```
48                                     /* First two digits have op code */
49                         if (iw.opc==0x25) {
50                             printf("%03X",0); /* Next comes 000 */
51                             fprintf(fid, "%03X\n",0); /* Next comes 000 */
52                         } else {
53                         if ((iw.opc==0x22)||(iw.opc==0x20)) {
54                           printf("%03X",iw.kk); /* Here comes aaa */
55                           fprintf(fid,"%03X\n",iw.kk); /* Here comes aaa
56      */
57                         } else {
58                         printf("%1X",iw.x); /* Next comes the desitination
59      register */
60                         fprintf(fid, "%1X",iw.x); /* Next comes the
61      desitination register */
62                         if (imm) {
63                             printf("%02X",iw.kk); /* Last two have kk or y0
64      */
65                 fprintf(fid, "%02X\n",iw.kk); /* Last two have kk or y0 */
66                         } else {
67                 printf("%02X",iw.y); /* Last two have kk or y0 */
68                 fprintf(fid, "%02X\n",iw.y); /* Last two have kk or y0 */
69                         }
70                         } } }}
71                         printf("\n"); /* 5 digits = line complete */
72                         /*fprintf(fid, "\n"); 5 digits = line complete */
73                     state=0; imm=0; offset=0;
74                     }
75      {HEX}           { strcpy(immis, yytext);
76                     vimm=h2i(immis[1])+16*h2i(immis[0]);
77                     if (pp) printf( "An hex: %s (%d)\n", yytext, vimm );
78                     else    { iw.kk=vimm; state=3; imm=1; }}
79      XOR|LOAD|STAR|INPUT|OUTPUT|ADD|ADDCY|SUB|SUBCY {
80                     if (pp)
81                 printf( "%d) 2 op ALU Instruction: %s opc=%2X\n",
82                             icount++, yytext, lookup_opc(yytext));
83                     else  {   state=1; iw.adr=icount++;
84                                     iw.opc=lookup_opc(yytext); }
85                     }
86      CALL|JUMP {
87                     if (pp)  printf( "%d) 1 op  Flow Instruction: %s\n",
88                                     icount++, yytext );
89                     else  {   state=2; iw.adr=icount++;
90                                     iw.opc=lookup_opc(yytext); }
91                     }
92      RETURN {   if (pp)  printf( "%d) 0 op Instruction: %s\n",
93                                     icount++, yytext );
94                     else  {   state=3; iw.adr=icount++;
95                         iw.opc=lookup_opc(yytext);}
96                     }
```

```
97    {REG}            { if (pp)  {printf( "An register: %s\n", yytext ); }
98                       else { state+=1; if (state==2) iw.x =
99                  h2i( yytext[1] ); if (state==3) iw.y=h2i( yytext[1] );}
100                  }
101   {NZ}             {if (pp) printf( "JUMP condition: %s\n", yytext );
102                     offset=0x14;}
103   {LABEL}          { if (pp) {printf( "A label: %s length=%d
104                     icount=%d\n", yytext , yyleng, icount);
105                     add_symbol(icount, yytext);}
106                  }
107   {GOTO}           { if (pp) printf( "A goto label: %s\n", yytext );
108                     else {state=3;
109                  sprintf(lblis,"%s:",yytext);iw.kk=lookup_symbol(lblis);}
110                  }
111   {COMMENT}        {if (pp) printf( "A comment: %s\n", yytext );}
112   {DELIMITER}      {if (pp) printf( "A delimiter: %s\n", yytext );}
113   [ \t]+           /* eat up whitespace */
114   .           printf( "Unrecognized character: %s\n", yytext );
115
116   %%
117
118   int yywrap(void) { return 1; }
119
120   int main(int  argc, char *argv[] )
121   { int k;
122
123     yyin = fopen( argv[1], "r" );
124     if (yyin == NULL ) { printf("Attempt to open file %s failed\n",
125                         argv[1]); exit(1); }
126     printf("Open file %s now...\n", argv[1]);
127     printf("--- First path though file ---\n");
128     yylex();
129     fclose(yyin);
130     pp=0;
131     printf("-- This is the psm2hex program with %d lines and %d
132             labels\n",icount,lcount);
133     icount=0;
134     printf("-- Copyright (c) Uwe Meyer-Baese\n");
135     list_symbols();
136     if (DEBUG) printf("--- Second path through file ---\n");
137     yyin = fopen( argv[1], "r" );
138     fid  = fopen("psm.hex","w");
139     yylex();
140     for (k=icount;k<4096;k++)
141       fprintf(fid, "00000\n");
142     fclose (fid);
143   }
144
145   /* define a linked list of symbols */
```

```
146    struct symbol {
147        char *symbol_name; int symbol_value; struct symbol *next;
148    };
149
150    struct symbol *symbol_list; /* first element in symbol list */
151
152    extern void *malloc();
153
154    int add_symbol(int value, char *symbol)
155    {
156        struct symbol *wp;
157        if(lookup_symbol(symbol) >= 0 ) {
158        printf("--- Warning: symbol %s already defined \n", symbol);
159        return 0;
160        }
161        wp = (struct symbol *) malloc(sizeof(struct symbol));
162        wp->next = symbol_list;
163        wp->symbol_name = (char *) malloc(strlen(symbol)+1);
164        strcpy(wp->symbol_name, symbol); lcount++;
165        wp->symbol_value = value;
166        symbol_list = wp;
167        return 1;    /* it worked */
168    }
169
170    int lookup_symbol(char *symbol)
171    {   int found = -1;
172        struct symbol *wp = symbol_list;
173        for(; wp; wp = wp->next) {
174          if(strcmp(wp->symbol_name, symbol) == 0)
175            {if (DEBUG) printf("-- Found symbol %s value is: %d\n",symbol,
176                           wp->symbol_value);
177                    return wp->symbol_value;}
178        }
179        if (DEBUG) printf("-- Symbol %s not found!!\n",symbol);
180        return -1;    /* not found */
181    }
182
183    int lookup_opc(char *opc)
184    { int k;
185      strcpy(opis, opc);
186      for (k=0; op_table[k].name !=0; k++)
187        if (strcmp(opc,op_table[k].name)==0) return (op_table[k].code);
188      printf("******* Ups, no opcode : %s --> exit \n",opc);exit(1);
189    }
190
191    void list_symbols()
192    {
193        struct symbol *wp = symbol_list;
194        printf("--- Print the Symbol list: ---\n");
```

```
195        for(; wp; wp = wp->next)
196          printf("-- Label: %s   line = %d\n",wp->symbol_name,
197                     wp->symbol_value);
198        printf("--- Print the Symbol done  ---\n");
199      }
200
201      /************* conversion hex to integer value ***********/
202      int h2i(char c) {
203        switch (c) {
204          case '1': return 1; break;
205          case '2': return 2; break;
206          case '3': return 3; break;
207          case '4': return 4; break;
208          case '5': return 5; break;
209          case '6': return 6; break;
210          case '7': return 7; break;
211          case '8': return 8; break;
212          case '9': return 9; break;
213          case 'A': case 'a': return 10; break;
214          case 'B': case 'b': return 11; break;
215          case 'C': case 'c': return 12; break;
216          case 'D': case 'd': return 13; break;
217          case 'E': case 'e': return 14; break;
218          case 'F': case 'f': return 15; break;
219          default : return 0; break;
220        }
221      }
```

The variable pp is used to decide if the preprocessing phase or the code generation second phase is running. Labels and variables are stored in a symbol table using the functions add_symbol and lookup_symbol. The op code values are defined in 20–24, followed by the function prototypes. The used pattern HEX, REG, NZ, DEIMITER, COMMENT, LABEL, and GOTO are specified (lines 32–38). Then the rule section follows in lines 39–116. Depending on the opcode, we have 0, 1, or two additional parameters. The even opcode is used for two register operations and the following odd for the 8-bit immediate. The routine main (lines 120–143) shows the two-pass operation. First labels are identified and in the second pass the output file augmented by 00000 to fill the 4K block memory is written. At the end, the utility functions are shown to add a symbol to the list (add_symbol: 154–168), the find a symbol in the list (lookup_symbol: 170–181), find the opcode for an instruction in the table (lookup_opc: 183–189), print a list of all symbols (list_symbols: 191–199), and the case insensitive conversion of one ASCII hex digit to the integer values (h2i: 202–221).

Here are the UNIX instructions to compile and run the code:

```
flex -opsm2hex.c psm2hex.l
gcc -o psm2hex.exe psm2hex.c
psm2hex.exe flash.psm
```

The fash.psm test program for the PICOBLAZE microprocessor looks as follows:

The PicoBlaze Test File 6.4: flash.psm

```
0    start:   INPUT  s3, 00   ; read switches
1    flash:   LOAD   s0, 20   ; start values loop
2             LOAD   s1, BC   ; counter has 3x8=
3             LOAD   s2, BE   ; 24 bits
4             OUTPUT s3, 00   ; write general LEDs
5    loop:    SUB    s0, 01   ; s0 -= 1
6             SUBCY  s1, 00   ; sub with carry
7             SUBCY  s2, 00   ; sub with carry
8             JUMP NZ, loop   ; count to zero
9             XOR    s3, FF   ; invert LEDs
10            JUMP flash      ; start all over
```

The output generated by psm2hex is shown next.

The Hex Output File 6.5: psm.hex

```
0    09300
1    01020
2    011BC
3    012BE
4    2D300
5    19001
6    1B100
7    1B200
8    36005
9    073FF
10   22001
11   00000
12   00000
     ...
```

For labels we store the instruction line the label occurs. An output of the symbol table for three labels will look as follows:

```
    ...
    --- Print the Symbol list:  ---
    -- Label: loop:   line = 5
    -- Label: flash:  line = 1
    -- Label: start:  line = 0
    --- Print the Symbol done   ---
    ...
```

We see that the label flash (defined in line 1) is used in line 10 of the program, and label loop (defined in line 5) is called in line 8 of the psm program. These psm.hex file can be used directly in Verilog together with the $readmemh instruction; see Sect. 4.5. In VHDL we would use the hex2prom utility discussed in Chap. 7 to convert it to a block RAM initialization file for VHDL.

6.3 Parser Development

From the program name YACC, i.e., yet another compiler-compiler [Joh75], we see that at the time YACC was developed, it was an often-performed task to write a parser for each new µP. With the popular GNU UNIX equivalent BISON, we have a tool that allows us to define a grammar. Why not use FLEX to do the job, you may ask? In a grammar we allow recursive expressions like a + b, a + b + c, a + b + c + d, etc. and if we use FLEX then for each algebraic expression it would be necessary to define the patterns and actions, which would be a large number even for a small number of operations and operands.

YACC or BISON both use the Bakus–Naur form (BNF) that was developed to specify the language Algol 60. The grammar rules in BISON use terminals and nonterminals. Terminals are specified with the keyword %token, while nonterminals are declared through their definition. YACC assigns a numerical code to each token, and it expects these codes to be supplied by a lexical analyzer such as FLEX. The grammar rule uses a look-ahead left recursive parsing (LALR) technique. A typical rule is written like

```
Expression   :   NUMBER '+' NUMBER     { $$ = $1 + $3; }
```

We see that an expression consists of a number followed by the add symbol and a second number can be reduced to a single expression. The associated action is written in parenthesis. Say in this case that we add element 1 and 3 from the operand stack (element 2 is the add operation) and push back the result on the value stack. Internally the parser uses an FSM to analyze the code. As the parser reads tokens, each time it reads a token it recognizes, it pushes the token onto an internal stack and switches to the next state. This is called a shift. When it has found all symbols of a rule it can reduce the stack by applying the action to the value stack and then reduction is applied to the parse stack. This is the reason why this type of parser is sometimes called a shift-reduce parser.

Let us now build a complete BISON specification around this simple addition rule. To do so we first need to know the formal structure of the BISON input file, which typically has the extension *.y. The BISON file has three major parts:

```
%{
 C header and declarations come here
%}
Bison definitions ...
%%
Grammar rules ...
%%
User C code ...
```

It is not an accident that this looks very similar to the FLEX format. Both original programs LEX and YACC were developed by colleagues at AT&T [LS75, Joh75] and the two programs work nicely together as we will see later. Now we are ready to specify our first BISON example add2.y

The YACC Simple Calculator Description 6.6: add2.y

```
1    /* Infix notation add two calculator */
2    %{
3    #define YYSTYPE double
4    #include <stdio.h>
5    #include <math.h>
6    void yyerror(char *);
7    %}
8
9    /* BISON declarations */
10   %token NUMBER
11   %left '+'
12
13   %% /* Grammar rules and actions follows */
14   program :    /* empty */
15              | program exp '\n' { printf("  %lf\n",$2); }
16              ;
17
18   exp        : NUMBER    { $$ = $1;}
19              | NUMBER '+' NUMBER      { $$ = $1 + $3; }
20              ;
21
22   %% /* Additional C-code goes here */
23
24   #include <ctype.h> int yylex(void)
25   { int c;
26   /* skip white space and tabs */
27   while ((c = getchar()) == ' ' || c == '\t');
28   /* process numbers */
29   if (c == '.' || isdigit(c)) { ungetc(c,stdin);
30                       scanf("%lf", &yylval); return NUMBER;
31   }
32   /* Return end-of-file */ if (c==EOF) return(0);
33   /* Return single chars */ return(c);
34   }
35
36   /* Called by yyparse on error */
37   void yyerror(char *s) { printf("%s\n", s); }
38   int main(void)    { return yyparse(); }
```

We have added the token NUMBER to our rule to allow us to use a single number as a valid expression. The other addition is the program rule so that the parser can accept a list of statements, rather than just one statement. In the C-code section we have added a little lexical analysis that reads in one character at a time and skips over whitespace. BISON calls the routine yylex every time it needs a token. BISON also requires an error routine `yyerror` that is called in case there is a parse error. The main routine for BISON can be short; a return `yyparse()` is all that is needed. Let us now compile and run our first BISON example.

```
bison -o -v add2.c add2.y
gcc -o add2.exe add2.c -lm
```

If we now start the program, we can add two floating-point numbers at a time and our program will return the sum, e.g.,

```
user: add2.exe
user: 2+3
add2: 5.000000
user: 3.4+5.7
add2: 9.100000
```

Let us now have a closer look at how BISON performs the parsing. Since we have turned on the −v compile option, we also get an output file that has the listing of all rules, the FSM machine information, and any shift-reduce problems or ambiguities. Here is the output file `add2.output`

```
Grammar
    0 $accept: program $end
    1 program: /* empty */
    2          | program exp '\n'
    3 exp: NUMBER
    4    | NUMBER '+' NUMBER
Terminals, with rules where they appear
$end (0) 0
'\n' (10) 2
'+' (43) 4
error (256)
NUMBER (258) 3 4
Nonterminals, with rules where they appear
$accept (6)
    on left: 0
program (7)
    on left: 1 2, on right: 0 2
exp (8)
    on left: 3 4, on right: 2
state 0
    0 $accept: . program $end
    $default  reduce using rule 1 (program)
    program  go to state 1
```

```
state 1
    0 $accept: program . $end
    2 program: program . exp '\n'
    $end     shift, and go to state 2
    NUMBER  shift, and go to state 3
    exp  go to state 4
state 2
    0 $accept: program $end .
    $default  accept
state 3
    3 exp: NUMBER .
    4     | NUMBER . '+' NUMBER
    '+'  shift, and go to state 5
    $default  reduce using rule 3 (exp)
state 4
    2 program: program exp . '\n'
    '\n'  shift, and go to state 6
state 5
    4 exp: NUMBER '+' . NUMBER
    NUMBER  shift, and go to state 7
state 6
    2 program: program exp '\n' .
    $default  reduce using rule 2 (program)
state 7
    4 exp: NUMBER '+' NUMBER .
    $default  reduce using rule 4 (exp)
```

At the start of the output file, we see our rules are listed with separate rule values. Then the list of terminals follows. The terminal NUMBER, for instance, was assigned the token value 258. These first lines are very useful if you want to debug the input file, for instance, if you have ambiguities in your grammar rules this would be the place to check what went wrong. More about ambiguities a little later. We can see that in normal operation of the FSM the shifts are done in states 1, 3, and 5, for the first number, the add operation, and the second number, respectively. The reduction is done in state 7, and the FSM has a total of seven states.

Our little calculator has many limitations, it can, for instance, not do any subtraction. If we try to subtract, we get the following message:

```
user: add2.exe
user: 7-2
add2: syntax error
```

Not only is our repertoire limited to adds, but also the number of operands is limited to two. If we try to add three operands our grammar does not yet allow it, e.g.,

```
user: 2+3+4
add2: syntax error
```

As we see the basic calculator can only add two numbers not three. To have a more-useful version we would add recursive grammar rules, the operations $*, /, -, \char`^$, and a symbol table that allows us to specify variables. This is left to an exercise at the end of this chapter.

We should now have the knowledge to write a more-challenging task. One would be a program 3ac.exe that generates three address assembler code from a simple C-like language. In [Nie04] all the steps to produce assembler code for a C-like language for a stack machine are given. [FH03, SF85, ASU88, Sta90] describe more-advanced C compiler designs. In [Par92] we find such code for a three-address machine. Since many modern RISC processor (e.g., ARM Cortex-A9, MICROBLAZE, or NIOS II discussed in later chapters) use this type of processor model lets us have a look at the BISON description for the three-address machine. For a three-address machine we would allow as input typical arithmetic expressions just as in the following example:

```
-- input file:   one.c:
r=(a+b*c)/(d*e-f);
```

We will need first a lexical analysis of the input file that can be achieved using the following FLEX file.

The FLEX Description 6.7: 3ac.l

```
1     /* FLEX Lexical Analyzer 3ac.l for three address code */
2     /* Author-EMAIL: Uwe.Meyer-Baese@ieee.org */
3     %{
4     #include <stdio.h>
5     #include "y3ac.h"
6     int yylval;
7     int symcount =0;
8     int nlines=0;
9     char sym_tbl[32][20];
10    %}
11
12    letter [A-Za-z]
13    digit   [0-9]
14    ident {letter}({letter}|{digit})*
15    number {digit}*
16    op     "+"|"-"|"*"|"/"|"("|")"|";"
17    ws      [ \t\n]+
18
19    %% /* Token and Actions */
20    \n          { nlines++;}
21    {ws}        ;
22    {ident}     { yylval = install(yytext); return ID; }
23    {number}    { yylval = install(yytext); return NUM; }
24    {op}        return yytext[0];
25    =           return ASSIGN;
26    .           {printf( "Unrecognized character: %s\n", yytext );}
27
28    %% /* User program section */
29
30    int yywrap(void) { return 1; } /** Needed by FLEX ***/
31
32    int install(char *ident)
33      { int i;i=0;
34        while (i <= symcount && strcmp(ident, sym_tbl[i]))
35          i++;
36        if (i<= symcount) return(i);
37        else {
38        symcount++;
39        strcpy(sym_tbl[symcount], ident);
40        return symcount;
41        }
42    }
```

We see that we now also can have variables that start with a letter and then can have multiple letters or numbers. All variables are stored in a symbol table. The C-code for the symbol table has been discussed in the FLEX code 6.5 psm2hex.l and can be found also as examples in the literature; see, for instance, [DS02], [Nie04], or [LMB95]. yytext and yylval are the text and value associated with each token, respectively. Table 6.2 shows the variables used in the FLEX↔BISON communication. The grammar for our more-advanced C-style grammar 3ac.y now looks as follows:

The BISON Description 6.8: `3ac.y`

```
 1   /* Yacc specification 3ac.y for three address code (3AC) */
 2   /* Author-EMAIL: Uwe.Meyer-Baese@ieee.org */
 3   %{
 4   #include <stdio.h>
 5   #include <stdlib.h>
 6   #include <string.h>
 7   int yylval;
 8   int symcount;
 9   char sym_tbl[32][20];
10   #ifndef YYSTYPE
11   #define YYSTYPE int
12   #endif
13   int  taccount =0, tcount =1595, Tmax;
14   #define MAXTACS 32
15   struct { char op; int a1, a2, a3;} tac[MAXTACS];
16   YYSTYPE make_tac(char op, YYSTYPE op1, YYSTYPE op2);
17   void yyerror (char *s);
18   %}
19
20   %left '+' '-'
21   %left '*' '/'
22
23   %token ID NUM
24   %token ASSIGN
25
26   %% /* Grammar rules and actions follows */
27   input:
28         | input stmt
29         ;
30   stmt  :  '\n'
31         |ID ASSIGN exp ';' { $$ = make_tac('=',$1,$3); Tmax=tcount; }
32         ;
33   exp   : exp '+' exp    { $$ = make_tac('+',$1,$3); }
34         | exp '-' exp    { $$ = make_tac('-',$1,$3); }
35         | exp '*' exp    { $$ = make_tac('*',$1,$3); }
36         | exp '/' exp    { $$ = make_tac('/',$1,$3); }
37         | '(' exp ')'    { $$ = $2;}
38         | ID             { $$ = yylval;}
39         | NUM            { $$ = yylval;}
40         ;
41
42   %% /* Additional C-code goes here */
43
44   YYSTYPE gettemp(void);
45   void list_tacs(void);
46   void list_table(void);
47
```

```
48    YYSTYPE gettemp(void)
49    { char str1[10]; int i,found;
50      tcount++;
51      sprintf(str1,"D.%d",tcount);
52      found= 1;
53      for (i=0;i<symcount;i++)
54        if (strcmp(str1,sym_tbl[i]) == 0)
55          found=i;
56      if (found!=-1)  {
57        return (found);
58      } else {
59        symcount++;
60        strcpy(sym_tbl[symcount],str1);
61        return (symcount);
62     }
63    }
64
65    YYSTYPE make_tac(char op, YYSTYPE op1, YYSTYPE op2)
66    { YYSTYPE new_a ;
67      if (op == '=') { new_a=op1;op1=0;
68      } else new_a = gettemp();
69      tac[taccount].op = op;
70      tac[taccount].a1 = op1;
71      tac[taccount].a2 = op2;
72      tac[taccount].a3 = new_a;
73      taccount++;
74      return(new_a);
75    }
76
77    void list_tacs(void)
78    { int i;
79      printf("\n Intermediate code:\n");
80      printf(" Quadruples         3AC\n");
81      printf(" Op Dst Op1 Op2\n");
82           /*(+,  3,   4,   5)    T1 <= x + b*/
83      for (i=0; i < taccount; i++) {
84        if (tac[i].op == '=') {
85          printf("(%c, %2d, %2d, --)
86    ",tac[i].op,tac[i].a3,tac[i].a2);
87          printf("%s  = %s\n", sym_tbl[tac[i].a3],sym_tbl[tac[i].a2]);
88        } else {
89          printf("(%c, %2d, %2d, %2d)
90    ",tac[i].op,tac[i].a3,tac[i].a1,tac[i].a2);
91          printf("%s = %s %c
92    %s\n",sym_tbl[tac[i].a3],sym_tbl[tac[i].a1],
93    tac[i].op,sym_tbl[tac[i].a2]);
94        }
95      }
96    }
```

```
97
98    void list_table(void)
99    {   int i;
100       printf("\n Symbol table:\n");
101       for(i=1; i <= symcount; i++)
102          printf("%2d : %s\n", i, sym_tbl[i]);
103       printf("\n");
104   }
105
106   void yyerror (char *s)
107   { printf("%s\n", s); }
108
109   /** Main program **/
110   main() {
111      do {
112         yyparse();
113      } while (!feof(stdin));
114       list_table();
115       list_tacs();
116       return 0;
117   }
118
```

Table 6.2 Special functions and variables used in the FLEX↔BISON communication; see Appendix A [DS02] for a full list

char *yytext	Token text
file *yyin	FLEX input file
file *yyout	FLEX file destination for ECHO
int yylength	Token length
int yylex(void)	Routine called by parser to request tokens
int yylval	Token value
int yywrap(void)	Routine called by the FLEX when end of file is reached
void yyparse();	The main parser routine
void yyerror(char *s)	Called by yyparse on error

There are several new things in this grammar specification we need to discuss. The specification of the terminals using %left and %right (lines 20, 21) ensures the right associativity. We prefer that $2 - 3 - 5$ is computed as $(2 - 3) - 5 = 6$ (i.e., left associative) and not as $2 - (3 - 5) = 4$, i.e., right associativity. For exponentiation ^ we use right associativity, since 2^2^2 should be grouped as 2^(2^2). The operands listed later in the token list have a higher precedence. Since * is listed after + we assign a higher precedence to multiply compared with add, e.g., $2 + 3 * 5$ is computed as $2 + (3 * 5)$ rather than $(2 + 3) * 5$. If we do not specify this precedence, the grammar will report many reduce-shift conflicts, since it does not know if it should reduce or shift if an expression like $2 + 3 * 5$ is found.

The grammar rules (lines 27–40) in the case of our complier are now written in a recursive fashion, i.e., an expression can consist of `expression operation expression` terms.

In the next major section, the additional C-code function and the main routine are listed. The `gettemp` (lines 48–63) will generate a new temporary variable of type `D.kkkk`, where `kkkk` is a 4 digit decimal number. This is a not standard C-style variable name but matches the GIMPLE style we will discuss later. The function `make_tac` (lines 65–75) will for arithmetic operation generate a new three address code entry, unless the operation was `'='` then no new temporary variable is requested. The functions `list_tacs` (lines 77–96) and `list_table` (lines 98–104) are used in the main program to print out the list of three address codes and the list of all symbols aka variables, respectively. Finally the `main()` program routine (lines 110–117) is specified that does the parsing of all input lines and when done print the variable table and the three address code.

Let us now briefly look at the compilation steps and FLEX↔BISON communication. Since we first need to know from BISON which kind of token it expects from FLEX, we run this program first, i.e.,

```
bison -y -d -oy3ac.c   3ac.y
```

This will generate the files `y3ac.c`, `y3ac.output`, and `y3ac.h`. The header file `ytab.h` contains the token values:

```
#/* Tokens.  */
#define ID 258
#define NUM 259
#define ASSIGN 260
```

Now we can run FLEX to generate the lexical analyzer. With

```
flex   -ol3ac.c        3ac.l
```

we will get the file `l3ac.c`. Finally, we compile both C source files and link them together in `3ac.exe`

```
gcc   yl3ac.c   y3ac.c   -o 3ac.exe
```

Now we can start our three-address code program with

```
3ac.exe  < one.c
```

We get the following output:
Symbol table:

```
 1 : r
 2 : a
 3 : c
 4 : b
 5 : D.1596
 6 : D.1597
 7 : d
 8 : e
 9 : D.1598
10 : f
11 : D.1599
12 : D.1600
```

Intermediate code:

```
Quadruples            3AC
Op Dst Op1 Op2
(*,   5,   3,   4)    D.1596 = c * b
(+,   6,   2,   5)    D.1597 = a + D.1596
(*,   9,   7,   8)    D.1598 = d * e
(-,  11,   9,  10)    D.1599 = D.1598 - f
(/,  12,   6,  11)    D.1600 = D.1597 / D.1599
(=,   1,  12,  --)    r   = D.1600
```

We have a working C-style code generator but with limited functionality. To make it a more useful compiler we would need to add control flow operation such as if and while, support for n-dimensional arrays, and function call. The development of a high-performance compiler usually requires an effort of 20–50 man-years, which is prohibitively large for most projects. That is the reason why the GCC retargetable compiler is used often. At the cost of less-optimized C compiler much shorter development time can be observed, if retargetable compilers like GNU GCC or LCC are used [FH03, Sta90]. We can use a shortcut in case we just need a three-address code front end compiler. Since most processor today use a three-address code, this has been a well-studied topic in particular in code tree generation and has become a favorite internal format of the GNU C/C++ compiler. It is an improved version of the SIMPLE tools [HDE92] that together with the GENERIC GNU tools is known as the GIMPLE three address code presentation [GNU18, Mer03]. We can generate this intermediate code for a test.c program as follows.

```
gcc -fdump-tree-gimple test.c
```

Let us demonstrate the program development using this intermediate GIMPLE format with a short example.

The Test Program 6.9 for GIMPLE: `test.c`

```
1   int main() {
2   int a,b,c,d,e,f,r;
3   r=(a+c*b)/(d*e-f);
4   }
```

The output of the intermediate code in GIMPLE now looks as follows

The Output Program 6.10 in GIMPLE Format: `test.c.004t.gimple`

```
1    main ()
2    {
3      int D.1596;
4      int D.1597;
5      int D.1598;
6      int D.1599;
7      int a;
8      int b;
9      int c;
10     int d;
11     int e;
12     int f;
13     int r;
14
15     D.1596 = c * b;
16     D.1597 = D.1596 + a;
17     D.1598 = d * e;
18     D.1599 = D.1598 - f;
19     r = D.1597 / D.1599;
20   }
```

We see the similarity to our 3AC Quadruples generated with `3ac.com` using BISON and a substantial simplification that allows us to use this intermediate code to generate assembler code for a target processor of our choice. The GIMPLE IR code beside full arithmetic operation supports also control constructs, function calls, and arrays/pointers. The verification is left to an exercise at the end of the chapter, see Exercise 6.27.

6.4 SW Debugger and Instruction Set Simulator

Before we download our designed microprocessor system to the FPGA, you may want to verify in software the program you have coded in assembler or C/C++. There are several different options that differ in additional effort and simulation speed and precision of the results. You may want to monitor registers and memory content during program execution and/or to modify these values, set breakpoint, etc. Let us briefly review our options:

- A *HDL simulator* such as MODELSIM that use the processor models HDL specification and run an application will allow us to monitor registers and memory
  content precisely. However, this HDL may get too complex such as for the ARM
  Cortex-A9, and may not always be available. This verification will also be the
  slowest of all with a couple of kilo simulation steps per second.
- For some processors so called *instruction set simulator* (ISS) are available. These
  are larger software packages that based on the µP architecture and the instruction
  set model in SW the behavior of the processor as close as possible. Often these
  simulators are only instruction accurate. For cycle accuracy a precise pipeline
  model of the processor including the memory and cache timing must be rebuild
  in software which is not a trivial task. Electronic system level tools such as LISA
  processor designer, nML, ASIP meister, or EXPRESSION are tools that have been
  develop to simplify processor and ISS development through architecture description languages, Fig. 6.2 shows the LISA processor designer generate ISS for the
  URSIC discussed in Chap. 1. Simulation speed of these ISS is substantial higher
  than HDL simulation in MODELSIM.
- *Software debugger* are also a popular option to verify a source program without
  having a hardware with your processor running, i.e., can be a cross compiler
  design. The GDB that comes with the GNU software tools is a good example. We

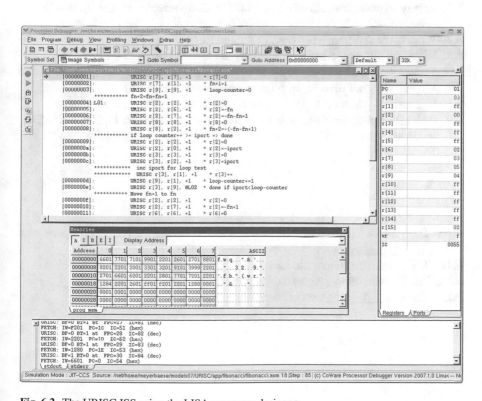

Fig. 6.2 The URISC ISS using the LISA processor designer

briefly discussed debugging in Chap. 5 on C/C++ program development. This allows you to go through you program in single steps or run up to a special breakpoint then monitor register, or variable/memory content.

- Most development tools also allow an *in-circuit debugging*. Here you use a special "monitor" program that let you download your program to the processor, run in single steps or until a breakpoint. This option will have the highest hardware requirement. Your system must be completely designed and up running before you can run such a program debugging. If you are working on a processor instruction set and/or architecture, this may not be a good option for you.

Review Questions and Exercises

Short Answer

6.1. Describe how a two-pass assembler works.
6.2. Explain the difference between relocatable versus absolute coding.
6.3. Briefly explain the following special function/variables in the FLEX↔BISON communication: `yytext`, `yylength`, `yyparse()`, and `yylex()`.
6.4. What is the difference between FLEX and BISON?
6.5. What is a difference between a native and a cross compiler?
6.6. Should we use FLEX only to build a general-purpose calculator?
6.7. Why are ISS not used in the microprocessor design phase?
6.8. What are requirements to use an in-circuit debugger to test a program.
6.9. Order the debug tools by simulation steps per seconds from high to low: HDL, in-circuit debugger, ISS.

Fill in the Blank

6.10. The `add2` calculator supports the following arithmetic operations _____
6.11. The `psm2hex` output file name is _____
6.12. The GIMPLE temporary variables have the format _____
6.13. The _____ was predecessor of the BISON program

True or False

6.14. _____ The add2 calculator was build using FLEX and BISON.
6.15. _____ BISON and FLEX have three major part: definition, rules, and user code
6.16. _____ The 3ac generated code is very similar to GNU GIMPLE code.
6.17. _____ The GIMPLE code is used to generate code for a stack machine.
6.18. _____ The cycle accurate ISS requires more system information than an instruction accurate ISS.
6.19. Given are the following FLEX expressions. Determine the pattern that match the expressions (YES). If the pattern does not match the expression use NO.

Expression	Matching pattern? (YES/NO)									
	a	A	ab	aa	AA	abc	aaaa	abcc	aabb	abccc
abc*										
abc+										
ab(c)?										
[a-zA-Z]										
((aa)|(bb))*										
a{1,3}										

6.20. Given is the following grammar: (R1) E → E + E (R2) E → E − E (R3) E → E/E (R4) E → (E) (R5) E → id. Determine if the following expressions can be generated with the above grammar:

   (i) a/(b + c)
   (ii) −a + b
   (iii) a − b − c − d
   (iv) (a + b) × c
   (v) a(b/c)
   (vi) a − (b + c)

6.21. Write leftmost derivations for (a − b)/c and provide the rules (R1)–(R5) from the Exercise 6.20.
6.22. Write rightmost derivations for p/(q + r) and provide the rules (R1)–(R5) from the Exercise 6.20.

Projects and Challenges

6.23. Extend the Verilog scanner to support all language elements found in `urisc.v`.

6.24. Modify the `vscanner` to count items in VHDL source file `urisc.vhd`.

6.25. Modify the `vscanner` to count items found in a C/C++ source code `add2.c`.

6.26. Extend the `add2` calculator with recursive grammar, *,−,^, and a symbol table for variables.

6.27. Write a test program for all 14 IR codes supported by GIMPLE:

      I. `x = a OP b`
     II. `*a = b OP c`
    III. `x = OP c`
    IV. `*a = x`
     V. `*a = f(args)`
    VI. `*a = *c`
   VII. `*a = (cast)b`
  VIII. `*a = &x`
    IX. `x = y`
     X. `x = &y`
    XI. `x = *c`
   XII. `x = f(args)`
  XIII. `x = (cast)b`
  XIV. `f(args)`

6.28. Build an instruction accurate ISS for the URISC processor and test it with a small loop example. Monitor PC, registers and I/O. Use a single step simulation.

6.29. Use the Niemann [Nie04] tutor (source files on book CD) to compile and test (; is return).

    I. Compile `calc1` and test with: $1 + 3 + 5$; $1 - 3 - 5$; $4/2$
   II. Compile `calc2` and test with $1 - 3 - 5$; $8/2$; $x = 8$; $y = 2$; x/y; $8/0$
  III. Compile `calc3a` and test with $x = 3 + 4*5 - 8/2$; print x
  IV. Compile `calc3b` test with $x = 3 + 4 * 5 - 8/2$; print x
   V. Compile `calc3g` and test graph output with $x = 3 + 4 * 5 - 8/2$

6.30. In [Nie04] add Boolean operations AND (use &), OR (use |), NOT (use ~) to `calc3` BISON and FLEX.
Test improved `calc3a` with `print 7 & 12; print 7| 12; x= ~7 + 1; print x` and improved `calc3b` and `calc3g` with `x = a | b & ~c;`

6.31. Explain the language limitations of Niemann `calc3` [Nie04] compared to ANSI-C.

6.32. Add Boolean operations AND (use &), OR (use |), NOT (use ~) to `c3ac.y` BISON and FLEX `3ac.1`.
Test improved `3ac.exe` with `a=7 & 12; b = 7| 12; c = ~7 + 1;` and `x = a | b & ~ c;`

References

[Alt04] Altera Netseminar Nios processor (2004), http://www.altera.com

[ASU88] A. Aho, R. Sethi, J. Ullman, *Compilers: Principles, Techniques, and Tools*, 1st edn. (Addison Wesley Longman, Boston, 1988)

[DS02] C. Donnelly, R. Stallman BISON: The YACC-compatible Parser Generator (2002), http://www.gnu.org

[FH03] C. Fraser, D. Hanson, A. Retargetable, C. Compilers, *Design and Implementation*, 1st edn. (Addison-Wesley, Boston, 2003)

[GNU18] GNU Compiler Collection (GCC) Internals: Chapter 12 GIMPLE https://gcc.gnu.org/onlinedocs/gccint/GIMPLE.html#GIMPLE

[HML02] A. Hoffmann, H. Meyr, R. Leupers, *Architecture Exploration for Embedded Processors with LISA*, 1st edn. (Kluwer Academic Publishers, Boston, 2002)

[HDE92] L. Hendren, C. Donawa, M. Emami, G. Gao, Justiani, B. Sridharan, Designing the McCAT compiler based on a family of structured intermediate representations, in *Proceedings of the 5th International Workshop on Languages and Compilers for Parallel Computing, no. 757 in Lecture Notes in Computer Science, New Haven, Connecticut*, (Springer-Verlag, Cham, 1992), pp. 406–420

[Joh75] S. Johnson YACC – Yet Another Compiler-Compiler," technical report no. 32, AT&T, (1975)

[Leu02] R. Leupers, *Code Optimization Techniques for Embedded Processors*, 2nd edn. (Kluwer Academic Publishers, Boston, 2002)

[LM01] R. Leupers, P. Marwedel, *Retargetable Compiler Technology for Embedded Systems*, 1st edn. (Kluwer Academic Publishers, Boston, 2001)

[LMB95] J. Levine, T. Mason, D. Brown, *LEX & YACC*, 2nd edn. (O'Reilly Media, Beijing, 1995)

[LS75] W. Lesk, E. Schmidt "LEX – a Lexical Analyzer Generator," Technical Report no. 39, AT&T, (1975)

[Mer03] J. Merrill, GENERIC and GIMPLE: A New tree representation for entire function. Proc GCC Developers Summit, 171–180 (2003)

[Nie04] T. Niemann A Compact Guide to LEX & Yacc (2004), https://www.epaperpress.com/lexandyacc/index.html

[Pax95] V. Paxson FLEX, Version 2.5: a fast scanner generator (1995), http://www.gnu.org

[Par92] T. Parsons, *Introduction to Compiler Construction*, 1st edn. (Computer Science Press, New York, 1992)

[SF85] A. Schreiner, H. Friedman, *Introduction to Compiler Construction with UNIX*, 1st edn. (Prentice-Hall, Englewood Cliffs, 1985)

[Sta90] R. Stallman Using and Porting GNU CC (1990), http://www.gnu.org

Chapter 7
Design of the PICOBLAZE Softcore Microprocessor

Abstract This chapter gives an overview of a popular 8-bit microprocessors in general and the PICOBLAZE in particular. PICOBLAZE is the most popular 8-bit softcore microprocessor from Xilinx that was designed by Ken Chapman. We will iteratively develop the PICOBLAZE architectures adding more and more features as we analyze the PICOBLAZE instruction set and its implementation in HDL.

Keywords PicoBlaze · Softcore · Hardware · Microcontroller · 2-address machine · Xilinx · 8-bitter · Ken Chapman programmable state machine (KCPSM) · TRISC2 · KCPSM6 · Instruction format · Scratchpad memory · LED toggle · Subroutine nesting · Vivado · ModelSim · Quartus · TerASIC · DE1 SoC · TimeQuest

7.1 Introduction

The 8-bitters have become the favorite controllers in embedded system design. One of the most important driving forces in the microcontroller market has been the automotive and home appliance areas. In cars, for instance, only a few high-performance microcontrollers are needed for audio or engine control; the other more than 50 microcontrollers are used in such functions as electric mirrors, air bags, speedometer, and door locking, to name just a few. 8-bitters sales are about 3 billion controllers per year, compared with 1 billion 4- or 16-/32-bit controllers. The 4-bit processors usually do not have the required performance, while 16- or 32-hit controllers arc usually too cxpensive. Xilinx PICOBLAZE fits right into these popular 8-bit applications and provides a nice and free-of-charge development platform. In this chapter, we discuss the hardware and in Chap. 8 the software tools for PICOBLAZE.

Several 8-bit FPGA softcores are available for many instruction sets like Intel's 8080 or 8051, Zilog's Z80, Microchip's PIC family, MOS Technology's 6502 (popular in early Apple and Atari computer), Motorola/Freescales 68HC11, or Atmel AVR microprocessors. At `www.edn.com/microprocessor` a full list of current controllers is provided.

Although Altera does not promote its own 8-bitter, the Altera Megafunction Partners Program (AMPP) partner supports several instruction sets, like 8081, Z80, 68HC11, PIC and 8051. For Xilinx devices, we find besides the PICOBLAZE also support for 8051, 68HC11, and PIC ISA (see Table 7.1).

It important to notice that low-level hardware optimization of the PICOBLAZE with only 177 four-input LUTs makes it the smallest and fastest 8-bitter for Xilinx devices. That's why we like to take a closer look in the following at this popular microprocessor.

The PICOBLAZE is one of many positive surprise stories we have witnessed in the FPGA field over the years. It is based on the Xilinx application Engineer named Ken Chapman programmable state machine (PSM) designs that basically is an FSM augmented by a program memory (see Chap. 1, Fig. 1.4). The PICOBLAZE microprocessor is a popular 8-bit microprocessor that has been around since 1990 running many thousand different applications. It's not a full featured CISC processor like Intel's Itanium processor with 592 million transistors but can still be very useful in embedded systems where small microcontroller task are more often needed than top of the line microprocessor performance for peripherals such as keyboard or LCD controller. The assembler/link/loader and VHDL code for the core are available royalty free. PICOBLAZE is optimized for Xilinx devices: the low-level LUT-based implementation of many functions like register file, scratch pad RAM, and call/return stack use the memory feature (aka distributed RAM) of Xilinx devices and make it hard to use an Altera device since LE-based RAM are not a feature of Altera FPGAs. The core is substantially smaller than other 8-bit softcores. At time of writing, we use the fifth generation of the PICOBLAZE called KCPSM6. There are a couple of features that have been constant over the last 25 year of PICOBLAZE such as:

- PICOBLAZE is a two-address machine, i.e., the logical and arithmetic operations are of the type sX <= sX □ sY.

Table 7.1 FPGA 8-bitter ISA support

μP name	Device used	LE/slices	BRAM/M9Ks	Speed (MHz)	Vendor
C8081	EP1S10-5	2061	3	108	CI
CZ80CPU	EP1C6-6	3897	–	82	CI
DF6811CPU	Stratix-7	2220	4	73	DCD
DFPIC1655X	Cyclone-II-6	663	N/A	91	DCD
DR8051	Cyclone-II-6	2250	N/A	93	DCD
Flip8051	Xc2VP4–7	1034	N/A	62	DI
DP8051	Spartan-III-5	1100	N/A	73	DCD
DF6811CPU	Spartan-III-5	1312	N/A	73	DCD
DFPIC1655X	Spartan-III-5	386	3	52	DCD
PICOBLAZE	Spartan-III	96	1	88	Xilinx

Vendors: DI = Dolphin Integration (France); CI = CAST Inc. (NJ, USA); DCD = Digital core design (Poland); N/A = Information not available

- PICOBLAZE has 7 instruction classes: arithmetic, interrupt, I/O, logical, program control, shift, and storage.
- All programs start at address zero.
- PICOBLAZE has around 30 instructions.

Over the years with changes in technology, the KCPSM had been evolved with slide modification (see Table 7.2): The first generation PICOBLAZE designed for the COOLRUNNER devices did not have distributed or block RAMs and to keep the footprint small it had only 256 × 16 bit program words and 8 × 8-bit registers. The second generation for Virtex-E devices now had 16 registers and stack depth 15. The third generations then later used distributed RAM for register file and call/return stack. This is the only KCPSM that had a 32-word register file. The improved block size now allowed 1024 × 18 program memory. The fourth generation then introduced an additional on-chip data memory called scratch pad RAM of 64 bytes; the register file was scaled to two banks of size 16 (see Fig. 7.1). Finally, the current

Table 7.2 PICOBLAZE feature and performance comparison of all 5 generations

Feature					
Generation	1.	2.	3.	4.	5.
Target device	CoolRunner	Virtex-E	Virtex-II	Spartan-3	Zynq,Vir-6
Assembler	ASM	KCPSM	KCPSM2	KCPSM3	KCPSM6
Program (max)	256 × 16	256 × 16	1K × 18	1K × 18	4K × 18
Registers 8 bits	8	16	32	16	2x16
Call Stack	4	15	31	31	30
#Instructions	26	26	26	30	39
Scratch Pad DRAM	N/A	N/A	N/A	64 Byte	64–256 Byte
MIPS (typical)	21	37	40–70	44–102	52–199
Reference	[X387]	[X213]	[X627]	KCPSM3 UG [Cha03]	KCPSM6 UG [Cha14]

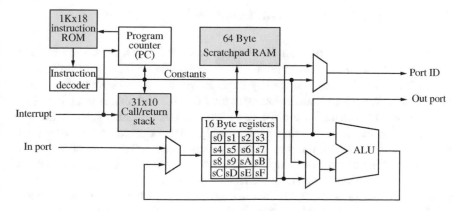

Fig. 7.1 The PICOBLAZE a.k.a. KCPSM3 core from Xilinx

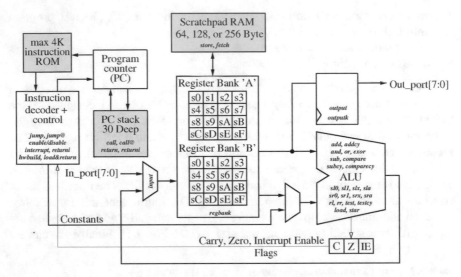

Fig. 7.2 The PICOBLAZE aka KCPSM6 core from Xilinx. Instruction (in italic) mainly associated with major hardware units. Substantial memory blocks (excluding registers) in gray

architecture KCPSM6 added 9 new instructions to the instruction set: START, COMPARECY, TESTCY, HWBUILD, JUMP@, call@, LOAD&RETURN, OUTPUTK, and REGBANK. We now have a total of 39 instructions, where most come in two flavors for the second operand, 8-bit immediate or another register, which essentially double the needed ISA code, such that almost all 6 bit OP code pattern are in used now (see Fig. 7.2). Most notable change to KCPSM3 is that now all register/register instructions are even, while all constant code uses odd operation code. Table 7.2 has the full details on the KCPSM history.

7.2　Overview KCPSM6 Instruction Set

If we intend to use a PICOBLAZE like microprocessor that can run on any FPGA or ASIC, we may want to consider taking advantage of the synthesizable Verilog code call PACOBLAZE [Koc06] designed by Pablo Bleyer Kocik. However, this is not the latest architecture since PACOBLAZE uses the KCPSM3 ISA that is not binary compatible with fifth generation PICOBLAZE KCPSM6, i.e., the latest version. Let us in the following therefore develop synthesizable HDL code for a small subset of PICOBLAZE instruction set. Let us call this microprocessor TRISC2 since it is a two-address machine and only a (tiny) RISC subset of KCPSM6 is implemented. But before we start the development of the HDL code, let us have a general look how the syntax for PICOBLAZE assembly instructions looks like. It is common practice to use the following coding abbreviation in the PICOBLAZE assembler coding:

Table 7.3 The 18-bit instruction bit coding format of KCPSM6

Format	6-bits	4-bits	4-bits	4-bits
R-TYPE	Opcode	sX 1. source/destination	sY 2. source	0 or (shift) code
I-TYPE	Opcode	sX 1. source/destination	8-bits immediate or indirect	
J-TYPE	Opcode	12-bits target address: aaa or NaN		

- aaa for the 12-bit address
- kk for the 8-bit immediate constants
- pp for the 8-bit port IDs
- p for a short 4-bit port ID version
- ss for the 8-bit scratch pad location aka data memory
- X for the 1. source and destination register coding
- Y for the 2. source/operand of arithmetic/logic operations

The KCPSM ISA (Table 7.3) has three basic 18-bit instruction formats:

- The register/register type, OP sX, sY type sX = sX □ sY or for indirect instructions OP@ (sX, sY) assembler coded if the register is used as indirect source
- The immediate type: OP sX, kk or pp, or OP kk, p if both are constant sX = kk
- The jump type: OP aaa, or without argument OP only.

Designing or just understanding a full instruction set of a modern RISC processor including assembler coding, flags, control, and data architecture can be overwhelming at first. We will therefore take a step-by-step approach by adding instructions to our PSM architecture as we go from a simple to a more complex CPU. The full instruction set of the PICOBLAZE is discussed in the next Chap. 8. Let our first example be a reading of the slider switch values and out putting these data to the LEDs we have on both boards. We will therefore need only three instructions:

1. 09xpp INPUT sX, pp instruction that read a value from port and store this in one of the 16 registers
2. 2Dxpp OUTPUT sX, pp instruction that takes the data from register X and stores this value in the output port
3. 22aaa JUMP aaa a jump instruction such that we can repeat the above forever

Now to finish our assembler program, we need to decide the register we like to use and the port number. Since our PICOBLAZE only needs single I/O ports, we can use zero or just ignore the port selection for our initial design. Since the instruction format contains many zeros, we may choose a none-zero register but that is up to you. If we use register 3, then the full program will look in 18-bit VHDL 2008 style HEX code as follows:

```
-- Program ROM definition and values for VHDL2008
TYPE MEMP IS ARRAY (0 TO 2) OF STD_LOGIC_VECTOR(17 DOWNTO 0);
CONSTANT prom : MEMP :=  18X"09300" , 18X"2D300" , 18X"22000"
```

In VHDL-1993 hex words must have a length of multiples of 4 bits, and we can therefore not build a 18-bit hex constant. We can use 6 octal digits ($3 \times 6 = 18$) or binary coding:

```
MEMP := ("001001001100000000","101101001100000000","10001000
0000000000")
```

As starting point for the synthesizable KCPSM6 design, we may use the Verilog PACOBLAZE or the TRISC0, a compact 8-bit stack machine type microprocessor introduced in [MB14]. The TRISC0 is a very compact HDL representation and seemed as evolving starting point a good choice. We can even reuse decoder, PC control, SRAM (aka scratch pad RAM), and part of the ALU.

Before we start coding our initial HDL design, let's add a few more ALU operations to our machine that it become a more useful microprocessor than just an I/O register controller. Let us assume our second task should be that we like to take the data we just read from the input port and like to toggle these data in around 1 second intervals, i.e., a LED should go on that was off and an illuminated LED should be off the next second. We have two problems to solve, (1) how do we toggle a register values since an INVERT is not part of the KCPSM6 instruction set and (2) how can we implement a 1 second long delay loop. Now for the first task, we can implement the INVERT operation with an XOR, since the Boolean rule says: x XOR $1 = x'$. So, if we add the instruction

4. 07xkk XOR sX, kk EXOR operation of register sX with constant, i.e., sX = sX XOR kk

   to the ALU operations, we should be able to toggle a whole register.

The second task is a little bit more challenging. Usually to implement delays with a loop, we would set a register to an initial value and count down (i.e., subtract 1) until the counter reaches zero. A single 8-bit counter period however will only require 256 clock cycles, which is pretty short at 50 MHz clock rate of our processor, so our counter needs to be about $\log_2(50 \times 10^6) \approx 24$ bit long for a 1 second interval. We will therefore need to implement a counter using three registers, and for the subtract, we need the SUB and the subtract with carry SUBCY from the KCPSM6 instruction set, i.e.,

5. 19xkk SUB sX, kk with C flag and Z flags based on the SUB operation
6. 1Bxkk SUBCY sX, kk with C flag and Z=Zold AND Znew

Our loop needs to continue until the 24 bit counter is zero, and we therefore need a conditional jump if not zero, i.e.,

7. 36aaa JUMP NZ, aaa jump to the address aaa if the zero flag is zero
The full assembler program now looks as follows (comments start with a semicolon and continue until the end of the line):

PSM Program 7.1: LED Toggle Using a 24-Bit Counter

```
 1          start:  INPUT s3, 00   ; read switches
 2          flash:  LOAD s0, 20    ; set loop counter values
 3                  LOAD s1, BC    ; counter has 3x8=
 4                  LOAD s2, BE    ; 24 bits
 5                  OUTPUT s3, 00  ; write general LEDs
 6          loop:   SUB s0, 01     ; s0 -= 1
 7                  SUBCY s1, 00   ; sub with carry
 8                  SUBCY s2, 00   ; sub with carry
 9                  JUMP NZ, loop  ; count to zero
10                  XOR s3, FF     ; invert LEDs
11                  JUMP flash     ; start all over
12
```

As can be seen, register 0, 1, and 2 are chained together to build a 24-bit counter, and register 3 is used for I/O. A single loop run takes now 4 clock cycles, and the counter is therefore set to $50,000,000/4 = 12,500,000_{10} = BEBC20_{16}$ for a 50 MHz clock of our TRISC2 (lines 2–4). When the counter reaches zero (line 10), we toggle the register s3 value using the XOR operation (line 11) and restart the counter by jumping to the line 2 instruction with label flash.

7.3   Initial PICOBLAZE Synthesizable Architecture

Let us now have a look at the initial HDL code of our reduced ISA PICOBLAZE aka TRISC2

VHDL Code 7.2: TRISC2 (Initial Design)

```
1     -- Title: T-RISC 2 address machine
2     -- Description: This is the top control path/FSM of the
3     -- T-RISC, with a single 3 phase clock cycle design
4     -- It has a two-address type ALU instruction type
5     -- implementing a subset of the KCPSM6 architecture
6     -- =====================================================================
7     LIBRARY ieee; USE ieee.std_logic_1164.ALL;
8
9     PACKAGE n_bit_int IS                   -- User defined types
10      SUBTYPE U8 IS INTEGER RANGE 0 TO 255;
11      SUBTYPE SLVA IS STD_LOGIC_VECTOR(11 DOWNTO 0); -- address prog. mem.
12      SUBTYPE SLVD IS STD_LOGIC_VECTOR(7 DOWNTO 0);  -- width data
13      SUBTYPE SLVD1 IS STD_LOGIC_VECTOR(8 DOWNTO 0); -- width data + 1 bit
14      SUBTYPE SLVP IS STD_LOGIC_VECTOR(17 DOWNTO 0); -- width instruction
15      SUBTYPE SLV6 IS STD_LOGIC_VECTOR(5 DOWNTO 0);  -- full opcode size
16      SUBTYPE SLV5 IS STD_LOGIC_VECTOR(4 DOWNTO 0);  -- reduced opcode
17      SUBTYPE SLV4 IS STD_LOGIC_VECTOR(3 DOWNTO 0);  -- register array
18    END n_bit_int;
19
20    LIBRARY work;
21    USE work.n_bit_int.ALL;
22
23    LIBRARY ieee;
24    USE ieee.STD_LOGIC_1164.ALL;
25    USE ieee.STD_LOGIC_arith.ALL;
26    USE ieee.STD_LOGIC_unsigned.ALL;
27    -- =====================================================================
28    ENTITY trisc2 IS
29    PORT(clk   : IN  STD_LOGIC; -- System clock (clk=>CLOCK_50)
30        reset : IN  STD_LOGIC; -- Active low asynchronous reset (KEY(0))
31        in_port : IN  STD_LOGIC_VECTOR(7 DOWNTO 0); -- Input port (SW)
32        out_port : OUT STD_LOGIC_VECTOR(7 DOWNTO 0));--Output port (LEDR)
33    END;
34    -- =====================================================================
35    ARCHITECTURE fpga OF trisc2 IS
36    -- Define GENERIC to CONSTANT for _tb
37      CONSTANT WA : INTEGER := 11;  -- Address bit width -1
38      CONSTANT WR : INTEGER := 3;   -- Register array size width -1
39      CONSTANT WD : INTEGER := 7;   -- Data bit width -1
40
41      COMPONENT rom4096x18 IS
42      PORT (clk   : IN STD_LOGIC;       -- System clock
43            reset : IN STD_LOGIC;       -- Asynchronous reset
44            pma   : IN STD_LOGIC_VECTOR(11 DOWNTO 0); -- Prog. memory adr.
45            pmd   : OUT STD_LOGIC_VECTOR(17 DOWNTO 0)); -- Prog. mem. data
46      END COMPONENT;
47
48      SIGNAL op6   : SLV6;
49      SIGNAL op5   : SLV5;
50      SIGNAL x, y, imm8 : SLVD;
51      SIGNAL x0, y0 : SLVD1;
52      SIGNAL rd, rs : INTEGER RANGE 0 TO 2**(WR+1)-1;
53      SIGNAL pc, pc1, imm12 : SLVA; -- program counter, 12 bit aaa
54      SIGNAL pmd, ir   : SLVP;
55      SIGNAL eq, ne, not_clk : STD_LOGIC;
56      SIGNAL jc        : boolean;
```

```
57        SIGNAL z, c, kflag : STD_LOGIC; -- zero, carry, and imm flags
58
59     -- OP Code of instructions:
60     -- The 5 MSBs for ALU and I/O operations (LSB is imm flag)
61        CONSTANT add     : SLV5 := "01000"; -- X10/1
62        CONSTANT addcy   : SLV5 := "01001"; -- X12/3
63        CONSTANT sub     : SLV5 := "01100"; -- X18/9
64        CONSTANT subcy   : SLV5 := "01101"; -- X1A/B
65        CONSTANT opand   : SLV5 := "00001"; -- X02/3
66        CONSTANT opxor   : SLV5 := "00011"; -- X06/7
67        CONSTANT load    : SLV5 := "00000"; -- X00/1
68        CONSTANT opinput : SLV5 := "00100"; -- X08/9
69        CONSTANT opoutput : SLV5 := "10110"; -- X2C/D
70     -- 6 Bits for all other operation
71        CONSTANT jump    : SLV6 := "100010"; -- X22
72        CONSTANT jumpz   : SLV6 := "110010"; -- X32
73        CONSTANT jumpnz  : SLV6 := "110110"; -- X36
74
75     -- Register array definition
76        TYPE RTYPE IS ARRAY(0 TO 15) OF SLVD;
77        SIGNAL s : RTYPE;
78
79     BEGIN
80
81        P1: PROCESS (reset, clk) -- FSM of processor
82        BEGIN
83           IF reset = '0' THEN
84              pc <= (OTHERS => '0');
85           ELSIF falling_edge(clk) THEN
86             IF jc THEN
87                pc <= imm12; -- any jumps that use 12 bit immediate aaa
88             ELSE
89                pc <= pc1;   -- Usual increment
90             END IF;
91           END IF;
92        END PROCESS p1;
93        pc1 <= pc + X"001";
94        jc <= (op6=jumpz AND z='1') OR (op6=jumpnz AND z='0') OR (op6=jump);
95
96        -- Mapping of the instruction, i.e., decode instruction
97        op6 <= ir(17 DOWNTO 12);  -- Full Operation code
98        op5 <= ir(17 DOWNTO 13);  -- Reduced Op code for ALU ops
99        kflag <= ir(12);          -- Immediate flag 0= use register 1= use kk;
100       imm8  <= ir(7 DOWNTO 0);   -- 8 bit immediate operand
101       imm12 <= ir(11 DOWNTO 0);  -- 12 bit immediate operand
102       rd <= CONV_INTEGER('0' & ir(11 DOWNTO 8));-- Index dest./1. src reg.
103       rs  <= CONV_INTEGER('0' & ir(7 DOWNTO 4));-- Index 2. source reg.
104       x <= s(rd); -- first source ALU
105       x0 <= '0' & x; -- zero extend 1. source
106       y <= imm8 when kflag='1'
107           else s(rs); -- second source ALU
108       y0 <= '0' & y; -- zero extend 2. source
109
110       prog_rom: rom4096x18       -- Instantiate a Block RAM
111       PORT MAP (clk   => clk,    -- System clock
112                 reset => reset,  -- Asynchronous reset
```

```
113                      pma    => pc,-- Program memory address
114                      pmd    => pmd);    -- Program memory data
115        ir <= pmd;
116
117        ALU: PROCESS (reset, clk, x0, y0, c, op5, op6, in_port)
118        VARIABLE res: STD_LOGIC_VECTOR(8 DOWNTO 0);
119        VARIABLE z_new, c_new : STD_LOGIC;
120        BEGIN
121          CASE op5 IS
122            WHEN add    =>    res  := x0 + y0;
123            WHEN addcy  =>    res  := x0 + y0 + c;
124            WHEN sub    =>    res  := x0 - y0;
125            WHEN subcy  =>    res  := x0 - y0 - c;
126            WHEN opxor  =>    res  := x0 XOR y0;
127            WHEN load   =>    res  := y0;
128            WHEN opinput =>   res  := '0' & in_port;
129            WHEN OTHERS =>    res  := x0; -- keep old
130          END CASE;
131          IF res = 0 THEN   z_new := '1'; ELSE z_new := '0'; END IF;
132          c_new := res(8);
133          IF reset = '0' THEN              -- Asynchronous clear
134            z <= '0'; c <= '0';
135            out_port <= (OTHERS => '0');
136          ELSIF rising_edge(clk) THEN
137            CASE op5 IS             -- Compute the new flag values
138              WHEN addcy | subcy  => z <= z AND z_new; c <= c_new;
139                              -- carry from previous operation
140              WHEN add | sub =>    z <= z_new; c <= c_new; -- Use new
141              WHEN OTHERS =>  z <= '0'; c <= '0';    -- reset all others
142            END CASE;
143            s(rd) <= res(7 DOWNTO 0); -- store alu result;
144            IF op5 = opoutput THEN out_port <= x; END IF;
145          END IF;
146        END PROCESS ALU;
147
148      END fpga;
```

We see in the coding first the package with the subtype definition (lines 9–18). We do not use generic definition and constant to simplify the timing verification as will become clear later. The ENTITY (lines 28–33) at this time only includes the four ports for input and output. We can monitor additional internal signals too in a functional simulation. The architecture part starts with general purpose signals, and then the operation code for the implemented instructions is listed as constant values (lines 59–73). The constant definitions allow very intuitive coding of the actions. The first PROCESS in the ARCHITECTURE body FSM hosts the finite state machine that is used to control the microprocessor (lines 81–94). Then *decoding* of the instruction word into separate components takes place such as the 2. operand selection for constant or register input (lines 96–108). The program memory is instantiated via external ROM block that can be generated with Xilinx tool kcpsm6.exe (download from www.Xilinx.com/picoblaze) or our psm2hex utility (see Sect. 6.2 FLEX file 6.1) and the program hex2prom.exe (lines 110–114) from the book CD that produces the *.hex file for Verilog $readmemh() and a complete VHDL file. All operations including LOAD that require an update of the register array are included in ALU PROCESS (lines 117–146). Extra test pins are assigned through the simulation script only. The design uses 84 ALMs, no M9K (VHDL coding), or two M9Ks (Verilog coding), no embedded multiplier, and we measure a registered performance of Fmax = 139.12 MHz using the TimeQuest slow 85C model when using Altera QUARTUS.

We are ready now to simulate the TRISC2 machine and the initial simulation steps are shown in Fig. 7.3. Figure 7.3 shows the VHDL MODELSIM simulation results, Verilog coding and VIVADO simulation can be found in Appendix A.

The simulation display shows the input signal for clock and reset, followed by local non I/O signals and finally the I/O signals for switches (i.e., in_port) and LEDs (i.e., out_port). After the release of the reset (active low), we see how the program counter (pc) continuously increases until the end of the loop. The pc changes with the falling edge and instruction word is updated with the rising edge (within the ROM). The value 5 is read from the input port in_port and put into register s3. Next, the counter-start values BE BC 20 are loaded. The register s3 value is loaded in the out_port register. The counter is decremented, and since the 24-bit value is not zero, the loop is repeated. This continues for many clock cycles until the 1 second simulation time is reached.

A "large-scale" simulation plot is shown in Fig. 7.4 and reveals that after 1 second (i.e., 1,000,000,000 ns), the counter reaches zero, and the bits are toggled from $05_{HEX} => 0000\ 0101_2$ to $FA_{HEX} => 1111\ 1010_2$.

This can also be observed after full compile and downloading the programming file to the FPGA board: The slider switch values toggle in a 1 second interval. Remember that the HDL I/O ports need to be mapped to the correct pins. In the Altera software when using DE boards, we typically start with SystemBuilder. exe utility and then modify the *.qsf file according to our I/O port names, i.e., CLOCK_50 become clk, SW becomes in_port, KEY(0) now is reset, and

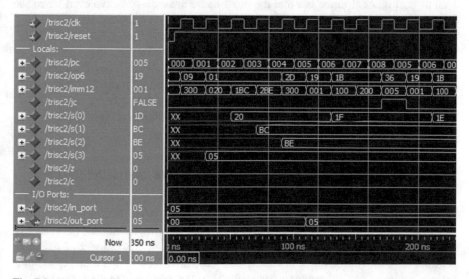

Fig. 7.3 First simulation steps for the toggle program showing 24-bit counter load and counting one loop iteration

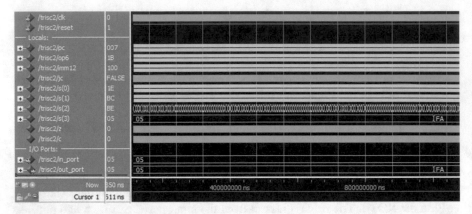

Fig. 7.4 Large-scale simulation that shows the toggle from 05 to FA

(a)   (b)

Fig. 7.5 The LED toggle in a 1 second intervals. (**a**) display of 5 in binary. (**b**) NOT 5 = FA display

LEDR is out_port. You need to change both assignments for logic level type and well as port name. Remember that the pins' name are case sensitive even when the VHDL is not. In the Xilinx Digilent Inc. ZyBo board, we typically start with the _ master.xdc file and substitute our clk, reset, in_port, and out_ port for the correct pins (i.e., clk, btn[0], sw, led).

Figure 7.5 shows the observation of the TerASIC DE FPGA board after full compile and download: The SW values toggle in a 1 second interval. A video (toggleMWT.MOV landscape for Windows Media Player and toggleQT.MOV portrait for Quick time player) can be found on the book CD of the first working design.

7.4   PicoBlaze Synthesizable Architecture with Scratchpad Memory

After this initial design, let us now add two more important features of a modern microprocessor: data memory and subroutine control. As most modern RISC processors, the PicoBlaze has a so-called load/store architecture that mean communication between the microprocessor, and his data memory (called scratch pad memory in the KCPSM architecture) is done via the register only, i.e., the ALU operations *cannot* access the data memory directly. This simplifies the ISA that we only need to support a STORE and a FETCH operation and no special ALU operation that use data memory. We add the two instructions:

8. 2Fxss STORE sX, ss using the 8 bit address ss from the scratch pad memory
9. 0Bxss FETCH sX, ss using the 8 bit address ss from the scratch pad memory

Since we like to use distributed or Block RAM, we use also a synchronous data RAM for stable design behavior. Since instruction aka PROM changes with the rising edge, the DRAM should use the falling edge. If we like to simplify array processing, then we should also add the indirect addressing options of the STORE and FETCH also, i.e.,

10. 2Exss STORE sX, (sY) using the 8 bit address found in the Y register to store X register values in the scratch pad memory
11. 0Axss FETCH sX, (sY) using the 8 bit address found in the Y register to load the scratch pad memory value into register X

However, synthesize tools usually require a rising edge, and we therefore implement an inverter gate not_clk <= NOT clk as clock signal for the HDL definition; this way the synthesis tool will use predefined Block/Distributed RAM. The HDL code otherwise is similar to Trisc0 data memory from [MB14], and we have:

VHDL Code 7.3: DRAM Data Definition and VHDL Behavior Code

```
 1      ......
 2      -- Scratch Pad memory definition
 3       TYPE MEMD IS ARRAY(0 TO 255) OF SLVD;
 4       SIGNAL dram : MEMD;
 5      BEGIN
 6      mem_ena <= '1' WHEN op6 = store ELSE '0';
 7                                    -- Active for store only
 8      not_clk <= NOT clk;
 9      scratch_pad_ram: PROCESS (reset, not_clk, y0)
10      VARIABLE idma : U8;
11      BEGIN
12        idma := CONV_INTEGER(y0); -- force unsigned
13        IF reset = '1' THEN            -- Asynchronous clear
14          dmd <= (OTHERS => '0');
15        ELSIF rising_edge(not_clk) THEN
16          IF mem_ena = '1' THEN
17            dram(idma) <= x;  -- Write to RAM at falling clk edge
18          END IF;
19          dmd <= dram(idma);  -- Read from RAM at falling clk edge
20        END IF;
21       END PROCESS;
```

We test the data memory by a sequence of writing three values (5,6,7) into the DRAM locations 1,2,3 followed by reading the same values from memory into register 2. The following listing shows the PSM test program:

PSM Program 7.4: DRAM Write Followed by Read `testdmem.psm`

```
1    start:   LOAD   s1, 05    ; load test data 5 in register 1
2             STORE  s1, 01    ; store data at memory location 1
3             LOAD   s1, 06    ; load test data 6 in register 1
4             LOAD   s2, 02    ; load address in register 2
5             STORE  s1,(s2)   ; store data using indirect address
6             FETCH  s3, 01    ; check data memory at address 1
7             FETCH  s3,(s2)   ; check data using indirect address
8             JUMP start       ; start all over
```

The PSM file can be compiled with the Xilinx `kcpsm6.exe` (download from `www.xilinx.com/picoblaze`) or our `psm2hex` utility from Chap. 6 and then converted to HDL with `hex2prom.exe`. Figure 7.6 shows the MODELSIM simulation results; Verilog coding and VIVADO simulation can be found in Appendix A.

We see in the simulation after reset the writing of values 05 from register s(1) into memory location dram(1) that can be seen after ca. 50 ns. Then the value 6 is written using the address specified in in register s(2) into memory location dram(2). After ca. 100 ns, the value 06 is available in dram(2). The immediate and indirect read from dram(1) and dram(2), respectively, is accomplished in the 130–170 ns time interval. The values are stored in register s(3). Then the unconditional jump follows such that the program starts from instruction zero.

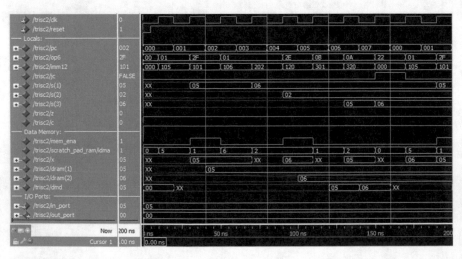

Fig. 7.6 TRISC2 data memory test routine. Immediate and indirect memory write are followed by reading the same locations back into register s3

7.5   PICOBLAZE Synthesizable Architecture with Link Control

A serious concern with small on-chip softcore processors such as the PICOBLAZE is usually program size. A substantial reduction in program length can often be achieved when the microprocessor provides architecture support for subroutines. This is illustrated by Fig. 7.7. In a program without subroutine, we may find program code that is executed several times, like a "wait for xx ms" task. Now if we place these instructions in a subroutine and allow for the microprocessor to memorize the location, we call this function and return to this program location after the subroutine is completed; we can substantially reduce our program code length. In the assembler language, we need two additional instructions: A CALL that allows the pc to jump to the subroutine and also to store the current pc value in a so called link register (lreg) and a RETURN instruction at the end of the subroutine such we can continue with program execution at the pc location we have stored in the lreg register. So, the price we pay are some extra registers and control hardware. In SW two additional instructions for each function call are executed, but we can substantially reduce the program size, as indicated by the free memory part of Fig. 7.7b.

We may even allow the subroutine to call another subroutine; this is called *nesting loops* and will require a little more housekeeping of the pc values that we store in a small stack. Nesting loops are illustrated by Fig. 7.7c. The KCPSM6 uses a stack with 30 locations, i.e., we can build highly nested code without the fear the stack may overflow. This is indeed a large stack compared with typical PDSPs that usually allow much less nesting such as typically 4 nested loops in the Analog Device ADSP21xx PDSP. From the implementation standpoint, we need to add another

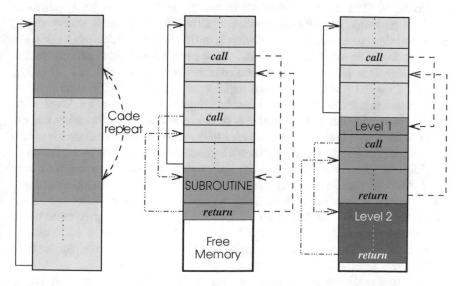

Fig. 7.7 Program organization with/without subroutine support. (**a**) In-line program without subroutine. (**b**) Program using subroutine. (**c**) Nesting of subroutines

register file (similar to the one we use for the ALU operations) to store the pc addresses and an up/down counter (lcount) that points to the current return address. To test our new hardware additions, we use a subroutine nesting and now use the previous DRAM memory access as task in the subroutines, i.e., we write three values (5, 6, 7) into the DRAM locations 1, 2, 3 followed by reading the same values from memory into register 2. However, the writing is now done in the nested subroutines. This is not a typical use of subroutines and only is used as a test case for demonstration of the pc modification necessary. The program listing for the subroutine test is shown next.

PSM Program 7.5: Subroutine Nesting Test Program

```
 1    start:  INPUT s1, 00  ; read switches
 2            CALL level1   ; call the subroutine 1
 3            FETCH s2, 01  ; check data memory at address 1
              FETCH s2, 02  ; check data memory at address 2
 4            FETCH s2, 03  ; check data memory at address 3
 5            OUTPUT s2, 00 ; write general LEDs
 6            JUMP start    ; start all over
 7    level3: STORE s1, 03  ; store to data memory location 3
 8            RETURN
 9    level2: STORE s1, 02  ; store to data memory location 2
10            ADD   s1, 01  ; load test data 7 in register 1
11            CALL level3   ; call the subroutine 3
              RETURN
12    level1: STORE s1, 01  ; store to data memory location 1
13            ADD   s1, 01  ; load test data 6 in register 1
14            CALL level2   ; call the subroutine 2
15            RETURN
```

To simplify the simulation using test bench, we may also consider adding additional ports to the ENTITY that automatically will show up in the simulation for both, function and timing simulation. Remember that not all signals are available after synthesis, and we only are guaranteed to see all signals in the I/O port list. On the other side when implementing using a development board, we do not want additional pins assigned to our local signals and we then should put a comment sign (-- in VHDL or // in Verilog) in front of the I/O port that the board logic is not confused. The VIVADO behavior VHDL simulation for the TRISC2 nesting subroutines is shown in Fig. 7.8. VIVADO was used just to demonstrate that our coding is device and/or vendor independent and can be synthesized with both tool sets. In order to demonstrate that our design work for many synthesis tools, the simulation shown is based on the Zynq test bench design.

Additional local seven signals for the link stack and the data memory have been added. All other signals are part of the I/O ports of the TRISC2 component and will be displayed by default in both functional and timing simulations. We write the three values (5, 6, 7) into the DRAM locations 1, 2, 3 followed by reading the same values from memory into register 2 (see s2_OUT[7:0] simulation 270 ns – 320 ns).

The register lcount shows the nesting level, while the first three register from the link stack lreg[0...2] show the pc return address values.

With these additional test ports, our final design is shown in our final TRISC2 HDL listing. An overview of the final implemented architecture is given in Fig. 7.9.

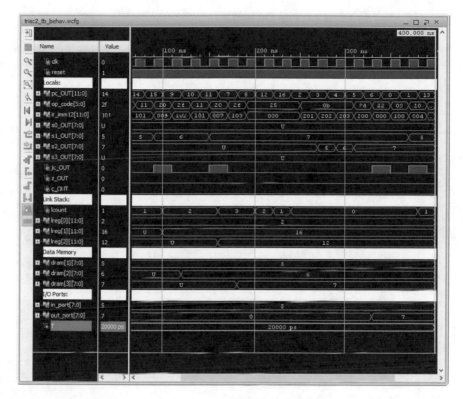

Fig. 7.8 VIVADO VHDL simulation of the TRISC2 nesting subroutines

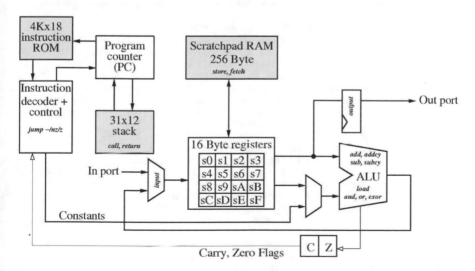

Fig. 7.9 The final implemented PICOBLAZE/KCPSM6 aka TRISC2 core. Instruction (in italic) mainly associated with major hardware units. Substantial memory blocks (excluding registers) in gray

VHDL Code 7.6: TRISC2 (Final Design)

```
1    -- Title: T-RISC 2 address machine
2    -- Description: This is the top control path/FSM of the
3    -- T-RISC, with a single 3 phase clock cycle design
4    -- It has a two-address type instruction word
5    -- implementing a subset of the KCPSM6 aka PicoBlaze v6 architecture
6    -- =================================================================
7    -- Xilinx modifications:
8    -- Use only STD_LOGIC or SLV and no generic in I/O
9    -- Modify code for jc_out (Boolean)
10   -- No conversion to/from integer using CONV_INTEGER(id)
11   -- or CONV_STD_LOGIC_VECTOR(id, width) required
12   -- =================================================================
13   LIBRARY ieee; USE ieee.std_logic_1164.ALL;
14
15   PACKAGE n_bit_type IS               -- User defined types
16     SUBTYPE U8 IS INTEGER RANGE 0 TO 255;
17     SUBTYPE SLVA IS STD_LOGIC_VECTOR(11 DOWNTO 0); -- address prog. mem.
18     SUBTYPE SLVD IS STD_LOGIC_VECTOR(7 DOWNTO 0);  -- width data
19     SUBTYPE SLVD1 IS STD_LOGIC_VECTOR(8 DOWNTO 0); -- width data + 1 bit
20     SUBTYPE SLVP IS STD_LOGIC_VECTOR(17 DOWNTO 0); -- width instruction
21     SUBTYPE SLV6 IS STD_LOGIC_VECTOR(5 DOWNTO 0);  -- full opcode size
22     SUBTYPE SLV5 IS STD_LOGIC_VECTOR(4 DOWNTO 0); -- reduced opcode size
23   END n_bit_type;
24
25   LIBRARY work;
26   USE work.n_bit_type.ALL;
27
28   LIBRARY ieee;
29   USE ieee.STD_LOGIC_1164.ALL;
30   USE ieee.STD_LOGIC_arith.ALL;
31   USE ieee.STD_LOGIC_unsigned.ALL;
32   -- =================================================================
33   ENTITY trisc2 IS
34    PORT(clk      : IN  STD_LOGIC; -- System clock
35         reset    : IN  STD_LOGIC; -- Active low asynchronous reset
36         in_port  : IN  STD_LOGIC_VECTOR(7 DOWNTO 0); -- Input port
37         out_port : OUT STD_LOGIC_VECTOR(7 DOWNTO 0) -- Output port
38   -- The following test ports are used for simulation only and
39   -- should be comment otherwise not to interfere with pin of the boards
40   --     s0_OUT   : OUT STD_LOGIC_VECTOR(7 DOWNTO 0);  -- Register 0
41   --     s1_OUT   : OUT STD_LOGIC_VECTOR(7 DOWNTO 0);  -- Register 1
42   --     s2_OUT   : OUT STD_LOGIC_VECTOR(7 DOWNTO 0);  -- Register 2
43   --     s3_OUT   : OUT STD_LOGIC_VECTOR(7 DOWNTO 0);  -- Register 3
44   --     jc_OUT   : OUT STD_LOGIC;    -- Jump condition flag
45   --     me_ena   : OUT STD_LOGIC;    -- Memory enable
46   --     z_OUT    : OUT STD_LOGIC;    -- Zero flag
47   --     c_OUT    : OUT STD_LOGIC;    -- Carry flag
48   --    pc_OUT    : OUT STD_LOGIC_VECTOR(11 DOWNTO 0); -- Program counter
49   --    ir_imm12  : OUT STD_LOGIC_VECTOR(11 DOWNTO 0);-- Immediate value
50   --    op_code   : OUT STD_LOGIC_VECTOR(5 DOWNTO 0)); -- Operation code
51   END;
52   -- =================================================================
53   ARCHITECTURE fpga OF trisc2 IS
54   -- Define GENERIC to CONSTANT for _tb
55     CONSTANT WA : INTEGER := 11;  -- Address bit width -1
56     CONSTANT WR : INTEGER := 3;   -- Register array size width -1
```

```
57       CONSTANT WD : INTEGER := 7;      -- Data bit width -1
58
59       COMPONENT rom4096x18 IS
60       PORT (clk   : IN STD_LOGIC;        -- System clock
61             reset : IN STD_LOGIC;        -- Asynchronous reset
62             pma   : IN STD_LOGIC_VECTOR(11 DOWNTO 0); -- Program mem.add.
63             pmd   : OUT STD_LOGIC_VECTOR(17 DOWNTO 0));--Program mem. data
64       END COMPONENT;
65
66       SIGNAL op6  : SLV6;
67       SIGNAL op5  : SLV5;
68       SIGNAL x, y, imm8, dmd : SLVD;
69       SIGNAL x0, y0 : SLVD1;
70       SIGNAL rd, rs : INTEGER RANGE 0 TO 2**(WR+1)-1;
71       SIGNAL pc, pc1, imm12 : SLVA; -- program counter, 12 bit aaa
72       SIGNAL pmd, ir   : SLVP;
73       SIGNAL eq, ne, mem_ena, not_clk : STD_LOGIC;
74       SIGNAL jc         : boolean;
75       SIGNAL z, c, kflag : STD_LOGIC; --zero, carry, and imm flags
76
77     -- OP Code of instructions:
78     -- The 5 MSBs for ALU operations (LSB is immidiate flag)
79       CONSTANT add     : SLV5 := "01000"; -- X10/1
80       CONSTANT addcy   : SLV5 := "01001"; -- X12/3
81       CONSTANT sub     : SLV5 := "01100"; -- X18/9
82       CONSTANT subcy   : SLV5 := "01101"; -- X1A/B
83       CONSTANT opand   : SLV5 := "00001"; -- X02/3
84       CONSTANT opor    : SLV5 := "00010"; -- X04/5
85       CONSTANT opxor   : SLV5 := "00011"; -- X06/7
86       CONSTANT load    : SLV5 := "00000"; -- X00/1
87     -- 5 bits for I/O and scratch Pad RAM operations
88       CONSTANT store   : SLV5 := "10111"; -- X2E/F
89       CONSTANT fetch   : SLV5 := "00101"; -- X0A/B
90       CONSTANT opinput : SLV5 := "00100"; -- X08/9
91       CONSTANT opoutput : SLV5 := "10110"; -- X2C/D
92     -- 6 Bits for all other operations
93       CONSTANT jump    : SLV6 := "100010"; -- X22
94       CONSTANT jumpz   : SLV6 := "110010"; -- X32
95       CONSTANT jumpnz  : SLV6 := "110110"; -- X36
96       CONSTANT call    : SLV6 := "100000"; -- X20
97       CONSTANT opreturn : SLV6 := "100101"; -- X25
98
99     -- Scratch Pad memory definition
100      TYPE MTYPE IS ARRAY(0 TO 255) OF SLVD;
101      SIGNAL dram : MTYPE;
102
103    -- Register array definition
104      TYPE RTYPE IS ARRAY(0 TO 15) OF SLVD;
105      SIGNAL s : RTYPE;
106
107    -- Link Register stack
108      TYPE LTYPE IS ARRAY(0 TO 30) OF SLVA;
109      SIGNAL lreg : LTYPE;
110      SIGNAL lcount : INTEGER RANGE 0 TO 30;
111
```

```
112    BEGIN
113
114      P1: PROCESS (op6, reset, clk) -- FSM of processor
115      BEGIN -- store in register ?
116          IF reset = '0' THEN
117             pc <= (OTHERS => '0');
118             lcount <= 0;
119          ELSIF falling_edge(clk) THEN
120             IF op6 = call   THEN
121                lreg(lcount) <= pc1; -- store link register
122                lcount <= lcount +1;
123             END IF;
124             IF op6 = opreturn THEN
125                pc <= lreg(lcount-1); -- Use next address after call/return
126                lcount <= lcount -1;
127             ELSIF jc THEN
128                pc <= imm12; -- any jumps that use 12 bit immediate aaa
129             ELSE
130                pc <= pc1;  -- Usual increment
131             END IF;
132          END IF;
133      END PROCESS p1;
134      pc1 <= pc + "000000000001";
135      jc <= (op6=jumpz AND z='1') OR (op6=jumpnz AND z='0')
136                                OR (op6=jump) OR (op6=call);
137
138      -- Mapping of the instruction, i.e., decode instruction
139      op6 <= ir(17 DOWNTO 12);  -- Full Operation code
140      op5 <= ir(17 DOWNTO 13);  -- Reduced Op code for ALU ops
141      kflag <= ir(12);          -- Immediate flag 0= use reg. 1= use kk;
142      imm8  <= ir(7 DOWNTO 0);    -- 8 bit immediate operand
143      imm12 <= ir(11 DOWNTO 0);    -- 12 bit immediate operand
144      rd <= CONV_INTEGER('0' & ir(11 DOWNTO 8));
145                               -- Index destination/1. source reg.
146      rs   <= CONV_INTEGER('0' & ir(7 DOWNTO 4)); -- Index 2. source reg.
147      x <= s(rd); -- first source ALU
148      x0 <= '0' & x; -- zero extend 1. source
149      y <= imm8 when kflag='1'
150          else s(rs); -- second source ALU
151      y0 <= '0' & y; -- zero extend 2. source
152
153      prog_rom: rom4096x18       -- Instantiate a Block RAM
154      PORT MAP (clk   => clk,   -- System clock
155                reset => reset,  -- Asynchronous reset
156                pma   => pc,    -- Program memory address
157                pmd   => pmd);   -- Program memory data
158      ir <= pmd;
159
160      mem_ena <= '1' WHEN op5 = store ELSE '0';  -- Active for store only
161      not_clk <= NOT clk;
162      scratch_pad_ram: PROCESS (reset, not_clk, y0)
163      VARIABLE idma : U8;
164      BEGIN
165        idma := CONV_INTEGER(y0); -- force unsigned
166        IF reset = '0' THEN        -- Asynchronous clear
167           dmd <= (OTHERS => '0');
```

```
168          ELSIF rising_edge(not_clk) THEN
169            IF mem_ena = '1' THEN
170              dram(idma) <= x;   -- Write to RAM at falling clk edge
171            END IF;
172            dmd <= dram(idma);   -- Read from RAM at falling clk edge
173          END IF;
174        END PROCESS;
175
176        ALU: PROCESS (op5, op6, x0, y0, c, in_port, dmd, reset, clk)
177        VARIABLE res: STD_LOGIC_VECTOR(8 DOWNTO 0);
178        VARIABLE z_new, c_new : STD_LOGIC;
179        BEGIN
180          CASE op5 IS
181            WHEN add      =>   res  := x0 + y0;
182            WHEN addcy    =>   res  := x0 + y0 + c;
183            WHEN sub      =>   res  := x0 - y0;
184            WHEN subcy    =>   res  := x0 - y0 - c;
185            WHEN opand    =>   res  := x0 AND y0;
186            WHEN opor     =>   res  := x0 OR y0;
187            WHEN opxor    =>   res  := x0 XOR y0;
188            WHEN load     =>   res  := y0;
189            WHEN fetch    ->   res  := '0' & dmd;
190            WHEN opinput  =>   res  := '0' & in_port;
191            WHEN OTHERS   ->   res  := x0; -- keep old
192          END CASE;
193          IF res = 0 THEN  z_new := '1'; ELSE z_new := '0'; END IF;
194          c_new := res(8);
195          IF reset = '0' THEN               -- Asynchronous clear
196            z <= '0'; c <= '0';
197            out_port <= (OTHERS -> '0');
198          ELSIF rising_edge(clk) THEN
199            CASE op5 IS            -- Compute the new flag values
200              WHEN addcy | subcy => z <- z AND z_new;
201                                    c <= c_new;
202                                  -- carry from previous operation
203              WHEN add | sub =>    z <= z_new; c <= c_new;
204                                         -- No carry
205              WHEN opor | opand | opxor =>  z <= z_new; c <= '0';
206                                         -- No carry; c=0
207              WHEN OTHERS =>    z <= z; c <= c; -- keep old
208            END CASE;
209            s(rd) <= res(7 DOWNTO 0); -- store alu result;
210            IF op5 = opoutput THEN out_port <= x; END IF;
211          END IF;
212        END PROCESS ALU;
213
214        -- Extra test pins:
215   --  pc_OUT <= pc; ir_imm12 <= imm12; op_code <= op6; -- Program
216   --  --jc_OUT <= jc; -- Control signals
217   --  jc_OUT <= '1' WHEN jc ELSE '0';  -- Xilinx modified
218   --  me_ena <= mem_ena; -- Control signals
219   --  z_OUT <= z; c_OUT <= c; -- ALU flags
220   --  s0_OUT <= s(0); s1_OUT <= s(1);      -- First two register elements
221   --  s2_OUT <= s(2); s3_OUT <= s(3);      -- Next two register elements
222
223   END fpga;
```

Compared with the initial PicoBlaze design, we see a few additions to the architecture and instruction set as well as a few modifications that allow to run the code under Quartus and Vivado. These Xilinx modifications are shown in the header (code lines 7–11). We see in the coding first the package with the subtype definition (lines 15–23). The ENTITY (lines 33–52) should for final synthesis only include ports for input and output used by the boards. Additional port can be used for verification, but not in the final design to avoid any damage to the boards. A couple of additional operations have been added to Trisc2 ISA and are listed as constant values (lines 77–97). The P1 first PROCESS in the ARCHITECTURE body FSM hosts now also the link stack control (lines 114–135). Then *decoding* of the instruction word into separate components takes place (lines 138–151). The program memory is instantiated via external ROM block that has been generated with the Xilinx kcpsm6.exe (download from www.Xilinx.com/picoblaze) or our psm2hex.exe and the program hex2prom.exe (lines 153–158). New is the scratch pad memory shown in lines 160–174. All operations including LOAD, INPUT, and OUPUT that require an update of the register array are included in ALU PROCESS (lines 176–212). Extra test pins are used during testing of new instructions but should not be used for the final synthesis that is downloaded to the board.

The synthesis results for the Trisc2 are shown in Table 7.4. The second and third columns show Xilinx Vivado synthesis results for VHDL and Verilog, respectively. The fouth and fifth columns show Altera Quartus synthesis results for VHDL and Verilog, respectively. In Xilinx Vivado the scratch pad RAM can be implemented with 1/2 RAMB36, i.e., one RAMB18 block as used in Verilog synthesis. Stack and register file each can be built with eight LUTs distributed RAM as in the VHDL synthesis. The ROM is often also implemented with LUTs since the size is small. We should see BRAM use for PROM as programs get longer. If distributed RAM can be used, we should see a substantial resource reduction as can be seen from comparison of VHDL and Verilog synthesis results. From the Altera VHDL synthesis report, we notice that due to the small program ROM only one

Table 7.4 Trisc2 synthesis results for Altera and Xilinx tools and devices

	Vivado 2015.1		Quartus 15.1	
Target device	Zynq 7K xc7z010t-1clg400		Cyclone V 5CSEMA5F31C6	
HDL used	VHDL	Verilog	VHDL	Verilog
LUT/ALM	192	299	323	278
Used as Dist. RAM	48	0	16	0
Block RAM RAMB18	0	0.5		
or M10K			1	3
DSP blocks	0	0	0	0
HPS	0	0	0	0
Fmax/MHz	113.64	85.47	95.37	98.35
Compile time:	3:35	3:40	4:45	4:58

Block RAM block has been used for register array, but no block RAM was used for DRAM, PROM, or the link stack. The ALM count is lower for the Verilog synthesis results: Here two of the 16×8 register arrays and the scratch pad RAM are built with M10K blocks reducing the ALM count. PROM again due to the small size is ALM-based.

We are finally having a working microprocessor with most of the KCPSM6 features implemented, such as register loading, logical, arithmetic operation, input and output, scratch pad memory, jump, and call. Missing from the complete instruction set are those group we did not need for our tasks so fare such as test and compare, shift and rotate, register bank selection, interrupt handling, and version control. This is left to the reader as exercises at the end of the chapter.

Review Questions and Exercises

Short Answer

7.1. Name two softcore 8-bit microprocessors each available for Altera and Xilinx.
7.2. What are maximum data and program memory size of the newest KCPSM6?
7.3. What is the purpose of the link control stack?
7.4. What are nested loops?
7.5. Why is the synthesis tool sometimes using Block RAM and sometimes LUT/ALM to implement program memory?

Fill in the Blank

7.6. The PICOBLAZE processor was developed by Application Engineer named_____ from Xilinx.
7.7. The first _____ bits of the PICOBLAZE opcode are used to encode the instruction.
7.8. The _____ utility is used by Altera/TerASIC to generate the I/O file for the pins of the project.
7.9. The constrain file with extension *. _____ are used by XILINX VIVADO to make pin assignments.
7.10. The scratch pad memory in the Xilinx PICOBLAZE can have _____ bytes.

True or False

7.11. _____ The PICOBLAZE is the vendor softcore with the smallest LE/Slice requirement.

7.12. _____ The PICOBLAZE KCPSM6 is the sixth generation of PICOBLAZE.

7.13. _____ The 8-bit microprocessor by Altera is called NIOS.

7.14. _____ The TRISC2 developed in this chapter implement the complete ISA of the PICOBLAZE.

Projects and Challenges

7.15. Doubling the Program PROM of the PICOBLAZE; how will this influence instruction coding, program length, hardware resource (LE/memory/DSP), and μP speed?

7.16. Double the Program PROM and modify the HDL of the TRISC2. Modify the JUMP from absolute to relative, and check your program with the `testd-nest` test program.

7.17. If we modify PICOBLAZE from a 8-bit data to 32-bit data processor; how will this influence instruction coding, program length, hardware resource (LE/memory/DSP), and μP speed?

7.18. Modify the TRISC2 to 32-bit data processor and verify with the flash program.

7.19. Use the `hex2prom` utility to identify instructions not supported in TRISC2 to run the factorial program.

7.20. Develop a TRISC2 program to implement a multiplication of two 8-bit numbers via repeated additions. Test you program by reading the SW values and computing q = x * x.

7.21. Add the missing instruction from the ISA to run the factorial program on the hardware. Test the single new instructions and then run the factorial program.

7.22. Develop a PSM assembler program to implement a left, right car turn signal and emergency light similar to the Ford Mustang 00X, 0XX, and XXX for a left turn where 0 is off and X is LED on. Use X00, XX0, and XXX for right turn signal. Use the slider switch for the left/right selection. For emergency light, both switches are on. The turn signal sequence should repeat once per second. See `mustangQT.MOV` or `mustangWMP.MOV` for a demo.

7.23. Develop a PSM program that has a random number generator, using ADD and XOR instructions. Display the random number on the LEDs. What is the period of your random sequence?

7.24. Develop a seven-segment display and use the toggle PSM to count up in seconds. This will require to modify the pin assignment that not only LEDs, but also the seven segments are driven by the PICOBLAZE output port.

7.25. Build a stopwatch that counts up in second steps. Use three buttons: start clock, stop or pause clock, and reset clock. Using the LEDs or the seven-segment display from the previous exercise 7.24 for display. See `stop_watch_ledQT.MOV` or `stop_watch_ledWMP.MOV` for a demo.

7.26. Use the pushbuttons to implement a reaction speed timer. Turn on one of four LEDs, and measure the time it takes until the associate button is pressed. Display the measurement after ten tries on LED or seven segment display.

7.27. Add an additional PICOBLAZE feature to the TRISC2 architecture or instruction such as test and compare, shift and rotate, register bank selection, interrupt handling, and version control. Add the HDL code and test with a PSM test program. Report the difference in synthesis (area and speed) compared to the original design.

References

[Cha03] K. Chapman, KCPSM3 Manual – PicoBlaze KCPSM3 8-bit Micro Controller for Spartan-3, Virtex-II and Virtex-II Pro, San Jose (2003)

[Cha14] K. Chapman, User Guide – PicoBlaze for Spartan-6, Virtex-6, 7-Series, Zynq and Ultra Scale Devices (KCPSM6), San Jose (2014)

[MB14] U. Meyer Base, *Digital Signal Processing with Field Programmable Gate Arrays*, 4st edn. (Springer, Heidelberg, 2014)., 930 pages

[Koc06] P. Kocik, PacoBlaze (2016), http://bleyer.org/pacoblaze/

[X387] Xilinx: PicoBlaze 8-bit microcontroller for CPLD devices, in Xilinx Application Note XAPP 387 San Jose (2003)

[X213] K. Chapman, PicoBlaze 8-bit microcontroller for Virtex-E and Spartan-II/IIE devices, in Xilinx Application Note XAPP 387 San Jose (2003)

[X627] K. Chapman, PicoBlaze 8-bit microcontroller for Virtex-II series devices, in Xilinx Application Note XAPP 627 San Jose (2003)

Chapter 8
Software Tools for the PICOBLAZE Softcore Microprocessor

Abstract This chapter gives an overview of a popular PICOBLAZE internal architecture from a programmer perspective only. In particular based on KCPSM6 architecture, the complete instructions set is discussed as well as all steps to develop program from C level, assembler level, or ISS view. This chapter does *not* require a detailed HDL knowledge as the previous chapter.

Keywords PicoBlaze · Software · Assembler · High-level language · 2-address machine · ISA coding · Ken Chapman Programmable state machine (KCPSM) · KCPSM6 · Instruction set simulator (ISS) · FIDEx · C-compiler · Pccomp · Recursive function · Loop function · psm2hex · Operation support · Data types · Assembler syntax rules · Small C compiler · Scratchpad RAM

8.1 Introduction

As an embedded software designer, you may be 100% satisfied with the PICOBLAZE initial architecture (and do not want to make any additions modification or have the HDL skills), and you may just need the details on the programming resources and assembler language details. This is the intend of this chapter: To have enough knowledge to take full advantage of the complete PICOBLAZE instructing set and development tools without being too concerned about the details of the HDL implementation. We will start with the programmer model and then talk about the assembler language details and finally have a look at the overall program development procedure.

The programming sources available in the KCPSM6 architecture can be summarized as follows:

- A $2^{12} \times$ 18-bit program memory (4K words)
- A 2-address machine with 2 banks of 16 registers each
- A 64-, 128-, or 256-byte data memory
- Flags for carry (C), zero (Z), interrupt (IE), and register bank (A/B)
- A 30-entry call/return stack for the pc

© Springer Nature Switzerland AG 2021   253
U. Meyer-Baese, *Embedded Microprocessor System Design using FPGAs*,
https://doi.org/10.1007/978-3-030-50533-2_8

The programming sources available in the synthesizable TRISC2 architecture developed in the last chapter can be summarized as follows:

- A $2^{12} \times 18$-bit program memory (4K words)
- A 2-address machine with 1 bank of 16 register
- A 256-byte data memory
- Flags for carry (C), zero (Z)
- A 31-entry call/return stack for the pc

There are three major professional development tools available to develop KCPSM6 and TRISC2 programs:

- The KCPSM6.exe assembler provided by Ken Chapman (www.xilinx.com/picoblaze)
- The instruction set simulator called FIDEx
- The C compiler for PICOBLAZE called PCCOMP

We had developed some experimental SW development tools in Chaps. 5 and 6 for PICOBLAZE, but these support only a reduced and not the full ISA of KCPSM6. In addition, the following utilities may be useful:

- The hex2prom.exe converts the HEX program file into a synthesizable VHDL file
- The psm2hex.exe program is a basic converter for TRISC2 written in FLEX to translate psm files into hex program files similar to the KCPSM6.exe assembler for the reduced instruction set

In the following we will have a closer look at these tools, including (not) implemented features, available documentations, and a few observations using these tools. Figure 8.1 gives an overview of how the different tools and utilities work together.

8.2   The KCPSM6 Assembler

The KCPSM6.exe can be downloaded free of charge from the Xilinx webpage (www.xilinx.com/picoblaze). It is a full-featured assembler that has a PSM file and a ROM_form.v(hd) as inputs and generates the necessary files to run PICOBLAZE in hardware. The KCPSM6 comes with a 124 pages user guide that has all details on each instruction, design flow recommendations, hardware reference information, and an information what is new compared to KCPSM3 [Cha03, Cha14]. It comes with some complete design examples such as UART, I2C, SPI, and XADC. For the assembler syntax we may consult an additional 1000+ lines reference syntax guide called all_kcpsm6_syntax.psm that can be used if in doubt about the assembler syntax and features. Finally, also a JTAG loader is provided that allows you to replace the current program ROM with a new program without a recompilation of the whole system which usually takes a substantial amount of time.

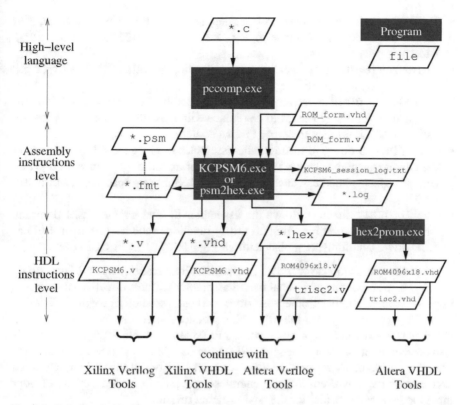

Fig. 8.1 Software development flow for Xilinx PicoBlaze and Trisc2

When writing PSM programs, we need to remember the following syntax rules and assembler directives:

- COMMENTS start with a semicolon and continue until the end of the line.
- INSTRUCTION and register names s0-sF are *not* case sensitive, e.g., sa=Sa=sA=SA.
- NAME such as labels, constant, strings, tables, or register name *are* case sensitive and can use the characters a-zA-Z0-9 and underscore, e.g., label start ≠ Start. A limitation is that a name cannot consist of hex values only or default register name s0-sF.
- The assembler is line but not space sensitive; you cannot write a single PSM instruction over multiple lines as in HDL or C/C++. Delimiter such as comma or semicolon need to be used to avoid error messages.
- Data and address values must match the required length, a hex constant kk such as 5 that need to be 8 bit long must be defined as 05, the zero cannot be omitted; 12-bit must use precisely 3 hex digits.
- It is possible to define 8-bit CONSTANT values placeholders, e.g., CONSTANT LED, 40.

- Constant values can be coded in Hex, decimal (use 'd) or binary (use 'b), char (use "") or as complement (use ~), e.g., `2B`, `43'd`, `00101011'b`, `"P"`, `~LED`.
- The default register names can be redefined using the `NAMEREG` directive, e.g., `NAMEREG s5, counter`.
- A `DEFAULT_JUMP` directive will replace the default `00000 => LOAD s0,s0` coding used for empty PROM to be replace with a `JUMP aaa` that the program can recover if it reaches an empty PROM address.
- The `ADDRESS` directive instructs the assembler to place the following code at the address specified. This is typically used for an interrupt service routine that start at a specified address, such as the end of the program space, e.g., `ADDRESS FF0`.
- The `INCLUDE` directive allows the assembler to load another PSM program. This is typically used for common utility functions such as 7 segment LED or LCD drivers, multiply, or global settings, e.g., `INCLUDE "util.psm."`

As we see the complete `KCPSM6.exe` (download from `www.Xilinx.com/picoblaze`) has some additional useful assembler features directives that should improve the readability of otherwise often verbose assembler programs. With an input file `project.psm`, the assembler produces three files `project.log`, `project.hex`, and `project.fmt`. If a ROM template file `ROM_form.vhd` is also present in the directory, then the PROM file `project.vhd` is also generated. The `*.log` file includes some additional useful statistics at the end on how often each instruction was used; it also generates a `*.fmt` formatted version of your input file you may consider for the next program revision.

Instructions can be grouped by alphabetical order, functionality, format, or operation coding. This is reviewed in the following. For complete details on each instruction, consult the "KCPSM User Guide" that comes with the distribution. The grouping by functionality let us build the following 12 groups:

1. Arithmetic (2×4): `ADD, ADDCY, SUB, SUBCY`
2. Logical (2×3): `AND, OR, XOR`
3. Jump (6): `JUMP`, conditional `JUMP Z/NZ/C/NC`, indirect `JUMP@`
4. Input and output (5): `INPUT/OUTPUT` direct, indirect, constant
5. Interrupt handling (4): `DISABLE/ENABLE  INTERRUPT, RETURNI DISABLE/ENABLE`
6. Register Bank Selection (2): `REGBANK A/B`
7. Register loading (4): `LOAD` and `STAR` direct or constant
8. Scratch Pad Memory (4): direct/indirect `STORE` and `FETCH`
9. Shift and Rotate (10): shift or rotate left/right using 0/1/arithmetic/carry
10. Subroutines (12): `CALL/RETURN` (un)conditional direct/indirect, `LOAD&RETURN`
11. Test and compare (8): `TEST` or `COMPARE` register/register, register/constant, with/without carry
12. Version control (1): `HWBUILD`

We count a total of 70 different op codes. Some share the same 6-bit MSB code such as all shift/rotate instruction (−9) and interrupts (−2), and ALU operation comes as R- and I-type instructions such that we from a syntax standpoint have 39 different instructions:

```
ADD, ADDCY, AND, CALL, CALL@, COMPARE, COMPARECY, DISABLE,
ENABLE, FETCH,  HWBUILD, INPUT, JUMP, JUMP@, LOAD, LOAD&RETURN,
OR,  OUTPUT, OUTPUTK, REGBANK , RETURN, RETURNI, RL, RR, SL0,
SL1, SLA, SLX, SR0, SR1, SRA, SRX, STAR, STORE, SUB, SUBCY, TEST,
TESTCY, XOR
```

Now finally let us have a look at the opcode PSM coding, format, operation type, and the function executed of all 70 instructions of KCPSM6 sorted by the opcode (Table 8.1). Coding abbreviation aaa, kk, ppp, p, ss, X, and Y were introduced in Sect. 7.1.

As can be seen from the table, the following six 6-bit opcodes are not used: 15, 23, 27, 2A, 33, 3B, and 3F. We see that most 6-bit opcodes are used in particular there is no easy way to add a multiply R/I pair to the set. Even if we just add the R-type (and use an extra LOAD for the I-format), that would require to redevelop the KCPSM6.exe (download from www.Xilinx.com/picoblaze) to support the new instruction since the assembler is not an open-source program.

8.3   Instruction Set Simulator for the PICoBLAZE

SW debugging and ISS have been reviewed in Chaps. 5 and 6. In-circuit debugging or the use of GDB is currently not available for PICoBLAZE, but ISS had been developed even for early PICoBLAZE version. Let us now have a look at a detailed example of ISS available for the PICoBLAZE processor. Xilinx does not provide an ISS for this processor, and we need to look at third-party products. For many years the pBLAZEIDE was the recommended ISS by Xilinx; however, the support has been discontinued and the newest PICoBLAZE v6 is not supported by the pBLAZEIDE ISS. But another ISS from Fautronix GmbH in Reutlingen, Germany, has become available that support the newest v6 PICoBLAZE. The licensing for the ISS is flexible ranging from a free license with 150 code line limit, over one processor @ E98 (≈$110 USD), to full license for E798 (≈$877 USD). PPT slide, tutorials, ISA help pages, and example files are free for download from https://www.fautronix.com/en/fidex. Figure 8.2 gives an overview of the ISS running a flash.psm example discussed in Chap. 7 that toggle the LEDs of the FPGA board in 1 second intervals. When using the FIDEx, we make several observations:

- The simulator is easily installed, setting up a project is simple, and running the first example is accomplished within a few minutes (No FPGA software, or FPGA board required).

Table 8.1 KCPSM6 ISA coding

OP-CODE	ASM coding	Format	Operation type	Operation description
00XY0	LOAD sX,sY	R	Register loading	Load register sX with register sY
01Xkk	LOAD sX,kk	I	Register loading	Load register sX with 8 bit constant
02XY0	AND sX,sY	R	Logical	sX = sX AND sY
03Xkk	AND sX,kk	I	Logical	sX = sX AND kk
04XY0	OR sX,sY	R	Logical	sX = sX OR sY
05Xkk	OR sX,kk	I	Logical	sX = sX OR kk
06XY0	XOR sX,sY	R	Logical	sX = sX XOR sY
07Xkk	XOR sX,kk	I	Logical	sX = sX XOR kk
08XY0	INPUT sX,(sY)	R	Input and output	Read indirect port (sY) store in sX
09Xpp	INPUT sX,pp	I	Input and output	Read port pp and store in sX
0AXY0	FETCH sX,(sY)	R	Scratch pad mem.	Indirect DRAM (sY) read/store in sX
0BXss	FETCH sX, ss	I	Scratch pad mem.	DRAM read sX = DRAM(ss)
0CXY0	TEST sX,sY	R	Test and compare	t=sX AND sY; update Z, C odd #1
0DXkk	TEST sX,kk	I	Test and compare	t=sX AND kk; update Z, C odd #1
0EXY0	TESTCY sX,sY	R	Test and compare	""; Z=Z_{new} AND Z_{old} C odd #1=C_{old}
0FXkk	TESTCY sX,kk	I	Test and compare	""; Z=Z_{new} AND Z_{old} C odd #1=C_{old}
10XY0	ADD sX,sY	R	Arithmetic	sX = sX + sY
11Xkk	ADD sX,kk	I	Arithmetic	sX = sX + kk
12XY0	ADDCY sX,sY	R	Arithmetic	sX = sX + sY + c
13Xkk	ADDCY sX,kk	I	Arithmetic	sX = sX + kk + c
14X00	SLA sX	R	Shift and rotate	Shift left add carry sX = {sX<<1, C}
14X02	RL sX	R	Shift and rotate	Rotate left sX = {sX<<1, MSB}
14X04	SLX sX	R	Shift and rotate	Double LSB: sX = {sX<<1, LSB}
14X06	SL0 sX	R	Shift and rotate	Shift left add 0: sX = {sX<<1, 0}
14X07	SL1 sX	R	Shift and rotate	Shift left add 1: sX = {sX<<1, 1}
14X08	SRA sX	R	Shift and rotate	Shift right add carry sX = {C,sX>>1}
14X0A	SRX sX	R	Shift and rotate	Double MSB sX = {MSB, sX>>1}
14X0C	RR sX	R	Shift and rotate	Rotate right sX = {LSB, sX>>1}
14X0E	SR0 sX	R	Shift and rotate	Shift right add 0: sX = {0, sX>>1}
14X0F	SR1 sX	R	Shift and rotate	Shift right add 1: sX = {1, sX>>1}
14X80	HWBUILD sX	R	Version control	Set sX=hw from ENTITY GENERIC
15				
16XY0	STAR sX,sY	R	Register loading	Copy sX to inactive bank sY
17Xkk	STAR sX,kk	I	Register loading	Store kk in inactive register sX
18XY0	SUB sX,sY	R	Arithmetic	sX = sX – sY
19Xkk	SUB sX,kk	I	Arithmetic	sX = sX – kk
1AXY0	SUBCY sX,sY	R	Arithmetic	sX = sX – sY – c
1BXkk	SUBCY sX,kk	I	Arithmetic	sX = sX – kk – c

(continued)

Table 8.1 (continued)

OP-CODE	ASM coding	Format	Operation type	Operation description
1CXY0	COMPARE sX,sY	R	Test and compare	test=sX-sY: zc=-1 if sX<sY zc=1- if sX=sY;zc=00 sX>sY
1DXkk	COMPARE sX,kk	I	Test and compare	test=sX-kk: zc=-1 if sX<kk zc=1- if sX=kk;zc=00 sX>kksY
1EXY0	COMPARECY sX,sY	R	Test and compare	test=sX-sY-C: C=C_{new}; Z=Z_{new} AND Z_{old}
1FXkk	COMPARECY sX,kk	I	Test and compare	test=sX-kk-C: C=C_{new}; Z=Z_{new} AND Z_{old}
20aaa	CALL aaa		Subroutines	Call subroutine PC=aaa
21Xkk	LOAD&RETURN sX,kk	I	Subroutines	Return from subroutine PC=stack(ptr) and load sX=kk
22aaa	JUMP aaa	J	Jump	Set program counter PC = aaa
23				
24XY0	CALL@ (sX,sY)	R	Subroutines	Call subroutine indirect PC={sX,sY}
25000	RETURN	J	Subroutines	Return from subroutine PC=stack(ptr)
26XY0	JUMP@ (sX,sY)	R	Jump	Set program counter PC={sX,sY}
27				
28000	DISABLE INTERRUPT	R	IRQ handling	Set flag IE = 0
28001	ENABLE INTERRUPT	R	IRQ handling	Set flag IE = 1
29000	RETURNI DISABLE	R	IRQ handling	IRQ routine complete return PC=stack(ptr) and set IE=0
29001	RETURNI ENABLE	R	IRQ handling	IRQ routine complete return PC=stack(ptr) and set IE=1
2A				
2Bkkp	OUTPUTK kk,p	J	Input and output	Write constant kk to port 0p
2CXY0	OUTPUT sX,(sY)	I	Input and output	Write sX to port specified in sY
2DXpp	OUTPUT sX,pp	I	Input and output	Set output port to sX
2EXY0	STORE sX,(sY)	R	Scratch pad mem.	Store in DRAM(sY)=sX
2FXss	STORE sX, ss	I	Scratch pad mem.	Store in DRAM(ss)=sX
30aaa	CALL Z, aaa	J	Subroutines	If flag Z=1 call subroutine at PC=aaa
31000	RETURN Z	J	Subroutines	Return from subroutine if flag Z=1
32aaa	JUMP Z, aaa	J	Jump	if flag Z=1 set PC=aaa
33				
34aaa	CALL NZ, aaa	J	Subroutines	If flag Z=0 call subroutine at PC=aaa
35000	RETURN NZ	J	Subroutines	Return from subroutine if flag Z=0
36aaa	JUMP NZ, aaa	J	Jump	If flag Z=0 set PC=aaa

(continued)

Table 8.1   (continued)

OP-CODE	ASM coding	Format	Operation type	Operation description
37000	REGBANK A	R	Bank selection	Use register bank A
37001	REGBANK B	R	Bank selection	Use register bank B
38aaa	CALL C, aaa	J	Subroutines	If flag C=1 call subroutine at PC=aaa
39000	RETURN C	J	Subroutines	Return from subroutine if flag C=1
3Aaaa	JUMP C, aaa	J	Jump	If flag C=1 set PC=aaa
3B				
3Caaa	CALL NC, aaa	J	Subroutines	If flag C=0 call subroutine at PC=aaa
3D000	RETURN NC	J	Subroutines	Return from subroutine if flag C=0
3Eaaa	JUMP NC, aaa	J	Jump	If flag C=0 set PC=aaa
3F				

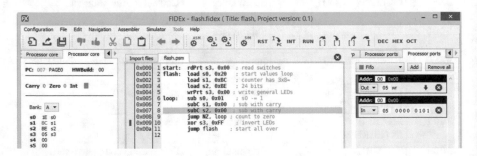

Fig. 8.2   The FIDEx ISS for Xilinx PICOBLAZE v6 running the `flash.psm` program

- The ISS updates registers (shown in the left panel) in each clock cycle, highlights the current code (shown in the center panel), lets you set the input port values (see right panel), and displays clock cycles and execution time (not shown). You can set breakpoints, run single steps, or until a breakpoint.
- Since FIDEx has the capability to simulate several ISA, it uses an internal unified assembler language. Some of the instruction coding do match the PICOBLAZE v6 syntax such as ADD, XOR, LOAD, JUMP, or SUB, while others have different coding such as addC vs. ADDCY, subC vs. SUBCY, rdPrt vs. INPUT, and wrPrt vs. OUPUT for FIDEx vs. Xilinx PSM code. FIDEx will automatically make this transformation for you when loading a Xilinx `*.psm file` into the ISS. When adding files to your project, always use `File->Import files ...` and as `import filter` use Xilinx; otherwise, you may see many errors when compiling your code directly without these conversions.
- Simulation speed is not very high. On an i7 PC (2.4 GHz;12 GB RAM; Win 8.1Pro), the simulation of 100,000 clock cycles took 2 minutes and 31 seconds, so if we want to see the output of the `flash.psm` program toggle after 1 second, we will need to wait $50 \times 10^6/10^5 \times 2.5m = 500 \times 2.5m \approx 20$ hours!

8.4   C-Compiler for the PICOBLAZE

Xilinx does not provide a C/C++ compiler for this processor, and we need to look at third-party products. For many years Xilinx supported through their webpage the download of the PICOBLAZE C-compiler PCCOMP written by Francesco Poderico. The PCCOMP came with an 18-page user manual, several working design examples, and was free of charge. The C-compiler was not written from scratch or using BISON, instead the highly successful "Small C-Compiler" initially published as a journal tutorial on compiler writing by Ron Cain [Cai80] in the 1980 and then later to be published in more professional manor as book written by James Hendrix [Hen90]. There are a couple of reasons why this seemed to be a good approach:

- The Small C-Compiler was developed of an Intel 8080/8086 μP, which as the PICOBLAZE is a two-address machine; GNU C/C++ compiler typically are developed and optimized for 3-address machines.
- The Small C-Compiler uses a C language subset that fit well with PICOBLAZE hardware: only char (8-bit) and int (16 bit) data types are supported; no 32-bit float or 64-bit double are supported.
- The Small C-Compiler had been successfully ported to other μP such as the Zilog Z80, VAX, or Motorola's 6809 microcontroller; see http://www.cpm.z80.de/small_c.html.

As the name "Small C-Compiler" suggests, the compiler we are talking about is not a full-featured ANSI C-Compiler that supports all possible coding styles; nevertheless, it has a substantial set of features that may be quite useful for software development. Let us go through the most important features: data types, operation support, control flow, and functions in the next section.

Data Types in PCCOMP

Much can be learned from the functionality of a compiler by looking at the supported keywords. That indeed can be also the good starting point for the evaluation of data types supported. There are 32 reserved words in ANSI C, and it is usually a good idea to order these by the 4 major classes: data types, data attributes, control flow, and storage class. Let us repeat these from our discussion in Chap. 5 and "strike through" not supported keywords in Small C:

1. Data types: char, ~~double~~, ~~enum~~, ~~float~~, int, void
2. Data attributes: ~~long~~, ~~short~~, ~~signed~~, ~~struct~~, ~~typedef~~, ~~union~~, unsigned
3. Control flow: break, case, continue, default, do, else, for, goto, if, return, switch, while
4. Storage class: ~~auto~~, ~~const~~, extern, ~~register~~, ~~static~~, ~~volatile~~

As we see from the data type, we should use `char` whenever possible since we have an 8-bit processor; `int` will require always 16-bit double word processing. The type `void` can be used for functional return types. Only a single data attributes `unsigned` is supported. Here is a list of our data types in PCCOMP:

Type	unsigned char	char	unsigned int	int
Range	0...255	-128...127	0....65535	-32768...32767

PCCOMP has a strict scheme when register or memory are used, and attribute cannot be used to force a use of register for a counter variable as in other C/C++ compilers. Identifiers allow underscore such that `char x, x_, _x, _x_;` will compile fine, but identifiers like `money$` or `P!NK` (special character) or `4you` (number first) will not compile. Identifiers are restricted to 38-character length, i.e., `int name456789012345678901234567890123467;` is a valid identifier but `int name4567890123456789012345678901234567890123456789012345678;` will not compile. The compiler also does not allow to build complicated union or struct data types keeping data handling simple. That will also make the -> and "." operator obsolete as we will see in the next section.

PCCOMP *Operation Support*

The operation support in PCCOMP is substantial as precedence Table 8.2 shows. If we compare this with Table 5.4 from Chap. 5, we see that all operations are supported except the -> and "." used for `struct` data types that are not supported.

A verification of each operation can be done with a small test sequence such as:

C Program 8.1: Verification of operations with test `opc.c`

```
 1   // Check the support operation
 2   char i, x, y; // char/integer 8 bits
 3   char a[11]; // array of numbers
 4   void main(){
 5      x=5; y=14;
 6    // Arithmetic test first
 7      a[0]= x + y;
 8      a[1]= y - x;
 9   // Boolean test next
10      a[5]= (x & y);
11      a[6]= (x | y);
12   // Shift test
13      a[9]= x << 2;
14   }
```

Table 8.2   Operation support in Pccomp

Precedence	Associativity	Operator	Description
15	L	() []	Function call, scope, array/member access
14	R	! ~ - * & sizeof (*type cast*) ++x --x	(Most) unary operations, size, and type casts
13	L	* / %	Multiplication, division, modulo
12	L	+ −	Addition and subtraction
11	L	<< >>	Bitwise shift left and right
10	L	< <= > >=	Comparisons: less-than, ...
9	L	== !=	Comparisons: equal and not equal
8	L	&	Bitwise AND
7	L	^	Bitwise exclusive OR (XOR)
6	L	|	Bitwise inclusive (normal) OR
5	L	&&	Logical AND
4	L	||	Logical OR
3	R	?:	Conditional expression
2	R	= += -= *= /= %= &= |= ^= <<= >>=	Assignment operators
1	L	,	Comma operator

After compilation of the C Code, we can import the `*.PSM` file in our ISS and verify the memory contend (Fig. 8.3). The first array a[0] starts at 0×32, and we have:

- a[0] = 5 + 14 = 19 (memory location 0×32)
- a[1] = 14-5 = 9 (memory location 0×33)
- a[5] = 0101_2 AND 1110_2 = 0100_2 = 4 (memory location 0×37)
- a[6] = 0101_2 OR 1110_2 = 1110_2 = 15 (memory location 0×38)
- a[9] = 5 << 2 = 5 * 4 = 20 (memory location 0×3B)

An extended operation test is left to Exercise 8.32 at the end of the chapter.

If you use the free FIDEx ISS, make sure your program does not produce more than 150 lines of PSM code.

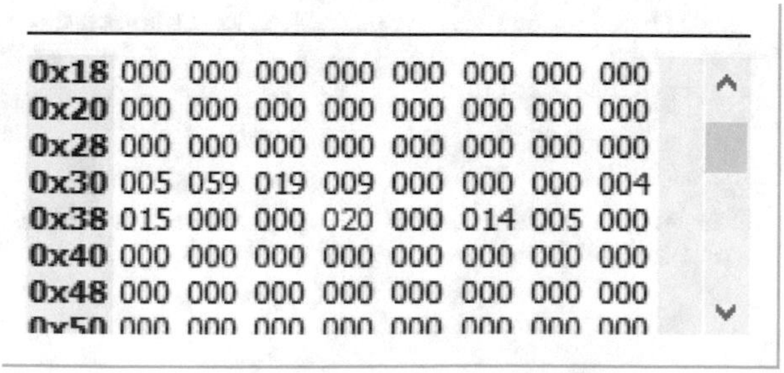

```
0x18 000 000 000 000 000 000 000 000   ^
0x20 000 000 000 000 000 000 000 000
0x28 000 000 000 000 000 000 000 000
0x30 005 059 019 009 000 000 000 004
0x38 015 000 000 020 000 014 005 000
0x40 000 000 000 000 000 000 000 000
0x48 000 000 000 000 000 000 000 000
0x50 000 000 000 000 000 000 000 000   v
```

Fig. 8.3 Memory contents after running the `opc.c` program 8.1 in ISS using `opc.psm`. Addresses are shown in hex, while the memory data displayed as decimal values

Control Flow Options in PCCOMP

One of the strengths of the Small C-Compiler is the support of all control instructions. You may argue why we have a while, for, and do loop when all three basically can do the same. But small C allows all three types. Close related with the question how to implement the loop is also the question what an efficient way is to access an array in a loop. A small example to build the sum of five array elements should show basic options.

C Program 8.2: Verification of loop function and array access with test loops.c

```
1    // Check supported loops and array access
2
3    char sum;// sum of the array
4    char i; // char/integer 8 bits
5    char a[5]; // array of numbers
6    char *ptr, *ptrlast; // address values
7
8    // Use only one loop type to count program length
9    //#define USEWHILE
10   //#define USEDO
11   #define USEFOR
12
13   void main(){
14
15     // Set test values
16     a[0]= 5;  a[1]= 10; a[2]= 15; a[3]=20; a[4]=25;
17     sum=0; // reset sum; should be 75 at the end
18
19   #ifdef USEWHILE
20     ptr =(char *) &a; i=0;
21     while (i<5) {
22     sum += a[i];
23     i++;
24     } // use standard array access: 98 lines PSM code
25   #endif;
26
27   #ifdef USEDO
28     ptr =(char *) &a; ptrlast = ptr+5;
29     do {          // increment array pointer: 92 lines PSM code
30     sum += *ptr++;
31     } while (ptr < ptrlast);
32   #endif;
33
34   #ifdef USEFOR
35     ptr =(char *) &a;  //increment array pointer: 89 lines PSM code
36     for (i=0; i<5; i++) sum += *ptr++;
37   #endif;
38   }
```

The first loop shows the use of the `while` loop together with standard array access (C code lines 20-24; PSM code length 98 lines). Next is a `do/while` sequence using post address increment access to the array (C code lines 28-31). Last is the `for` loop with a post address increment access to the array (C code line 35-36). Compiler directives are used that only one loop type is active, and we can measure the different program length. While the overall PSM codes are similar in length, it turns out that the last `for` loop type with address increment gives the shortest PSM code. The code required 418 clock cycles to complete in FIDEx ISS. The shortest program code is not necessary the fasted to compute the array sum. If we use a "loop unrolling," do not use a loop and write five instructions:

```
//Loop unrolling --> needs 115 lines/228 clock cycles
ptr = (char *) &a;
sum += *ptr++;
sum += *ptr++;
sum += *ptr++;
sum += *ptr++;
sum += *ptr;
```

This will complete the simulation in 228 clock cycles. The unrolled loop however has a longer PSM code of 115 code words. Again, the Small C-Compiler shows some substantial support of loops as well as the method to access an array.

Functions Call in PCCOMP

The use of functions in C/C++ compiler not only builds a hierarchy in your code and makes it easier to maintain and reuse you code, but for PCCOMP the fact that the use of functions reduces the code size is an even more important feature. The KCPSM6 has 31 PC level stack depth so multilevel subroutine calls are possible in assembler. However, in modern compiler, the data communication with a function is usually done through a data stack. While in early PICOBLAZE versions we had no on-chip memory and therefore no data stack, the use of functions was basically impossible. For scratch pad RAM free PICOBLAZE, the PCCOMP had required that the memory data were provided via the I/O port, which would be a substantial additional hardware effort if used this way. With the introduction of the scratchpad memory, the function call has become a smooth and easy way to run. You may be surprised that our Small C-Compiler even can process recursive function calls without a problem. Let us try this with a small example.

C Program 8.3: Recursive function call verification with `fact.c`

```
1    /* Factorial using STD loop and recursive calls
2     * Author: Uwe Meyer-Baese
3     */
4    unsigned char fact(unsigned char n);
5    unsigned char N, f, r, i;
6
7    void main()
8    {
9        N= 5; f=1;
10       for (i=1; i<=N; i++) f = f * i;
11       r= fact(N);
12
13   }
14   unsigned char fact(unsigned char n)
15   {
16     if (n==0)
17        return 1;
18     else
19          return (n * fact (n-1));
20   }
```

We use the `unsigned` data type (lines 4 and 5) that we can use the faster and shorter unsigned multiplier routine and since we have only unsigned data anyway. We see the conventional computation of the factorial using the `for` loop in line 10. We would expect for $N = 5$ as result $f = 1 \times 2 \times 3 \times 4 \times 5 = 120$. Now in the second approach, we use a recursive function call using the recursive function shown in lines 14 to 20. Notice how the function calls itself again in line 19 with `fact (n-1)`. Now let us verify if our Small C-Compiler can indeed handle such a sophisticated task. The memory for the variable is allocated as follows:

```
.....
        #equ      _N            ,      0x3f
        #equ      _f            ,      0x3e
        #equ      _r            ,      0x3d
       #equ      _i            ,      0x3c
                           ......
```

Therefore, the memory space below `3c` is available as data stack when calling the function recursively. The following snapshot (Fig. 8.4) shows the memory data and the `pc` stack in the deepest loop (see Fig. 8.4a) and at program end; see Fig 8.4b. Clearly at program end, both factorial values 120 (memory location `0x3e` and `0x3d`) and the `pc` stack should be empty; see Fig. 8.4b. As the recursive calls are executed the `pc` stack growths as well and the data stack is expanded to smaller memory location (Fig. 8.4b).

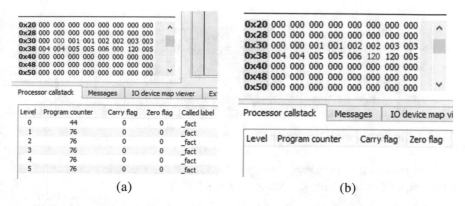

Fig. 8.4 Recursive factorial computation in PCCOMP. (**a**) Level 5 call stack. (**b**) Result (i.e., 120 in red) after recursive call is complete

Additional PCCOMP Features and Recommendations

Compiler directive can be used in PCCOMP, for instance, as global switches as shown in the `loops.c` example above with the preprocessor directives `#ifdef`, `#ifndef`, `#else`, or `#endif`. We can also include additional files such as often used system functions using the `#include` directive. If we like to integrate some low-level assembler code in our ANSI C code, we can do this with the `#asm` and `#endasm` preprocessor directives. Below is our flash example (i.e., toggle the LED in second intervals) coded in C that shows the use of the `#include` and `#asm` directives.

C Program 8.4: Preprocessor directives example file `cflash.c`

```
1    // Short I/O example in PCC
2    #include "lib\io.c"
3    char   data, j;
4    unsigned int i, s;
5    void main(){
6       data = inchar(0); // read data from Input
7    #asm
8    loop:
9    #endasm
10      s=0;
11      outchar(0,data);   // display reading
12      for (j=100;j>0;j--)
13      for (i=10000;i>0;i--) s += i; // waste time
14      data = data ^ 0xFF; // xor
15   #asm
16   jump loop
17   #endasm
18   }
```

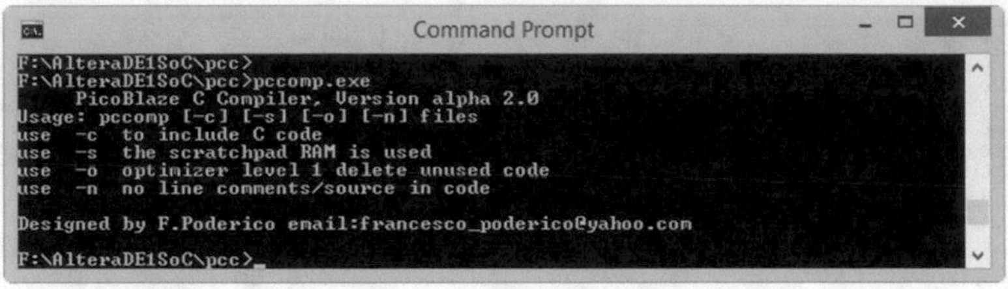

Fig. 8.5 The Pccomp compiler command line options

User manual (see `Pccomp_manual.pdf`) includes a couple more useful recommendations when using the Pccomp such as:

- Global variable definitions should be preferred to local to reduce code size.
- Only one-dimensional arrays are supported.
- Function arguments must be scalar, i.e., cannot be vectors.
- The Pccomp scratchpad memory has 64 bytes.
- Data and function stack share the 64-byte memory.
- Always compile with option -s for KCPSM6 code.

Figure 8.5 shows the compiler option when starting `pccomp.exe`

It is recommended to use a dos command window that you can turn on the switch -s for using the scratchpad RAM.

This concludes to overview on Pccomp.

Review Questions and Exercises

Short Answer

8.1. What are the 12 instruction groups in KCPSM6 assembler?

8.2. Name five addressing modes used by PicoBlaze.

8.3. How will the PROM size (total bits) change if you increase the address bit by one; decrease the address bits by one bit; increase the data bits by 1; and decrease the data bits by one bit?

8.4. PicoBlaze has no special "clear register." What instructions can be used to set a register to zero?

8.5. If you like to double the program size, what consequence does this have to the ISA and assembler coding?

8.6. What PicoBlaze architecture is supported by ISS pBlazeIDE and FIDEx?

8.7. Why is a unified assembler code used by pBlazeIDE and FIDEx?

8.8. Why is the Small C-Compiler by Cain/Hendrix a good choice as starting point for the PicoBlaze C compiler?

8.9. Why are not all ANSI C keywords supported by the Pccomp?

Fill in the Blank

8.10. The following op code are not in use in KCPSM6: _____

8.11. The PICOBLAZE operation SUBCY   sF,s9 will be encoded in hex as _____

8.12. The PICOBLAZE operation LOAD   sA,BB will be encoded in hex as _____

8.13. The PICOBLAZE operation FETCH   s2,03 will be encoded in hex as _____

8.14. The PICOBLAZE operation ADD   s1,01 will be encoded in hex as _____

8.15. The PICOBLAZE operation _____ will be encoded in hex as 22222.

8.16. The PICOBLAZE operation _____ will be encoded in hex as 2D500.

8.17. The PICOBLAZE operation _____ will be encoded in hex as 0AAB0.

8.18. The PICOBLAZE operation _____ will be encoded in hex as 09912.

8.19. The simulation of 100,000 ISS cycles on FIDEx typically takes about _____ minutes.

True or False

8.20. _____ The basic KCPSM6 assembler code line format is
label: instruction rd, rs; comment

8.21. _____ The Xilinx KCPSM6 HDL code will compile fine also with QUARTUS tools.

8.22. _____ The FIDEx uses other assembler coding for SUBCY, INPUT, and OUTPUT.

8.23. _____ The FIDEx is a C/C++ simulator for the PICOBLAZE.

8.24. _____ The FIDEx is a free PICOBLAZE simulator up to 150 lines of code.

8.25. _____ The PCCOMP supports char, int, and float but no double data type.

8.26. _____ The PCCOMP supports identifiers definitions x, _x_, _x, and x_.

8.27. _____ The PCCOMP has char range 0...255 and unsigned char range -128...127.

8.28. _____ In PCCOMP the short int uses 16 bits and int type 32 bits.

8.29. _____ The PCCOMP does not support. and → operators since struct is not supported.

8.30. _____ The factorial program can be run on the TRISC2 microprocessor without any hardware modifications.

Projects and Challenges

8.31. Modify the C program 8.1 `opc.c` that each expression output is equal to array index, e.g., `a[5]= (x&y)+1;`.

8.32. Test (extend `opc.c`) the following operations if supported by PCCOMP and verify via ISS:

```
i.   a[2]= x * y - 68;
ii.  a[3]= y / x + 1;
iii. a[4]= y % x;
iv.  a[7]= (x ^ y) - 4;
v.   a[8] = (~x) + 14;
vi.  a[10]= (y >> 1) + 3;
```

8.33. Develop a KCPSM6 assembler program to implement a left, right car turn signal and emergency light similar to the Ford Mustang: 00X, 0XX, and XXX for a left turn where 0 is off and X is LED on. Use X00, XX0, and XXX for right turn signal. Use the slider switch for the left/right selection. For emergency light both switches are on. The turn signal sequence should repeat once per second. See `mustangQT.MOV` or `mustangWMP.MOV` for a demo.

8.34. Develop a KCPSM6 assembler program that has a random number generator, using `add` and `xor` functions. Display the random number on the LEDs.

8.35. Develop a KCPSM6 assembler program that works as a stopwatch that counts up in second steps. Use three buttons: start clock, stop or pause clock, and reset clock. Using the LEDs or the seven-segment display for display. See `stop_watch_ledQT.MOV` or `stop_watch_ledWMP.MOV` for a demo.

8.36. Develop a KCPSM6 assembler program for a reaction speed timer. Use the pushbuttons to implement. Turn on one of four LEDs, and measure time it takes until the associate button is pressed. Display the measurement after ten tries on LED or seven-segment display.

8.37. Develop a PCC program to implement a left/right car turn signal and emergency light similar to the Ford Mustang: 00X, 0XX, and XXX for a left turn where 0 is off and X is LED on. Use X00, XX0, and XXX for right turn signal. Use the slider switch for the left/right selection. For emergency light both switches are on. The turn signal sequence should repeat once per second. See `mustangQT.MOV` or `mustangWMP.MOV` for a demo.

8.38. Develop a PCC program that has a random number generator, using `add` and `xor` functions. Display the random number on the LEDs.

8.39. Develop a PCC program that works as a stopwatch, which counts up in second steps. Use three buttons: start clock, stop or pause clock, and reset clock. Using the LEDs or the seven-segment display for display. See `stop_watch_ledQT.MOV` or `stop_watch_ledWMP.MOV` for a demo.

8.40. Develop a PCC program for a reaction speed timer. Use the pushbuttons to implement. Turn on one of four LEDs, and measure time it takes until the associate button is pressed. Display the measurement after ten tries on LED or seven-segment display.

References

[Cai80] R. Cain, A small C compiler for the 8080's. Dr. Dobb's J, Vol 5 April-May, 5–19 (1980)

[Cha03] K. Chapman, *KCPSM3 Manual -- PicoBlaze KCPSM3 8-Bit Micro Controller for Spartan-3, Virtex-II and Virtex-II Pro* (San Jose, Xilinx Inc., 2003)

[Cha14] K. Chapman, *User Guide -- PicoBlaze for Spartan-6, Virtex-6, 7-Series, Zynq and Ultra Scale Devices (KCPSM6)* (San Jose, Xilinx Inc., 2014)

[Hen90] J.E. Hendrix, *Small C Compiler, M & T Books* (M&T, Redwood City) (1990)

Chapter 9
Altera Nios Embedded Microprocessor

Abstract This chapter gives an overview of the Altera Nios microprocessor system design, its architecture, and instruction set. It starts with a brief Nios history followed by ISA architecture and interesting design features such as custom instructions.

Keywords Nios · DMIPS · Nios II · TRISC3N · JTEG debugger · Video graphics array (VGA) · World clock · Qsys · Dhrystone · Platform designer · Top-down-design · Bottom-up-design · Basic computer · Floating-point · Custom IP · Square root · Nios Instruction set architecture · Subroutine nesting · DE1-SoC · Compiler optimization

9.1   Introduction

Altera's highly successful Nios embedded microprocessor has set a standard for soft-core embedded processors since its introduction in the fall of 2000. It is a configurable, general-purpose RISC processor that can be easily integrated with user logic and programmed into any Altera FPGA with sufficient logic resources. Initially the Nios processor was a two-address (i.e., RA ← RB □ RB; see Fig. 1.8a) pipelined RISC architecture machine with 16-bit instructions with a user-selectable 16- or 32-bit data path; see Fig. 9.1a. It also featured a maximum of 512 registers with a sliding window of 32 visible registers to the programmer at one time. The processor came with a SOPC library of many standard soft peripherals that are configurable for a wide variety of applications. One of the main reasons why Altera's Nios development systems have been a huge success is based on the fact that besides a fully functional microprocessor, also all the necessary software tools including a GCC-based C/C++ compiler is generated at the same time when the IP block parametrization takes place in the GUI. The Nios had great success with over 10,000 systems sold to over 3000 different customers in the first 3 years since its introduction. We can reason that the great success has been a surprise even to Altera since initially SOPC builder generated the (non-encrypted) VHDL or Verilog source code of the Nios processor and it's peripheral, highly unusual for an IP block today [A04].

© Springer Nature Switzerland AG 2021
U. Meyer-Baese, *Embedded Microprocessor System Design using FPGAs*,
https://doi.org/10.1007/978-3-030-50533-2_9

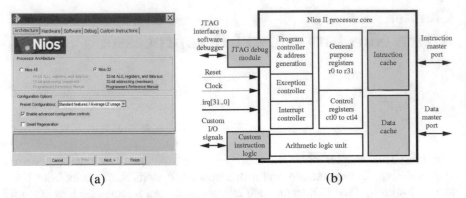

Fig. 9.1 Overview on embedded microprocessor. (**a**) The 16-/32-bit first-generation Nios configuration GUI. (**b**) The Nios II architecture (gray blocks are optional)

The second-generation processor called Nios II introduced in 2004 was more in line with other standard 32-bit processors. The number of registers now was fixed to 32 instructions and data length 32 bits, it was now a three-address machine (i.e., rC ← rA □ rB; see Fig. 1.8b), and all files now were encrypted. Figure 9.1b gives an overview of the architecture. The Nios II is available in three versions: economic (/e), standard (/s), and fast (/f). All Nios II include:

- 32 × 32 Registers file
- Arithmetic logic unit (ALU)
- Interface to custom instruction logic
- Exception controller
- Internal interrupt controller
- Memory management unit (MMU)
- Tightly coupled memory interfaces for instructions and data

And the following optional functional units:

- JTAG debug module
- 0.5–64 Kbyte instruction cache memories (/s and /f)
- Branch prediction unit (/s and /f)
- Hardware multiplier (/s and /f)
- Hardware divider (/s and /f)
- 0.5–64 Kbyte data cache memories (/f)
- Barrel shifter (/f)
- Dynamic branch prediction (/f)

The /s and /f versions only allow additional features such as data/instruction caches and hardware multiplier and divider at the cost of additional resources. Floating-point is available for all three Nios II flavors. Floating-point operation support requires the addition of a custom instruction IP that is discussed in Sect. 9.4. At the time of writing, only the /e version is license-free; /s and /f need a purchased

Table 9.1 Instruction latency for Nios II gen 2. [AI16]

Instruction type	Nios II gen 2/e	Nios II gen 2/f
ALU instructions (e.g., `add`, `cmplt`)	6	1
Combinational CIP	6	1
Branch taken	6	2
Branch mispredicted	6	4
`call`, `jmpi`	6	2
`jmp`, `ret`	6	3
Load word	6 + Avalon transfer	1+
32 × 32 multiply	11	1
Divide	>100	35
Shift/rotate	7–38	1–32
All other instructions	6	1

Table 9.2 DMIPS performance of popular embedded processors

μP name	Device used	Speed (MHz)	DMIPS measured	Hard/Soft Core
Nios II/e	Stratix	330	50	S
MICROBLAZE	Spartan-3(-4)	85	68	S
MICROBLAZE	Virtex-II PRO-7	150	125	S
V1 ColdFire	Stratix	145	135	S
ARM Cortex-M1	Stratix	200	160	S
Nios II/s	Stratix	270	170	S
ARM922T	Excalibur	200	210	H
MIPS32	Stratix	290	300	S
Nios II/f	Stratix	290	340	S
PPC405	Virtex-4 FX	450	700	H
ARM Cortex-A9	Arria/Cyclone/Zynq	800	4000	H

license. The pipelining in the Nios II processor has a substantial impact on the performance. Table 9.1 gives an overview of popular instruction class for the two versions in the newest generation called Nios II gen 2.

Embedded microprocessors are often benchmarked via DMIPS programs originally written in Ada programming language by R. Weicker (see Chap. 5). DMIPS source code was in the past one of the example software projects in the "old" Nios II IDE but is no longer part of the current Eclipse tool. The DMIPS depends on the systems (cache) architecture, external memory interface, and processor speed so often is provided as best case DMIPS/MHz; see Table 5.19. The Nios Reference Documentation [AI17a] includes the data along with typical resource and performance data for different FPGA families.

As can be seen from Table 9.2, the DMIPS performance of the Nios II / f version is among the best for softcore processors.

Recently the Nios II has been updated: The SOPC system has been replaced with QSYS[1] that enables better control of bus structures in multiprocessor designs, and the "classic" Nios II with three versions is replaced by Nios II gen 2 with /e and /f only and has a few minor improvements such as full 32-bit address space, user-define cache bypass, and improved Qsys interface. The /e is a size-optimized free IP (ca. 0.15 DMIPS/MHz, one pipeline stage), while the /f is the performance-optimized RISC (ca. 1.16 DMIPS/MHz; six pipeline stages) and requires a paid license [AI17b].

The Nios II system can be designed by combining the Nios II with the needed components, and we call this the *bottom-up* design approach. Alternatively, we can use one of the starting-point systems provided by the FPGA or board vendors and modify this design as needed. We would call this approach the *top-down* method, and since this is usually a little easier, we will start with this top-down design approach.

9.2   Top-Down Nios II System Design

The majority of FPGA boards come with a starting-point design for quick evaluation purpose. Since vendor wants to demonstrate all the great features a board has, this starting-point designs are usually not minimum systems, instead usually most if not all peripheral components are accessed in the start-up design to make the system more appealing and demonstrate the large number of functions of the board. Since these systems are configured such that all peripheral components are functionally fine, the mistakes of a wrong configuration of a peripheral driver are less likely in the top-down design approach. We simply remove all design components that are not needed in the system target application and/or modify the existing component as needed.

For the majority of TerASIC/Altera DE board (DE-Nano, DE2, DE2–115, etc.), the AUP provide two starting-point designs: a "Basic Computer" system that has a few basic I/O in use, a larger memory, and a JTAG interface. A full-featured "Media Computer" extends the basic version to include the more complex I/O such as VGA, USB, Audio, LCD, IrDA, etc. (if available). For the DE1-Nano-SoC and DE1 SoC board, only one full-featured starting-point design is provided that includes the ARM and two Nios II processors and the majority of I/Os (SDRAM, LEDs, seven-segment, switches, buttons, PS2, 4 × JTAG, IrDA, 4 × timer, ADC, Audio, VGA, Video). Three quite sophisticated bridges (64-bit FPGA→HP, 128-bit HPS → FPGA, and 32-bit lightweight HPS → FPGA interface) are used such that both ARM and Nios II can access the majority of I/O components. Missing are only the I^2C FPGA peripheral and Gigabit Ethernet, MicroSD, and two-port USB on the HPS/ARM side. Nevertheless, the system is large and the possible applications huge; see Fig. 9.2 for the included IP.

---

[1] QSYS has recently been renamed by Intel to PLATFORM DESIGNER.

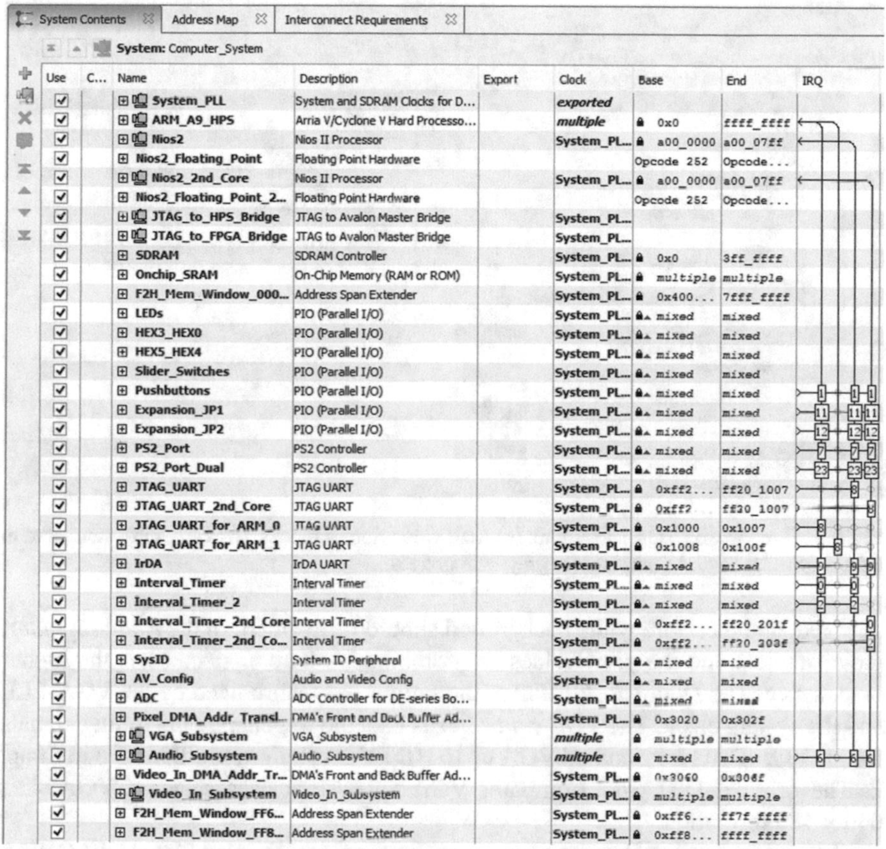

System Contents ⊠ | Address Map ⊠ | Interconnect Requirements ⊠

System: Computer_System

Use	C...	Name	Description	Export	Clock	Base	End	IRQ
✓		⊞ System_PLL	System and SDRAM Clocks for D...		*exported*			
✓		⊞ ARM_A9_HPS	Arria V/Cyclone V Hard Processo...		*multiple*	0x0	ffff_ffff	
✓		⊞ Nios2	Nios II Processor		System_PL...	a00_0000	a00_07ff	
✓		⊞ Nios2_Floating_Point	Floating Point Hardware			Opcode 252	Opcode...	
✓		⊞ Nios2_2nd_Core	Nios II Processor		System_PL...	a00_0000	a00_07ff	
✓		⊞ Nios2_Floating_Point_2...	Floating Point Hardware			Opcode 252	Opcode...	
✓		⊞ JTAG_to_HPS_Bridge	JTAG to Avalon Master Bridge		System_PL...			
✓		⊞ JTAG_to_FPGA_Bridge	JTAG to Avalon Master Bridge		System_PL...			
✓		⊞ SDRAM	SDRAM Controller		System_PL...	0x0	3ff_ffff	
✓		⊞ Onchip_SRAM	On-Chip Memory (RAM or ROM)		System_PL...	multiple	multiple	
✓		⊞ F2H_Mem_Window_000...	Address Span Extender		System_PL...	0x400...	7fff_ffff	
✓		⊞ LEDs	PIO (Parallel I/O)		System_PL...	mixed	mixed	
✓		⊞ HEX3_HEX0	PIO (Parallel I/O)		System_PL...	mixed	mixed	
✓		⊞ HEX5_HEX4	PIO (Parallel I/O)		System_PL...	mixed	mixed	
✓		⊞ Slider_Switches	PIO (Parallel I/O)		System_PL...	mixed	mixed	
✓		⊞ Pushbuttons	PIO (Parallel I/O)		System_PL...	mixed	mixed	
✓		⊞ Expansion_JP1	PIO (Parallel I/O)		System_PL...	mixed	mixed	
✓		⊞ Expansion_JP2	PIO (Parallel I/O)		System_PL...	mixed	mixed	
✓		⊞ PS2_Port	PS2 Controller		System_PL...	mixed	mixed	
✓		⊞ PS2_Port_Dual	PS2 Controller		System_PL...	mixed	mixed	
✓		⊞ JTAG_UART	JTAG UART		System_PL...	0xfff2...	ff20_1007	
✓		⊞ JTAG_UART_2nd_Core	JTAG UART		System_PL...	0xfff2	ff20_1007	
✓		⊞ JTAG_UART_for_ARM_0	JTAG UART		System_PL...	0x1000	0x1007	
✓		⊞ JTAG_UART_for_ARM_1	JTAG UART		System_PL...	0x1008	0x100f	
✓		⊞ IrDA	IrDA UART		System_PL...	mixed	mixed	
✓		⊞ Interval_Timer	Interval Timer		System_PL...	mixed	mixed	
✓		⊞ Interval_Timer_2	Interval Timer		System_PL...	mixed	mixed	
✓		⊞ Interval_Timer_2nd_Core	Interval Timer		System_PL...	0xfff2...	ff20_201f	
✓		⊞ Interval_Timer_2nd_Co...	Interval Timer		System_PL...	0xfff2...	ff20_203f	
✓		⊞ SysID	System ID Peripheral		System_PL...	mixed	mixed	
✓		⊞ AV_Config	Audio and Video Config		System_PL...	mixed	mixed	
✓		⊞ ADC	ADC Controller for DE-series Bo...		System_PL...	mixed	mixed	
✓		⊞ Pixel_DMA_Addr_Transl...	DMA's Front and Back Buffer Ad...		System_PL...	0x3020	0x302f	
✓		⊞ VGA_Subsystem	VGA_Subsystem		*multiple*	multiple	multiple	
✓		⊞ Audio_Subsystem	Audio_Subsystem		*multiple*	mixed	mixed	
✓		⊞ Video_In_DMA_Addr_Tr...	DMA's Front and Back Buffer Ad...		System_PL...	0x3060	0x306f	
✓		⊞ Video_In_Subsystem	Video_In_Subsystem		System_PL...	multiple	multiple	
✓		⊞ F2H_Mem_Window_FF6...	Address Span Extender		System_PL...	0xff6...	ff7f_ffff	
✓		⊞ F2H_Mem_Window_FF8...	Address Span Extender		System_PL...	0xff8...	ffff_ffff	

Fig. 9.2 DE1 SoC Computer starting-point design

To demonstrate the use of the system, let us assume the implemented VGA subsystem resolution with $320 \times 240 \times 16$-bit RGB does not meet exactly our project requirements and we like to use a $640 \times 480 \times 8$-bit gray resolution. The AUP provides a substantial number of IP blocks available for VGA processing as shown in Fig. 9.3a. The original processing of the VGA data is organized as follows: The *VGA Pixel DMA* unit stores and retrieves to and from memory without molesting the Nios II. It specifies addressing modes, frame resolution, and pixel format. It controls the Dual-Clock-FIFO. The *Dual-Clock FIFO* helps to transfer a stream between two clock domains. Then the data run through the *VGA resampler* that converts the 16-bit color code (5-bit red, 6-bit green, 5-bit blue) to the 30 bit data needed by the 3×10 bit RGB DAC ADV7123KSTZ140; see Fig. 2.7 in Chap. 2. The on-chip memory is not large enough to store a full-size color VGA 640×480, so a *VGA Pixel scaler* is used to duplicate each pixel in row and column direction such the original 320×240 pixels are displayed as 640×480. The VGA also has a *Character Buffer* (size 80×60) that allows ASCII charters to be displayed in "white" on the VGA monitor. The color and font type can unfortunately not be

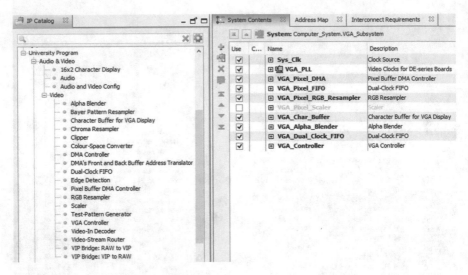

Fig. 9.3 Overview on VGA subsystem. (**a**) AUP video library blocks. (**b**) VGA subsystem used in the DE1 SoC modified Computer for grayscale images

changed, and therefore white background should be avoided. In the *Alpha Blender*, the characters and the pixel values are combined, where character has the higher precedence. The *Dual Clock FIFO* at the output is synchronized by the VGA-PLL and the system clock. The *VGA controller* is capable to generate the timing signals for the VGA DAC for standard VGA up to HDTV resolution; see Table 2.8 in Chap. 2. The original DE1 SoC Computer VGA subsystem uses standard VGA, i.e., 640×480 resolution.

Now let us assume that we want to replace the 16-bit color 320×240 QVGA with a 8-bit gray full-size VGA format, i.e., 640×480 to improve the display resolution, for instance, for fractal display or a dual clock. There are a couple of modifications we need to apply to the DE1 SoC Computer system and the VGA subsystem: First we need to adjust the on-chip memory such that we can store the 640×480 pixels. In the original system, the pixel buffer DMA controller uses the X/Y addressing mode, i.e., each pixel can easily address with

```
pixel_ptr = FPGA_ONCHIP_BASE + (row << 10) + (col << 1);
```

to keep the Nios II processing simple. Now each line aka row with 16 bit $\times$ 320 pixels needs one M10K embedded memory blocks or 240 for the whole QVGA image. The Cyclone V SoC 5CSEMA5F31C6 used for the DE1 SoC has total of 397 memory blocks, and already 350 are needed for the whole starting-point Computer system. The QVGA uses 256 M10K blocks or 262,144 Bytes of these. If we continue using the X/Y mode, we would need an additional $480-256 = 224$ M10Ks, but only 47 are not in use. If we now give up on the X/Y addressing mode and use

```
pixel_ptr = FPGA_ONCHIP_BASE + (row * 640) + col;
```

We have a more compact image storage at the cost of a multiply instead of shift operations in the pixel address computation. This is beneficial to our memory requirements since we now need 307,200 bytes, or 300 M10Ks, to store our 8-bit gray full-size VGA image with 640 × 480 resolution. We will need to adjust the onchip_SRAM component accordingly. The *Pixel Buffer DMA Controller* needs to be changed to 8-bit grayscale, consecutive addressing, and 640 × 480 resolution. The *Pixel FIFO* now has 8 bits and one-color plane. Next, we would remove or disable the VGA Pixel scaler of the VGA subsystem design and adjust the internal bus connections accordingly, i.e., output of resampler is now connected to the alpha blender input. The sys clock, video clocks, VGA char buffer, alpha blender, output FIFO, and VGA controller remain unchanged; see Fig. 9.3b. Lastly we remove the audio and video subsystems since they are not in use in our application and will free some resources and reduce compile time. The modified system can be found on the book CD under the directory /NiosII/DE1_SoC_TopDown. Table 9.3 shows the required Qsys changes. Table 9.4 compares and summarizes synthesis results and some key features of the original computer system and our modified system.

As example design (see app_C/CLOCK/alalog_clock2x2.c), we have implemented a world clock with two time zones; see Fig. 9.4. A four-fractal display is left to the reader as project Exercise 9.53.

The full set of example projects for the top-down system includes CLOCK, FLASH, FRACTAL_COLOR, FRACTAL, GREY, MOVIE_COLOR, and MOVIE_GREY and is located in the NiosII/DE1_SoC_TopDown/app_C folder on the CD.

9.3 Bottom-Up Nios II System Design

In the past the majority of DE boards (DE2, DE2–70, DE2–115, DE0-Nano) came with two starting-point designs: a complete systems that integrates almost all of the components/features available (called *Media Computer*) and a small footprint

Table 9.3 Qsys parameter changes for the VGA 640 × 480 × 8-bit gray system of the DE1 SoC. All other block parameters are unchanged

Block	Parameter	Original system	Modified system
On-chip SRAM	Total memory size	262,144 Bytes	307,200 Bytes
VGA Pixel DMA	Addressing mode	X-Y	Consecutive
	Width	320	640
	Height	240	480
	Color space	16-bit RGB	8-bit grayscale
VGA Pixel FIFO	Color bits	16	8
RGB Resampler	Incoming format	16-bit RGB	8-bit grayscale
VGA Pixel Scaler	Use	Enable	Disable
Audio Subsystem	Use	Enable	Disable
Video In Subsystem	Use	Enable	Disable

Table 9.4 Overview of original and modified system of the DE1 SoC

	Original computer system	Top-down modified system
ALM	23,178	18,617
DSP blocks	15	2
PLLs	3	2
M10K blocks	350	366
Compile time (min:sec)	58:23	40:32
Features	With audio and video subsystem	No audio and video subsystem
	VGA 320 × 240 × 16-bit RGB	VGA 640 × 480 × 8-bit gray

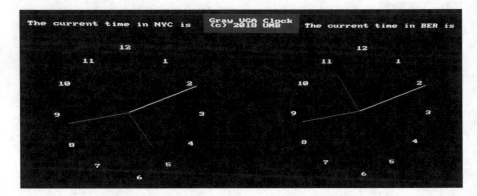

Fig. 9.4 World clock example using VGA 640 × 480 × 8-bit gray resolution

starting-point design with simple I/O only (switches, buttons, LEDs) called *Basic Computer*. Recently Altera has stopped to provide the Basic Computer systems since these are typically build with a bottom-up tutorial such as the "Introduction to Altera Qsys System Integration Tool" and the "Using the SDRAM on Altera's DE1 SoC Board with VHDL/Verilog Designs." Let us have a brief look at the major steps in putting together a Basic Computer system.

We first need to decide on the features and components we like to include in our system. It should have the Nios II/e with the JTAG interface for debugging via a host computer since this would not require a paid license. As basic components, we included all LEDs, switches, and buttons we can find on the board and a timer. The timer is included such that we can do measurements needed for DMIPS, FMIPS, or MWIPS scores. Because the on-chip memory is small we will immediately start to integrate the SDRAM. Since on DE board the SDRAM leads the Nios clock by 3 ns we need also a PLL or the University Program DE series clock IP. Seven-segment displays, system ID, and on-chip memory that are also part of the typical Basic Computer are left to the reader as project Exercise 9.57.

Now let us get started with the initial design. We will first add all desired components and then assign address space, interrupts, export names, and wiring. If not mentioned otherwise, we will use the components with the default settings. We assume that the (free) University IP packages for the Nios II have been installed. Here are the steps in the order on how we like to organize the system.

Memory Profile	Timing

Data Width
Bits: 16

Architecture
Chip select: 1
Banks: 4

Address Width
Row: 13
Column: 10

Generic Memory model (simulation only)
☑ Include a functional memory model in the system testbench

Memory Size = 64 MBytes
33554432 x 16
512 MBits

Memory Profile	Timing

CAS latency cycles::   ○ 1   ○ 2   ◉ 3

Initialization refresh cycles:	2	
Issue one refresh command every:	15.625	us
Delay after powerup, before initialization:	100.0	us
Duration of refresh command (t_rfc):	70.0	ns
Duration of precharge command (t_rp):	20.0	ns
ACTIVE to READ or WRITE delay (t_rcd):	20.0	ns
Access time (t_ac):	5.5	ns
Write recovery time (t_wr, no auto precharge):	14.0	ns

(a)   (b)

Fig. 9.5 SDRAM configuration of the DE1 SoC Computer. (a) Memory Profile. (b) Timing

Use the *New Project Wizard...* to define a project with name *DE1_SoC_Basic_Computer*[2] and select as device 5CSEMA5F31C6. If you use another board, use the correct device, and the instruction may need adjustments in a couple of values in particular the SDRAM timing. Now start *Tools→Qsys*, and on the left panel, i.e., the

IP Catalog, right click the component or use the [➕ Add...] button to add the following components:

- *University Program→Clock→System and SDRAM Clocks for DE-Series Boards.* The *Reference Clock* should be 50 MHz. Set *Desired Clock* to 100 MHz and *Board* to DE1 SoC. Rename the component to `clocks`.
- *Processor and Peripherals→Embedded Processors→Nios II Processor.* Choose the Nios II/e. Later we will also set the CPU reset and exception vectors to `sdram` starting at 0 and 0x20, respectively.
- *Memory Interfaces and Controllers→SDRAM→SDRAM Controller.* Since configuration and size of the SDRAM on the DE are different, the controller parameters for the particular board have to be customized carefully. You may want to check the provided tutorial and the provided (Media) Computer for precise setting. Memory and Timing data for the DE1 SoC Computer are shown in Fig. 9.5. These differ a little from the parameter in the SDRAM tutorial.
- *Interface Protocols→Serial→JTAG UART* and keep the default settings.
- *Processor and Peripherals→Peripherals→PIO (Parallel I/O)* and change the *Basic Setting* to 10 bits and *Direction* Input. Rename the component to `switches`.
- *Processor and Peripherals→Peripherals→PIO (Parallel I/O)* and change the *Basic Setting* to 10 bits and *Direction* Output. Rename the component to `leds`.

---

[2]As a general recommendation, try to avoid space and special characters in path and file names. Many TCL/TK scripts will not work properly if you use special characters or spaces such as a path that includes "My Documents."

- *Processor and Peripherals→Peripherals→PIO (Parallel I/O)* and change the *Basic Setting* to 4 bits and *Direction* Input and the other setting as in Fig. 9.6a. Rename the component to `pushbuttons`.
- *Processor and Peripherals→Peripherals→Interval Timer* and change the settings as in Fig. 9.6b. Rename the component to `timer`.

Now that we have all components, we start assigning address space, interrupts, export names, and wiring. Export the following. Double click *Base* and enter the address such that the I/O components match the address used in the DE1 SoC Nios Computer system, as shown in Table 9.5.

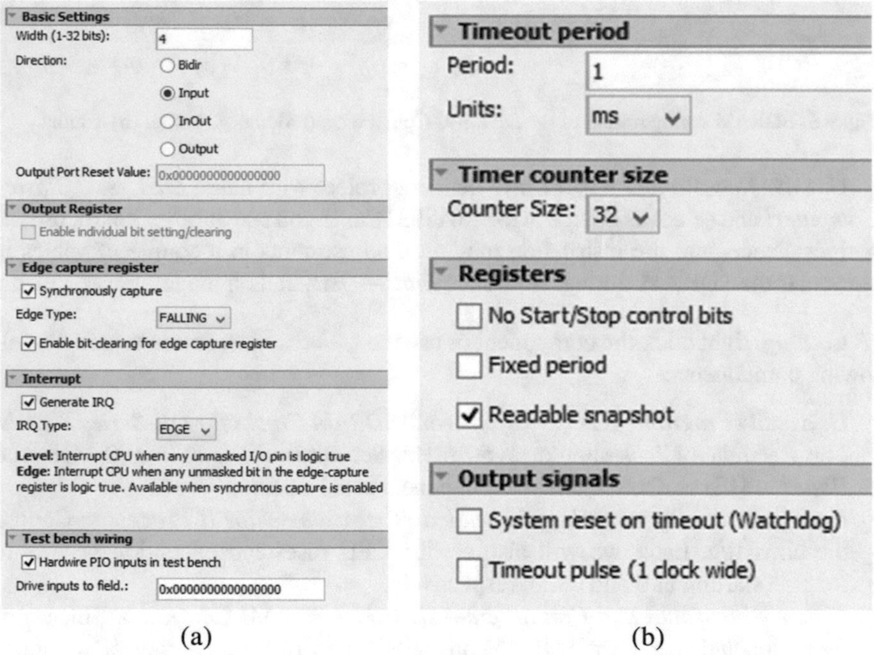

(a)   (b)

Fig. 9.6 DE1 SoC Basic Computer configuration. (**a**) Push button. (**b**) Timer

Table 9.5 DE1 SoC Basic Computer address map

Component	Base	End
jtag_uart	0xff20_1000	0xff20_1007
leds	0xff20_0000	0xff20_000f
nios2_gen2	0x04000_0800	0x0400_0fff
pushbuttons	0xff20_0050	0xff20_005f
sdram	0x0000_0000	0x03ff_ffff
switches	0xff20_0040	0xff20_004f
timer	0xff20_2000	0xff20_201f

Table 9.6 DE1 SoC Basic
Computer export signals

Component	Name	Export
clocks	ref_clk	clk
clocks	ref_reset	reset
clocks	sdram_clk	sdram_clk
sdram	wire	sdram_wire
switches	external_ connection	switches
leds	external_ connection	leds
pushbuttons	external_ connection	pushbuttons

The interrupts should be used as follows: `timer` should use IRQ 0, `pushbut-tons` IRQ 1, and `jtag_uart` the IRQ 5. Now we need to export a couple of signals such that we can make the correct connections in the HDL code. We need to do this for a total of seven connections: clocks, SDRAM, switches, LEDs, and push buttons. Use the export name as shown in Table 9.6 such that it matches the HDL code later.

Finally we now implement the connection between the components by clicking at the connection points on the left as shown in Fig. 9.7. For the majority of components, two to three connections are not made (e.g., `sdram_clk` and `instruction_master`). You should save the whole Qsys system under the name `nios_system`. Then select from the menu *Generate→Generate HDL...* or click *Generate HDL ...* in the lower Qsys right corner. The *Generation Menu* will pop up. Select your preferred synthesis language and make sure the output directory is as desired in your project. After save system is complete (click ok), the system generation starts. This may take a few seconds for our basic computer; the DE1 SoC Computer may need substantially longer time. Close the window after generation is complete.

Before we can compile the system (so that we can download the bit stream to the FPGA), we need to add a HDL wrapper and the FPGA pin list with driving strength. This procedure is substantially simplified if we use the same names (e.g., `LEDR` for red LEDs; `SW` for switches; `KEY` for push buttons, etc.) that are used by DE1 SoC Computer. Such a `*.qsf` file can be generated with the `DE1SoC_SystemBuilder.exe` program from the board CD-ROM. Before you exit the Qsys editor, you may want to have a look at the instantiation template for our Nios II system. You find these under the Menu *GenerateArial Unicode MSShown Instantiation Template....* It has the component definition (VHDL) and the instantiation example (VHDL and Verilog). You will need to add the basic module/entity definition with the names as defined in the `*.qsf` file. In the QUARTUS panel, open a new file with *File→New... → Design Files* and then *Verilog HDL File* or *VHDL File.* You may want to check that the Nios system is added to your project under *Project→Add/Remove Files in Project...* before you close the Qsys editor which can be done using *File→Exit.*

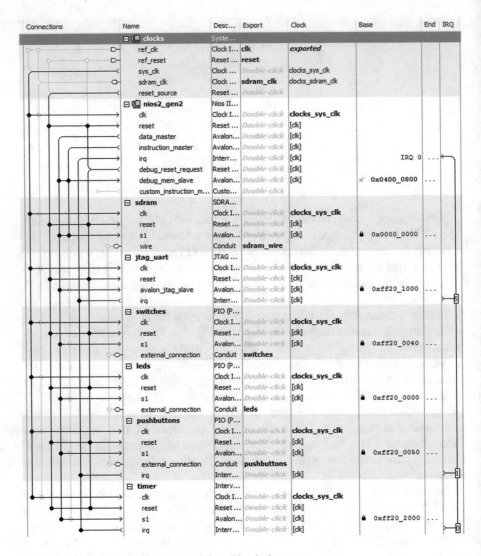

Fig. 9.7   DE1 SoC Basic Computer starting-point design

VHDL Code 9.1: The DE1 SoC Basic Computer Top Level

```vhdl
1   --  =======================================================
2   -- Implements a simple Nios II system for the DE1-SoC board.
3   -- Inputs: SW9-0 are parallel port inputs to the Nios II system
4   -- CLOCK_50 is the system clock
5   -- KEY is the system push button
6   -- Outputs: LEDR9-0 are parallel port outputs
7   LIBRARY ieee;
8   USE ieee.STD_LOGIC_1164.ALL;
9   USE ieee.STD_LOGIC_unsigned.ALL;
10  -- =======================================================
11  ENTITY DE1_SoC_Basic_Computer IS
12    PORT (
13      CLOCK_50   : IN STD_LOGIC;
14      KEY        : IN STD_LOGIC_VECTOR (3 DOWNTO 0);
15      SW         : IN STD_LOGIC_VECTOR (9 DOWNTO 0);
16      LEDR       : OUT STD_LOGIC_VECTOR (9 DOWNTO 0);
17      DRAM_DQ    : INOUT STD_LOGIC_VECTOR (15 DOWNTO 0);
18      DRAM_ADDR  : OUT STD_LOGIC_VECTOR (12 DOWNTO 0);
19      DRAM_BA    : OUT STD_LOGIC_VECTOR (1 DOWNTO 0);
20      DRAM_CAS_N, DRAM_RAS_N, DRAM_CLK : OUT STD_LOGIC;
21      DRAM_CKE, DRAM_CS_N, DRAM_WE_N : OUT STD_LOGIC;
22      DRAM_UDQM, DRAM_LDQM : OUT STD_LOGIC);
23  END ENTITY;
24  -- =======================================================
25  ARCHITECTURE fpga OF DE1_SoC_Basic_Computer IS
26
27  COMPONENT nios_system IS
28  port (
29    clk_clk    : IN    STD_LOGIC   := 'X';          -- clk
30    leds_export : OUT  STD_LOGIC_VECTOR(9 DOWNTO 0); -- export
31    pushbuttons_export : IN STD_LOGIC_VECTOR(3 DOWNTO 0)
32                                 := (others => 'X'); -- export
33    reset_reset  : IN   STD_LOGIC  := 'X';       -- reset
34    sdram_clk_clk   : OUT   STD_LOGIC;       -- clk
35    sdram_wire_addr : OUT   STD_LOGIC_VECTOR(12 DOWNTO 0);  -- addr
36    sdram_wire_ba   : OUT   STD_LOGIC_VECTOR(1 DOWNTO 0);   -- ba
37    sdram_wire_cas_n : OUT  STD_LOGIC;                -- cas_n
38    sdram_wire_cke  : OUT   STD_LOGIC;                -- cke
39    sdram_wire_cs_n : OUT   STD_LOGIC;                -- cs_n
40    sdram_wire_dq   : INOUT STD_LOGIC_VECTOR(15 DOWNTO 0)
41                                 := (others => 'X'); -- dq
42    sdram_wire_dqm  : OUT   STD_LOGIC_VECTOR(1 DOWNTO 0);   -- dqm
43    sdram_wire_ras_n : OUT  STD_LOGIC;                -- ras_n
44    sdram_wire_we_n : OUT   STD_LOGIC;                -- we_n
45    switches_export : IN    STD_LOGIC_VECTOR(9 DOWNTO 0)
46                                 := (others => 'X')); -- export
47    END COMPONENT nios_system;
48
49  BEGIN
50
51  u0 : COMPONENT nios_system
52    port map
53    (clk_clk            => CLOCK_50,             -- clk.clk
```

```
54    leds_export            => LEDR(9 DOWNTO 0),    -- leds.export
55    pushbuttons_export => NOT KEY(3 DOWNTO 0),    -- pushbuttons.export
56    reset_reset            => '0',                -- reset.reset
57    sdram_clk_clk          => DRAM_CLK,           -- sdram_clk.clk
58    sdram_wire_addr        => DRAM_ADDR,          -- sdram_wire.addr
59    sdram_wire_ba          => DRAM_BA,            -- .ba
60    sdram_wire_cas_n       => DRAM_CAS_N,         -- .cas_n
61    sdram_wire_cke         => DRAM_CKE,           -- .cke
62    sdram_wire_cs_n        => DRAM_CS_N,          -- .cs_n
63    sdram_wire_dq          => DRAM_DQ,            -- .dq
64    sdram_wire_dqm(1)      => DRAM_UDQM,          -- .dqm(1)
65    sdram_wire_dqm(0)      => DRAM_LDQM,          -- .dqm(2)
66    sdram_wire_ras_n       => DRAM_RAS_N,         -- .ras_n
67    sdram_wire_we_n        => DRAM_WE_N,          -- .we_n
68    switches_export        => SW(9 DOWNTO 0));    -- switches.export
69
70  END ARCHITECTURE;
```

The file listing 9.1 shows the instantiation of the Nios system. We start with a header explaining I/O ports. Lines 27–47 contain the component definition followed by component instantiation in lines 51–68. We use a NOT KEY assignment since the push button is active low and the reset is active high and therefore holds to zero all the time.

After the HDL wrapper compiles (syntax error free) we can run a full compile of the system. Do not forget to add the pin assignment *.qsf first. After the successful run, we can test the system by downloading a test program such as the nios_flash.c from the CD. The listing is shown next.

C Program 9.2: Flash Program for the Nios II Basic Computer

```
1    // ==============================================================
2    // This program demonstrates use of basic ports in the DE1-SoC board
3    //
4    // It performs the following:
5    //   1. displays the SW switch values on the red LEDR
6    //   2. toggle the LEDR every second
7    //   3. if KEY[3..0] is pressed, uses the SW switches as the pattern
8    // ==============================================================
9    #define LEDR_BASE      0xFF200000
10   #define SW_BASE        0xFF200040
11   #define KEY_BASE       0xFF200050
12   #define TIMER_BASE     0xFF202000
13   #define Fcpu           100000000
14
15   int main(void)
16   {
17     /* Declare volatile pointers to I/O registers (volatile means that
18      * I/O load and store instructions will be used to access
19      * these pointer locations, instead of regular memory loads
20      * and stores)
21      */
22   volatile int *red_LED_ptr   = (int *) LEDR_BASE;  // Red LED address
23   volatile int *SW_switch_ptr = (int *) SW_BASE;    // Slider switch
24   volatile int *KEY_ptr       = (int *) KEY_BASE;  // Pushbutton KEY
25   volatile int * interval_timer_ptr = (int *) TIMER_BASE; // Timer
26
27   int high_half, counter, User_Time=0;
28   int SW_value, KEY_value;
29
30     SW_value = *(SW_switch_ptr); // Read the initial SW switch values
31     while(1)  {
32       /* Set the interval timer to 32 bit max */
33       *(interval_timer_ptr + 1) = 0x8;
34                         // Set STOP=1, START = 0,CONT = 0, ITO = 0
35       *(interval_timer_ptr + 0x2) = 0xFFFF;
36       *(interval_timer_ptr + 0x3) = 0x7FFF;
37       *(interval_timer_ptr + 1) = 0x4;
38                         // Set STOP=0, START = 1, CONT = 0, ITO = 0
39
40       *(red_LED_ptr) = SW_value;       // Light up the red LEDs
41       KEY_value = *(KEY_ptr);          // Read the pushbutton KEY
42       if (KEY_value != 0)              // Check if any KEY was pressed
43       {  SW_value = *(SW_switch_ptr);  // Read the SW slider switch
44        while (*KEY_ptr);}             // Wait for pushbutton KEY release
45
46       // Use the timer to find out when 500ms are over
47       User_Time=0;
48       while (User_Time < 500) {
49         // Make a counter snapshot by writing a dummy value to snap1
50         *(interval_timer_ptr + 0x4) =  0;
51   // Read the 32-bit counter snapshot from the 16-bit timer registers
52     high_half = *(interval_timer_ptr + 0x5) & 0xFFFF;
53     counter=(*(interval_timer_ptr + 0x4) & 0xFFFF)|(high_half << 16);
```

```
54   // Compute User Time is in ms
55      User_Time = (0x7FFFFFFF - counter) / (Fcpu/1000);
56                              // Clock cycles divided by CPU frequency
57      }
58      SW_value = SW_value ^ 0xFFFF; // Build complement
59    }
60   }
```

Note since we have used the same address map (lines 9–12) as in the DE1 SoC Computer, we can at any time also test our program with the precompiled Nios II system. The program loads the SW values and displays the values on the red LEDs (line 30). The display toggles in 1 Hz between on and off. If we change the SW value, we will not see an immediate update in the LED pattern. Only after we press any of the KEYs (line 41–42), i.e., push button, we see the new SW pattern.

Note also that we have only used ANSI C statements, i.e., no BSP package functions. In this way our program compiles much faster. The price to pay is that we need to have some more detailed knowledge about the timer registers. The register of the timer are shown in Fig. 9.8. There are six control bits for the counter. The RUN bit indicates if the counter is running (i.e., counting down). If it reaches zero, the TO bit is set. START/STOP is used to continue/suspend the counting operation. CONT is the wraparound flag. With CONT set to 1, if it reaches zero, counting continues. ITO is used to generate interrupts. With STOP=0, START=1, CONT=0, and ITO=0, we start the counter, and with STOP=1, START=0, CONT=0, and ITO=0, we stop the counter. The timer registers have 16 bits each, so loading a 32-bit value into the counter takes two clock cycles. Since the counter runs with CPU clock to compute the time, we need to divide the counter values by the processor clock. Since we like to have ms counter, we multiply by 1000. This concludes the discussion of the DE1 SoC Basic Computer. In the next section, we will add a custom IP to the Basic Computer.

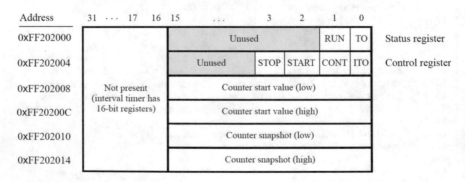

Fig. 9.8 Interval timer register

9.4 Custom Instruction Nios II System Design

Modern softcore processors allow a close integration of custom logic often called *custom intellectual property* (CIP) within the microprocessor through peripheral buses such Avalon or AIX. If the core is tightly coupled within the CPU, such a core is then called *custom instruction* (CI) without the long delay through external bus operations. Typical examples are image processing co-processors that offload continue display operation (discussed in Sect. 10.4 for MICROBLAZE design) or custom bitwise operation (see Chap. 11 on a bit-reverse example for ARM). Floating-point operation support is also often implemented as CIP and is discussed next for Nios II. There are however a few considerations related to the processor and algorithms which should be taken into account:

- The most improvement through a CI is expected for a "slow" processor such as the Xilinx PICOBLAZE or the Nios II/e. For an ARM Cortex-A9 or Nios II/f, the improvement may be less significant. A highly pipelined processor may actually be run slower with a CI.
- Large improvements can be expected for algorithms only if a hardware implementation is fast and compact. Bitwise operations like a bit-reverse needed in DCT or FFT or switch boxes in cryptographic algorithms are good examples that should result in large improvements with a CI. On the other hand, a 256-point FFT was improved by only 45–77% when using a butterfly processor [S04, MSC06, M07]. The large design effort in a CI may then not be justified for such a small improvement obtained for custom arithmetic circuits.

Before we start with the details of the CIP design, let us briefly review floating-point number format and available HDL source code and a typical implementation example.

Floating-Point Number Formats and Operations

Floating-point systems were developed to provide high resolution over a large dynamic range. Floating-point systems can often provide a solution when fixed-point systems, with their limited dynamic range, fail. Floating-point systems, however, bring a speed and complexity penalty. Most microprocessor floating-point systems comply with the published single- or double-precision IEEE floating-point standard [I85, I08], while FPGA-based systems can also employ custom formats [SAA95]. We will therefore discuss in the following the standard 32- and 64-bit floating-point formats, called `float` and `double` in ANSI C; see Table 9.7 for format bit allocation, bias, and range.

Floating-point arithmetic blocks are available from several "intellectual property" providers, FPGA vendors via LPM block by Altera, and have recently been included in the VHDL-2008 standard.

Table 9.7 IEEE floating-point 754–2008 standard interchange formats

	Short	Single	Double	Extended
Word length	16	32	64	128
Mantissa	10	23	52	112
Exponent	5	8	11	15
Bias	15	127	1023	16,383
Range	$2^{16} \approx 6.4 \times 10^4$	$2^{128} \approx 3.8 \times 10^{38}$	$2^{1024} \approx 1.8 \times 10^{308}$	$2^{16384} \approx 10^{4932}$

A standard normalized floating-point word consists of a sign-bit s, exponent e, and an unsigned (fractional) normalized mantissa m, arranged as follows:

Sign bit s	Exponent e	Unsigned mantissa m

Algebraically, a (normalized) floating-point word is represented by

$$X = \left(-1\right)^s \times 1.m \times 2^{e-\text{bias}} \tag{9.1}$$

Note that this is a signed magnitude format. The "hidden" one in the mantissa is not present in the binary coding of the normalized floating-point number. If the exponent is represented with E bits, then the bias is selected to be

$$\text{bias} = 2^{E-1} - 1 \tag{9.2}$$

To illustrate, let us determine the decimal value 9.25 in 32-bit floating-point binary representation. The bias is 127, and after normalization to a $1.m$ format, i.e., $f = 9.25_{10} = 1001.01_2 = 1.00101 \times 2^{130-127}$, we have the following binary coding:

Binary		
s	E	M
0	100 0001 0	001 0100 000 000 0000 0000

The 32-bit hex number becomes $f = 4114\ 0000_{\text{HEX}}$. The HEX code can also be displayed using the %X printf() format in ANSI-C. In MATLAB we can display single and double floats in Hex with %tX and %bX format, respectively.

The IEEE standard 754–2008 for binary floating-point arithmetic [I08] also defines some additional useful special numbers to handle, for instance, overflow and underflow. The exponent $e = E_{\max} = 1 \ldots 1_2$ in combination with zero mantissa $m = 0$ is reserved for ∞. Zeros are coded with zero exponent $e = E_{\min} = 0$ and zero mantissa $m = 0$. Note that due to the signed magnitude representation, plus and minus zero are coded differently. There are two more special numbers defined in the 754 IEEE standard, but these additional representations are most often not supported in FPGA floating-point arithmetic. These additional numbers are denormals and NaNs (not a number). With denormalized numbers, we can represent numbers smaller than $2^{E\min}$

by allowing the mantissa to represent numbers without the hidden one, i.e., the mantissa can represent numbers smaller than 1.0. The exponent in denormals is coded with $e = E_{min} = 0$, but the mantissa is allowed to be different from zero. NaNs have proven useful in software systems to reduce the number of "exceptions" that are called when an invalid operation is performed. Examples that produce such "quiet" NaNs include:

- Addition or subtraction of two infinities, such as $\infty - \infty$
- Multiplication of zero and infinite, e.g., $0 \times \infty$
- Division of zeros or infinities, e.g., $0/0$ or ∞/∞
- Square root of negative operand

In the IEEE standard, 754 for binary floating-point arithmetic NaNs are coded with exponent $e = E_{max} = 1...1_2$ in combination with a nonzero mantissa $m \neq 0$. Table 9.8 shows the five major floating-point coding including the special numbers.

There are four supported rounding modes for floating-point type which are rounding-to-nearest-even (i.e., the default), rounding-to-zero (truncation), rounding to-∞ (round up), and round-to-negative-∞ (round down). In MATLAB the equivalent rounding functions are round(), fix(), ceil(), and floor(), respectively. The only small difference between the MATLAB and IEEE 754 modes is the rounding-to-even-nearest for numbers with $0.5_{10} = 0.1_2$ fractional part. Only if the integer LSB is one do we round up – otherwise, we round down; 32.5 is rounded down, but 33.5 is rounded up in the rounding-to-even-nearest scheme. Table 9.9 shows an example rounding that may occur in floating-point format.

It is interesting to observe that the default operation rounding-to-nearest-even is the most complicated scheme to implement and the rounding-to-zero is not only the cheapest but also may be used to reduce an undesired gain in the processing since we always round to zero, i.e., the amplitude does not grow due to rounding.

Table 9.8 The five major coding types in the 754–1985 and updated 754–2008 IEEE binary floating-point standard

Sign s	Exponent e	Mantissa m	Meaning
0/1	All-zeros	All-zeros	± 0
0/1	All-zeros	Nonzero	Denormalized $(-1)^s \times 0.m \times 2^{e\text{-bias}}$
0/1	$1 < e < E_{max}$	M	Normalized $(-1)^s \times 1.m \times 2^{e\text{-bias}}$
0/1	All-ones	All-zeros	$\pm\infty$
–	All-ones	Nonzero	NaN

Table 9.9 Rounding examples for the four floating-point types

Mode	32.5	33.25	33.5	33.75	−32.5	−32.25
Rounding-to-nearest-even	32	33	34	34	−32	−32
Rounding-to-zero	32	33	33	33	−32	−32
Rounding-to-∞	33	34	34	34	−32	−32
Rounding-to-nearest-even	32	33	33	33	−33	−33

Although the IEEE standard 754–1985 for binary floating-point arithmetic [I85] is not easy to implement with all its details, such as four different rounding modes, denormals, or NaNs, the early introduction in 1985 of the standard helped as it has become the most adopted implementation for microprocessors. The parameters of this IEEE single and double format can be seen from Table 9.7. Due to the fact that already single-precision 754 standard arithmetic designs will require substantial device resources, we find that sometimes FPGA designers do not adopt the IEEE 754 standard and define a special format. Shirazi et al. [SAA95], for instance, have developed a modified format to implement various algorithms on their custom computing machine called SPLASH-2, a multiple-FPGA board based on Xilinx XC4010 devices. They used an 18-bit format so that they can transport two operands over the 36-bit wide system bus of the multiple-FPGA board. The 18-bit format has a 10-bit mantissa, 7-bit exponent, and a sign bit and can represent a range of 3.7×10^{19}.

Floating-Point Operations HDL Synthesis

Implementing floating-point operations in HDL can become a labor-intensive task if we try to build our own complete library including all necessary operations and conversion functions. Luckily, at least for VHDL, we have seen the introduction of a very sophisticated library for operations and functions that can be used. This is part of the VHDL-2008 standard, and a library with over 7 K lines of VHDL code that are compatible with VHDL-1993 has been provided by David Bishop and can be downloaded from https://github.com/FPHDL/fphdl. Since most vendors support only a subset of the standard VHDL language, small modified versions are available on that web page that have been tested for Altera, Xilinx, Synopsys, Cadence, and MENTORGRAPHICS (MODELSIM) tools. The new floating-point standard is documented in Appendix G (pp. 537–549) of the VHDL-2008 LRM, and several textbooks now cover this new floating-point data types and the operations too [A08, P10, R11]. To use the library at a minimum in VHDL-1993, we would write

```
LIBRARY ieee_proposed;
USE ieee_proposed.fixed_float_types.ALL;
USE ieee_proposed.float_pkg.ALL;
```

The library allows us to use standard operators like we use for INTEGER and STD_LOGIC_VECTOR data types:

Arithmetic	+, -, *, /, ABS, REM, MOD
Logical	NOT, AND, NAND, OR, NOR, XOR, XNOR
Comparison	=, /=, >, <, >=, <=
Conversion	TO_SLV, TO_SFIXED, TO_FLOAT
Others	RESIZE, SCALB, LOGB, MAXIMUM, MINIMUM

There are also a few predefined constant values. The six values are zero = zerofp, NaN = nanfp, quite NaN = qnanfp, ∞ = pos_inffp, −∞ = neg_ inffp, and −0 = neg_zerofp. Predefined types of length 32, 64, and 128 bit as in the IEEE standards 854 and 754 [I85, I08] are called FLOAT32, FLOAT64, and FLOAT128, respectively.

Assuming now we like to implement a floating-point number with one sign, 6-bit exponent, and five fractional bits, we would define

```
SIGNAL a, b : FLOAT(6 DOWNTO -5);
SIGNAL s, p : FLOAT(6 DOWNTO -5);
```

and operations can be specified simply as

```
s <= a + b;
p <= a * b;
```

The code is short since left and right sides use the same data type. No scaling or resizing is required. However, the underlying arithmetic uses the default configuration setting that can be seen from the fixed_float_types library file. The rounding style is round_nearest, denormalize and error_check are set to true, and three guard bits are used. The minimum hardware effort on the other end will happen if we set rounding to round_zero (i.e., truncation), denormalize and error_check false, and guard bits to 0, so basically the opposite to the default setting. Most operations in VHDL-2008 float type are also available in a function form, e.g., for arithmetic function, we can use ADD, SUBTRACT, MULTIPLY, DIVIDE, REMAINDER, MODULO, RECIPROCAL, MAC, and SQRT. Then it is much easier to modify the rounding style and guard bits:

```
r <= SQRT(arg=>y,  -- Should be the "cheapest" design
          round_style => round_zero,
          guard => 0,
          check_error => FALSE,
          denormalize => FALSE);
```

The left and right operands are specified first, followed by the four synthesis parameters that should give the smallest number of ALMs. Note that the IEEE VHDL-2008-1076 (p. 540) says "guard_bits" not "guard" as has been used in the library written by David Bishop.

Comparison can also be used as a function call (similar names as in FORTRAN) via EQ, NE, GT, LT, GE, and LE. For scaling the function, SCALB(y,n) implements the operation $y \times 2^n$ with a reduced hardware effort compared to normal multiplication or divide. MAXIMUM, MINIMUM, square root SQRT, and multiply-and-add MAC are additional functions that can be useful.

Now let us discuss how these functions may work in a small ALU; see Exercise 9.41. Since most simulators so far not fully support the new data types such as a negative index in arrays, it seems to be a good approach that we stay with the standard STD_LOGIC_VECTOR as I/O type. The library provides so-called bit-preserving conversion functions that just redefine the meaning of the bits in the STD_LOGIC vector to a sfixed or float type of the same length. Such a conversion is done by the VHDL preprocessor and should not consume any hardware resources. On the other hand, we also need a value-preserving operation if we do a conversion between the sfixed and float types. This conversion will preserve the value of our data, but that will indeed require substantial hardware resources.

The VHDL-2008 library allows us to write efficient compact code for any float specified. The only disadvantage is that the overall speed of such a design will not be very high since substantial arithmetic is used, but no pipelining is implemented. FPGA vendor usually provide predefined floating-point blocks in 32- and 64-bits that are highly pipelined. Xilinx offers the LOGICORE floating-point IP, and Altera has a full set of LPM functions. The LPM blocks can be used as graphical block or instantiated from the component library that can be found under *quartus → libraries → vhdl → altera_mf → altera_mf _components.vhd*.

Let us take a brief look at the pipeline requirements to achieve high throughput. Table 9.10 lists the minimum number of pipeline stages in Altera's LPM; see altera_mf_components.vhd. Figure 9.9 shows a comparison of the throughput that can be achieved when using the VHDL-2008 library and the LPM blocks. The benefit of the pipelining in the LPM blocks is clearly visible. In order to measure the registered performance Fmax with the VHDL-2008 library, registers were added to the input and output ports, but no pipelining inside the block has been used.

Figure 9.10 shows the synthesis results for all seven basic building blocks in 32- and 64-bit width. The solid line shows the required ALMs with default VHDL-2008 settings, while the dashed lines show the results when synthesis options are set to minimum hardware effort, i.e., rounding set to round_zero (i.e., truncation), denormalize and error_check false, and guard bits to 0. Conversion between fixed and FLOAT32 requires about 100 ALMs. We see that basic math operations +,-,* require about 300 ALMs with standard setting and twice that for the divide. The SQRT function for 64 bits is the largest of all blocks with over 1 K ALMs when using LPM blocks. The IEEE 64-bit sqrt() could not be synthesized due to unsupported 64-bit divider size in QUARTUS.

Table 9.10 Minimum pipelining required in the Altera LPM block library for 32-bit and 64-bit floating-point data

	32-bits	64-bits
Int to FP	6	6
FP to int	6	6
Add/sub	7	7
Multiply	7	7
Divide	6	10
sqrt	16	30

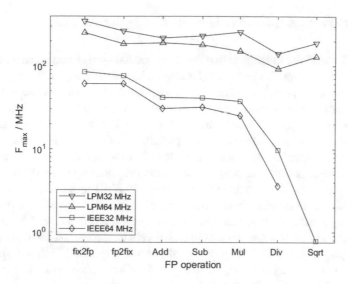

Fig. 9.9 Speed data for VHDL-2008 operations and LPM blocks from Altera. The IEEE 64-bit `sqrt()` could not be synthesized due to unsupported divider size in QUARTUS

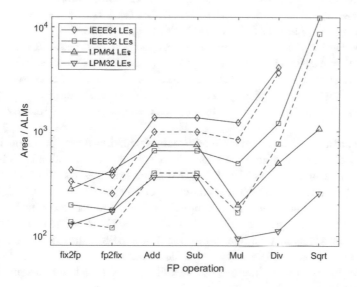

Fig. 9.10 Size data for VHDL-2008 operations and LPM blocks from Altera for default setting (solid line) and minimum hardware effort (dashed line). The IEEE 64-bit `sqrt()` could not be synthesized due to unsupported divider size in QUARTUS

Floating-Point Algorithms Design Example: FECG

In Chap. 5 we briefly mentioned that some embedded system may need a substantial number of floating-point operations. Examples include high-quality image processing, mechatronic systems, communication protocols, or biomedical signal processing. A modern compiler such as GCC can indeed emulate these large dynamic range floating-point operations via software algorithms, but this will require a substantial amount of instruction cycles for each operation. For a Nios II/e system (we had designed in the previous section), the cycle count measured for the basic floating-point operations is rather large, and the number of floating-point operations for a 100 MHz processor is then only in the range of 105 FMIPS; see Fig. 5.7. Let us now see how this floating-point performance impacts a typical biomedical algorithm. As a typical algorithm, let us assume we have sampled a multichannel FECG signal at 250 Hz/second and now use the first five channels to reconstruct the fetus ECG and the beats-per-minute (BPM). Figure 9.11 shows some typical FECG signals.

To compute the fetus BPM, we use the following six-step algorithm (partial MATLAB code):

```
x = ascii2fp(i);          %%%% 1) Convert int->float
s = sum(x);               %%%% 2) Calculate DC of input
x = x-sum(x)/length(x);   %%%% 2) remove DC part
Cxx = cov(x', 1);         %%%% 3) Calculate the covariance matrix
[E, D] = eig(Cxx);        %%%% 4) Eigenvector and Eigenvalues
V=D^-.5 * E';             %%%% 5) Compute reconstruction matrix and
z=V*x;                    %%%%      apply reconstruction
bpm(z)                    %%%% 6) Calculate the BPM and S/N
```

There are a substantial number of add/mul/sub floating-point operations involved and a few divides and square roots too. The total calculation with a Nios II/e from the previous chapter will need 161,803,090 clock cycles or for 100 MHz CPU we get would require a total of 1.6 seconds for completion of the FECG algorithm. However, a real-time system that computes a new BPM value for each incoming value would require a 4 ms computation time, i.e., 400 times faster. If we relax the real-time constrains that we like to provide a new BPM estimate at each beat, we will need a 2 Hz rate or 0.5 second or a factor 3.2 faster calculation. Let us in the following discuss how to improve our FECG monitor by adding custom IP to our Nios II processor.

Custom Floating-Point Hardware Options for Nios II

For floating-point acceleration via custom IP of the Nios system, we have basically two options: we may use any of the previously discussed VHDL blocks and integrate these into our Nios system, or we may use the custom floating-point IP

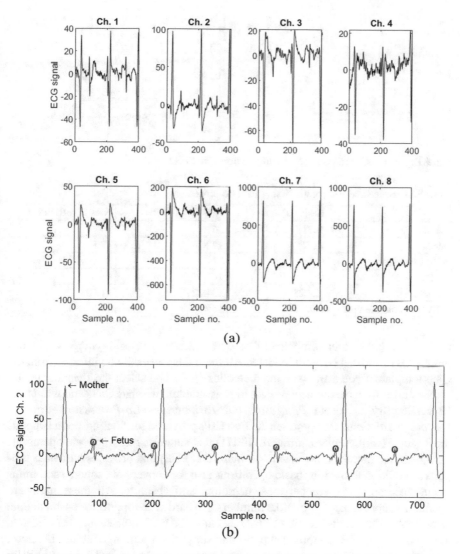

Fig. 9.11 (**a**) First 400 points or 1.6 s of the eight-channel DaISy signals. (**b**) The DaISy leads two noninvasive fetal ECG Database, FQRS detected (red circle). Superposition of two 3 s signals: large peaks (mother ECG) and small peaks (fetus ECG)

provided for us by Altera in the Qsys IP Catalog. Since the latter is much simpler, let us start with the IP Catalog and discuss how to add an HDL component later (Fig. 9.12).

Before we add the CIP to our system, we may want to create a second system that we can later compare performance. To this end we first rename the Altera Basic Computer to our new project name DE1_SoC_Custom_IP. You need to change the name and contents for the *.qsf, *.vhd, and *.qpf files. You may also need to copy the component directory called nios_system to your project. The HDL

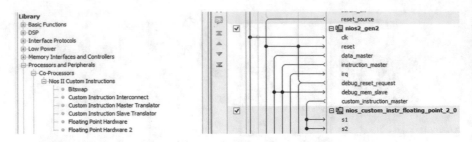

Fig. 9.12 DE1 SoC Basic Computer configuration with FPH2

Table 9.11 Speed and size comparison of FPH co-processors

	Basic Computer	Nios II with FPH1	Nios II with FPH2
ALMs	1071	3636	2059
Memory	11 K	12 K	27 K
M10K blocks	6	21	9
DSP 27 × 27	0	1	4
Fmax/ MHz	124.25	134.12	119.03
QUARTUS compile time	6:03	10:45	7:35

is placed with the other component files under `nios_system` synthesis submodules by QSYS and can be modified later. All these CIs can easily be inferred from the QSYS template if you start QSYS and then click *New...* and select the *Template* menu.

Now after we have our new project files assembled, we start QSYS and within the IP catalog try to locate *Processor and Peripherals→Co-Processors→Nios II Custom Instructions*. Here you see the two CIPs provided for floating-point support: the legacy *Floating Point Hardware* (FPH1) that supports the four basic operations +,*,-,/ and the new *Floating Point Hardware 2* (FPH2) that provides hardware supports in addition to the four basic operations also for conversion, square root, minimum, maximum, six comparisons, negation, and absolute. All these hardware operations support only 32-bit float numbers. No hardware support for 64-bit double is available. This second version FPH2 has improved performance and reduced area at the cost of limited subnormal support and simplified rounding. Table 9.11 shows a speed and size comparison of our Basic Computer extended by FPH1 and FPH2. A drawback with FPH2 is that it requires a BSP to fully support all functions, FPH1 can be compiled using standard ANSI C program settings with additional compiler option `-mcustom-fpu-cfg=60-2` (where 60–2 is used when the floating-point divider is enabled and 60–1 is used when you don't use a custom floating-point divider). The linker option `-lm` is always required when using floating-point operations. You will notice that in particular the generation of the newlib C library will require substantially more compile time for FPH2 than using an ANSI C program.

Let us now do some basic arithmetic computation with the three systems. We like to test the operations needed in our FECG algorithms, i.e., FP INT conversions, +,*,-,/ and `sqrt()` functions. Since `sqrt()` from the `math.h` library in the original definition is a 64-bit double operations, we also included calculations for the

32-bit `sqrtf()` introduced in the C99 standard that most likely has enough precision for many applications. If you read carefully the documentation for FPH2, then you will notice that GCC 4.7.3 has some trouble to map the custom IP to the internal function. This can be solved by forcing GCC to use the CIP via

```
#define my_sqrtf(A)    __builtin_custom_fnf(ALT_FPCI2_FSQRTS_N, (A))
```

The ALT constant is defined in an additional header file you need to include in your coding via

```
#include "altera_nios_custom_instr_floating_point_2.h"
```

With FPH2 the `sqrtf()` performance is substantially higher, while `sqrt()` still has the same performance. Figure 9.13 shows the comparison of the three systems. We clearly see an improvement for the four basic operations for both CIPs. For FPH2 we see additional improvements for conversion and the `sqrtf()` function. Finally, we measure the performance of the three systems for the FECG algorithm. The total computation time for one frame BPM calculation is now 515 ms with −O1 and for maximum compiler optimization 497 ms with −O3 so the desired 2 Hz performance could be achieved. Using the Altera Monitor Program, the −O3 performance is a little cumbersome when using the BSP. In the ANSI C, we could simply modify the compiler option, which is not possible when using BSP. Here are the steps to get the −O3 fast options with the Altera Monitor Program:

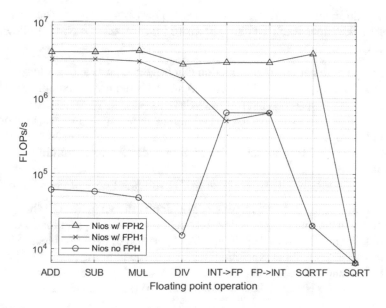

Fig. 9.13 The Nios floating-point performance for basic operations

1. Set up your C source Program to use the BSP style, and download your Nios system to the board.
2. Select *Actions→Generate Device Drivers (BSP)*.
3. Modify the `Makefile` in the BSP directory, and replace -O1 by -O3.
4. Run *Actions→Compile* to generate the top-level `Makefile` (this may take substantial time).
5. From your installed QUARTUS programs, start the tool called *"Nios II 15.1 Command Shell,"* and use **cd** to move to the source code folder that contains the top-level `Makefile`.
6. Edit the top-level Makefile, and replace -O1 by -O3 and remove -g option.
7. Delete the generated `*.elf` and `*.srec` files.
8. Run the top-level `Makefile` by typing `make` in your Nios II 15.1 Command Shell.
9. Use `nios2-elf-objcopy -O srec fmips.elf fmips.srec` to generate the `*.srec` file.
10. Use *Actions→Load* to run the `*.srec` program on your Nios II.

Now we have a working Nios II system with hardware support for basic 32-bit floating-point operations. Let us now in the following see how we may also add hardware support for other operations such as double-precision operations by using one of our predesigned HDL blocks.

Creation and Integration of Custom Logic Block

There are different types of CI in the Nios II QSYS environment. The basic ports (inputs A, B, output RESULT) are shown in Fig. 9.14a. The Nios II allows five types of CI:

- The pure combination circuit function that has three ports and no clock
- The fixed multicycle that adds a clock input and is required to complete the computation after a specified number of cycles
- The variable multicycle that uses an additional done port to indicate the completion of the operation
- The extended CI that has an 8-bit port n that allows one to multiplex different output ports
- The internal register file type CI that has 32 words for each of the 3 I/O ports

The outline of the port and naming convention is shown in Fig. 9.14b.

Let us try in the following to add 64-bit double-precision support for the `sqrt()` operation. We may use the IEEE 1076–2008 operation or the LPM block. Since the IEEE block comes without pipelining, the overall speed of the Nios II will be very low <1 MHz and the gain through the reduced instruction count will make all other instructions of the Nios II take much more time. The LPM block on the other side

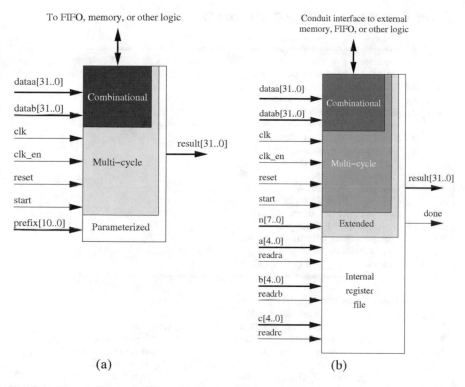

Fig. 9.14 Nios II CIP. (**a**) Adding custom logic to the Nios ALU. (**b**) Physical ports for the custom logic block

will have a substantial latency from pipelining (30 or 57 clock cycles for 64-bit double precision), but since it is highly pipelined, the overall speed of the Nios II will not be reduced much. In fact from Fig. 9.9, we see that the double-precision sqrt() operations can run with over 128 MHz, such that our Nios II processor can still run at 100 MHz as before. Let us go quick through the steps to set up a new component in the Qsys IP Library. Since writing to our core is much faster than reading the pipelined delay output, we will need to design a variable multicycle CIP. The Altera UG for custom instruction has a similar CRC example, so you may want to check out the CRC tutorial from the UG [AI17b] as an additional source of information. We start again with the Basic Computer and change the name and contents for the *.qsf, *.vhd, and *.qpf files. We also copy the component directory called nios_system to your project. To see also a Verilog example, this time we will use a Verilog design entry. We modify the *.vhd Basic Computer to a Verilog file and make appropriate replacements such as ENTITY→module, IN → input, OUT → output, N DOWNTO K port → [N:K], and component mapping from P=>Q → .P(Q), etc. Here is the new top-level Verilog file DE1_SoC_CustomIPv.v

Verilog Code 9.3: The DE1_SoC_CustomIPv.v Top Level

```verilog
1    //=====================================================================
2    // Implements a simple Nios II system for the DE1-SoC board.
3    // Inputs: SW9-0 are parallel port inputs to the Nios II system
4    // CLOCK_50 is the system clock
5    // KEY0 is the system reset
6    // Outputs: LEDR9-0 are parallel port outputs
7    // =====================================================================
8    module DE1_SoC_CustomIPv (
9        input CLOCK_50,
10       input [3:0] KEY,
11       input [9:0] SW,
12       output [9:0] LEDR,
13       inout [15:0] DRAM_DQ,
14       output [12:0] DRAM_ADDR,
15       output [1:0] DRAM_BA,
16       output DRAM_CAS_N, DRAM_RAS_N, DRAM_CLK,
17       output DRAM_CKE, DRAM_CS_N, DRAM_WE_N,
18       output DRAM_UDQM, DRAM_LDQM);
19   // =====================================================================
20
21     nios_system UUT
22             (.clk_clk(CLOCK_50),          // clk.clk
23              .leds_export(LEDR[9:0]),// leds.export
24              .pushbuttons_export(~ KEY[3:0]), // pushbuttons.export
25              .reset_reset( 1'b0),         // reset active high
26              .sdram_clk_clk(DRAM_CLK),    // sdram_clk.clk
27              .sdram_wire_addr(DRAM_ADDR),    // sdram_wire.addr
28              .sdram_wire_ba(DRAM_BA),        // .ba
29              .sdram_wire_cas_n(DRAM_CAS_N),  // .cas_n
30              .sdram_wire_cke(DRAM_CKE),      // .cke
31              .sdram_wire_cs_n(DRAM_CS_N),    // .cs_n
32              .sdram_wire_dq(DRAM_DQ),        // .dq
33              .sdram_wire_dqm({DRAM_UDQM, DRAM_LDQM}),
34              .sdram_wire_ras_n(DRAM_RAS_N), // .ras_n
35              .sdram_wire_we_n(DRAM_WE_N),   // .we_n
36              .switches_export(SW[9:0])); // switches.export
37
38   endmodule
```

Before we start adding the CIP to QSYS, you may want to develop the HDL wrapper for your CIP that meets the QSYS requirements. Such templates can be found in the appendix of legacy custom instruction UG [AI08, AI15]. As an alternative you may use as starting point the CRC tutorial file CRC_Custom_Instruction.v from the crc_hw folder. For the variable multicycle CIP with one operand, we need the inputs clk, reset, dataa, n, clock_en, and start and the two output ports done and result. The input n may have 8 bits, and dataa and result ports are 32 bit. All other ports are 1 bit each. For our 64-bit square root, we first use the IP Catalog to generate our LPM block. Select *Tools→IP Catalog → Library→Basic Functions→Arithmetic→ALTFP_SQRT*, click *Add...*, use your project home directory, name it fsqrt64, select Verilog as file type, and then click *OK*. The *MegaWizard Plug-In Manager* will open for *ALTFP_SQRT*, and in the General panel, select Double precision (64 bits) and latency 30 and then click *Next*. Do not add any extra ports and click *Next* twice. In the *Summary* panel, select *Instantiation template* file fsqrt64_inst.v so you can easily use the new component and then click *Finish*. Now complete your template for the CIP. Copy your new component, and connect it to the local nets; see lines 44–47 below. You will need to add one register to store your LSB inputs for n=0, MSBs for n=1 (see lines 70–78 below) and each one register for your outputs when n=2 (LSBs) and n=3 (MSBs), see lines 80–88 below. Writing to our CIP is not time critical, and we immediately issue a done impulse. The rising edge of the done impulse is used by Nios to continue with other operations. Since the fsqrt64 LPM block needs 30 clock cycles to complete, we use a small counter (lines 53–68 below) to issue the done aka output data valid signal after 30 clock cycles whenever one of the input values changed. We store the component output result in our output register in clock cycle 32, issue a done impulse at counter clock cycle 33, and stop counting after the counter reaches 34. If the counter has reached 34 and we receive a read request, we set the counter back to 30 to perform storing the correct part LSB/MSB in the output register and issue another done impulse. Overall the wrapper for our CIP now looks as follows.

Verilog Code 9.4: The CIP HDL Wrapper for the SQRT64

```
1     //====================================================================
2     // Verilog Custom Instruction Template for Internal Register Logic
3     // Ports as required for multi-cycle or extended multicycle
4     `define DEBUG_PRINT
5     module fp_sqrt64(
6       clk, // CPU system clock
7       reset, // CPU master asynchronous active high reset
8       clk_en, // Clock-qualifier
9       start, // Active high signal used to specify that inputs are valid
10      done, // Active high signal used to notify the CPU result is valid
11      n, // N-field selector (required for extended)
12      dataa, // Operand A (always required)
13      datab, // Operand B (optional)
14      //======== Add some test ports:
15    `ifdef DEBUG_PRINT
16      done_delay_out, sqrt_out, cnt_out,
17    `endif
18      result); // Result (always required)
19      input clk;)
20      input reset;
21      input clk_en;
22      input start;
23      output done;
24      input[7:0] n;
25      input[31:0] dataa;
26      input[31:0] datab;
27      //======== Add some test ports:
28    `ifdef DEBUG_PRINT
29      output done_delay_out;
30      output [63:0] sqrt_out;
31      output [4:0] cnt_out;
32    `endif
33      output [31:0]result;
34
35    // ==========================================================
36      reg [63:0] data;
37      wire [63:0] sqrt;
38      reg [31:0] r;
39      reg [6:0] cnt = 0;
40
41    // Use the n[7..0] port as a select SIGNAL on a multiplexer
42    // to select the value to feed result[31..0]
43
44    //  altfp_sqrt  pipeline range: single 16-28; double 30-57
45      fsqrt64  fsqrt64_inst (
46      .clock(clk), .data(data), .result(sqrt)
47      );
48
49      wire done_delay = (cnt==7'd33)? 1 : 0;
50      wire ready = (n>8'h01) ? done_delay : start;
51      assign done = ready;
52
53      always @ (posedge clk or posedge reset) // Pipeline counter
```

```
54    if (reset)  begin
55       cnt = 0;
56    end else begin
57        if ((start) && ((n==8'h02) || (n==8'h03)) &&
58                   (clk_en) && (cnt==7'd34)) begin // decrement counter
59            cnt = 7'd29;
60        end
61        if ((start) && ((n==8'h00) || (n==8'h01))
62                   && (clk_en)) begin // reset counter for new values
63            cnt = 7'd0;
64        end
65        if (cnt < 7'd34) begin // count until 34 is reached
66            cnt = cnt + 7'd1;
67        end
68    end
69
70    always @ (posedge clk or posedge reset) // get the input data
71      if (reset) begin
72        data = 32'h0;
73      end else begin
74        if ((n==8'h00)&& (clk_en))
75          data[31:0] = dataa;
76        else if ((n==8'h01) && (clk_en))
77          data[63:32] = dataa;
78      end
79
80    always @ (posedge clk or posedge reset) // write the results
81      if (reset) begin
82        r = 32'h0;
83      end else begin
84        if ((n==8'h02)&& (clk_en) && (cnt==7'd31))
85          r = sqrt[31:0];
86        else if ((n==8'h03)&& (clk_en) && (cnt==7'd31))
87          r = sqrt[63:32];
88      end
89
90    assign result = r; // connect to outputs
91
92    //=== Test port assignments:
93    `ifdef DEBUG_PRINT
94    assign done_delay_out = done_delay;
95    assign sqrt_out = sqrt;
96    assign cnt_out = cnt;
97    `endif
98    endmodule
```

For debugging of the core, we have added a few additional ports, which conveniently can be disabled in the final design by removing `define   DEBUG_PRINT. Since a LPM block is involved, we will use the timing simulation and not a functional simulation of the Verilog output file *.vo from QUARTUS. Such a simulation will require a full compile in QUARTUS but overall simplifies the MODELSIM simulation. Figure 9.15 shows the simulations of our CIP.

We see that first input LSBs and MSBs are stored in registers. The output LSB results for reading are available after 350 ns and the MSBs after 425 ns. The test

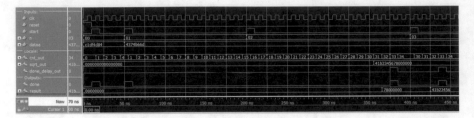

Fig. 9.15 Verilog simulation of CIP for 64-bit `sqrt` LPM function with port names as required by QSYS CIP

data can be verified with MATLAB `sprintf()` using the `%bX` format for 64-bit floats. We are now ready to add our new custom IP in Qsys. Implementing a Nios II custom instruction hardware uses the following steps:

1. To open the component editor, first start *Tools→Qsys* and load the *nios_system*. This should be the same system we had developed for the Basic Computer, i.e., a Nios with `SDRAM`, `UART`, `switches`, `leds`, `pushbuttons`, and `timer`. Now on the left IP Catalog panel, click *New...* and the *Component Editor* will pop up.
2. In the *Component Type* panel, specify *Name* and *Display name* as `fp_sqrt64`; you may also add a *Description* and *Created by* entry.
3. Move to the *Files* panel, and add there under *Synthesis Files* your wrapper file `fp_sqrt64.v` and the LPM component `fsqrt64.v`. Make sure the `fp_sqrt64.v` has the *Top-level File* as *Attribute*. Now click *Analyze Synthesis Files* button. When done, click *Close*.
4. Configuring the custom instruction parameter type is not required for our CIP since we have no parameters.
5. To set up the custom instruction interfaces, select *View→Interfaces*. Note that this panel does not show up in the default GUI. As *Type* select *Custom Instruction Slave* and *Operands*: 1.
6. To configure the custom instruction signal types, select *View→Signals*. This panel is also not part of the initial GUI. For all nine signals, the interface type in the second column must be *nios_custom_instruction_slave*. The *Signal Type* entry in the third column must match the *Name* entry in the first column. If you have all data correct, go back to your *Interface* panel, and click *Remove Interfaces With No Signals*. The clock and reset entry should disappear from the *Interfaces* panel. Now you should also see the *Access Waveform* and under *Parameters* that *Clock cycles* is 0 and that *Clock cycle type* has changed to *Variable*, which are the correct settings for our CIP. Finally the two panels for *Interfaces* and *Signals* should look as the one shown in Fig. 9.16.
7. Review that all data are correct by clicking on *nios_custom_instruction_slave* in the *Signal & Interfaces* panel. Save the new CIP and add the new custom instruction to your Nios System; see Fig. 9.17. Make the connection from the `Nios II gen2` processor to your new component. The Opcode of your component should be ok with 0.
8. Now generate the new Nios system for Verilog source code and run a full compile in the Intel QUARTUS Prime software.

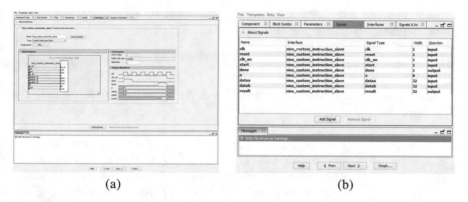

(a)   (b)

Fig. 9.16 The two new component editor panels not included in the original GUI. Nios II CIP board tests. (**a**) Interfaces. (**b**) Signals

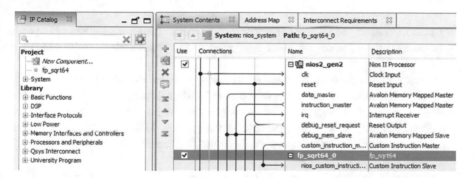

Fig. 9.17 New CIP available now from the IP catalog

You should see that the initial ca. 1071 ALMs of the Basic Computer with the SQRT64 will increase to about 2146 ALMs after a full compilation. Make sure that the Nios II still can run at the requested 100 MHz by checking the TIMEQUEST results. Start then the Altera Monitor Program to configure a new BSP type project with your new DE1_SoC_CustomIPv Nios system. Let us in the following first test the new CIP with two short examples before we continue to measure the speed-up achieved with our 64-bit SQRT CIP.

A function prototype of our new CIP including the required constant definition is placed in the system.h file as part of the generated BSP and can be used verbatim via #include or copied to your ANSI C source file. Since the default function name is rather long, we redefine the function call into a shorter version and place this new definition at the beginning of our source file:

```
#define ALT_CI_FP_SQRT64_0_N 0x0
#define ALT_CI_FP_SQRT64_0_N_MASK ((1<<8)-1)
#define my_sqrt64(n, A) \

__builtin_custom_ini(ALT_CI_FP_SQRT64_0_N+(n&ALT_CI_FP_
SQRT64_0_N_MASK),(A))
```

Note how our four CIP instructions are identified via an offset to the opcode base. The standard double-precision square root call

```
d[1] = sqrt(d[0]);
```

can now be replaced when using the CIP with the following four instructions aka function calls.

```
k = my_sqrt64(0, i[0]);   // Write to CIP LSBs
k = my_sqrt64(1, i[1]);   // Write to CIP MSBs
i[2] = my_sqrt64(2, 0);   // Read the SQRT64 LSBs
i[3] = my_sqrt64(3, 0);   // Read the SQRT64 MSBs
```

The integer representation allows us also to display the binary/hex value in ANSI C via

```
printf("LSBs: %08d_10=%08X_hex\n", i[0], (unsigned int) i[0]);
printf("MSBs: %08d_10=%08X_hex\n", i[1], (unsigned int) i[1]);
```

As test data we use a trivial example with $4*4 = 16$ that can also be implemented by reading the slider switch values of the board and an advanced example with $x*x$, where $x = 0 \times 12345678$. For $k = 4$ the 64-bit hex representation is 40100000_00000000, and for x the hex presentation is 41B23456_78000000. The test program sqrt64.c will produce the following test results:

```
================== repeat with 2. number ==================
with x=0x12345678(hex)=>FP: 41B23456_78000000
gives x*x=4374B66D_C1DF4D84 hex
write to  CIP: LSBs   : -1042330236_10=C1DF4D84_hex
write to  CIP: MSBs   : 1131722349_10=4374B66D_hex
read from CIP: LSBs   : 2013265920_10=78000000_hex
read from CIP: MSBs   : 1102197846_10=41B23456_hex
```

If we assume our data are stored in a double array, then we need to split up the double floating-point number in 32-bit MSBs and 32-bit LSBs (unsigned) integers. We cannot use a casting operation for this split, but in Chap. 5 we discussed a convenient way for this type of task using pointer arithmetic. The pointer of the double and int array is aligned such that we can access the 64-bit double or the two 32-bit int data. Our arrays and pointers should be defined as follows:

```
double d[4];    double *d_ptr;
int i[8];       int *i_ptr;
```

Here is the memory alignment and the final test data values in memory of the example:

Output:0x12345678		Input: 0x12345678[2]		Output: 4.0		Input: 16.0	
$41B23456_{16}$	78000000_{16}	$C1DF4D84_{16}$	$4374B66D_{16}$	40300000_{16}	00000000_{16}	40100000_{16}	00000000_{16}
MSBs	LSB	MSBs	LSB	MSBs	LSB	MSBs	LSB
d[3]		d[2]		d[1]		d[0]	
i[7]	i[6]	i[5]	i[4]	i[3]	i[2]	i[1]	i[0]

Below is the ANSI C coding including restoring the output data as 64-bit double number and data read/write I/O that runs at the end of the test program in an infinity loop:

```
SW_value = *(SW_switch_ptr);       // Read the SW slider switch
d[0] = (double) SW_value;          // Initial double: 25 or 100
d_ptr = & d[0];                    // Get starting address
i_ptr = (int *) d_ptr;             // Point to same address
i[0] = (unsigned int) *i_ptr;      // First 32 bits
i[1] = (unsigned int) *(i_ptr+1);  // Second 32 bits
k = my_sqrt64(0, i[0]);            // Write to CIP LSBs
k = my_sqrt64(1, i[1]);            // Write to CIP MSBs
i[2] = my_sqrt64(2, 0);            // Read the SQRT64 LSBs
i[3] = my_sqrt64(3, 0);            // Read the SQRT64 MSBs
d_ptr = & d[1];                    // Get starting address
i_ptr = (int *) d_ptr;             // Point to same address
*(i_ptr) = i[2];                   // Place LSBs in double d[1]
*(i_ptr+1) = i[3];                 // MSBs in double d[1]
*(red_LED_ptr) = (int) d[1];       // Display at LEDs
```

First we read the SW value and convert to double. Then we split our double word into two ints using aligned pointer. It follows the four CIP function calls. Finally, we place the two integers in the memory location such that we get a 64-bit double precision number. Finally, we display the results on the red LEDs. Figure 9.18 shows examples for input $5^2 = 25$ and $10^2 = 100$. The switch values show the input to the CIP, while the red LEDs show the output result after square root.

For each of the test program sections, we measure the number of clock cycles used. These data are shown in Table 9.12 along with the synthesis data of the initial design (i.e., Basic Computer) and the CIP design. Depending if we assume the data are 64-bit addition pointer arithmetic is needed or the data are available as integer array, we get a speed-up factor of 41 and 69, respectively.

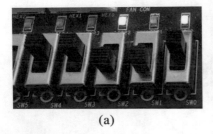 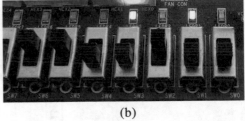

(a)   (b)

Fig. 9.18 Nios II CIP board tests. (**a**) Twenty-five SW = 25_{10} = 11001_2 result red LED = 5_{10} = 101_2. (**b**) One hundred SW = 100_{10} = 1100100_2 result red LED = 10_{10} = 1010_2

Table 9.12 Speed and size comparison of Nios systems with/without SQRT64 CIP

	Basic Computer	Nios II with SQRT64
ALMs	1071	2146
Memory	11 K	11 K
M10K blocks	6	6
DSP 27x27	0	0
Fmax/ MHz	124.25	110.3
QUARTUS compile time	6:03	6:49
Float sqrtf() clock cycles	4879	
Double sqrt() clock cycles	13,640	195
Double sqrt() with pointer arithmetic	–	331
Speed improvement (Factor)	41–69	

9.5   Looking Under the Hood: Nios II Instruction Set Architecture

In the majority of projects, we do not need to have precise knowledge of the inner working of our machine aka CPU, but in a couple of instance like writing driver for new components such as a JTAG controller, such knowledge is beneficial sometimes even necessary to know the nitty-gritty of the CPU architecture such as pipe lining, registers, or ANSI C compiler expectations.

It is usually pretty tiresome to go through each instruction, addressing modes, registers, etc. one-by-one, so let us make this study a little bit more fun by designing our own tiny RISC Nios II (TRISC3N) that implements a reduced set of the Nios II Instruction Set Architecture (ISA). We will not attempt to implement all instructions but a useful subset that allows us to write simple tasks such as a flash program, memory access, or (nested) function calls such that we can study the hardware-software interface between Nios II and our GCC.

So, let's get started how registers are organized in Nios II. Some have special meaning for the GCC compiler, and these registers get special names. The Nios II has thirty-two 32-bit registers. The first register r0 is always zero and has the same name. The second register (at) is used as assembler temporary. Register r2 and r3 may be used as function return values and r4–r7 as function arguments. Registers r8–r23 are general-purpose registers. Registers r24–r31 are special registers for the compiler and should not be used for general purpose. The compiler uses often r24=et as exception temporary register, r26=gp as global pointer register, r27=sp as stack pointer, and r31=ra as return address for function calls. Less frequently used are r25=bt break temporary, r29=ea exception return address, and r30=ba break return address.

Nios II uses three instruction types aka categories. These are the immediate type (I-type), the register type (R-type), and the jump type (J-type). These allow us to encode and decode each instruction also by hand; see Exercise 9.15–9.22. All instructions are 32-bit long, and the three formats are shown in Table 9.13.

Table 9.13 The three 32-bit instruction coding format used by Nios II

Bits:	31...27	26...22	21...17	16..11	10...6	5...0
Format	5-bits	5-bits	5-bits	6-bits	5-bits	6-bits
R-TYPE	A	B	C	OPX	IMM5	OP
I-TYPE	A	B	IMM16			OP
J-TYPE	IMM26					OP

All three types specify the operation (OP) in the six LSBs 5...0. The J-type includes a 26-bit long IMM26 constant for instructions like jmpi or call that require a large constant value. Most of the I-type instructions are of the type r(B)=r(A) □ IMM16. B specifies one of the 32 destination registers and A the index of the first source. For r(B) often the abbreviation rB is used. Since all data are 32 bits in size the IMM16 values are zero or sign extended. Some operations like orhi use IMM16 in the 16 MSBs. All R-type instructions are coded with OP=0x3A. The individual operations are encoded with additional 6 bits in the eXtended code OPX in bits 16...11. The majority of R-type instruction uses the r(C)=r(A) □ r(B) style as typical for the three-address machine. A few instructions use IMM5 to define a small constant as needed for instant in the immediate shift operation slli rC = rA << IMM5. Out of the possible $2^6 = 64$ codes Nios II uses 46 operation codes and 42 eXtended OPX codes. Other than PICoBLAZE the Nios II ISA has been using the identical OP code for over 15 years [A03], adding only jmpi and four new control instructions. Let us have a look briefly at the available instruction classes. The grouping by functionality lets us build the following eight groups:

- Data transfer, i.e., load and store, are available in byte, 2 byte (half word), and 4 byte aka word size. The instructions ending with io will bypass the cache, and for unsigned operand a "u" is included (16): ldb, ldbio, ldbu, ldbuio, stb, stbio, ldh, ldhio, ldhu, ldhuio, sth, sthio, ldw, ldwio, stw, and stwio.
- The ten logical instructions contain four R-type, and, or, xor, and nor, and six IMM16 type, andi, ori, xori, andhi, orhi, and xorhi, where the addition "h" indicates a use of IMM16 as MSBs.
- We count ten arithmetic operations, two I-type addi and muli and eight R-type, add, sub, mul, div, divu, mulxss, mulxuu, and mulxsu with u/s indicating (un)signed operands.
- Nios II has 12 comparisons operations; 6 (cmpeq (==), cmpne (!=), cmpge (>=), cmpgeu (>= unsigned), cmplt (<), cmpltu (< unsigned)) are R-type and another 6 using IMM16: cmpeqi, cmpnei, cmpgei, cmpgeui, cmplti, and cmpltu. The "u" indicates unsigned operands.
- We count nine shift and rotate operations. R-type operation names and meanings are as in VHDL: rol, ror, sll, sra, and srl. There are also four I-type operations: roli, slli, srai, and srli.

- Nios II has six program control operations. Two use IMM26 from J-type (call and jmpi); three use one register (callr, jmp, and ret); and br uses IMM16.
- We count six conditional branches. Naming is similar to the FORTRAN comparison operation bge, bgeu, blt, bltu, beq, and bne. With "u" indicating unsigned.
- Nios II has 19 other control instruction that often deal with pipeline (flash) control, coding the R-Type, or custom instructions: trap, eret, break, bret, rdctl, wrctl, flushd, slushi, initd, initi, flushp, sync, initda, flushda, rdprs, wrprs, nextpc, R-type, and custom.

The assembler also has 19 pseudo-instruction you may see in example assembler programs such as five different move instructions, nop, or subi that do not have a direct hardware implementation but simplify reading code and are replaced by equivalent operation that has the same effect for the CPU.

A full description of all instructions can be found in the processor reference manual. Make sure you download the correct one matching your installed software [AI16]. Now let us have a closer look at the instructions we like to implement in HDL. In the table we use $\sigma(\text{IMM}16)$ as shortcut for a sign extension. Sorted by increasing operation code, Table 9.14 shows opcode, assembler coding, format, operation group, and a brief operation description.

We try to use similar CONSTANT/PARAMETER coding. Only for logic operation an op... is added since AND, OR, and XOR are VHDL keywords. Now let us recode our PICOBLAZE LED flash example in Nios II assembler language. Here is the program listing:

ASM Program 9.5: LED Toggle Using a Nios II Assembler Coding

```
1           .text                        /* executable code follows */
2         .global    _start
3   _start: /* initialize base addresses of parallel ports */
4           orhi    r1, r0, 0xFF20   /* MSBS for SW slider and LED base */
5           ldwio   r2, 0x40(r1)     /* load slider switches */
6   flash: stwio    r2, 0(r1)        /* write to red LEDs */
7   /*      movia   r3, 25000000        delay counter */
8           orhi    r3, r0, 0x2FB    /* HEX: 17D0000+7840=17D77840 */
9           addi    r3, r3, 0x7840   /* DEC: 24969216+30784=2500000 */
10          addi    r4, r0, 0x1      /* place 1 in register 4 */
11  loop:   sub     r3, r3, r4       /* decrement counter r3 */
12          bne     r3, r0, loop     /* check if counter is zero */
13          xori    r2, r2, 0x3FF    /* complement all bits */
14          jmpi    flash            /* start another loop */
```

Starting label for Nios II assembler is always the _start label. The I/O address are listed in address_map_nios2.h, and we see that the slider switch at 0xFF200040 and red LEDs at address 0xFF200000 share the same 16-bit

Table 9.14 Nios II ISA coding examples. OP and OPX code with OP = 3A (Hex)

OP	ASM coding	Format	Operation group	Operation description
00	`call label`	J	Prog. control	ra = PC + 4; PC = {PC $_{31:28}$, IMM26<<2}
01	`jmpi label`	J	Prog. control	PC = {PC $_{31:28}$, IMM26<<2}
04	`addi rB,` `rA, IMM16`	I	Arithmetic	rC = rA + σ(IMM16)
06	`br label`	I	Prog. control	PC = PC + 4 + σ(IMM16)
0C	`andi rB,` `rA, IMM16`	I	Logical	rB = rA AND {0x0000,IMM16}
14	`ori rB,` `rA, IMM16`	I	Logical	rB = rA OR {0x0000,IMM16}
15	`stw rB,` `offset(rA)`	I	Data transfer	MEM[rA+ σ(IMM16)] = rB
17	`ldw rB,` `offset(rA)`	I	Data transfer	rB = MEM[rA+ σ(IMM16)]
1C	`xori rB,` `rA, IMM16`	I	Logical	rB = rA OR {0x0000,IMM16}
1E	`bne`	I	Conditional branch	If (rA = rB) THEN PC = PC + 4+ σ(IMM16) ELSE PC = PC + 4
26	`beq`	I	Conditional branch	If (rA! = rB) THEN PC = PC + 4+ σ(IMM16) ELSE PC = PC + 4
34	`orhi rB,` `rA, IMM16`	I	Logical	rB = rA OR {IMM16,0x0000}
35	`stwio rB,` `offset(rA)`	I	Data transfer	MEM[rA+ σ(IMM16)] = rB
37	`ldwio rB,` `offset(rA)`	I	Data transfer	rB = MEM[rA+ σ(IMM16)]
3A		R	Others	R_type use OPX code below
OPX	**ASM coding**	**Format**	**Operation type**	**Operation description**
05	`ret`	R	Prog. control	PC = ra
0D	`jmp rA`	R	Prog. control	PC = rA
0E	`and rC,` `rA, rB`	R	Logical	rC = rA AND rB
16	`or rC, rA,` `rB`	R	Logical	rC = rA OR rB
1E	`xor rC,` `rA, rB`	R	Logical	rC = rA XOR rB
31	`add rC,` `rA, rB`	R	Arithmetic	rC = rA + rB
39	`sub rC,` `rA, rB`	R	Arithmetic	rC = rA – rB

MSBs. We load first the slider switch values (lines 4–5) and write this to the red LED outputs. For I/O components we use the ...io style of the data transfer instruction to skip the data cache. The Altera Monitor translates the 32-bit pseudo-loading instruction movia for the delay counter value 25,000,000 into register r3 (lines 8–9) by a series of or high and add immediate instructions. We could also have used the high (orhi) and low (or) instruction instead as recommended by Nios Tutorial [A15]. The instructions from lines 11–12 implement the delay loop: counting down r3 until it reaches zero. Then all bits in register 2 are inverted. This could have been done with a nor instruction also. In line 14 we jmpi back to line 6 (label flash) to start another loop. We see that the I-type instruction is used very often. We have only used one R-type (sub) and one J-type (jmpi) instruction. Encoding and decoding of these instructions in HEX code is left as an exercise at the end of the chapter; see Exercises 9.15–9.22.

Now let us have a look at the VHDL simulation in Fig. 9.19. The simulation shows the input signal for clock and reset, followed by local non-I/O signals and finally the I/O signals for switches and LEDs. After the release of the reset (active low), we see how the program counter (pc) continuously increases until the end of the loop. The pc changes with the falling edge, and instruction word is updated with the rising edge (within the ROM). The value 5 is read from the input port in_port and put into register r2. The register r2 value is loaded in the out_port register. Next decrement 1 is loaded in register r4, and the counter start value 0x017D7840 is loaded into register r3 in two steps: first the MSBs and then the LSBs. The counter is decremented by 1 in two clock cycles, and since the 32-bit value is not zero, the loop is repeated. This continues for many clock cycles until the 1 second simulation time is reached and the output toggles.

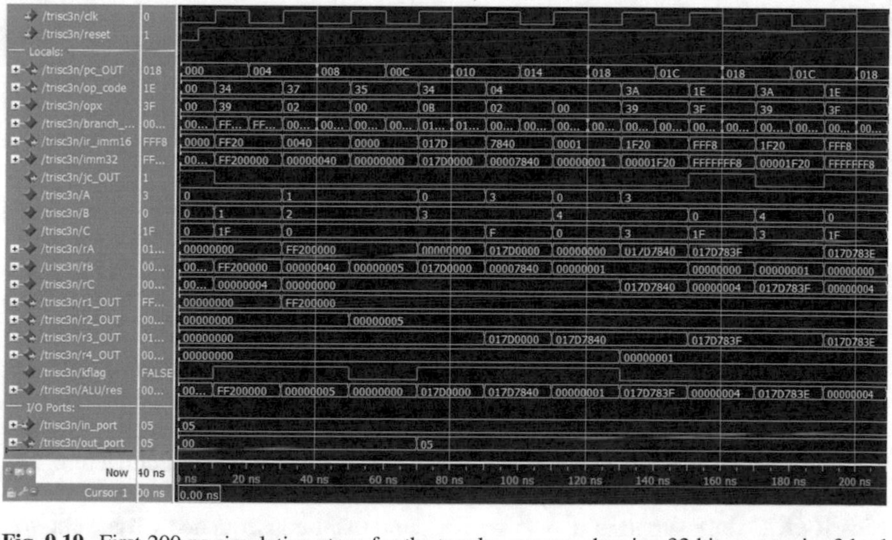

Fig. 9.19 First 200 ns simulation steps for the toggle program showing 32 bit counter in r3 load and counting down ...40,...3F,...3E,...

Now with basic I/O of Nios II understood, let us have a look on how our GCC organizes our data for our Nios II processor. In order to have similar GCC memory addressing and assembler code we need for our HDL, we make a small modification to our Basic Computer first: since our HDL will have only 4 K address, we first modify our DE1 SoC Basic Computer and replace the DRAM by a 4K×32 on-chip memory and recompile the system.

On-Chip Memory (RAM or ROM)
altera_avalon_onchip_memory2

▼ Size	
Data width:	32 ∨
Total memory size:	16384 bytes

The memory is now much smaller and will no longer be able to host things like a `printf()`; for our short test programs the 4 K word size should be sufficient.

In ANSI C code, we can place out variable/data definitions in the initial section of our code, making the data "global" available in the main section as well as any of the functions we may have. The other option is that we can place the variable definition within the main code, i.e., after `main()`. Let us have a look on how GCC is handling these two cases. Here is a little ANSI C test program for this.

DRAM Program 9.6: DRAM Write Followed by Read

```
1    volatile int g;// Volatile allows us to see the full assembler code
2    volatile int garray[15];
3    int main(void) {
4      volatile int s;    // Declare volatile tells compiler do
5      volatile int sarray[14]; // not try optimization
6      s = 1; // Memory location sp+0
7      g = 2; // Memory location gp+0
8      sarray[0] = s; // Memory location array[0] is sp+4
9      garray[0] = g; // Memory location array[0] is gp+4
10     sarray[s] = 0x1357; // use variable; requires *4 for byte address
11     garray[g] = 0x2468; // use variable; requires *4 for byte address
12   }
```

The assembler output code of interest in the Altera monitor is shown in Fig. 9.20. We notice that global scalar as well as global arrays are indexed by GCC using the global pointer aka register `gp=r26`. The local data defined within our `main()` sections are indexed via the stack pointer `sp=r27`. All data (and program) access is done in increments of bytes, and that is why our first integer variable has address `0(sp)`, the second `4(sp)`, third `8(sp)`, etc. The Nios II bus architecture is internally designed to have separate data and program buses aka Harvard architecture, that can be connected to (one or more) tightly coupled data memories. However, since our Nios II GCC only supports one `.text` section, as a consequence we have a von Neumann memory organization, i.e., program and data share the same memory. The

			_start:		
	0x00000000	06C00034	**orhi**	sp, zero, 0x0	
	0x00000004	DED00004	**addi**	sp, sp, 0x4000	
	0x00000008	DEF6303A	**nor**	sp, sp, sp	
	0x0000000C	DEC001D4	**ori**	sp, sp, 0x7	
(a)	0x00000010	DEF6303A	**nor**	sp, sp, sp	
	0x00000014	06800074	**orhi**	gp, zero, 0x1	
	0x00000018	D6A22804	**addi**	gp, gp, -0x7760	
	0x0000001C	06000034	**orhi**	et, zero, 0x0	
	0x00000020	C6023B04	**addi**	et, et, 0x8EC	
	0x00000024	00800034	**orhi**	r2, zero, 0x0	
	0x00000028	10804604	**addi**	r2, r2, 0x118	
	0x0000002C	1000683A	**jmp**	r2	

			main:		
	0x00000030	DEFFF104	**addi**	sp, sp, -0x3C	
			volatile int s; // Declare volatile tells compiler do		
			volatile int sarray[14]; // not try optimization		
			s = 1; // Memory location sp+0		
	0x00000034	00800044	**addi**	r2, zero, 0x1	
	0x00000038	D8800035	**stwio**	r2, 0(sp)	
			g = 2; // Memory location sp+0		
	0x0000003C	00800084	**addi**	r2, zero, 0x2	
	0x00000040	D0A00335	**stwio**	r2, -32756(gp)	
			sarray[0] = s; // Memory location array[0] is sp+4		
	0x00000044	D8800037	**ldwio**	r2, 0(sp)	
	0x00000048	D8800135	**stwio**	r2, 4(sp)	
			garray[0] = g; // Memory location array[0] is sp+4		
	0x0000004C	D0A00337	**ldwio**	r2, -32756(gp)	
	0x00000050	00C00034	**orhi**	r3, zero, 0x0	
	0x00000054	18C22C04	**addi**	r3, r3, 0x8B0	
	0x00000058	18800035	**stwio**	r2, 0(r3)	
(b)			sarray[s] = 0x1357; // use variable; requires *4 for byte address		
	0x0000005C	D8800037	**ldwio**	r2, 0(sp)	
	0x00000060	1085883A	**add**	r2, r2, r2	
	0x00000064	1085883A	**add**	r2, r2, r2	
	0x00000068	D885883A	**add**	r2, sp, r2	
	0x0000006C	10800104	**addi**	r2, r2, 0x4	
	0x00000070	0104D5C4	**addi**	r4, zero, 0x1357	
	0x00000074	11000035	**stwio**	r4, 0(r2)	
			garray[g] = 0x2468; // use variable; requires *4 for byte address		
	0x00000078	D0A00337	**ldwio**	r2, -32756(gp)	
	0x0000007C	1085883A	**add**	r2, r2, r2	
	0x00000080	1085883A	**add**	r2, r2, r2	
	0x00000084	1885883A	**add**	r2, r3, r2	
	0x00000088	00C91A04	**addi**	r3, zero, 0x2468	
	0x0000008C	10C00035	**stwio**	r3, 0(r2)	
			}		
	0x00000090	DEC00F04	**addi**	sp, sp, 0x3C	
●	0x00000094	F800283A	**ret**		

(c)		+0x0	+0x4	+0x8	+0xc
	0x00003FB0	00000000	00000000	00000000	00000001
	0x00003FC0	00000001	00000001	00001357	00000000

(d)		+0x0	+0x4	+0x8	+0xc
	0x000008A0	00000000	00000480	00000480	00000002
	0x000008B0	00000002	00000000	00002468	00000000

Fig. 9.20 TRISC3N data memory ASM test routine. Local and global memory write are followed by reading the same locations. (**a**) Starting sequence. (**b**) main() program. (**c**) Memory at sp (**d**) memory at gp

data memory can only use the address space after program section. Starting address of gp is somewhere after the code sections, while sp use a starting address that is at the end of our physical memory decreased by the number of bytes needed to host our data. These values are set by GCC at the beginning of the program; see Fig. 9.20a.

We ran the testdem.c program until it reaches ret and then monitor the memory content. At the end of our physical memory, we see (Fig. 9.20c) the local values indexed by sp. We see that sarray[1]=0x1357 can be found at address

0x3FC8, i.e., close to the end 0x4000, and garray[2]=0x2468 at address
0x8B8 somehow after the program code. This should also be the memory location
we should monitor in a HDL simulation.

As a third example, let us check what instructions are needed for (nested) functions calls. Again, we start with a short ANSI C example and then see what GCC
requires as instructions. Here is our C code.

3 Level Nesting Program 9.7: Nesting Function Test Code

```
1    void level3(int *array, int s1) {
2        s1 += 1;
3        array[3] = s1;
4        return;
5    }
6    void level2(int *array, int s1) {
7        s1 += s1;
8        array[2] = s1;
9        level3(array, s1); // call level 3
10       return;
11   }
12   void level1(int *array, int s1) {
13       s1 += 1;
14       array[1] = s1;
15       level2(array, s1); // call level 2
16       return;
17   }
18   int main(void) {
19     volatile int s1, s2, s3;
20     int array[11];
21     s1 = 0x1233; // Memory location sp+0
22     while(1)  {
23       level1(array, s1);
24       s1 = array[1];
25       s2 = array[2];
26       s3 = array[3];
27     }
28   }
```

As in our PICOBLAZE studies, we use also three levels of loops and only local, i.e.,
sp indexed variables. The starting sequence setting up sp, gp, and er and jumping
to main() via r2 is similar to the previous example, so we skip this section. In the
function call, we include one array and one scalar. During the call we increment the
scalar and assign the new value to our array. We use an initial value of 0×1233 such
that it can be easily identified in memory. Figure 9.21a shows the assembler code for
the three functions, while Fig. 9.21b shows assembler code for the main() code.

Other than PICOBLAZE, the Nios II does not have a pc stack to host return
addresses for loops. We only have one register r31=ra to store a return address. If
we just call one function that would be sufficient (see level 3 in Fig. 9.21a). But if
we call another function during our first function call, we need to store the ra value
into memory (see lines 5C–60) and restore ra at the end of the function (lines
70–74). To index the memory location for the loop address, the stack register sp is

```
                  void level3(int *array, int sl) {                          int main(void) {
                      sl += 1;                                               main:
                  level3:                         0x0000007C  DEFFF104          addi   sp, sp, -0x3C
0x00000030  29400044    addi   r5, r5, 0x1        0x00000080  DFC00E15          stw    ra, 56(sp)
                      array[3] = sl;                                            volatile int sl, s2, s3;
0x00000034  21400315    stw    r5, 12(r4)                                       int array[11];
0x00000038  F800283A    ret                                                     sl = 0x1233; // Memory location sp+0
                                                  0x00000084  00848CC4          addi   r2, zero, 0x1233
                      return;                     0x00000088  D8800035          stwio  r2, 0(sp)
                  }                                                             while(1) {
                  void level2(int *array, int sl) {                               level1(array, sl);
                  level2:                         0x0000008C  D9400037          ldwio  r5, 0(sp)
0x0000003C  DEFFFF04    addi   sp, sp, -0x4        0x00000090  D9000304          addi   r4, sp, 0xC
0x00000040  DFC00015    stw    ra, 0(sp)          0x00000094  000005C0          call   0x00000017 (0x0000005C: level1)
                      sl += sl;                                                   sl = array[1];
0x00000044  294B883A    add    r5, r5, r5         0x00000098  D8800417          ldw    r2, 16(sp)
                      array[2] = sl;              0x0000009C  D8800035          stwio  r2, 0(sp)
0x00000048  21400215    stw    r5, 8(r4)                                         s2 = array[2];
                      level3(array, sl); // call level 3   0x000000A0  D8800517  ldw    r2, 20(sp)
0x0000004C  00000300    call   0x0000000C (0x00000030: level3)  0x000000A4  D8800135  stwio  r2, 4(sp)
                      return;                                                     s3 = array[3];
                  }                               0x000000A8  D8800617          ldw    r2, 24(sp)
0x00000050  DFC00017    ldw    ra, 0(sp)          0x000000AC  D8800235          stwio  r2, 8(sp)
0x00000054  DEC00104    addi   sp, sp, 0x4                                     }
0x00000058  F800283A    ret                       0x000000B0  003FF606          br     -0x28 (0x0000008C)

                  void level1(int *array, int sl) {
                  level1:
0x0000005C  DEFFFF04    addi   sp, sp, -0x4
0x00000060  DFC00015    stw    ra, 0(sp)
                      sl += 1;
0x00000064  29400044    addi   r5, r5, 0x1
                      array[1] = sl:
0x00000068  21400115    stw    r5, 4(r4)
                      level2(array, sl); // call level 2
0x0000006C  000003C0    call   0x0000000F (0x0000003C: level2)
                      return;
}
0x00000070  DFC00017    ldw    ra, 0(sp)
0x00000074  DEC00104    addi   sp, sp, 0x4
0x00000078  F800283A    ret
                           (a)                                                   (b)
```

Fig. 9.21 TRISC3N nesting loop ASM test routine. Local and global memory write are followed by reading the same locations. (**a**) Three-level functions. (**b**) main() program

used. Since sp is also used to index our local variables, a copy of sp is placed in r4 (line 0x90) to index our variables during function calls. At the end we will have in the memory region addressed by sp (with increasing index) the variable values and below the sp address the return address used for the function call. We run our main program loop one time and then monitor the memory content around sp=0x3FC0. After one loop iteration, we will observe:

At location 0x3FC0 we find sl=0x1234 and array[1] at 0x3FD0. At address 0x3FBC we see the return address 0x98 for the call to first level1 function and 0x70 as the return address for the call to level2 within the level1 function.

These three short programs will be our test bench data for the HDL simulation that follows next.

HDL Implementation and Testing

The PICOBLAZE design trisc2.vhd or trisc2.v should be our starting point for the tiny Nios II design we call trisc3n since Nios II is a three-address machine. The HDL code is shown next.

VHDL Code 9.8: TRISC3N (Final Design)

```
1     -- Title: T-RISC 3 address machine
2     -- ================================================================
3     -- Title: T-RISC 3 address machine
4     -- Description: This is the top control path/FSM of the
5     -- T-RISC, with a single 3 phase clock cycle design
6     -- It has a 3-address type instruction word
7     -- implementing a subset of the Nios II architecture
8     -- ================================================================
9     LIBRARY ieee; USE ieee.std_logic_1164.ALL;
10
11    PACKAGE n_bit_type IS                    -- User defined types
12      SUBTYPE U8 IS INTEGER RANGE 0 TO 255;
13      SUBTYPE U12 IS INTEGER RANGE 0 TO 4095;
14      SUBTYPE SLVA IS STD_LOGIC_VECTOR(11 DOWNTO 0); -- Address prog. mem.
15      SUBTYPE SLVD IS STD_LOGIC_VECTOR(31 DOWNTO 0); -- Width data
16      SUBTYPE SLVP IS STD_LOGIC_VECTOR(31 DOWNTO 0); -- Width instruction
17      SUBTYPE SLV6 IS STD_LOGIC_VECTOR(5 DOWNTO 0);  -- Full opcode size
18    END n_bit_type;
19
20    LIBRARY work;
21    USE work.n_bit_type.ALL;
22
23    LIBRARY ieee;
24    USE ieee.STD_LOGIC_1164.ALL;
25    USE ieee.STD_LOGIC_arith.ALL;
26    USE ieee.STD_LOGIC_unsigned.ALL;
27    -- ================================================================
28    ENTITY trisc3n IS
29     PORT(clk      : IN  STD_LOGIC; -- System clock
30          reset    : IN  STD_LOGIC; -- Active low asynchronous reset
31          in_port  : IN  STD_LOGIC_VECTOR(7 DOWNTO 0); -- Input port
32          out_port : OUT STD_LOGIC_VECTOR(7 DOWNTO 0); -- Output port
33    -- The following test ports are used for simulation only and should be
34    -- comments during synthesis to avoid outnumbering the board pins
35          r1_OUT    : OUT SLVD;  -- Register 1
36          r2_OUT    : OUT SLVD;  -- Register 2
37          r3_OUT    : OUT SLVD;  -- Register 3
38          r4_OUT    : OUT SLVD;  -- Register 4
39          sp_OUT    : OUT SLVD;  -- Register 27 aka stack pointer
40          ra_OUT    : OUT SLVD;  -- Register 31 aka return address
41          jc_OUT    : OUT STD_LOGIC;   -- Jump condition flag
42          me_ena    : OUT STD_LOGIC;   -- Memory enable
43          k_OUT     : OUT STD_LOGIC;   -- constant flag
44          pc_OUT    : OUT STD_LOGIC_VECTOR(11 DOWNTO 0); -- Program counter
45          ir_imm16  : OUT STD_LOGIC_VECTOR(15 DOWNTO 0); -- Immediate value
46          imm32_out : OUT SLVD;               -- Sign extend immediate value
47          op_code   : OUT STD_LOGIC_VECTOR(5 DOWNTO 0)   -- Operation code
48              );
49    END;
50    -- ================================================================
51    ARCHITECTURE fpga OF trisc3n IS
52    -- Define GENERIC to CONSTANT for _tb
53      CONSTANT WA : INTEGER := 11;  -- Address bit width -1
54      CONSTANT NR : INTEGER := 31;   -- Number of Registers -1
55      CONSTANT WD : INTEGER := 31;   -- Data bit width -1
56      CONSTANT DRAMAX : INTEGER := 4095; -- No. of DRAM words -1
```

```
57          CONSTANT DRAMAX4 : INTEGER := 16383; -- No. of DRAM bytes -1
58
59          COMPONENT rom4096x32 IS
60          PORT (clk   : IN STD_LOGIC;        -- System clock
61                reset : IN STD_LOGIC;        -- Asynchronous reset
62                pma   : IN STD_LOGIC_VECTOR(11 DOWNTO 0); -- Program mem. add.
63                pmd   : OUT STD_LOGIC_VECTOR(31 DOWNTO 0));--Program mem. data
64          END COMPONENT;
65
66          SIGNAL op, opx  : SLV6;
67          SIGNAL dmd, pmd, dma : SLVD;
68          SIGNAL imm5 : STD_LOGIC_VECTOR(4 DOWNTO 0);
69          SIGNAL sxti, imm16 : STD_LOGIC_VECTOR(15 DOWNTO 0);
70          SIGNAL imm26 : STD_LOGIC_VECTOR(25 DOWNTO 0);
71          SIGNAL imm32 : SLVD;
72          SIGNAL A, B, C : INTEGER RANGE 0 TO NR;
73          SIGNAL rA, rB, rC : SLVD := (OTHERS => '0');
74          SIGNAL ir, pc, pc4, pc8, branch_target, pcimm26 : SLVP;-- PCs
75          SIGNAL eq, ne, mem_ena, not_clk : STD_LOGIC;
76          SIGNAL jc, kflag : boolean; -- jump and imm flags
77          SIGNAL load, store, read, write : boolean; -- I/O flags
78
79     -- OP Code of instructions:
80     -- The 6 LSBs IW for all implemented operations sorted by op code
81       CONSTANT call    : SLV6 := "000000"; -- X00
82       CONSTANT jmpi    : SLV6 := "000001"; -- X01
83       CONSTANT addi    : SLV6 := "000100"; -- X04
84       CONSTANT br      : SLV6 := "000110"; -- X06
85       CONSTANT andi    : SLV6 := "001100"; -- X0C
86       CONSTANT ori     : SLV6 := "010100"; -- X14
87       CONSTANT stw     : SLV6 := "010101"; -- X15
88       CONSTANT ldw     : SLV6 := "010111"; -- X17
89       CONSTANT xori    : SLV6 := "011100"; -- X1C
90       CONSTANT bne     : SLV6 := "011110"; -- X1E
91       CONSTANT beq     : SLV6 := "100110"; -- X26
92       CONSTANT orhi    : SLV6 := "110100"; -- X34
93       CONSTANT stwio   : SLV6 := "110101"; -- X35
94       CONSTANT ldwio   : SLV6 := "110111"; -- X37
95       CONSTANT R_type  : SLV6 := "111010"; -- X3A
96
97     -- 6 bits for OP eXtented instruction with OP=3A=111010
98       CONSTANT ret     : SLV6 := "000101"; -- X05
99       CONSTANT jmp     : SLV6 := "001101"; -- X0D
100      CONSTANT opand   : SLV6 := "001110"; -- X0E
101      CONSTANT opor    : SLV6 := "010110"; -- X16
102      CONSTANT opxor   : SLV6 := "011110"; -- X1E
103      CONSTANT add     : SLV6 := "110001"; -- X31
104      CONSTANT sub     : SLV6 := "111001"; -- X39
105
106    -- Data RAM memory definition use one BRAM: DRAMAXx32
107      TYPE MTYPE IS ARRAY(0 TO DRAMAX) OF SLVD;
108      SIGNAL dram : MTYPE;
109
110    -- Register array definition 32x32
111      TYPE REG_ARRAY IS ARRAY(0 TO NR) OF SLVD;
112      SIGNAL r : REG_ARRAY;
```

```
113
114      BEGIN
115
116        P1: PROCESS (op, reset, clk) -- FSM of processor
117        BEGIN -- update the PC
118          IF reset = '0' THEN
119            pc <= (OTHERS => '0'); pc8 <= (OTHERS => '0');
120          ELSIF falling_edge(clk) THEN
121            IF jc THEN
122              pc <= branch_target ; -- any jumps that use immediate
123            ELSE
124              pc <= pc4;  -- Usual increment by 4 bytes
125              pc8 <= pc + X"00000008";
126            END IF;
127          END IF;
128        END PROCESS p1;
129        pc4 <= pc + X"00000004"; -- Default PC increment is 4 bytes
130        pcimm26 <= pc(31 DOWNTO 28) & imm26 & "00";
131        jc <= (op=beq AND rA=rB) OR (op=R_type AND (opx=ret OR opx=jmp))
132           OR (op=bne AND rA/=rB) OR (op=jmpi) OR (op=br) OR (op=call);
133
134        branch_target <= pcimm26 WHEN (op=jmpi OR op=call)
135                        ELSE   r(31) WHEN (op=R_type AND opx=ret)
136                        ELSE   rA WHEN (op=R_type AND opx=jmp)
137                        ELSE   imm32+pc4; -- WHEN (op=beq OR op=bne OR op=br)
138
139        -- Mapping of the instruction, i.e., decode instruction
140        op   <= ir(5 DOWNTO 0);        -- Operation code
141        opx <= ir(16 DOWNTO 11);       -- OPX code for ALU ops
142        imm5  <= ir(10 DOWNTO 6);      -- OPX constant
143        imm16 <= ir(21 DOWNTO 6);      -- Immediate ALU operand
144        imm26 <= ir(31 DOWNTO 6);      -- Jump address
145        A   <= CONV_INTEGER('0' & ir(31 DOWNTO 27)); -- Index 1. source reg.
146        B   <= CONV_INTEGER('0' & ir(26 DOWNTO 22));
147                                               -- Index 2. source/des. register
148        C   <= CONV_INTEGER('0' & ir(21 DOWNTO 17));-- Index destination reg.
149        rA <= r(A); -- First source ALU
150        rB <= imm32 WHEN kflag ELSE r(B); -- Second source ALU
151        rC <= r(C);   -- Old destination register value
152     -- Immediate flag 0= use register 1= use HI/LO extended imm16;
153        kflag <= (op=addi) OR (op=andi) OR (op=ori) OR (op=xori)
154                            OR (op=orhi) OR (op=ldw) OR (op=ldwio);
155        sxti <= (OTHERS => imm16(15)); -- Sign extend the constant
156        imm32 <= imm16 & X"0000" WHEN op=orhi
157              ELSE sxti & imm16; -- Place imm16 in MSbs for ..hi
158
159        prog_rom: rom4096x32        -- Instantiate a Block RAM
160        PORT MAP (clk   => clk,    -- System clock
161                  reset => reset,  -- Asynchronous reset
162                  pma   => pc(13 DOWNTO 2),-- Program memory address 12 bits
163                  pmd   => pmd);   -- Program memory data
164        ir <= pmd;
165
166        dma <= rA + imm32;
167        store <= ((op=stw) OR (op=stwio)) AND (dma <= DRAMAX4);-- DRAM store
168        load <= ((op=ldw) OR (op=ldwio)) AND (dma <= DRAMAX4); -- DRAM load
```

```
169    write <= ((op=stw) OR (op=stwio)) AND (dma > DRAMAX4); -- I/O write
170    read <= ((op=ldw) OR (op=ldwio)) AND (dma > DRAMAX4);  -- I/O read
171    mem_ena <= '1' WHEN store ELSE '0';   -- Active for store only
172    not_clk <= NOT clk;
173    ram: PROCESS (reset, dma, not_clk) -- Use one BRAM: 4096x32
174    VARIABLE idma : U12 := 0;
175    BEGIN
176      idma := CONV_INTEGER('0' & dma(13 DOWNTO 2));-- uns/skip 2 LSBs
177      IF reset = '0' THEN        -- Asynchronous clear
178        dmd <= (OTHERS => '0');
179      ELSIF rising_edge(not_clk) THEN
180        IF mem_ena = '1' THEN
181          dram(idma) <= rB;  -- Write to RAM at falling clk edge
182        END IF;
183        dmd <= dram(idma);   -- Read from RAM at falling clk edge
184      END IF;
185    END PROCESS;
186
187    ALU: PROCESS (op,opx,rA,rB,rC,in_port,dmd,reset,clk,load,read)
188    VARIABLE res: SLVD;
189    BEGIN
190      res := rC; -- keep old/default
191      IF (op=R_type AND opx=add) OR (op=addi) THEN res:=rA + rB; END IF;
192      IF op=R_type AND opx=sub THEN res := rA - rB; END IF;
193      IF (op=R_type AND opx=opand) OR (op=andi) THEN res := rA AND rB;
194                                                                  END IF;
195      IF (op=R_type AND opx=opor) OR (op=ori) OR (op=orhi) THEN
196                                           res := rA OR rB; END IF;
197      IF (op=R_type AND opx=opxor) OR (op=xori) THEN res := rA XOR rB;
198                                                                  END IF;
199      IF load THEN res := dmd; END IF;
200      IF read THEN res := X"000000" & in_port; END IF;
201      IF reset = '0' THEN            -- Asynchronous clear
202        out_port <= (OTHERS => '0');
203        FOR k IN 0 TO NR LOOP -- Need to reset at least r(0)
204          r(k)  <= X"00000000";
205        END LOOP;
206      ELSIF rising_edge(clk) THEN -- Nios has no zero or carry flags !
207        IF op=call THEN -- Store ra for operation call
208          r(31) <= pc8; -- Old pc + 1 op after return
209        ELSIF kflag AND B>0 THEN -- All I-type
210          r(B) <= res;
211        ELSIF C > 0 THEN
212          r(C) <= res; -- Store ALU result (default)
213        END IF;
214        IF write THEN out_port <= rB(7 DOWNTO 0); END IF;
215      END IF;
216    END PROCESS ALU;
217
218    -- Extra test pins:
219    pc_OUT <= pc(11 DOWNTO 0);
220    ir_imm16 <= imm16;
221    imm32_out <= imm32;
222    op_code <= op; -- Program
223    --jc_OUT <= jc; -- Control signals
224    jc_OUT <= '1' WHEN jc ELSE '0';  -- Xilinx modified
```

```
225      k_OUT <= '1' WHEN kflag ELSE '0';  -- Xilinx modified
226      me_ena <= mem_ena; -- Control signals
227      r1_OUT <= r(1); r2_OUT <= r(2);       -- First two user registers
228      r3_OUT <= r(3); r4_OUT <= r(4);       -- Next two user registers
229      sp_OUT <= r(27); ra_OUT <= r(31);      -- Compiler registers
230
231    END fpga;
```

We can use a similar I/O interface as for the PICOBLAZE but would not monitor register 0 since it is always zero (lines 28–49). For the memory we will monitor location around the sp values. A few additional flag, such as the IMM16 flag, called kflag will also be monitored. Operation codes are now all 6 bits, and pc, data, and address width need to be adjusted to 32 bits. Since VHDL-1993 does not allow coding 6-bit hex values, we use a binary coding of OP and OPX but also show the hex code for comparison with the assembler code (lines 79–104). Without the link register stack, the FSM processor controller will be a little shorter (lines 116–128). pc increment counting is byte-wise, so each standard instruction increases the program counter by 4 (line 129). For the branch target, we now have more choices: it can come from the IMM26 directly for instructions such as jmpi or call, relative to pc (br, bne, or beq), from r31=ra register for the ret instruction, or from any register for jmp (lines 134–137).

Decoding the instruction is pretty straightforward since we have no exception of the three instruction formats. A, B, and C get the $3 \times 5 = 15$ MSBs, while OP uses the six LSBs (lines 140–148). We have three constants of size 5, 16, and 26 to identify. The first ALU source will always be rA=r(A); the second ALU source may be a register or the (sign extended or zero appended) IMM16 value. The kflag identifies all instructions that use IMM16 or have rB as destination, e.g., ldwio (lines 145–151). Next comes the program memory (lines 159–164) that instantiates a 4 K word program ROM with 32 bit for each word. Next comes the data RAM with the control signal and the IO components. Since we do not plan to include a cache, the standard word and io type instructions are handled the same way. The DRAM (lines 173–185) is a negative edge synchronous memory. The number of words depends on the global constant DRAMAX. For a small footprint we would use something like DRAMMAX=255; if you like an easy transition for ANSI C programs, you may use the same size as the program ROM, i.e., 4095.

Now the ALU coding follows (lines 187–216). The R-type and I-type ALU instructions are combined to allow for resource sharing since only the second operand is different but the operation the same. All logic, arithmetic, and data transfer are coded that write to a register. R-type instructions write to r(C), while I-type write to register r(B). ra=r(31) is written for the call operation. Register write are controlled by the rising edge clock. Register zero is not written. All registers are set to zero when reset is active. Finally output assignments are made to some additional test ports (lines 218–229); these should be used for simulation only and need to be disabled before device synthesis to avoid overloading of the I/O pins of the FPGA.

The simulation for the memory access example discussed in program listings 9.6 is shown in Fig. 9.22. The memory values shown in the simulation match the memory content of the monitor program from Fig. 9.20. The variables have been mapped by the compiler using the gp and sp pointers to DRAM addresses as follows:

Variable	s	g	sarray[0]	garrray[0]
DRAM address (decimal)	4080	555	4080	556

The major events of the DRAM test simulation are s=1 @ 140ns; g=2 @ 180 ns; sarray[0]=s @ 220 ns; garray[0]=g @ 280 ns; sarray[1]=0x1357 @ 420 ns; and garray[2]=2468 @ 540 ns.

The HDL simulation for the nesting loop code from Program 9.7 is shown in Fig. 9.23. The address values in the range of sp match the assembler debug display in Table 9.15. Special attention should be given to the return address register r(31)=ra for the loops.

The major events of the loop behavior simulation can be summarized as follows: s1=0x1233 @ 320 ns; ra=0x98 @390 ns; ra=98 stored at DRAM(4079) @ 420 ns; ra=70 @ 490 ns; ra=0x70 stored at DRAM(4078) @ 520 ns; ra=50 @

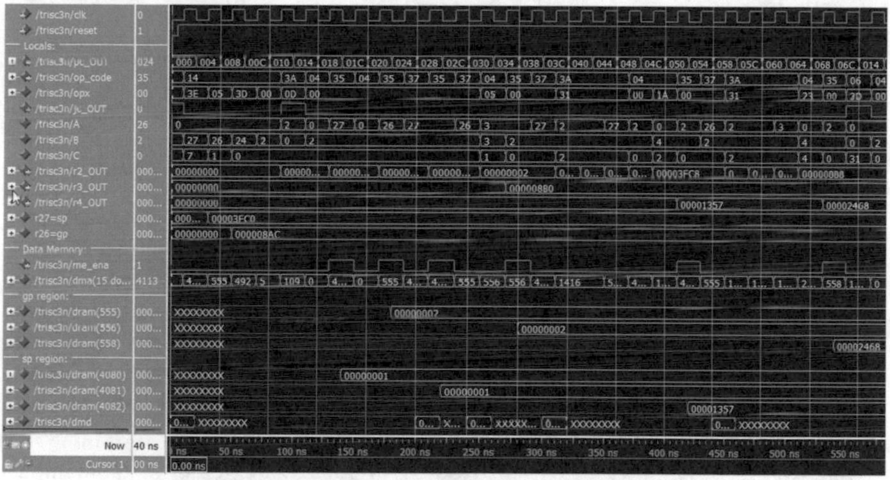

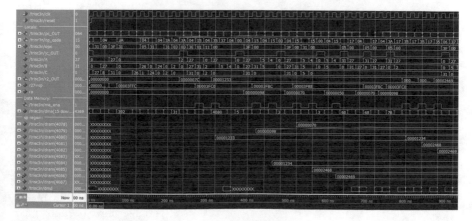

Fig. 9.22 Simulation shows the TRISC3N memory access

Fig. 9.23 Simulation shows the TRISC3N nesting loop behavior

Table 9.15 The memory contends the `nesting.c` program run

	+0×0	+0×4	+0×8	+0×c
0×00003FB0	00000000	00000000	00000070	00000098
0×00003FC0	00001234	00002468	00002469	00000000
0×00003FD0	00001234	00002468	00002469	00000000

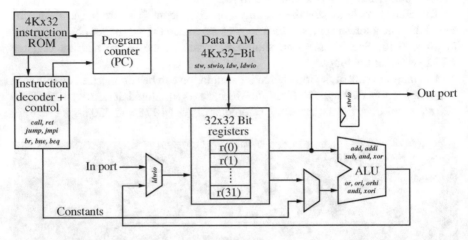

Fig. 9.24 The final implemented tiny Nios II aka TRISC3N core. Instruction (in italic) mainly associated with major hardware units. Substantial memory blocks (excluding registers) in gray

590 ns; loading `pc` with `ra=0x50` @ 640 ns; loading `pc` with `ra=70` @ 700 ns; and loading `pc` with `ra=0x98` @ 760 ns. The subroutines are entered as follows: `level1` @ 380 ns, `level2` @ 480 ns, and `level3` @ 580 ns. The final DRAM values for `s1`, `s2`, and `s3` are located at `DRAM(4080)`, `DRAM(4081)`, and `DRAM(4082)`, respectively.

Finally, Fig. 9.24 shows the overall TRISC3N architecture together with the implemented instructions. Implementation of additional instructions is left to the reader as exercise at the end of the chapter; see Projects 9.58–9.60.

Synthesis Results for TRISC3N

The synthesis results for the TRISC3N are shown in Table 9.16. The second and third columns show Xilinx VIVADO synthesis results for VHDL and Verilog, respectively. Columns four and five show Altera QUARTUS synthesis results for VHDL and Verilog, respectively. The major differences in the four synthesis results come from the method on how the Program ROM is implemented. Since our programs are small, the synthesis tool has in three out of four cases mapped the ROM to logic cells and not BRAM. Only Verilog QUARTUS version uses block RAM for RAM and ROM requiring twice as many block RAMs, which seemed not a good choice in this

Table 9.16 TRISC3N synthesis results for Altera and Xilinx tools and devices

	VIVADO 2016.4		QUARTUS 15.1	
Target device	Zynq 7 K xc7z010t-1clg400		Cyclone V 5CSEMA5F31C6	
HDL used	VHDL	Verilog	VHDL	Verilog
LUT/ALM	778	791	515	1420
ROM as BRAM	0	0	0	0
Block RAM RAMB36	4	4	–	–
or M10K	–	–	16	32
DSP Blocks	0	0	0	0
HPS	0	0	0	0
Fmax/MHz	62.5	66.6	59.22	42.53
Compile time	3:02	3:02	5:26	5:18

case. The Altera needs four times the number of block RAMs due to the smaller size of the embedded memory blocks. Compile time is reasonable, and speed is over 50 MHz. The VIVADO on ZyBo may need a clock divider by 2, but the design worked fine even with the 125 MHz system clock.

We finally have a working microprocessor with 21 out of the 88 Nios II operations implemented. Major units such as register array, logical, arithmetic operation, input and output, data memory, jump, and call are included. Missing from the complete instruction set are those group we did not need for our tasks so far such as comparisons, conditional branches, shift and rotate, multiply and divide, interrupt handling, and version control. Conditional branches are needed for ANSI C control language elements such as if or switch. Working with 64-bit number such as long long int may also be interesting to implement since Nios II ALU does not have carry or overflow flags. This is left to the reader as exercises at the end of the chapter (Project 9.58).

Review Questions and Exercises

Short Answer

9.1. What are the main differences between Nios and Nios II?
9.2. Name three features of the Nios II /f version that are not part of the Nios II /e.
9.3. When would you use the Nios II/e?
9.4. What are the eight instruction groups in Nios II assembler?
9.5. Name five addressing modes used by Nios II. For each mode specify an instruction example.
9.6. How are nested loops implemented by GCC for Nios II?
9.7. Name five advantages of the TRISC3N HDL design approach compared to Nios II QSYS design.

9.8. Nios II has no `subi`, `nop`, or clear register instructions. Define "pseudo-instructions" using the instructions from TRISC3N that accomplish these tasks.

9.9. Give Nios II architecture examples for the RISC design principles: (a) simplicity favors regularity; (b) smaller is faster; and (c) good designs demand good compromise.

Fill in the Blank

9.10. The last _____ bits of the Nios II instruction opcode are used to encode the instruction.

9.11. The Nios II is a _____ address machine.

9.12. The TRISC3N uses _____ instructions of the Nios II ISA.

9.13. The Nios development system sold _____ times in the first 3 years after introduction.

9.14. The three immediate constants in the Nios II instructions are of length _____.

9.15. The NIOS II operation `beq  r4,  r0,  0x18` will be encoded in hex as _____.

9.16. The NIOS II operation `andi  r3,  r4,  0x1` will be encoded in hex as _____.

9.17. The NIOS II operation `sub  r17,r16,r21` will be encoded in hex as _____.

9.18. The NIOS II operation `jmpi  0x00000009C` will be encoded in hex as _____.

9.19. The NIOS II operation _____ will be encoded in hex as 0xD8800B17.

9.20. The NIOS II operation _____ will be encoded in hex as 0xF800283A.

9.21. The NIOS II operation _____ will be encoded in hex as 0x000008C0.

9.22. The NIOS II operation _____ will be encoded in hex as 0x00BFFFF4.

True or False

9.23. _____ The first Nios generations had a 16-bit and a 32-bit data path option.

9.24. _____ In the bottom-up design approach we assemble components one-at-a-time that are needed.

9.25. _____ The top-down design method is more complex than the bottom-up method.

9.26. _____ The Altera QSYS system only allows Nios II/e and II/s designs; for Nios II/f, the SOPC builder must be used.

9.27. _____ The Altera monitor program compiler messages show up in the Info & Errors window.

9.28. _____ The Nios II/e core needs less logic resources than the Nios II/f core.

9.29. _____ The Altera monitor program default start shows five windows: Disassembly, Memory, Registers, Terminal, and Info & Errors.

9.30. _____ The Altera floating-point FPH1 implements hardware for *,+,-,/ and sqrt().

9.31. _____ The Altera FPH2 CIP has more floating-point functions but requires less logic resources than FPH1.

9.32. _____ The `bsp.h` file of the Altera Monitor program contains the I/O address of the component.

9.33. _____ The Nios II microprocessor can be programmed in assembler and C/C++.

9.34. _____ To download the Nios system to the DE2 board, we need QUARTUS; the Altera monitor program cannot be used for that.

9.35. _____ ANSI C code that uses the `float` data type must use also a FPH custom IP.

9.36. Identify standard features (**T**rue/**F**alse) for the Nios II/e/s/f cores in the table below.

Feature	Nios II/e	Nios II/s	Nios/f
RISC architecture			
I-Cache 8 KBYTE			
D-Cache 2 KBYTE			
Dynamic branch prediction			
Two M10K + cache			
JTAG level 1			
Barrel shifter			

Projects and Challenges

9.37. Run an example from *app_software_nios2_asm* folder from the *University_program→ Computer_Systems*, and write a short essay (1/2 page) what the examples are about and what I/O components are used.

9.38. Run an example from *app_software_nios2_C* folder from the *University_program→ Computer_Systems*, and write a short essay (1/2 page) what the examples are about and what I/O components are used.

9.39. Develop an assembler or ANSI C program to implement a left, right car turn signal and emergency light similar to the Ford Mustang: 00X, 0XX, and XXX for a left turn where 0 is off and X is LED on. Use X00, XX0, and XXX for right turn signal. Use the slider switch for the left/right selection. For emergency light, both switches are on. The turn signal sequence should repeat once per second. See `mustangQT.MOV` or `mustangWMP.MOV` for a demo.

9.40. Write an ANSI C program that implements a running light. The speed of the running light should be determined by the SW values, and any change should be displayed at the Altera Monitor Terminal. See `led_coutQT.MOV` or `led_countWMP.MOV` for a demo.

9.41. Develop an ANSI C programs that lights up one LED only and moves to the left or right by pushing button 4 or 3, respectively.

9.42. Repeat the previous exercise with a three-color bar on the VGA monitor.

9.43. Develop an assembler or ANSI C program that has a random number generator, using `ADD` and `XOR` instructions. Display the random number on the LEDs. What is the period of your random sequence?

9.44. Build a stopwatch, which counts up in second steps. Use three buttons: start clock, stop or pause clock, and reset clock. Use the LEDs or the seven-segment display from the previous exercise for display. See `stop_watch_ledQT.MOV` or `stop_watch_ledWMP.MOV` for a demo.

9.45. Use the push buttons to implement a reaction speed timer. Turn on one of the four LEDs, and measure the time it takes until the associate button is pressed. Display the measurement after ten tries on LED or seven-segment display.

9.46. Write ANSI C program to display all characters A–Z on a seven-segment display. Take special attention for characters such as X, M, and W. You may use a bar to indicate double length, e.g., $m = \bar{n}; w = \bar{v}$. Produce a short video that shows all characters on the seven-segment display. See `ascii4sevensegment.MOV` for a demo.

9.47. Use the character set from the last exercise to generate a running text on the seven-segment display using the current month and year and/or our name or your initials if your name is too long. See `idQT.MOV` and `idWMP.MOV` from the CD for a demo.

9.48. Write ANSI C program to reproduce the ASCII from Fig. 5.2 for all printable character. (Decimal 32…127) on the VGA display and the Altera Terminal window; see Fig. 5.2b.

9.49. Write an ANSI C program for the DE1 SoC Computer that draws a circle on the VGA monitor using Bresenham circle algorithm: start with $y = 0$; $x = R$. Move in y direction one pixel in each iteration. Decrement x if error $|x^2 + y^2 - R^2|$ is smaller for $x' = x\text{-}1$; otherwise, keep x. Use symmetry to complete all four segments of the circle.

9.50. Use the Bresenham algorithm from the last exercise to draw a filled circle in a specified color.

9.51. Write an ANSI C program for the DE1 SoC Computer that draws an ellipse on the VGA monitor using standard floating-point arithmetic: $x = a*\cos([0:0.1:\pi/2$

]);$y = b*\sin([0:0.1:\pi/2])$. Use symmetry to fill the ellipse with horizontal lines. Is the ellipse completely filled? What is the exercise with this type of algorithm?

9.52. Write an ANSI C program for the DE1 SoC Computer that draws an ellipse on the VGA monitor using Bresenham algorithm: start with $y = 0$; $x = R$. Move in y direction one pixel in each iteration. Decrement x if error $|x^2/a^2 + y^2/b^2 - R^2|$ is smaller for $x' = x-1$; otherwise, keep x. Use symmetry to fill the ellipse with horizontal lines. Is the ellipse completely filled? Compare this algorithm to the version in the previous exercise, and name advantages and disadvantages.

9.53. Write an ANSI C program for the color DE1 SoC Computer or the gray DE1 SoC Top-Down Computer that implements a display of the four fractals (use range $x,y = -1.5..1.5$). The complex iteration equations are as follows:

(a) Mandelbrot: $c = x + i*y$; $z = \mathbf{0}$
(b) Douday rabbit: $z = x + iy$; $c = -0.123 + j0.745$
(c) Siegel disc: $z = x + i*y$; $c = -0.391-j0.587$
(d) San marco: $z = x + i*y$; $c = -0.75$

Find the number of iteration (and code in unique color) $z = z.*z + c$, such that $|z|^2 > 5$; see Fig. 9.25. The name or the fractal should be displayed in the center of the VGA screen. See 4fractals.MOV for a demo.

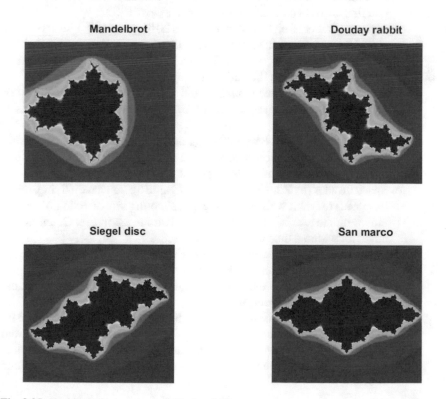

Fig. 9.25 Four fractal examples from Project 9.53

9.54. Write an ANSI C program for the DE1 SoC Computer that implements a zoom of a Mandelbrot fractal. Compute the fractal at 25 different zoom levels, and then display the 25 images in a series. The number of the frames should be displayed in the center of the VGA screen. How much of the data memory in percent was used for the images? See `MandelbrotZoomColor.MOV` and `MandelbrotZoomGrey.MOV` for a demo.

9.55. Write an ANSI C program for the DE1 SoC Computer that implements a game such as:

(a) Tic-Tac-Toe (by Christopher Ritchie, Rohit Sonavane, Shiva Indranti, Xuanchen Xiang, Qinggele Yu, Huanyu Zang 2016 game 3/6/8; 2017 game 9/10/11): Use two players. Enter selecting with slider switches and the push button to submit your selection for player 1 (button 3) and player 2 (button 0). Alternatively toggle by default, and display which player turn it is on LED or seven-segment. For a DE115 you can use slider switches 0–8 for player 1 and 9–17 slider switches for player 2.

(b) Reaction Time (by Mike Pelletier 2016 game 4): Light up one of the three squares on the screen, and measure the time until the correct button was pressed.

(c) Minesweeper (Sharath Satya 2016 game 7): In a 3x3 field, three bombs are hidden. You need to find the six boxes that do not contain a bomb. If you find all six, you win; otherwise, player loses.

(d) Pong (by Samuel Reich and Chris Raices 2017 game 4/5): Use one player only. Use one direction to move the ball and the push button for moving the paddle on the other axis. Set a timer to count how long the game runs until the player misses the ball. Ball may be a small square.

(e) Simon Says (by Melanie Gonzalez and Kiernan Farmer 2016 game 1/2): Repeat the three-color sequence of VGA square with the push buttons 1, 2, and 3.

(f) Box Chaser (by Ryan Stepp 2017 game 7): A big box moves randomly over the display, and you need to avoid getting your small box hit by this big box by moving it to the left or right with the push buttons.

(g) Space Invades (by Zlatko Sokolikj 2017 game 6): You have a flying object that needs to be hit with your firing gun shooting in y direction. Your gun moves on the x axis, so a total of three buttons are in use. Counting the target hit within a time frame gives the score.

(h) Flappy Bird (by Javier Matos 2017 game 3): Pressing a button moves your bird in y direction, while your bird continuously drops. Obstacles such as wall with a hole move from right to left. Game is over when the bird hits the wall or ground.

(i) Bounce (by Kevin Powell 2016 game 5): A three-color bar is moved by two push buttons to the left or right. A ball bounces in y direction and changes color whenever the ball reaches the top. The ball color must match the bar color when the ball hits the ground; otherwise, the game is over.

For some of these games, you will find more detailed description in Chap. 11, Sect. 11.2. All games have been designed by students within a semester project competition in 2016 and 2017. Write a brief description of your implementation including Computer System, I/O component used, and game rule. Submit a short video of a game demonstration.

9.56. Write an ANSI C program for the DE1 SoC Computer that implements a game such as:

    (a) Four in a Row: Select the row with the slider switch and then use a push button for submitting your choice; see Fig. 9.26a for an example.

    (b) Tetris: Let different elements move slowly in y direction; see Fig. 9.26b for an example.

    (c) Sudoku: Place easy, normal, or hard starting configurations on the board. Then allow with three entry, row, column, and number (1–9), to make your submission; see Fig. 9.26c for an example.

9.57. Extend the bottom-up design from Sect. 9.3 to include seven-segment displays, system ID, and on-chip memory that are also part of the typical Basic Computer. Test the new component with short ANSI C programs.

9.58. Run short sequence of ANSI C code to identify the missing instruction in TRISC3N set for:

    (a) Handling `long long int`, i.e., 64 bits for add, sub, multiply, and divide.

    (b) Add the missing instructions, and test the HDL with a short assembler program.

9.59. Run short sequence of ANSI C code to identify the missing instruction in TRISC3N set for:

    (a) `if` and `switch` statements,

    (b) The factorial example with for loop

    (c) The factorial example with recursive function call

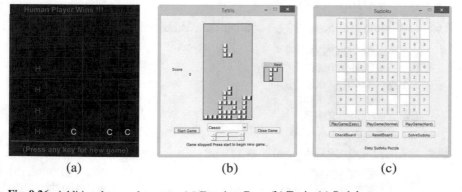

          (a)   (b)   (c)

Fig. 9.26 Additional example games. (**a**) Four in a Row. (**b**) Tetris. (**c**) Sudoku

9.60. Add additional Nios II features to the TRISC3N architecture, i.e., instruction such as:

   (a) 12 comparisons
   (b) 9 shift and rotate
   (c) 4 missing condition branches

   Add the HDL code and test with a short assembler test program. Report the difference in synthesis (area and speed) compared to the original design.

9.61. Consider a floating-point representation with a sign bit, E = 7-bit exponent width, and M = 10 bits for the mantissa (not counting the hidden one).

   (a) Compute the bias using (9.2).
   (b) Determine the (absolute) largest number that can be represented.
   (c) Determine the (absolutely measured) smallest number (not including denormals) that can be represented.

9.62. Using the result from the previous Exercise:

   (a) Determine the representation of $f_1 = 9.25_{10}$ in this (1,7,10) floating-point format.
   (b) Determine the representation of $f_2 = 10.5_{10}$ in this (1,7,10) floating-point format.
   (c) Compute $f_1 + f_2$ using floating-point arithmetic.
   (d) Compute $f_1 * f_2$ using floating-point arithmetic.
   (e) Compute f_1/f_2 using floating-point arithmetic.

9.63. For the IEEE single-precision format, determine the 32-bit representation of:

   (a) $f_1 = -0$
   (b) $f_2 = \infty$
   (c) $f_3 = -0.6875_{10}$
   (d) $f_4 = 10.5_{10}$
   (e) $f_5 = 0.1_{10}$
   (f) $f_6 = \pi = 3.141593_{10}$
   (g) $f_7 = \sqrt{3}/2 = 0.8660254_{10}$

9.64. Convert the following 32-bit floating-point number into a decimal number:

   (a) $f_1 = $ 0x44FC6000
   (b) $f_2 = $ 0x49580000
   (c) $f_3 = $ 0x40C40000
   (d) $f_4 = $ 0xC1AA0000
   (e) $f_5 = $ 0x3FB504F3
   (f) $f_6 = $ 0x3EAAAAAB

9.65. Design a 32-bit floating-point ALU with a select signal sel to choose one of the following operations: (0) fix2fp, (1) fp2fix, (2) add, (3) sub, (4) mul, (5) div, (6) reciprocal, and (7) power-of-2 scale. Test your design using the simulation shown in Fig. 9.27.

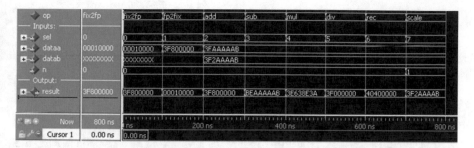

Fig. 9.27 MODELSIM RTL simulation of eight functions of the floating-point arithmetic unit fpu. 0001.0000 in sfixed become 0x3F800000 in hex code for FLOAT32. The arithmetic test data are 1/3 = 0x3EAAAAAB and 2/3 = 0x3F2AAAAA

9.66. Develop a seven-segment display using the flash program listing 9.37 to count up in seconds. This will require to modify the pin assignment that not only LEDs but also the seven-segment are driven by the TRISC3N output port.

9.67. Write an ANSI C program for the DE1 SoC Computer that test a component:

   (a) AUDIO-IN: Use an amplitude tracking to determine sags or swells in a power signal.

   (b) AUDIO-OUT: Implement an "echo" that changes with SW settings.

   (c) USB: Check the device for R/W errors and measure maximum transfer rate.

   (d) InDA: Allow an optical device to transmit data.

   (e) PS2: Input data from keyboard and display on VGA or LED.

   (f) VGA: Let "EuPSD" fly over the VGA monitor.

   (g) SD Card: Check memory R/W and measure maximum transfer rate.

   (h) Custom Instruction: Implement S-Boxes used in the DES algorithm.

   (i) Ethernet: Try to communicate via Ethernet cable to another device.

   (j) JTEG-UART: Measure file transfer rate from/to the Nios II.

   (k) G-sensor: Build a tremor sensor.

   (l) Video: Edge detection or resampler.

9.68. Modify the HDMI CIP from Chap. 10 to be used as standalone text terminal for the VGA port. If you have DE1 SoC, use the available VGA port. You will also need a VGA cable. The VGA uses the same timing as HDMI for row and column control but analog output signals; see Chap. 2. You will also need to forward the internal H/V sync signals to the output. The VGA port contains three DACs for this purpose. Study the master *.QSF for required ports and names. You will use the eight MSBs for red, green, and blue. For the ports use the following assignment: reset=KEY[0]; WE=KEY[3]; DATA = SW[7:0]; Pixel→VGA_B=VGA_G=VGA_R; LEDR[9:0]= {V[4:0], H[4:0]}; (keep SW[1:0]=font; CLOCK_125=clock 50 MHz). Keep QSF output names. Bonus: Use the text from the next exercise for display. Add your name in the banner.

9.69. Write a short MATLAB or C/C++ program to generate a new memory initialization file for the VGA encoder. Check the file `eupsd_hex.mif` for required format. Use as new text:

   (a) Your home country constitution
   (b) Your favorite poem
   (c) Your favorite joke
   (d) Your favorite song text

9.70. Develop a `srec2prom` utility that reads a Motorola `*.srec` file (SREC file format; see, for instance, [A01]) as generated by GCC in the Altera Monitor program. Generate a complete VHDL file that includes to code as constant values or for Verilog a `*.mif` table that can be loaded with a `$read-memh()`. Test your program with the Nios II `flash.c` example.

9.71. Identify missing instructions from TRISC3N set that are needed if you use the `srec2prom` from the previous exercise. Implement the missing instruction in HDL and run the `srec2prom` generated program.

9.72. Develop a TRISC3N program to implement a multiplication of two 32-bit numbers via repeated additions. Test your program by reading the SW values and computing $q = x * x$.

9.73. Develop a TRISC3N program, and add the HDL to implement a `mul` multiplication of two 32-bit numbers via array multiply. Test your program by reading the SW values and computing $q = x * x$.

References

[A01]   Altera, Converting .srec Files to .flash Files for Nios Embedded Processor Applications. White Paper (2001)
[A03]   Altera, *Nios II Processor Reference Handbook 5.0*. San Jose (2003)
[A04]   Altera, Net seminar Nios processor. (2004), http://www.altera.com
[A08]   P. Ashenden, *The Designer's Guide to VHDL*, 3rd edn. (Morgan Kaufman Publishers, Inc., San Mateo, 2008)
[A15]   Altera, *Introduction to the Altera Nios II Soft Processor 15.1*. San Jose (2015)
[AI08] Altera/Intel, *Nios II Custom Instruction User Guide 8.0*. San Jose (2008)
[AI15] Altera/Intel, *Nios II Custom Instruction User Guide 15.1*. San Jose (2015)
[AI16] Altera/Intel, *Nios II Gen2 Processor Reference Guide*. San Jose (2016)
[AI17a] Altera/Intel, *Nios II Performance Benchmarks 17.1*. San Jose (2017)
[AI17b] Altera/Intel, *Nios II Custom Instruction User Guide 17.1*. San Jose (2017)
[I85]   IEEE, Standard for binary floating-point arithmetic. IEEE Std. **754-1985**, 1–14 (1985)
[I08]   IEEE, Standard for binary floating-point arithmetic. IEEE Std. **754-2008**, 1–70 (2008)
[MSC06] U. Meyer-Baese, D. Sunkara, E. Castillo, E.A. Garcia, Custom instruction Set NIOS-Based OFDM Processor for FPGAs, in Proc. SPIE Int. Soc. Opt. Eng. Orlando (2006), pp. 6248o01–15
[M07]   U. Meyer-Baese, *Digital Signal Processing with Field Programmable Gate Arrays*, 3rd edn. (Springer, Berlin, 2007) 774 pages

[P10]   V. Pedroni, *Circuit Design and Simulation with VHDL* (The MIT Press, Cambridge, MA, 2010)

[R11]   A. Rushton, *VHDL for Logic Synthesis*, 3rd edn. (John Wiley & Sons, New York, 2011)

[S04]   D. Sunkara, Design of Custom Instruction Set for FFT using FPGA-Based Nios processors, Master's thesis, Florida State University (2004)

[SAA95]   N. Shirazi, P. Athanas, A. Abbott, Implementation of a 2-D Fast Fourier Transform on an FPGA-based custom computing machine. Lect. Notes Comput. Sci. **975**, 282–292 (1995)

Chapter 10
Xilinx MICROBLAZE Embedded Microprocessor

Abstract This chapter gives an overview of the Xilinx MICROBLAZE microprocessor system design, its architecture, and instruction set. It starts with a brief ISA architecture overview followed by ZyBo interesting system design options and custom instructions.

Keywords MICROBLAZE · ZyBo · Xilinx · Vivado · TRISC3MB · High definition multimedia interface (HDMI) · SDK · Basic computer · Custom intellectual property (CIP) · TMDS · DMIPS · Pmod · Top-down-design · Bottom-up-design · Floating-point performance · Instruction set architecture · Subroutine nesting · Text Font Design · ModelSim · MatLab

10.1 Introduction

The Xilinx 32-bit softcore processor MICROBLAZE was for many years a Harvard 32-bit RISC processor with three or five pipeline stages and separate data and instruction buses, which is available since the EDK was first published in October 2002 [X05, X08]. The standard key features of the MICROBLAZE core can be summarized as follows:

- MICROBLAZE area optimized is a three-pipeline-stage core with 1.03 DMIPS per MHz.
- MICROBLAZE performance optimized is a five-pipeline-stage core with branch optimizations delivering 1.38 DMIPS/MHz.
- The ALU, shifter, and 32×32 register file are standard.

Optional items (see Fig. 10.1) that can be included at configuration time are:

- Barrel shifter, array multiplier, and divider
- Single precision floating-point unit for add, subtraction, multiply, divide, and comparison
- Data cache from 2 to 64 KB
- Instruction cache from 2 to 64 KB

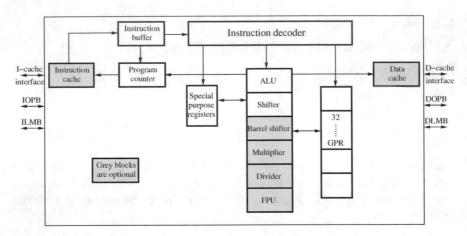

Fig. 10.1 The architecture of the Xilinx on MICROBLAZE Embedded Microprocessor (grey blocks are optional)

The five-stage pipeline is executed in the following steps: (1) fetch, (2) decode, (3) execute, (4) memory access, and (5) write-back. The data and instruction caches have a direct mapped cache architecture. The cache can be access in blocks of four or eight words. One or more Block-RAMs are used to store the data, while an additional Block-RAM is used to store the tag data. A study of typical memory and cache configurations for the MICROBLAZE can be found in the literature [M14, F05].

While the architecture and the peripherals have been updated and refined over the years, recently we have seen two substantial additions:

- An 8-stage pipeline version has been added in 2016.
- A 64-bit MICROBLAZE implementation has been added in 2018.

MICROBLAZE system can be designed by combining the processor with the needed components, and we call this the *bottom-up* design approach. Alternatively, we can use one of the starting point systems provided by the FPGA or board vendor and modify this design as needed. We would call this approach the *top-down* method, and since this is usually a little easier, we will start with this top-down design approach.

10.2   Top-Down MICROBLAZE System Design

On the ZyBo board, a substantial number of I/O are directly wired to the ARM aka PS and cannot be used immediately for the programmable logic (PL), i.e., the MICROBLAZE. This is easiest seen by looking how the FPGA is wired to the peripheral component; see Fig. 10.2. The ZyBo FPGA is a ZC7Z010-1CLG400C, i.e., it uses a ball grid array package (the pin looks like half a soccer ball) arranged in a 20 × 20 array. From the device package file, we can identify the 400 pins [D14] and we have:

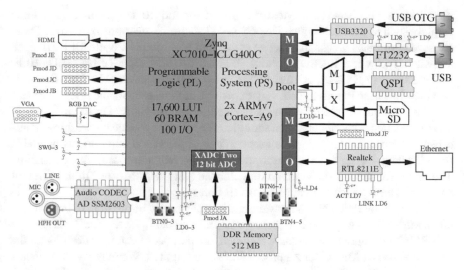

Fig. 10.2 ZyBo peripheral connection to PS (ARMv7 Cortex A9) and programmable logic (PL), i.e., MICROBLAZE

- The PS ARM MIO group 500 with 18 pins connected directly to LED4, SPI Flash, and Pmod JF.
- The PS ARM MIO group 501 with 40 pins connected directly to ENET, USB, SD-Card, UART 1, BTN 4 + 5.
- The PS Double Data Rate (DDR) DRAM group with 75 pins.
- User PL I/O group 34 with 50 high voltage range pins.
- User PL multifunction I/O group 35 with 50 pins.
- Not connected are 25 pins and 17 special dedicated pins.
- The remaining pins are power pins.

This leaves the VGA/HDMI ports, audio CODEC, 4 LEDs, 4 switches, 4 buttons, the XADC, and several 12 pin Pmods for the PS, i.e., the MICROBLAZE can utilize. Figure 10.2 gives an overview how peripherals are connected. We notice in particular that the MICROBLAZE does not have access to a substantial off-chip memory such as the DDR, and also a UART communication is missing. We can still run a couple of experiments such as DMIPS rating and measure time via the general-purpose input/output (GPIO) LEDs, but a UART to print results on the PC terminal would be helpful. We can try to refigure the ARMv7 MIO that the MICROBLAZE can send data over the UART (you may find tutorials on YouTube for that), but this seemed a lot of trouble and not a solution easy to be transferred to other systems or boards. Much more practical would be to add the USBUART Pmod module (available for about 10 E/USD from Digilent Inc.) to your system for communication.

While the number of available peripherals may seem small compared to the DE1 SoC board, these changes dramatically when we consider to invest a few USD/Euro in some additional Pmod modules. Remember the ZyBo board is much cheaper than the DE1 SoC, so the overall system cost after adding a couple of Pmod modules should still be reasonable. Digilent Inc. (in Germany sold by Trenz GmbH) Pmod modules are very low cost. Most are priced around 10 USD/Euro; only some displays such as the MultiTouch Display (74 USD/Euro) are more expensive. There are several of these blocks that can be connected with simple Vivado GPIO IP such as switches (SWT; 5 E/USD), buttons (BTN; 8 E/USD), eight LEDs (8LD; 10 E/USD), 2xSSD (7 E/USD), or 16x2 LCD (31 E/USD), and VGA (9 E/USD) that uses two Pmods. These Pmods are not very expensive as you see from the price estimates in USD/Euro but may require some advanced project planning. For the majority of more sophisticated peripheral, you should make sure that you have a working IP aka driver available. Table 10.1 gives an overview of the Pmod components available from Digilent, a brief function description, and approximate cost at time of writing. The driver for these blocks can be downloaded from GitHub or Digilent free of charge and easily added to your Vivado IP library.

With so many Pmod modules available, it is physically impossible to build a top-down MicroBlaze system that incorporates all Pmods. In fact, since the ZyBo has an ARMv7 microprocessor already included, the number of example designs for the ZyBo that uses a MicroBlaze processor is quite limited. We may try to recycle a design from another board, add our ZyBo board files, and master pin file instead. But since the available number of peripherals is small, we can go right ahead and just use a Vivado MicroBlaze template and add the peripheral yourself. We would call this a bottom-up approach, and it is explained in the next section.

10.3   Bottom-Up MicroBlaze System Design

Putting together a MicroBlaze FPGA system is a multistep process but with most designs steps supported by *Block* and *Connection Automation* not too difficult. Let us have a look at the major steps putting together a MicroBlaze Basic Computer

Table 10.1   Pmod blocks with available drivers with estimated price in USD/Euro

Block	Description	Price	Block	Description	Price
ACL	3-axis accelerometer	15	JSTK	2-axis joystick (smaller)	21
ACL2	3-axis accelerometer	18	JSTK2	2-axis joystick	17
AD1	Two 12-bit ADC	31	KYPD	16-button keypad	10
AD2	Four 12-bit ADC	21	MTDS	Multitouch display	74
ALS	Light sensor	17	NAV	9-axis barometer	30
AMP2	Audio amplifier	10	OLED	128 × 32 OLED	16
BT2	Bluetooth interface	27	OLEDrgb	96 × 64 RGB OLED	21
CLS	16 × 2 LCD serial	31	R2R	R2R resistor DAC	5
DA1	Four 8-bit DAC	21	RTCC	Real-time clock	9
DPG1	Pressure sensor	29	SD	SD card slot full size	10
ENC	Rotary encoder	7	SF3	32 MB flash	11
GPS	Receiver module	42	TC1	Thermocouple wire	20
GYRO	3-axis gyroscope	21	TMP3	Temperature sensor	7
HYGRO	Humidity and temperature sensor	15	WIFI	802,11b Wi-Fi	20

Parts*Parts* Parts*

system. We get started with a usual RTL project: Start *Vivado* and then click *Create New Project*. Choose your desired project location and then click *Next*. Use *RTL Project type* and click *Next* a few times until you can select *Parts* and *Boards*. Select the *ZyBo* board to get the correct I/O components and locations. Since we plan to use also some of the Pmod ports, add the ZyBo_Master.xdc also as constrain file to you project. The XDC file and board file called master.zip can be downloaded from GitHub or www.digilentinc.com. You should install the board files in the Vivado path under ...data/boards/board_files.

Now start *IP Integrator → Create Block Design*, and click the 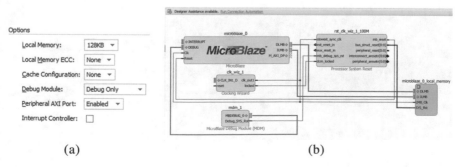 button such that you can place a new library component of the *MicroBlaze* in the block design. Tip: Use the search for *Micro*, then click on the component with your left mouse button one time, and then use the *Enter* on your keyboard. On the top of your Diagram, you should see *Designer Assistance* available and click the blue <u>Run Block Automation</u> link. You will be asked to specify major parameters such as local memory size, caches, and interrupt controller. Since our PL resources are limited, we only use (largest possible) local memory such we can run some non-trivial programs without cache memory (see Fig. 10.3a). When block automation has finished Vivado has added the components: *Clocking Wizard, MDM, Processor System Reset* and the 128 KB *local_memory*, see Fig. 10.3b. The clocking Wizard produces a 100 MHz output frequency that is used by MicroBlaze and the AXI components. If you have a predesigned system, you can use *IP Integrator → Open Block Design* of the predesigned system, and then after you double-click the *MicroBlaze* systems, you will be able to modify processor parameter, add floating-point co-processors, etc. later, but you cannot modify the local memory size easily any more.

After you have successfully configured your MicroBlaze, you may want to add some simple I/O such as switches, LEDs, UART, and a timer to your Basic Computer.

Right click on the *Block Diagram* and select or right click and select *Add IP...* look for *AXI GPIO* and click return. Repeat this one more time so that you have two *AXI GPIO*. Double-click the AXI GPIO and set *Board Interface* to the I/O, we like to use *leds_4bits* and *sws_4bits*. Then click *OK* to close the *AXI GPIO* window. Keep the default setting for the timer. We will use the button btn[0] as reset for

(a)   (b)

Fig. 10.3 The MicroBlaze instantiation: (**a**) Main parameter (**b**) MicroBlaze after block automation run

our MICROBLAZE. Use 📑 two more times and use *Add IP...* to look for the *AXI Timer* and the *Uartlite* if you have added the Pmod USBUART as recommended earlier (see Fig. 10.6a). Double-click the *AXI Uartlite* and set *Baud Rate* to 115200 since this is the default rate in the Software Developer's Kit (SDK) terminal. The initial Uartlite rate 9600 will work too, but you will then need to change the Baud rate every time you use SDK. You should see at the top of the *Diagram* that *Designer Assistance available*. Click the Run Connection Automation highlighted and under-lined in blue. The processor system reset, AXI switch, and I/O ports will be added and all connections are routed. Also inside the IP, an update of the *IP Configuration* within the *AXI GPIOs* takes place. You may need to add a constant value 1 block and connect this constant to the ext_reset_n of the *Processor System Reset* block. You will need to update the ZyBo_Master.xdc to reflect the port name VIVADO choose for you. You can also force renaming by dragging the I/O from the board window to your block design window IP. Typical names are sys_clock for the 125 MHz input clock and reset_rtl for the reset of the *Clocking Wizard* IP. For the USBUART, you should be aware that the functionality for input and output changes. What is an output for the *Uartlite* IP is an input for the *USBUART Pmod* and vice versa. We used Pmod JE (see Fig. 10.6a) but any other (except MIO or XADC) Pmod will work too. The setting in the XDC should be done as follows:

Uartlite	Pmod	Pin of Pmod JE
uart_rtl_txd	RXD	JE1
uart_rtl_rxd	TXD	JE2

The RTS (pin 0) and CTS (pin 3) ports of the *UASBUART Pmod* and the inter-rupt pins of *Uartlite* are not used. We do not need two resets so we made the external processor reset a constant. You may want to rename the *AXI GPIOs* to reflect the ports used and finally use *Validate Design* 🗹 to check for connection errors and the 🔄 button or right click and select *Regenerate Layout* to have a nice aligned *Diagram* of your design. Figure 10.4 shows the final layout of the *Diagram*.

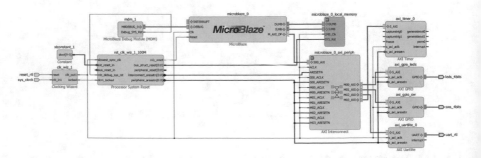

Fig. 10.4 The MICROBLAZE Basic Computer design with added timer, UART, and 2 GPIO

Now we are ready to translate our design into a bitstream that can be downloaded to the FPGA. To compile the graphic design, we need to build a HDL wrapper first. In the *Block Design* panel, right click on design_1.bd and select *Create HDL Wrapper ...* and choose the VIVADO *auto update* option. Now under the *Program and Debug* panel, select *Generate Bitstream*. This may take some time to complete.

It is now time to start with the program development. SDK is more or less a standalone tool, and in VIVADO we use *File → Export → Export Hardware...* to get started. Do not forget to select the *Include bitstream* option before export and then click *OK*. Next select *File → Launch SDK* and click *OK*. Later you can start SDK directly as a standalone package using a double-click on ![SDK] on your desktop. SDK will start with a welcome message listing the device, design tool, and all IP blocks (see Fig. 10.5a). Since we have new components, it seemed to be a good idea to take advantage of the SDK feature to generate a peripheral test program for us that tests all components, but only the components we have added to the system. Select *File → New → Application Project* and choose a *Project name* such as test_IO click *NEXT>*, and then as *Templates*, choose *Peripheral Tests* then click *Finish*. You will get test routines for the LED and switches, but not for the USBUART since SDK is smart enough to use this IP as STDOUT. If you did not download the system bitstream to the FPGA with VIVADO, you can do that in SDK too. Just press the ![icon] button or use *Xilinx Tools → Program FPGA*. To add a terminal window, click on the ![icon] icon in the middle SDK terminal panel; your USBUART (most likely) has the last COM port in the list. Use the default parameter (i.e., *Baud Rate* 115200) and specify COM port and *Timeout (sec)* 5 (see Fig. 10.5b). If you have not modified the *Uartlite* default Baud rate, you need to set it to 9600. Both works fine just make sure you use the same settings for the SDK terminal and the IP. You can run the test and

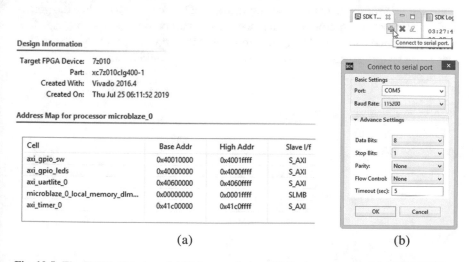

(a)   (b)

Fig. 10.5 The SDK initial steps: (**a**) Welcome window. (**b**) Setting up terminal for USBUART

you will see success message in the Terminal window and a very short running light on the four green LEDs of the ZyBo board. You will get a project folder and a second folder with ..._*bsp* extension that includes system files and ANSI C service routines. Have a look at the main routine test_IO → src → testperiph.c and the associate test programs in the same folder. These contain many useful functions you may need in your program development. For instance, for each of the XGpio component, SDK provides an *Initialize* function and a *DataDirection* function. For the leds, we would, for instance, use:

```
XGpio_Initialize(&leds, XPAR_AXI_GPIO_LEDS_DEVICE_ID);
XGpio_SetDataDirection(&leds, 1, 0x0);
```

The DEVICE_IDs and device names can we look-up in xparameters.h under the bsp include files. In the right panel, just double-click xparameters.h, and the file will open up for you. We can then display an integer count value at the LEDs using

```
XGpio_DiscreteWrite(&leds, 1, count);
```

For our new AXI Timer, we would use initially: XTmrCtr_Initialize(), XTmrCtr_SelfTest(), XTmrCtr_SetOptions(), XTmrCtr_SetResetValue(), XTmrCtr_Reset, and XTmrCtr_Start(). For our measurements, we would make calls before and after the code segment to XTmrCtr_GetValue(). The advantage of using the AXI Timer is that we can use 99% the same code for our measurement with ARM or MICROBLAZE, eliminating any error that may occur when using the internal ARM vs. the AXI Timer.

A typical short SW and LED test that continues display a counter pattern on ZyBo and a message in the terminal now looks as the Listing 10.1 below. The SW value can be used to increase the speed of the LED counter. We modify the generated testperiph.c program and place the first part around the main() section, while the second part should be placed in front of the line print("---Exiting main---\n\r"). Here are the code segments that should be added.

ANSI C Code 10.1: The LED Test Routine for USBUART

```
1    ...
2    #define LOOP_ITERATIONS 2000000
3
4    int main()
5    {
6        XGpio sw, leds;    // from xgpio.h struct
7        int inc, count=0;
8        volatile unsigned int i = 0;
9    ...
10
11   //initialize SW/LEDs and set data direction to I/O
12   XGpio_Initialize(&leds, XPAR_AXI_GPIO_LEDS_DEVICE_ID);
13   XGpio_SetDataDirection(&leds, 1, 0x00000000);
14   XGpio_Initialize(&sw, XPAR_AXI_GPIO_SW_DEVICE_ID);
15   XGpio_SetDataDirection(&sw, 1, 0xFFFFFFFF);
16
17   while(1){
18   //output count value on the LEDs
19       XGpio_DiscreteWrite(&leds, 1, count);
20       count+=1;
     //waiting for some time; use SW for increment
         inc = XGpio_DiscreteRead(&sw, 1);
         for(i = 0; i < LOOP_ITERATIONS; i+=inc+1){}
         xil_printf("New LED value = %d\n",count);
     }
     print("---Exiting main---\n\r");
```

Short video of the ZyBo counter light and the SDK can be found in the Videos folder book CD under PmodUART_ZyBo.MOV and PmodUART_SDK.MOV. You will notice the continuous change in the green LEDs of the ZyBo as well as the blinking reds LED of the *USBUART* Pmod every time the MicroBlaze print out aka sends a line to the terminal. Snapshot of the videos can be seen in Fig. 10.6.

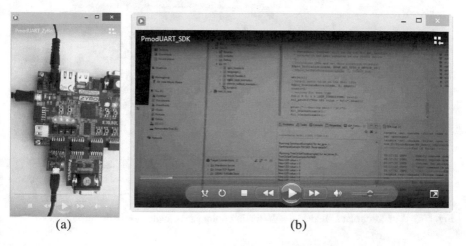

(a)   (b)

Fig. 10.6 The SDK peripheral test: (**a**) ZyBo with USBUART. (**b**) SDK terminal display

After we have the initial setup working, we can take on a more sophisticated task such as measuring the floating-point operation performance of our MICROBLAZE Basic Computer we had discussed briefly in Chap. 5, Sect. 5.7 and in more details in Chap. 9, Sect. 9.4.

Figure 10.7 shows the comparison of the systems with and without floating-point hardware support. We clearly see an improvement (factor over 200) for the four basic operations for FPH extension at the cost of about twice the number of LUTs. The compiler option −O3 improves a factor 3 compared to −O0 (default). Conversion time is also substantial improved. We can see from the measurements that both, Nios II and MICROBLAZE, also improve the 32-bit square root sqrtf() but the 64-bit sqrt() should be avoided if possible.

Table 10.2 shows speed and size comparison of our Basic Computer and the extended floating-point hardware (FPH) requirement. Memory requirements, speed, and compile time remain similar, only the number of LUTs is 54% higher for the version with floating-point hardware support.

We just saw that it is much more convenient (and precise) to display measurement data over the USBUART Pmod than measuring the time of a routine with a stopwatch between turning on and off a LED. If we want to use the VGA or HDMI port to display a text information, then we would need to develop a *custom intellectual property* (CIP) for the ports. This will be discussed in the next section.

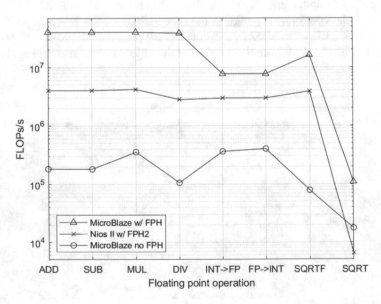

Fig. 10.7 The MICROBLAZE and Nios II floating-point performance for basic operations

Table 10.2 Speed and size comparison of FPH co-processors

	MicroBlaze Basic Computer	MicroBlaze Basic Computer with FPH
LUTs	1759	2722
LUTRAM	139	146
36 K Blocks	32	32
DSPs	0	2
Fmax/ MHz	125.7	121.21
Vivado Submodule compile time	44:44	48:45
Vivado Syn. + impl. compile time	10:36	2:58

10.4 Custom Instruction MicroBlaze System Design

Modern sophisticated softcore embedded processors such as the MicroBlaze processor have substantial integer and floating-point processing power to cover a wide range of applications. However, there are a couple of tasks even the most powerful embedded processor will have difficulty to handle. Think, for instance, of the large amount of data a video displays such as VGA or HDMI which continuously needs to be provided. In the lowest resolution 640 × 480 × 8-bit, a HDMI controller needs to deliver 250 Mb/s, i.e., twice the speed our MicroBlaze can run with! Even the fastest microprocessor overburden, and substantially performance decrease is expected when the microprocessor has to pick up these data from memory and forward it to the HDMI ports or VGA DACs. Here a so-called custom intellectual property (CIP) is beneficial. These are FPGA-based custom logic blocks closely coupled with the microprocessor. Such a CIP system was discussed in detail in Chap. 9 for the Nios II to improve floating-point performance. Let us in the following have a closer look at the design of a HDMI CIP for the MicroBlaze since we not really have a monitor display with our ZyBo so far.

Before we start with the details of the CIP design, let us briefly review the DVI and HDMI standards.

DVI and HDMI Display Options Implementation

For many years, the (cathode-ray tube) CRT monitor was the dominant technology. The CRT monitor uses X/Y plates to guide an electric beam with the associated analog VGA control signals. VGA were discussed in Chap. 2 (see Fig. 2.4). Since a couple of years now, digital aka LCD monitors have been replacing CRT monitors due to technology advantages such as higher resolution, less power dissipation, flat screen architecture, etc. to name just a few. Most digital LCD monitors although digital display physically usually still have VGA inputs and need to sample the analog RBG signals to display on the digital LCD. Since most image/video are stored by a processor in DRAM memory, we basically first need to convert our stored image via a DAC to an analog VGA signal, and then the LCD uses a ADC to convert it back to digital signals for display on a LCD screen. A better solution seemed therefore to transmit directly the digital signal from our CPU to the LCD monitor. To simplify the transition from analog to digital video technology, the digital visual interface (DVI) standard was developed by the digital display working group (DDWG). Initial members of DDWG were Intel, Silicon Image, Compaq, Fujitsu, HP, IBM, and NEC.

The DVI standard continued to support the VGA signals but also had Plug & Play ports and the high-speed Transition Minimized Differential Signaling (TMDS) included that translates the 8-bit RGB into high-speed differential signals that run first through a transition minimization and DC balancing.

VGA, DVI, and as we see later HDMI basically use the same row/column timing developed for the CRT monitor, i.e., besides the row/column display periods, we also have horizontal and vertical pause sections that allowed the electric beam to return to the original location and to implement "back porch" synchronization impulses. HDMI uses this back-porch time interval to code audio signal and multimedia features. The image of the lowest resolution of 640×480 pixels then actually needs a total of 800×525 pixel clock cycles to complete, leaving 45 lines and 160 horizontal pixel clock cycles to code audio and synchronization. The standard frame rate is 60 Hz, but 100 Hz, 120 Hz, and also 30 Hz are permitted.

Let us in the following have a brief look at the TMDS signaling since it is used verbatim also in HDMI coding, the dominant video standard today, e.g., most recent ZyBo board no longer support VGA and only have HDMI ports.

TMDS Encoding and Decoding

To reduce the radio interference in the first step, TMDS tries to reduce the number of transitions. The algorithm determines if a transformation via XOR or XNOR gates will be more beneficial. An additional 9 bit is added to indicate which transition was used. Here are two examples:

01010101 → 00110011 via XOR: $Q(n) = Q(n - 1) \oplus D(n); D(0) = D(0)$
10101010 → 11001100 via XNOR: $Q(n) = Q(n - 1) \odot D(n); Q(0) = D(0)$

With MSB on the left. Encoded in bit 9 is if XOR (1) or XNOR (0) was used. If the number of one's in the input word is larger than 4 or 4 and $d(0) = 0$, then XNOR is used otherwise XOR. For an all zero or all one string, the transformation will not change the data:

00000000 → 00000000 using XOR bit(9) = 1:
11111111 → 11111111 using XNOR bit(9) = 0:

In the second step, a DC balancing is performed to avoid very long runs of zeros or ones. If the string was flipped, this is encoded with a one in bit 10 otherwise zero.

The TMDS decoder is substantially simpler to implement than the encoder, since the encoder needs first to make statistical analysis before deciding on bit 9 and 10. The decoder simply checks bit 10, if a complement of all bits is required, and then applies the XOR or XNOR based on bit 9. Figure 10.8b shows the decoder logic.

The encoder is a little bit more complex than the decoder. Based on the number of one's in the input data D, we apply the XOR or XNOR operation. If the number of 1's is larger than the number of zero or both the same and $D(0) = 0$, then XNOR otherwise the XOR is used to compute q_m. If no valid video data are available, then two control bits $C0 = $ hsync and $C1 = $ vsync are used to code control sequences (not shown in Fig. 10.8) that are later used to build the horizontal and vertical synchronization signals within the blue channel only. Based on the count N of one's in q_m and the accumulated disparity from previous pixel, the decision to flip the string is computed and the disparity counter updated. The flip is implemented if previous disparity was zero or count is zero and $q_m(8) = 0$, or both have the same sign. The update of the disparity counter is a little bit more complex and summarized in Table 10.3.

Here we determine the differences between the number of one's (N1Q) and zero's (N0Q) in DISP = N1Q-N0Q = 8-2N1Q since for 8-bit strings we have N1Q = 8-N0Q.

We are now ready to start with the HDL coding. The decoder is simple and left as a project design (see Exercise 10.64). For the encoder, we find a couple of sources, ranging from textbooks [P10], over application notes [X460, X495] to online resources [H19]. Here is a possible solution in Verilog following closely the scheme from Table 10.3:

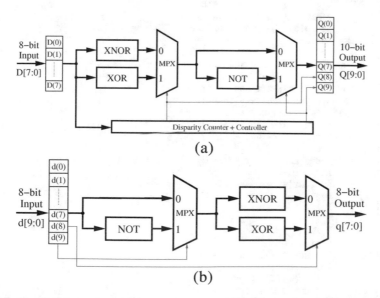

Fig. 10.8 The TMDS signals (**a**) encoder. (**b**) Decoder (without control sequences)

Verilog Code 10.2: TMDS Encoder

```
1    // =====================================================================
2    // IEEE STD 1364-2001 Verilog tmds_encoder.v
3    // Encoding 8-bit input to 10 bit transition minimized code
4    // =====================================================================
5    module tmds_encoder (
6        input CLK,        // Pixel clock
7        input RESET,      // Synchronous reset
8        input DE,         // Data enable
9        input [7:0] PD,   // Pixel data
10       input [1:0] CTL,  // Control data
11       output reg [9:0] Q_OUT); // Output data
12   // =====================================================================
13   // Compute number of 1s in pixel data
14   wire [3:0] N1D = PD[0] + PD[1] + PD[2] + PD[3] + PD[4] + PD[5]
15                                                     + PD[6] + PD[7];
16   // Compute internal vector Q_M
17   wire USE_XNOR = (N1D > 4'd4) || (N1D == 4'd4 && PD[0]==1'b0);
18   wire [8:0]  Q_M = USE_XNOR ?
19                {1'b0, ~(Q_M[6:0] ^ PD[7:1]), PD[0]} : // use XNOR
20                {1'b1,   Q_M[6:0] ^ PD[7:1], PD[0]}; // use XOR
21
22   // Compute number of 1s in Q_M data
23   wire [3:0] N1Q = Q_M[0] + Q_M[1] + Q_M[2] + Q_M[3] + Q_M[4] + Q_M[5]
24                                                     + Q_M[6] + Q_M[7];
25   wire [6:0] N1N0 = 2*N1Q - 8; // i.e., N1Q - N0Q;
26   reg INV_Q;
27   reg [6:0] CNT, CNT_NEW;
28   reg [9:0] Q_CTL, Q_DATA;
29
30   // determine the control tokens
31      always @*
32        case (CTL)
33        2'b00   :  Q_CTL <= 10'h354;
34        2'b01   :  Q_CTL <= 10'h0AB;
35        2'b10   :  Q_CTL <= 10'h154;
36        default :  Q_CTL <= 10'h2AB;
37      endcase
38
39   // Compute output and new disparity counter value
40      always @(*) begin
41      INV_Q <= 1'B0; CNT_NEW <= 0;
42   // Update the disparity counter
43      if (CNT==0 || N1N0==0)
44        if (Q_M[8]==1'B0) begin
45          CNT_NEW <= CNT - N1N0; INV_Q <= 1'B1;
46        end else
47          CNT_NEW <= CNT + N1N0;
48      else
49        if (CNT[6] == N1N0[6]) // same sign?
50        begin
51          INV_Q <= 1'B1;
52          if (Q_M[8]==1'b0)
53            CNT_NEW <= CNT - N1N0;
54          else
55            CNT_NEW <= CNT - N1N0 + 2;
56        end else
57          if (Q_M[8]==1'b0)
58            CNT_NEW <= CNT + N1N0 - 2;
```

```
59                  else
60                      CNT_NEW <= CNT + N1N0;
61          end
62      // Compute the q_data vector
63          always @*
64          Q_DATA <= INV_Q ? {1'B1, Q_M[8], ~Q_M[7:0]} :
65                            {1'B0, Q_M[8],  Q_M[7:0]} ;
66
67      // store counter and q_out in registers
68          always @(posedge CLK) begin
69          if (RESET == 1'b1) begin
70              CNT = 1'b0; Q_OUT <= 10'h000;
71          end else  // PIXEL DATA (DE=1) OR CONTROL DATA (DE=0) ?
72              if (DE == 1'b1) begin
73                  CNT <= CNT_NEW;
74                  Q_OUT <= Q_DATA;
75              end else begin
76                  CNT <= 0;
77                  Q_OUT <= Q_CTL;
78              end
79          end
80
81      endmodule
```

Table 10.3 Update of disparity counter in TMDS encoder

DISP = 0 OR cnt(t-1) = 0	$q_m(8)$	sign(cnt(T-1)) = sign(DISP)	Counter update
T	1	–	cnt(t) = cnt(t-1) + DISP
T	0	–	cnt(t) = cnt(t-1)-DISP
F	–	T	cnt(t) = cnt(t-1)-DISP + 2
F	–	F	cnt(t) = cnt(t-1) + DISP-2

The input port of the encoder has three control signals, clock (CLK), reset (RESET), and data enable (DE), and two vectors, 8-bit pixel data (PD) and the two bits control vector (CTL) for horizontal and vertical synchronization. The data output (Q_OUT) is 10-bit long. The encoder logic follows closely Fig. 3.5 [D99] of the DDWG document. We start with counting the number of one (N1D) in the input data (line 14–15). Based on the number of one's, we decide if the XNOR or XOR gate is used for transition minimization (lines 17–18) and compute Q_M. We then compute the number of one's and zero's in Q_M, where we use the fact that number of one's plus number of zero's together give the number of bits in the string. Based on the control bits, we then compute the four different control token Q_CTL (lines 30–37). In the next always block, the inversion flag and the disparity counter update according to Table 10.3 is computed (line 39–61). Based on the inversion flag, we compute Q_DATA in original or complement form and adding the decision bit in the MSB position, such that we have a 10-bit vector. In the last always block (line 67–79), we store our (updated) counter value and results in registers. If no data are processed (DE = 0), we forward one of the four control tokens to the output Q_OUT.

We now run some simulations with our files. First, we run short simulation to verify the control token and the correct update of the counter and DC balancing by using a value ff as input. With a 00 as input, the output becomes 100 and for ff the output is 0ff. The 5516 = 010101012 is transformed to 13316 = 1001100112. The DC balancing is demonstrated with a long ff input sequence. Notice how the output toggles between 0ff and 200, i.e., avoiding long ff sequences (see Fig. 10.9).

In a second simulation (see Fig. 10.10) we verify that encoder and decoder work for all 256 input data. Since the counter uses feedback, this is not covering all possible states, but at least 50%. If we make a precise counting, we will find that not $2 \times 256 = 512$ pattern but exactly 460 patterns are possible. There are a few extra control tokens such as hsync and vsync. HDMI differs from DVI that it uses preamble control token for video guard band and data island guard band [X460]. Since it is hard to monitor that many numbers, we can select the "analog" display in MODELSIM and the triangle function of the input to the encoder should match the triangle function in the output data of the decoder (see Fig. 10.10). Input values increase linear from 00 to ff entering the encoder, the same triangular output values of the decoder are shown in the simulation.

One last item we have not discussed is the second part of the name TMDS. Since HDMI is a high-speed serial transmission, it is sensitive to DC drifts and peak/impulse noise. To this end, a differential encoding is used for R, G, B, and the clock. This requires special I/O hardware since only ¼ bit jitter is permitted [TI07]. Even at the lowest resolution with a TMDS rate of $25 \times 10 = 250$ MHz, this is translated into a maximum allowed jitter of $\frac{1}{4} 1/(250 \times 10^6) = 1$ ns which is lower than most FPGA LUT delays. ZyBo has a special OBUFDS hard macro to access these high-speed differential I/O (see Fig. 10.14 outputs).

Let us in the following use the TMDS encoder to build a complete HDMI encoder that can then be packaged as CIP for design reuse. As starting point for HDMI encoder, any HDL description of VGA encoder would be helpful that can be found in the literature [P10, M14].

HDMI Encoder

The HDMI encoder includes also audio data encoding, so it is a little denser coding than VGA. HDMI allows compressed or uncompressed audio data with rates from 32 kHz, 44.1 KHz, 48 kHz, 88.2 KHz to 192 kHz using 2-channel L-PCM compression with a IEC60958 packet structure. The order of a complete line for a 640 × 480 frame including audio data is shown in Fig. 10.11. It processes the data in the following order:

- Video guard band (2 pixels)
- Active video data (640 pixels): Each 8-bit pixel is translated with TMDS encoder into 460 unique 10-bit symbols for all three channels, red, green, and blue.
- Data island preamble: Advanced indication that audio data follows

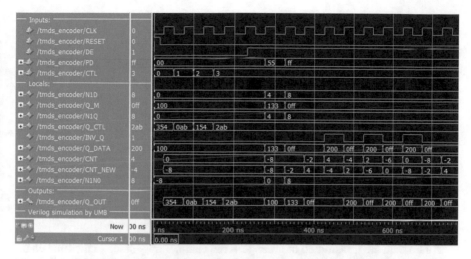

Fig. 10.9 The TMDS encoder detail view

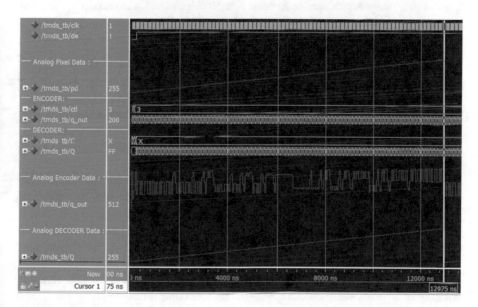

Fig. 10.10 The TMDS encoder and decoder overall behavior for data 0...255

- Data island guard band leading (2 pixels): Allows synchronization for audio data between transmitter and receiver.
- Active audio data: Audio data are encoded as 10-bit TERC4 symbols TREC4 uses 16 unique 10-bit characters. Transmission is on green and red channels only since blue is used for video synchronization.
- Data island guard band tailing (2 pixels).
- Video preamble: Indication that video data will follow.

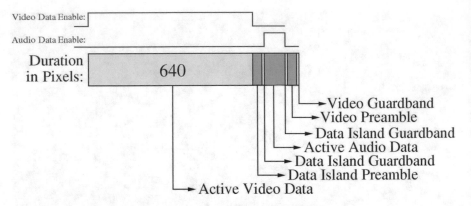

Fig. 10.11 The HDMI line coding with video and audio data [X460]

Audio data are transmitted in green and red channel only since in the blue channel hsync and vsync signals are coded. Following the initial 640-pixel data we have in the "beam return" phase in each line, 16 clock cycles front porch, 96-pixel long synchronization, and 48 clock cycles back porch. For the vertical signaling, we have first 480 lines data then 10 lines front porch, 2 lines synchronization, and 33 lines back porch. The horizontal synchronization signal is therefore active between column 656 and 752, and the vertical for line 490 and 491, if the counter starts at zero. HDMI is coded with a two-bit control vector combined from C1 = vsync and C0 = hsync. The four 10-bit codes can be seen in Listing 10.2 for code line 30–37 and are only active during horizontal and vertical synchronization.

Besides the standard RGB signaling, HDMI also supports YC_BC_R in 4:4:4 and 4:2:2 coding. For the 4:2:2 coding, Y can have 10- or 12-bits using C_B and C_R downsampled by 2.

The Data Display Channel (DDC) is an optional HDMI output but required for HDMI in most cases. The DDC transmits, for instance, the encryption key for copy protection. The DDC provides in 28 bytes information about supported and default video formats, RGB colorimetry, aspect ratio, scaling, and repetition factor. The audio info frame also has 28 bytes and includes audio, channel count, sampling frequency. The signaling for DDC is the I^2C bus. The HDMI standard also includes an optional CEC bus with a slow 400 bits/second bus speed that is used for interoperability of devices, setup tasks, or infrared remote control usage.

The HDMI standard has seen a couple of updates over the years. The first HDMI standard 1.0 introduced in 2002 had maximum data rate of 3.56 GB and a character rate of 165 MHz. Version 1.3 in year 2009 added pixel coding with 10, 12, 16 bits; higher refresh rate of 240 Hz; and increased resolution to maximum of 1440 p and 4 K. Composition of RGB into 4:2:0 YC_BC_R were added. Version 1.4 added Ethernet data channel of 100 Mbits/sec and 3D video to the standard. In version 2.0 (2013) and 2.1 (2017) image size of 8 K: 7680 × 4320 and encoding in 16b/18b were added. Maximum data rate is now at 42.6 Gbits/s and character rate at 1.26 GHz; maximum audio channels were increased to 32.

Last but not least it should be mentioned that HDMI is not a free standard if you plan the sell HDMI systems. It requires a license payment of $10 K for companies that produce more than 10 K units and $5 K/year and $1 per unit for low volume. HDMI founders is a similar group as DDWG with members, Hitachi, Sanyo, Philips, Silicon Image, Sony, Technicolor, and Toshiba.

Before we start with HDL requirements for the CIP, let us briefly discuss how we may reduce the large memory requirement usually associated with image display.

Text Terminals and Font Design

The image processing even with low resolution usually comes with a substantial memory requirement. In the lowest resolution, a RGB images needs $640 \times 400 \times 3 \times 8$-bit or 7.37 Mbits. On the ZYBO, we have 60 BRAM each with 36 Kbit capacity. The color VGA image would need 7.37 M/36 K = 204 blocks, so too many for our ZYBO board. However, in the majority of embedded applications, it would be sufficient to just display text and not color images. That will substantially reduce our memory requirement if we just store the required ASCII values for the characters and would allow us to use the HDMI monitor to display text message such as DMIPS or FPMIPS measurement. So, let us briefly talk about how to develop u IIDMI text terminal for our ZYBO and make a little field trip in the area of text font designs.

Character fonts can and have been built with dot matrix of 5×7, 6×16, 8×8, etc. pixels for LCDs and monitors. The 8×8 size has the distinguish advantage that we can use the three LSBs of the row/column counter to address the character and therefore simplify the design essential.

Now for the typeface aka fonts family, we have many different choices. Just check your favorite word processing program and you see the many choices. Textbooks (such as this one you are reading right now) typically use serif fonts such as Times New Roman (e.g., EμPSD) or Courier New (e.g., EμPSD) we use in program listing in this book. These serif fonts have cute little ticks at the end of each stroke to make them look more appealing.

For computer displays, with a limited number of pixels, fonts without serif aka san serif fonts such as Arial (e.g., EμPSD) are more popular. The Arial font also is often used in PPT slides. The Commodore PC made this 8×8 san serif typewriter font popular for monitor display. We may also make modification to fonts such as boldface or Italic aka cursive or may use a modern looking font such as the Future font. Figure 10.12 shows examples of four fonts we like to use.

We should support all standard ASCII character, i.e., 127 characters should be sufficient (see Chap. 5, Fig. 5.2). We may want to add some often-used special symbols such as ®, ©, or μ to our set in the nonstandard character places. If we now try by hand to design all 127 characters in 8×8 pixel resolution, we will be quite busy for a long time. A more convenient way would be to design the fonts using a font editor such as `PixelFontEdit` by Richard Prinz. We have used Version 2.7.0 of

`PixelFontEdit` that can be downloaded free of charge from https://www.min. at/prinz/o/software/pixelfont to help with the font design. The font editor provides several font templates. We use four fonts and add three special symbols, ®, ©, or μ at locations 5, 6, and $7F_{16} = 127$. Figure 10.13a shows a character Z example. To design a character, you simply just click on the pixel, and the 8×8 pixel code is immediately computed. Row and column separation space/lines should also be provided within the characters for easy reading. This pixel code are stored as constant values in a `*.h` file to be used in an ANSI C compiler (see Fig. 10.13b). A short MATLAB script allows us to read in the `*.h` file and produce the binary memory initialization file we need for Verilog `$readmemb()` functions. You should notice how the zero/one pattern also allows us to verify the correct coding of our character Z. Since Z has ASCII code 90 and each letter needs 8 lines, we find the letter Z in our MIF file starting after line 90*8 = 720 (see Fig. 10.13c).

For our size 8×8 font, the 640×480-pixel monitor is able to display 640/8 = 80 characters in one row, and we will have 480/8 = 60 rows. We will need $80 \times 60 \times 7 = 33{,}600$ bit or one Zynq BRAM to store the data. Here is the Verilog description of the character memory.

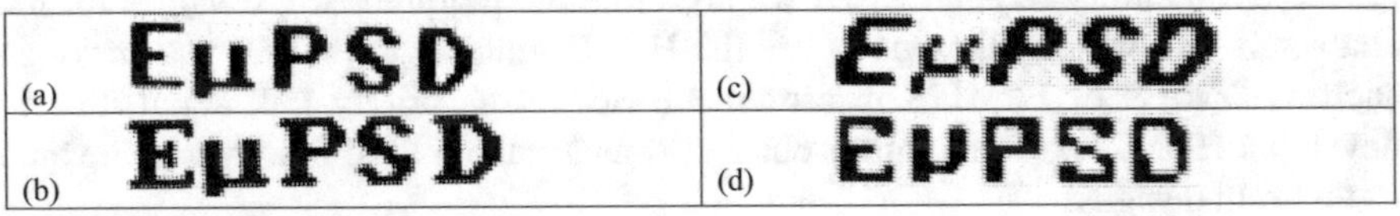

Fig. 10.12 Font examples from `PixelFontEdit` with modifications. (**a**) San serif example, APEAUS; (**b**) serif examples, HERCULES; (**c**) italic example, HERCITAL; (**d**) modern example, SPACE8

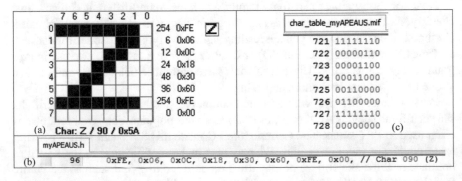

Fig. 10.13 The Font design using `PixelFontEdit` by Richard Prinz. (**a**) Character editing via mouse. (**b**) ANSI C table generation. (**c**) MIF file conversion for Verilog

HDL Code 10.3: Character RAM Description with Initialization

```verilog
1   //==================================================================
2   // IEEE STD 1364-2001 Verilog file: img_ram.v
3   // Store the 80x60 character array; use initialization file
4   //==================================================================
5   module img_ram
6   #(parameter DATA_WIDTH=8, parameter ADDR_WIDTH=13)
7     (input   CLOCK,                      // System clock
8      input WE,                          // Write enable
9      input [DATA_WIDTH-1:0] DATA,       // Data input
10     input [ADDR_WIDTH-1:0] ADDRESS,    // Read/write address
11     output [DATA_WIDTH-1:0] Q);        // Data output
12  //==================================================================
13  // Declare the RAM variable
14     reg [DATA_WIDTH-1:0] ram[2**ADDR_WIDTH-1:0];
15
16  // Variable to hold the registered read address
17     reg [ADDR_WIDTH-1:0] addr_reg;
18
19     initial
20     begin
21       $readmemh("eupsd_hex.mif", ram);
22     end
23
24     always @ (posedge CLOCK)
25     begin
26       if (WE)     // Write
27         ram[ADDRESS] <= DATA;
28
29       addr_reg <= ADDRESS; // Synchronous memory, i.e. store
30     end                    // address in register
31
32     assign Q = ram[addr_reg];
33
34   endmodule
```

The Code 10.3 is implementing character memory. The port description (lines 7–11) is followed by the memory array and address register (lines 13–17). Here we use an initialization file eupsd_hex.mif with $80 \times 60 = 4800$ lines that shows the sentence "EµPSD w/ Altera ® and Xilinx ®FPGAs by © 2019 Dr. Uwe Meyer-Baese" in each line and to the right a test of all characters in the font. Notice the three special symbols ®, ©, and µ used in the test sentence. A small MatLab script is beneficial to generate the test data. Listing 10.3 follows the coding of the synchronous RAM (lines 24–32). The movie HDMI4color4fonts. MOV shows the test sentence in different colors and fonts.

For each font, we will need $128 \times 8 \times 8 = 8192$ bits or one BRAM. Depending how the four font ROMs and character memory are synthesized, we will need 2 to 5 BRAM in total. This compares nicely with the over 200 BRAM a standard 640×480-pixel color image would need. Here is the Verilog description for the font memory.

HDL Code 10.4: Character ROM LUT for 4 Fonts

```
1    //=====================================================================
2    // IEEE STD 1364-2001 Verilog file: char_rom.v
3    // ROM LUT for the 4 fonts
4    //=====================================================================
5    // Initialize the ROM with $readmemb. For Vivado
6    // use the MIF files extension. Without this file,
7    // this design will not compile. See Verilog
8    // LRM 1364-2001 Section 17.2.8 for details on the
9    // format of this file.
10   module char_rom
11   #(parameter DATA_WIDTH=8, parameter ADDR_WIDTH=10)
12     (input  CLK,                        // System clock
13      input [1:0] SW, // Font selection
14      input [6:0] ADDRESS, // Address input
15      input [2:0] ROW, COL, // row/column address
16      output wire Q); // Data output single bit
17   //=====================================================================
18   // Declare the ROM variable
19     reg [DATA_WIDTH-1:0] rom0[2**ADDR_WIDTH-1:0];
20     reg [DATA_WIDTH-1:0] rom1[2**ADDR_WIDTH-1:0];
21     reg [DATA_WIDTH-1:0] rom2[2**ADDR_WIDTH-1:0];
22     reg [DATA_WIDTH-1:0] rom3[2**ADDR_WIDTH-1:0];
23     reg [DATA_WIDTH-1:0] word0,word1,word2,word3,word;
24     wire [ADDR_WIDTH-1:0] address_row;
25     reg [2:0] col_d, col_dd;
26
27     initial
28     begin
29       $readmemb("char_table_myAPEAUS.mif", rom0); //Arial san serif font
30       $readmemb("char_table_myHERCULES.mif", rom1); // Times serif font
31       $readmemb("char_table_myHERCITAL.mif", rom2); // Italic serif font
32       $readmemb("char_table_mySPACE8.mif", rom3); // Fun/SPACE type font
33     end
34
35     assign address_row = {ADDRESS , ROW};// concardination table address
36
37     always @ (posedge CLK)  begin// read table value
38       word0 <= rom0[address_row];
39       word1 <= rom1[address_row];
40       word2 <= rom2[address_row];
```

```
41        word3 <= rom3[address_row];
42     end
43
44     always @*
45       case (SW[1:0])
46         2'b00   :  word <= word0;
47         2'b01   :  word <= word1;
48         2'b10   :  word <= word2;
49         default :  word <= word3;
50       endcase
51
52     always @ (posedge CLK) begin // delay for column values
53         col_d <= COL;
54         col_dd <= col_d;
55     end
56
57     // pick the bit needed
58     assign Q = word[~col_dd];
59
60   endmodule
```

The char ROM HDL hosts the four font LUTs. The HDL code starts with port definitions such as clock (CLK), followed by the font selection switch SW. 7 bits are used to access the ASCII char and 3 bits each for rows and columns. The output is just a single bit for each clock cycle. Next follows the internal registers and wire definitions (lines 19–25). The initial section (lines 27–33) is used to load the four fonts into the ROM memory. Each font has its own memory initialization file. In VIVADO, it is highly recommended to use the extension MIF for this type of file. Modern FPGAs have only synchronous memory, so we load based on the address and row one word from memory into a register (lines 37–42). The next always block (lines 44–50) contains the 4 to 1 multiplexer based on the setting of the SW input. The following always block (lines 52–55) implements a 2-clock cycle delay for the column selection. Since we write on the monitor from left to right, we need to return first the MSB. That is why the output assignment (line 58) uses a complement in the column addressing. The movie HDMI4color4fonts.MOV toggle through four colors and four fonts.

We now have all necessary information together to build our 80 × 60-character HDMI encoder.

HDMI Encoder in HDL

The major blocks of our HDMI encoder are shown in Fig. 10.14. The major blocks are

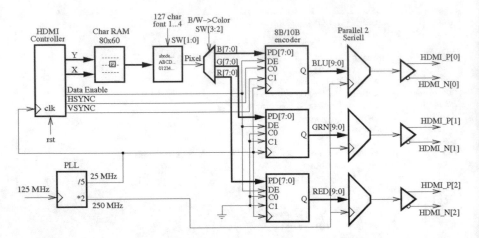

Fig. 10.14 The HDMI encoder overview

- Clock generation: A PLL is used to translate the 125 MHz system clock to 25 MHz symbol rate and the 250 MHz pixel clock rate.
- Control signal generation: Counters are needed to determine valid image data (data enable), row/column of character, vsync, and hsync.
- Character and font storage: We have a 80×60 character RAM storage and four $127 \times 8 \times 8$ ROM font memories to design in two separate HDL files.
- 3×TMDS encoder: We need thee TMDS encoder for the three colors to translate the 8-bit raw data into 10-bit transition minimized and DC balanced strings. The 10-bit parallel data are translated into a serial bit stream and should drive 250 MHz differential outputs.

With the major blocks discussed, we can take a look at the HDL code next.

HDL Code 10.5: HDMI Encoder in Verilog

```
1   //=============================================================
2   //   PORT declarations
3   //=============================================================
4   module HDMI(
5   input  CLOCK_125,  // CLOCK 125 MHz
6   output [3:0] LEDR, // 4 LEDs
7   input  [3:0] KEY,  // Push Buttons
8   input  [3:0] SW,   // Slider switches
9   input  [7:0] X,    // X coordinate
10  input  [7:0] Y,    // Y coordinate
11  input  [7:0] D,    // Data value
12  output HDMI_OUT_EN,
13  output [2:0] HDMI_D_P, HDMI_D_N,
14  output HDMI_CLK_P, HDMI_CLK_N
15  /////// Extra test ports
16  //output [9:0] TMDS_RED_out
17  );
18  //=============================================================
19  // REG/WIRE declarations
20  //=============================================================
21  wire RESET = KEY[0]; // btn[0] active high reset on ZYBO
22  wire  iCLOCK_25, CLOCK_25; // The 25 MHz clock
23  wire  iCLOCK_250, CLOCK_250; // The 250 MHz clock
24  wire CLOCK_25_n; // Complement 25 MHz
25  reg DLY_RST; // Delay reset for PLL signal
26  reg [19:0] CONT; // Delay counter
27  assign HDMI_OUT_EN = 1; // ZYBO is source
28  //=============================================================
29  //   Reset Delay Timer
30  //=============================================================
31  always@(posedge CLOCK_125 or posedge RESET)
32  begin
33    if (RESET) begin
34      CONT  <= 20'H00000;
35      DLY_RST <= 1;
36    end
37    else
38    //if (CONT < 20'hFFFFF) // Normal operation
39    if (CONT < 20'h0003C) // Simulation short cut
40      CONT    <= CONT + 1;
41    else
42      DLY_RST  <=  0;
43  end
44  //=============================================================
45  //   PLL to generate the 25 and 250 MHz clock
46  //=============================================================
47  wire LOCKED, FB_CLK;  // Clock feedback and lock signal
48  PLLE2_BASE #(
```

```
49        .BANDWIDTH("OPTIMIZED"), // OPTIMIZED, HIGH, LOW
50        .CLKFBOUT_MULT(8), // Multiply value for all CLKOUT, (2-64)
51        .CLKFBOUT_PHASE(0.0), // Phase offset in degrees, (-360.0-360.0).
52        .CLKIN1_PERIOD(8.0), // Input clock period in ns.
53     // CLKOUT0_DIVIDE - CLKOUT5_DIVIDE: Divide amount each CLKOUT (1-128)
54        .CLKOUT0_DIVIDE(4),
55        .CLKOUT1_DIVIDE(40),
56     // CLKOUT0..5_DUTY_CYCLE: Duty cycle for each CLKOUT (0.001-0.999).
57        .CLKOUT0_DUTY_CYCLE(0.5),
58        .CLKOUT1_DUTY_CYCLE(0.5),
59     // CLKOUT0..5_PHASE: Phase offset for each CLKOUT (-360.000-360.000).
60        .CLKOUT0_PHASE(0.0),
61        .CLKOUT1_PHASE(0.0),
62        .DIVCLK_DIVIDE(1), // Master division value, (1-56)
63        .REF_JITTER1(0.01), // Reference input jitter in UI, (0.000-0.999).
64        .STARTUP_WAIT("FALSE")) // Delay until PLL Locks, ("TRUE"/"FALSE")
65     PLLE2_BASE_inst (
66        // Clock Outputs: 1-bit outputs: User configurable clock outputs
67        .CLKOUT0(iCLOCK_250), // 1-bit output: CLKOUT0
68        .CLKOUT1(iCLOCK_25), // 1-bit output: CLKOUT1
69        // Feedback Clocks: 1-bit (each) output: Clock feedback ports
70        .CLKFBOUT(FB_CLK), // 1-bit output: Feedback clock
71        .LOCKED(LOCKED), // 1-bit output: LOCK
72        .CLKIN1(CLOCK_125), // 1-bit input: Input clock
73        // Control Ports: 1-bit (each) input: PLL control ports
74        .PWRDWN(0), // 1-bit input: Power-down
75        .RST(0), // 1-bit input: Reset
76        // Feedback Clocks: 1-bit (each) input: Clock feedback ports
77        .CLKFBIN(FB_CLK)); // 1-bit input: Feedback clock
78        // End of PLLE2_BASE_inst
79
80     BUFG BUFG_250(.I(iCLOCK_250), .O(CLOCK_250));
81     BUFG BUFG_25 (.I(iCLOCK_25), .O(CLOCK_25));
82     //=====================================================================
83     //  Sync signal generation
84     //=====================================================================
85     reg [9:0] COUNTER_X, COUNTER_Y;
86     reg H_SYNC, V_SYNC, DATA_ENABLE;
87     always @(posedge CLOCK_25)
88        if (DLY_RST) begin
89          COUNTER_X  <= 0; COUNTER_Y  <= 0;
90          DATA_ENABLE <= 0; H_SYNC <= 0; V_SYNC <= 0;
91        end else begin
92          // Sync signal are code withing the control BLU
93          DATA_ENABLE <= (COUNTER_X>=2) && (COUNTER_X<642)
94                              && (COUNTER_Y<480); // Allow two memory access
95          COUNTER_X <= (COUNTER_X==799) ? 0 : COUNTER_X+1;
96          if (COUNTER_X==799) COUNTER_Y <=(COUNTER_Y==524)? 0 : COUNTER_Y+1;
```

```
97          H_SYNC <= (COUNTER_X>=658) && (COUNTER_X<754); // delay by two
98          V_SYNC <= (COUNTER_Y>=490) && (COUNTER_Y<492);
99        end
100    //================================================================
101    //  Load image data using X/Y Counters
102    //================================================================
103    reg [7:0] RED, GRN, BLU;
104    //// Address array is linear:
105    wire [18:0] ADDR = COUNTER_X[9:3] + COUNTER_Y[9:3]*80;
106    wire WE = ((COUNTER_X==(X<<3)) && (COUNTER_Y==(Y<<3)) && KEY[3]);
107    wire [7:0] DATA = D; // { 4'h3, SW};
108    wire [7:0] INDEX;
109    wire PIXEL;
110    img_ram img_data_inst (
111     .ADDRESS(ADDR[12:0]),
112     .WE(WE),
113     .DATA(DATA),
114     .CLOCK (CLOCK_25),
115     .Q(INDEX));
116    ////// Load the binary 8x8 character font values
117    char_rom Char_inst(
118     .SW(SW[1:0]),
119     .ADDRESS (INDEX[6:0]),
120     .CLK (CLOCK_25),
121     .COL(COUNTER_X[2:0]),
122     .ROW(COUNTER_Y[2:0]),
123     .Q(PIXEL));
124    // Map color data to (default) gray scale
125    wire [7:0] GRAY_DATA = (PIXEL) ? 8'hFF : 8'h00;
126    always @(posedge CLOCK_25)
127      if (DLY_RST) begin
128        RED  <= 0; GRN <= 0; BLU <= 0;
129      end else begin
130        if (SW[2]) RED <= 8'h00; else RED <= GRAY_DATA;
131        if (SW[3]) GRN <= 8'h00; else GRN <= GRAY_DATA;
132        BLU <= GRAY_DATA;
133      end
134    //================================================================
135    //  TMDS encoder instantiations
136    //================================================================
137    wire [9:0] TMDS_RED, TMDS_GRN, TMDS_BLU ;
138    tmds_encoder B_inst(.CLK(CLOCK_25), .RESET(DLY_RST), .PD(BLU),
139               .CTL({V_SYNC,H_SYNC}), .DE(DATA_ENABLE), .Q_OUT(TMDS_BLU));
140    tmds_encoder G_inst(.CLK(CLOCK_25), .RESET(DLY_RST), .PD(GRN),
141                        .CTL(2'b00), .DE(DATA_ENABLE), .Q_OUT(TMDS_GRN));
142    tmds_encoder R_inst(.CLK(CLOCK_25), .RESET(DLY_RST), .PD(RED),
143                        .CTL(2'b00), .DE(DATA_ENABLE), .Q_OUT(TMDS_RED));
144    //================================================================
```

```
145   // Parallel to serial may need the OSERDESE2 for high CLOCK rates
146   //====================================================================
147   reg [3:0] TMDS_MOD10=0;  // modulus 10 counter
148   reg [9:0] TMDS_SHIFT_RED=0, TMDS_SHIFT_GRN=0, TMDS_SHIFT_BLU=0;
149   reg TMDS_SHIFT_LOAD=0;
150   //always @(posedge CLOCK_250) TMDS_SHIFT_LOAD <= (TMDS_MOD10==4'd9);
151
152   always @(posedge CLOCK_250)
153     if (DLY_RST) begin
154       TMDS_SHIFT_LOAD <= 0;
155       TMDS_SHIFT_RED <= 0; TMDS_SHIFT_GRN <= 0;
156       TMDS_SHIFT_BLU <= 0; TMDS_MOD10 <= 0;
157     end else
158     begin
159       TMDS_SHIFT_RED <= TMDS_SHIFT_LOAD ? TMDS_RED : TMDS_SHIFT_RED[9:1];
160       TMDS_SHIFT_GRN <= TMDS_SHIFT_LOAD ? TMDS_GRN : TMDS_SHIFT_GRN[9:1];
161       TMDS_SHIFT_BLU <= TMDS_SHIFT_LOAD ? TMDS_BLU : TMDS_SHIFT_BLU[9:1];
162
163       TMDS_MOD10 <= (TMDS_MOD10==4'd9) ? 4'd0 : TMDS_MOD10+4'd1;
164       TMDS_SHIFT_LOAD <= (TMDS_MOD10==4'd9);
165     end
166   //====================================================================
167   // Generate the differential output signals with hard macro
168   //====================================================================
169   OBUFDS OBUFDS_BLU(.I(TMDS_SHIFT_BLU[0]), .O(HDMI_D_P[0]),
170                                            .OB(HDMI_D_N[0]));
171   OBUFDS OBUFDS_GRN(.I(TMDS_SHIFT_GRN[0]), .O(HDMI_D_P[1]),
172                                            .OB(HDMI_D_N[1]));
173   OBUFDS OBUFDS_RED(.I(TMDS_SHIFT_RED[0]), .O(HDMI_D_P[2]),
174                                            .OB(HDMI_D_N[2]));
175   OBUFDS OBUFDS_clock(.I(CLOCK_25), .O(HDMI_CLK_P), .OB(HDMI_CLK_N));
176
177   assign LEDR[0] = CLOCK_25;
178   assign LEDR[1] = CLOCK_250;
179   assign LEDR[2] = DLY_RST;
180   assign LEDR[3] = LOCKED;
181   assign TMDS_RED_out = TMDS_RED;
182
183   endmodule
```

The HDMI encoder starts with the port descriptions (lines 5–14). Name and size of the port have been chosen so that we can make an easy assignment to the CIP ports needed in the next section. Port descriptions are followed by some additional wire and register definition (lines 21–27). The first logic block (lines 31–43) contains a reset delay for the whole system so that the monitor can complete his self-check first. Next, we instantiate the PLL block (lines 48–77). Input is the 125 MHz system clock and output the 25 MHz character rate and the 250 MHz pixel rate needed by the TMDS encoder. We use the BUFG macro to tell VIVADO to use a global clock network for the two new generated clock signals (lines 80,81). The next logic block (lines 85–99) contains the major counter and control logic of the HDMI encoder. It generates the X and Y counters, data enable, the vertical synchronization (V_SYNC), and horizontal synchronization (H_SYNC) signals. The following block (lines 110–115) instantiates the image RAM that was discussed in Listing 10.3 earlier. The output of the image RAM is the input to the character ROM aka font memory instantiated in lines 117–123. Character ROM LUT was discussed in HDL Listing 10.4 earlier. The two LSBs of SW are used to select one of the four fonts. Since our ROM is providing just zero and one values, we use the slider switch bit 3 and 4 to add some color to the display. If RGB all have the same data, we have a gray or in our case black/white. With SW(2), we set the red signal to zero, and with SW(3), we can set all green values to zero. We will always have a black background, but now besides R + G + B = white (00), also G + B = cyan aka aqua blue (01), R + B = magenta (10), and blue (11) can be selected.

CIP Interface for HDMI Encoder

The process to build a CIP for the HDMI encoder is a multistep process. A good example on the procedure is provided by Digilent Inc. in the "Creating a Custom IP core using IP Integrator" tutorial online. Here is the outline of the steps:

1. Create a New RTL project first. Alternatively you can continue using a copy of the Basic Computer system from the last section. Make sure the *Target language* in *General Project Settings* is set to *Verilog* that the appropriate template for our HDMI component is generated.
2. Go to *Tools → Create and Package New IP...* click *Next* and then select *Create a new AXI4 peripheral* and click *Next*. Under *Peripheral Detail* choose a descriptive name such as my_hdmi_cip starting with my... such you can easily find you component later in the IP library; also check the box for *Overwrite existing* and click *Next*. As *AXI interface Type*, choose the default *Lite*, use default *Data Width* 32. Click *Next*, and under the *Create Peripheral* panel choose *Edit IP* and click *Finish*.
3. A complete new copy of VIVADO will pop up.
4. Expand Design Sources my_hdmi_cip_v1_0 and double-click my_hdmi_cip_v1_0_S00_AXI_inst or right click it and select *Open*. There are three locations we typically need to modify the HDL file:

   - Add user parameter (top of file; nothing for HDMI)
   - Add user ports (located just after parameter section in the code)
   - Add user logic (end of file)

```
            // Users to add parameters here
        // nothing to add for HDMI
            // User parameters ends
    ...

            // Users to add ports here
        output HDMI_OUT_EN,
        output [2:0] HDMI_D_P, HDMI_D_N,
        output HDMI_CLK_P, HDMI_CLK_N,
            // User ports ends
    ...

    // Add user logic here
    wire [3:0] LEDR;
    HDMI HDMI_inst(
     .CLOCK_125(S_AXI_ACLK),     // CLOCK 125 MHz
    .LEDR(LEDR), //4 LEDs (not connected in final design)
     .KEY({slv_reg0[3], 2'h0, ~S_AXI_ARESETN}),
                    // Push Buttons Key(0) reset KEY(3)=WE
    .SW(slv_reg0[7:4]),    // [7:6] color; [5:4] fonts
    .X(slv_reg0[15:8]),     // X coordinate
    .Y(slv_reg0[23:16]),    // Y coordinate
    .D(slv_reg0[31:24]),    // Data value
    .HDMI_OUT_EN(HDMI_OUT_EN),
    .HDMI_D_P(HDMI_D_P),
    .HDMI_D_N(HDMI_D_N),
    .HDMI_CLK_P(HDMI_CLK_P),
    .HDMI_CLK_N(HDMI_CLK_N));
    // User logic ends
```

5. Now modify the wrapper file my_hdmi_cip_v1_0.v. There are now four locations we typically need to modify the HDL file:

   • Add user default parameter (top of file; nothing for HDMI)
   • Add mapping for user ports (locate just after parameter section)
   • Instantiation parameter of new ports (within …inst(..))
   • Add mapping of user logic (end of file; nothing for HDMI)

```
                    // Users to add parameters here
                    // nothing to add for HDMI
                    // User parameters ends
          ...

                    // Users to add ports here
                     output HDMI_OUT_EN,
                     output [2:0] HDMI_D_P, HDMI_D_N,
                     output HDMI_CLK_P, HDMI_CLK_N,
                    // User ports ends
          ...

                    ) my_ip_text4hdmi_v1_0_S00_AXI_inst (
                   .HDMI_OUT_EN(HDMI_OUT_EN),
                   .HDMI_D_P(HDMI_D_P),
                   .HDMI_D_N(HDMI_D_N),
                   .HDMI_CLK_P(HDMI_CLK_P),
                   .HDMI_CLK_N(HDMI_CLK_N),
                    .S_AXI_ACLK(s00_axi_aclk),
          ...

                    // Add user logic here
                    // nothing to add for HDMI
                    // User logic ends
```

6. After you have instantiated the HDMI encoder, VIVADO marks the missing source code component with a question mark `? HDMI_inst - HDMI`. Right click on the entry and add the appropriate HDL source file. After you added HDMI.v, VIVADO will then ask for img_ram.v, char_rom.v and tmds_encoder.v. You will need to locate these files and add this to the project. Do not forget to add the four font files and default monitor text initialization file *.MIF to the project too.

7. Now the core design is complete, and we need to package the core so that we can reuse the CIP in our designs. Click on the *Package IP – my_hdmi_cip* tap and make sure that for selection *Compatibility*, the *zynq* is available. Click on Merge changes from Customization Parameters Wizard such that the new files and ports are recognized by VIVADO. Click on *File Groups* and make sure all files, listed in Fig. 10.15a, are included in the CIP. Under *Ports and Interfaces*, the new HDMI signals should be listed (see Fig. 10.15b). If you click on *Customize GUI*, you should see the new ports included in the CIP symbol (see Fig. 10.15c). Finally click on *Review and Package* panel, and select the *Re-Package IP* button. Answer *Yes* for closing the project. The 2. VIVADO window will close and we are back to our original design.

8. Now we can integrate *my_hdmi_cip* into our MICROBLAZE system. It highly recommended to use a copy of the Basic Computer from the last section or another working design. If you work on a revision and your IP block is locked, you need to run the *Upgrade IP* first. If it still locked, then you may try to close VIVADO and then start again. You can just instantiate a copy from the new CIP via

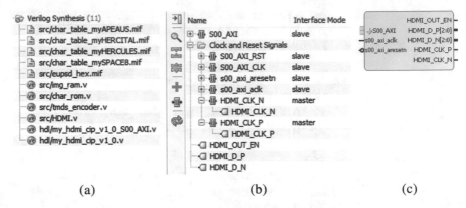

(a)   (b)   (c)

Fig. 10.15 The CIP Package IP (**a**) Files in the project. (**b**) The CIP ports. (**c**) The new GUI

, search for my... and use <u>Run Connection Automation</u>. The finals Block Design including the CIP is shown in Fig. 10.16, where the new CIP is in the upper right corner. We will need to create all ports in BD and update the XDC file. Look for the ##HDMI Signals section. You will need to enable the following 9 ports: HDMI_CLK_N, HDMI_CLK_P, HDMI_D_N[0], HDMI_D_P[0], HDMI_D_N[1], HDMI_D_P[1], HDMI_D_N[2], HDMI_D_P[2], and HDMI_OUT_EN.

9. It is now time to start with the program development. In VIVADO use *File → Export → Export Hardware...* to get started. Do not forget to select the *Include bitstream* option before export then click *OK*. Next select *File → Launch SDK* and click *OK*. Since we have a new component, it seemed to be a good idea to take advantage of the SDK feature to generate a peripheral test program for us that test all component. Select *File → New → Application Project* and choose a *Project name* such as test_IO and click OK and then as *Template* choose *Peripheral Tests*. When you run the test, you will see success message in the Terminal window for each peripheral and a very short running light on four green LEDs of the ZyBo board. We can then modify the main routine test_ MY_SWAP → src → testperiph.c to measure performance or add the LED counter.

10. Finally, let us test the HDMI encoder on the FPGA. Typically, a monitor today has at least three inputs, VGA, DVI, and HDMI. If your HDMI input is already in use by your computer, you may use the DVI (together with a cheap hardwired HDMI→DVI adapter). If you download the VIVADO design to the ZyBo, you should be able to see the HDMI test picture (see Fig. 10.17a). If not make sure your design, board, and cable are set up ok. Next you may try to print some information on your monitor such as DMIPS or FPMIPS results. The code below (Program Listing 10.6) uses two short ANSI C functions that allow you to print a line using one of the four fonts and your choice of color too.

Now we should remember how the communication with the CIP should be assigned to our 32-bit vector.

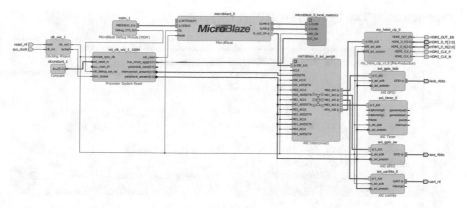

Fig. 10.16 Final Basic Computer system in Vivado with HDMI CIP included

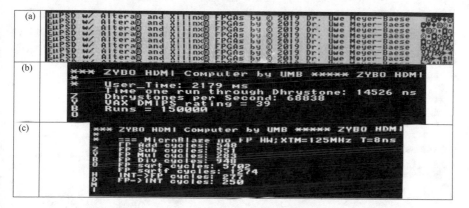

Fig. 10.17 The HDMI text terminal examples (**a**) First eight lines of default display. (**b**) The DMIPS measurement results. (**c**) Floating-point basic operation time measurements

Bits	[31:24]	[23:16]	[15:8]	[7:6]	[5:4]	[3]	[0]
Function	Data	Y	X	Color	Font	WE	Reset

A convenient way to put together the 32-bit vector is to use four 8-bit masks and shifts. We may want to keep the slider switches from the ZyBo board as selection for color and font. Two utility functions seemed useful. The first function HDMI_ text() prints a string starting at any X/Y location. We can then use coding like

```
char banner[40] = "*** ZyBo HDMI Computer by UMB ***\0";
HDMI_text (23, 3, banner);
```

The monitor would start in column 23 and line 3 to display the text string. The second function HDMI_cls() can be used as clear screen function. We may want to add a banner at the border of the monitor to make it more appealing. Here are the ANSI C function for the task.

Program 10.6: Two Utilities for HDMI Display

```
1    //===============================================================
2    // Subroutine to send a string of text to the HDMI monitor
3    //===============================================================
4    void HDMI_text(int x, int y, char *text_ptr)
5    {
6      volatile int i;
7      u32 sv, w0, w1, w2, w3, val, xx=x;
8
9      /* assume that the text string fits on one line */
10     while ( *(text_ptr) ) {
11       sv =  Xil_In32(SWITCHES);       // sv= color + font
12       w0= mask0 & ((sv << 4) + 8);  // set WE=1
14       w1= mask1 & (xx << 8);
15       w2= mask2 & (y << 16);
16
16       i = *(text_ptr);  // print one char at a time
17       text_ptr++;
18       w3= mask3 & (i << 24); // Char ASCII 0=48 dec
19       val= w3 + w2 + w1 + w0;
20       Xil_Out32(HDMI, val);
21       xx++;
22       for(i=0;i<(125000000/9)/30; i++); // wait WE 1sec/30 Hz
23     }
24   }
25   //===============================================================
26   // Subroutine to clear the first ymax lines; add banner to monitor
27   //===============================================================
28   void HDMI_cls(int ymax)
29   {
30     volatile int i;
31     u32 sv, w0, w1, w2, w3, val, x, y;
32
33     for (y=0;y<ymax;y++)
34       for (x=0;x<80;x++) {
35         sv =  Xil_In32(SWITCHES);      // serif font= SW=01;
36         w0= mask0 & ((sv << 4) + 8);  // color=BW=00; set WE=1
37         w1= mask1 & (x << 8);
38         w2= mask2 & (y << 16);
39         if ((x==0)||(x==79)||(y==0)||(y==59))
40           i=banner[(x+y)%32];
41         else
42           i=32; // space char dec=32
43         w3= mask3 & (i << 24); // Char ASCII 0=48 dec
44         val= w3 + w2 + w1 + w0;
45         Xil_Out32(HDMI, val);
46         for(i=0;i<(125000000/9)/30; i++); // wait 1sec/60 Hz
47     }
48   }
```

Synthesis Results for HDMI Encoder

We may also want to know what the CIP will cost us in terms for additional logic resources. That would be mainly the block RAM and the LUT needed for the encoder logic.

Overall the number of LUT increases by 242 LUT and 4 BRAMs as can be seen from Table 10.4. The compile time for the IP block is substantial, but since this is a onetime effort and is not needed often when using smart synthesis that only recompiles if changes occur, the 5-minute synthesis and implementation compile time are more realistic after small changes to the design.

10.5   Looking Under the Hood: MICROBLAZE Instruction Set Architecture

Looking under the hood of your car is usually a pretty impressive thing to do. You will see a shiny engine and all the energy, know-how, and love the engineers have put into the car that power your ride every day. For microprocessors we do a similar thing: We can have a look at the inner working of our machine aka CPU and will be impressed usually with the amount of architecture and details that goes into a microprocessor design. It may even be argued that building a highly pipelined 700 million transistor CPU is not less complicated than anything else that has been built in the world you can think of, e.g., pyramids. Anyway, such a precise knowledge of the CPU is a benefit in a couple of instances like writing driver for new components such as a JTAG controller. Such knowledge is beneficial sometimes even necessary to know the nitty-gritty of the CPU architecture such as pipelining, registers, or ANSI C compiler expectations.

It is usually pretty tiresome to go through each instruction, addressing modes, registers, etc. one-by-one – so let us make this study a little bit more fun by designing

Table 10.4 MICROBLAZE Basic Computer (BC) synthesis results for device Zynq 7 K xc7z010t-1clg400

	BC no HDMI encoder	BC with HDMI encoder
LUT	1759	2001
LUT as RAM	139	139
Block RAM RAMB36	32	36
DSP blocks	0	0
PLL	0	1
HPS	0	0
Slack at 125 MHz sys clock	0.518 ns	0.454 ns
IP compile time	44:44:36	1:02:10
Synthesis + implementation compile time	5:58	5:42

our own tiny MICROBLAZE we will call TRISC3MB that implements a reduced set of
the MICROBLAZE Instruction Set Architecture (ISA). We will not attempt to implement all instructions, but a useful subset that allows us to write simple tasks such as
a memory access, (nested) function calls, or flash program, such we can study the
hardware-software interface between MICROBLAZE and our GCC. So, let's get
started how registers are organized in MICROBLAZE. Some have special responsibil-
ity and these registers get special names by the GCC compiler. The MICROBLAZE
has 32 32-bit registers. The first register r0 is always zero. Writing to r0 will not
be executed and can indeed be consider or used as NOP operation. The GCC uses r1
as stack pointer (sp) and often build images of sp in register r19; therefore we
should watch both during memory access. MICROBLAZE does not use a global
pointer register and uses instead a symbol/memory table for global variable locations. Register r15 is used as return address for subroutines. Other dedicated registers are r2, r13, r14, r16, r17, and r18 used by the compiler and should not be
used for general purpose, e.g., r14 hosts the interrupt return address, and r16 has
address for traps used by the debugger.

MICROBLAZE uses two instruction types aka categories. These are the three-
register type (type A) and the instructions that uses an immediate (type B). This
allows us to encode and decode each instruction also by hand (see Exercise
10.17–10.24). All instructions are 32-bit long and the format is shown in Table 10.5.

Both types specify the operation code in the first six bits 0...5. The type B
includes a 16-bit long IMM16 constant for any instruction that requires a large constant value. Most of the type B instructions are of the type r(D) = r(A) □
IMM16. D specifies one of the 32 destination registers and A the index of the first
source. For r(D) often the abbreviation rD is used. Some operations like addi
require a 32-bit constant; then IMM16 is sign extended. If a true 32-bit constant is
required, then the imm instruction preceding the addi (or any other . . . i instruction) will load 16 bits in the registers MSBs.

Not all instructions follow precise the two formats then only $2^6 = 64$ instructions
could be coded but the MICROBLAZE UG984 shows in Table 2.6 [X18] that 148
instructions are available. All branch instruction for instance are coded with opcode
0x27. The individual conditions for the 2×6 branches such as EQ, NE, LT,..., GE
are encoded within bits 6...10.

The majority of type A instructions use the r(D)=r(A) □ r(B) style as typical for the three-address machine. But the floating-point compare instructions, for
instance, use also bits 25...27 for seven different compare conditions. This way a
single opcode implements seven different instructions. The MICROBLAZE ISA has
also evolved over the years: the UG081 [X08] from 2008 shows 129 instruction,
while our VIVADO 2016.4 [X18] UG984 shows 148 instructions.

Table 10.5 The two 32-bit instruction coding formats used by MICROBLAZE

Bits	0...5	6...10	11...15	16..20	21...31
Size	6-bits	5-bits	5-bits	5-bits	11-bits
TYPE A	Opcode	Destination D	Source A	Source B	0x000
TYPE B	Opcode	Destination D	Source A	IMM16	

Let us have a look briefly at the available instruction classes. The grouping by functionality let us build the following five groups:

- We count 19 arithmetic instructions: Three B type `addi`, `muli`, and `rsubi` and eight R-type, `add`, `clz`, `mul`, `idiv`, `mulh`, `mulhu`, `mulhsu`, and `rsub` with `u` indicating unsigned operands. Eight floating-point operations are defined, `fadd`, `frsub`, `fmul`, `fdiv`, `fcmp`, `flt`, `fint`, and `fsqrt`, that would require that FP hardware is enabled
- The 13 logical and shift instructions contain the following A type, `and`, `andn`, `bs`, `or`, `sra`, `src`, `srl`, and `xor` and five IMM16 type `andi`, `andni`, `ori`, `xori`, `bsi`. The barrel shift operation `bs` and `bsi` would require addition hardware support enabled at synthesis.
- We count 21 branches and conditions. We have two unconditional (`br` and `brk`) and six conditional branches. Naming is similar to the FORTRAN comparison operations `beq`, `bge`, `bgt`, `ble`, `blt`, and `bne`. Same operations "i" indicating IMM16 use are `beqi`, `bgei`, `bgti`, `blei`, `blti`, `bnei`, `bri`, and `brki`. There are four return operations: `rtbd`, `rtid`, `rted`, and `rtsd` and one compare `cmp`.
- Data transfers, i.e., load and store, are available in byte, 2 byte (half word) and 4 byte aka word size. The instructions for unsigned operand a "u" is included, `lbu`, `lhu`, `lw`, `sb`, `sh`, `sw`, `lwx`, and `swx`, and for the first 6, the IMM16 equivalent operations `lbui`, `lhui`, `lwi`, `sbi`, `shi`, `swi`. The imm instruction stores 16-bit in MSBs. There are some addition instructions that deal with register manipulations and stream interface: `get`, `getd`, `mbar`, `mfs`, `msrclr`, `msrset`, `mts`, `put`, `putd`, `swapb`, and `swaph`. We count a total of 26 load/store operations.
- MicroBlaze has seven other special instructions that deal with cache memory, (`wdc` and `wic`), sign extension (`sext16` and `sext8`), and matching byte pattern: `pcmpbf`, `pcmpeq`, and `pcmpne`.

Some of the instructions come with additions flag for refinement. The unconditional branch `br` instructions, for instance, use 3 flags: D flag from bit 11 for indication of branch delay slot use, A flag in bit 12 for absolute addressing, and L flag from bit 13 for link such that a total of 6 branch operation use the same operation code. A full description of all instructions can be found in the processor reference manual, i.e., User Guide 984 [X18]. Make sure you download the correct one matching your installed software. Now let us have a closer look at the instructions we like to implement in HDL. In the table, we use σ(IMM16) as shortcut for a sign extension. These are listed in Table 10.6 along with opcode, assembler coding, format, and operation description. Some of the instructions can be combined if we use the third bit as immediate flag. Branch instruction may use a branch delay slot when bit D is set.

We try to use similar CONSTANT/PARAMETER coding for HDL and assembler.

Since in SDK GPIOs require usually three function calls (see Listing 10.1), let us first have a look at the data organization by GCC for our MicroBlaze. In ANSI C code, we can place out variable/data definitions in the initial section of our code, making the data "global" available in the main section as well as any of the functions we may have. The other option is that we can place the variable definition within the main code. Let us have a look at how GCC is handling these two cases. Here is a little ANSI C test program for this.

DRAM Program 10.7: DRAM Write Followed by Read

```
1    volatile int g;// Volatile allows us to see the full assembler code
2    volatile int garray[15];
3    int main(void) {
4      volatile int s;   // Declare volatile tells compiler do
5      volatile int sarray[14]; // not try optimization
6      s = 1; // Memory location sp+
7      g = 2; // Memory location gp+
8      sarray[0] = s; // Memory location sarray[0] is sp+4
9      garray[0] = g; // Memory location garray[0] is gp+4
10     sarray[s] = 0x1357; // use variable; requires *4 for byte address
11     garray[g] = 0x2468; // use variable; requires *4 for byte address
12   }
```

Table 10.6 MICROBLAZE ISA subset implemented in HDL. D,A,L flags in bits 11,12,13. For condition branch and return D in bit 6

OP	ASM coding	Extra cond.	Operation type	Operation description
000000	add rD, rA, rB		Arithmetic	rD = rA + rB
000010	addc rD, rA, rB		Arithmetic	rD = rA + rB + C
000100	addk rD, rA, rB		Arithmetic	rD = rA + rB [Keep C]
000110	addkc rD, rA, rB		Arithmetic	rD = rA + rB + C [Keep C]
000101	cmpu		Branch	rD = rB-rA; rD(MSB) = (rA > rB)?
001000	addi rD, rA, IMM		Arithmetic	rD = rA + σ(IMM16)
001010	addic rD, rA, IMM		Arithmetic	rD = rA + σ(IMM16) + C
001100	addik rD, rA, IMM		Arithmetic	rD = rA + σ(IMM16) [Keep C]
001110	addikc rD, rA, IMM		Arithmetic	rD = rA + σ(IMM16) + C [Keep C]
100000	or rD, rA, rB		Logical	rD = rA OR rB
100010	xor rD, rA, rB		Logical	rD = rA XOR rB
100110	br rB	DAL = 000	Branch	PC = PC + rB
100110	bra rB	DAL = 010	Branch	PC = rB
100110	brd rB	DAL = 100	Branch	PC = PC + rB [delay]
100110	brad rB	DAL = 110	Branch	PC = rB [delay]

(continued)

Table 10.6 (continued)

OP	ASM coding	Extra cond.	Operation type	Operation description
100110	brld rD, rB	DAL = 101	Branch	rD = PC; PC = PC + rB [delay]
100110	brald rD, rB	DAL = 111	Branch	rD = PC; PC = rB [delay]
100111	beq, bne, blt, ble, bgt, bge	D = 0, B8–10: 000..100	Branch	If (cond) THEN PC = PC+ σ(IMM16)
100111	beqd, bned, bltd, bled, bgtd, bge	D = 1, B8–10: 000..100	Branch	If (cond) THEN PC = PC + σ(IMM16) [delay]
101000	ori rD, rA, IMM		Logical	rD = rA OR σ(IMM16)
101010	xori rD, rA, IMM		Logical	rD = rA xOR σ(IMM16)
101100	imm		Data Trans.	rI = {IMM16,0x0000}
101110	bri	DAL = 000	Branch	PC = PC+ σ(IMM16)
101110	brai	DAL = 010	Branch	PC = σ(IMM16)
101110	brid	DAL = 100	Branch	PC – PC+ σ(IMM16) [delay]
101110	braid	DAL = 110	Branch	PC = σ(IMM16) [delay]
101110	brlid	DAL = 101	Branch	rD = PC;PC=PC + σ(IMM16) [delay]
101110	bralid	DAL = 111	Branch	rD = PC; PC = σ(IMM16) [delay]
	rtsd		Branch	If (rA = rB) THEN PC = PC + 4 + σ(IMM16) ELSE PC = PC + 4
101111	beqi, bnei, blti, blei, bgti, bgei	D = 0, B8–10: 000..100	Branch	If (cond) THEN PC = PC + σ(IMM16)
101111	beqid, bneid, bltid, bleid, bgtid, bgeid	D = 1, B8–10: 000..100	Branch	If (cond) THEN PC = PC + σ(IMM16) [delay]
110010	lw rD, rA, IMM		Data Trans.	rD = MEM[rA + rB]
110110	sw rD, rA, rB		Data Trans.	MEM[rA + rB] = rD
111110	swi rD, rA, IMM		Data Trans.	MEM[rA + σ(IMM)] = rD
111010	lwi rD, rA, IMM		Data Trans.	rD = MEM[rA + σ(IMM16)]

The assembler output code of interest in the SDK elf file is shown in Fig. 10.18. We notice that global scalar and arrays are indexed by GCC using a global symbol table directly (see line 28). The local data defined within our main() sections are indexed via the stack pointer sp=r1. A copy of the stack pointer r1 is also placed

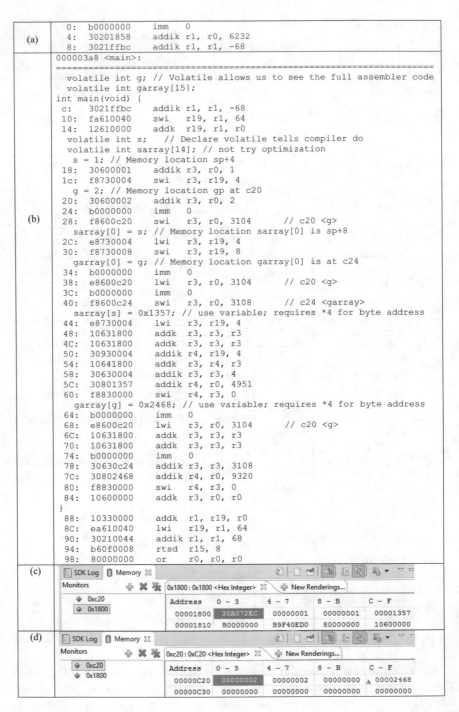

Fig. 10.18 TRISC3MB data memory ASM test routine. Local and global memory write are followed by reading the same locations. (**a**) Starting sequence. (**b**) main () program. (**c**) Memory at sp (**d**) memory for global variable (at gp)

in register r19 (see line 14). All data (and program) access is done in increments of bytes and that is why our first integer variable has address 0 (sp), the second 4 (sp), third 8 (sp) , etc. The MICROBLAZE bus architecture is internally designed to have separate data and program busses aka Harvard architecture, which can be connect to (one or more) tightly coupled data memories. However, since our MICROBLAZE GCC only supports one .text section, as a consequence we have a von Neuman memory organization, i.e., program and data share the same memory. The data memory can only use the address space after program section. Starting address of the global data is somewhere after the code sections, while sp use a starting address that is at the end of our physical memory decreased by the number of bytes needed to host our data. These values are set by GCC at the start of the program (see Fig. 10.18a).

We ran the mb_testdmem.c program until it reaches ret and then monitor the memory content. At the end of our physical memory, we see (Fig. 10.18c) the local values indexed by sp. We see that sarray[1] with value 0x1357 can be found at address 0x180C. As word address for sarray[1], we have $180C_{16}/4 = 603_{16} = 1539_{10}$. We find garray[2] (i.e., value 0x2468) at address 0xC2C somehow after the program code. As word address for garray[2], we get $C2C_{16}/4 = 30B_{16} = 779_{10}$. This should also be the memory location we should monitor in a HDL simulation (see Fig. 10.22).

As a second example, let us check what instructions are needed for (nested) function calls. Again we start with a short ANSI C example and then see what GCC requires as instructions. Here is our C code.

3 Level Nesting Program 10.8: Nesting Function Test Code

```
1    void level3(int *array, int s1) {
2        s1 += 1;
3        array[3] = s1;
4        return;
5    }
6    void level2(int *array, int s1) {
7        s1 += s1;
8        array[2] = s1;
9        level3(array, s1); // call level 3
10       return;
11   }
12   void level1(int *array, int s1) {
13       s1 += 1;
14       array[1] = s1;

15       level2(array, s1); // call level 2
16       return;
17   }
18   int main(void) {
19     volatile int s1, s2, s3;
20     int array[11];
21     s1 = 0x1233; // Memory location sp+0
22     while(1)  {
23        level1(array, s1);
24        s1 = array[1];
25        s2 = array[2];
26        s3 = array[3];
27     }
28   }
```

As in our PICOBLAZE studies, we use here also three levels of loops and only local, i.e., sp indexed variables. The starting sequence setting up sp and jumping to main() via r2 is different from the previous example since we want to preserve the numbering of the code lines. We include therefore after setting up the stack pointer a substantial number (total of 231) of NOPs (implemented as or r0,r0,r0), such that the main() program and the functions start at exactly the same program memory addresses in SDK and the HDL simulation. This ensures that call and return address stored in the return address register (i.e., r15) match in ANSI C debugger and HDL. The first initial instructions are also a good place to check out the different branch instructions. We have pc relative (br or bri), absolute (brai), and register indirect branches (bra). In addition, we can test the functionality of the branch delay slot. Here the instruction immediately following the branch is also executed. In the listing in Fig. 10.19a line 008 below, we see that loading sp=r1 with the value $6348_{10} = 18CC_{16}$ is in the delay slot. So, if we monitor r1, we will see if the delay slot is working.

In the function call we include one array and one scalar. During the call we increment the scalar and assign the new value to our array. We use an initial value of 0x1233 such that it can be easily identified in memory. Fig. 10.19a shows the assembler code for the level 2 and 3 functions, while Fig. 10.19b shows assembler code for level 1 and the main() code.

```
000: b0000000   imm   0
004: b81804b4   braid   4b4
008: 302018CC   addik  r1, r0, 6348
00c: 80000000   or    r0, r0, r0
...
3a4:  80000000 or   r0, r0, r0
  void level3(int *array, int s1){
3a8:  3021fff8 addik r1, r1, -8
3ac:  fa610004 swi  r19, r1, 4
3b0:  12610000 addk r19, r1, r0
3b4:  f8b3000c swi  r5, r19, 12
3b8:  f8d30010  swi  r6, r19, 16
   s1 += 1;
3bc:  e8730010  lwi  r3, r19, 16
3c0:  30630001  addik r3, r3, 1
3c4:  f8730010  swi  r3, r19, 16
   array[3] = s1;
3c8:  e873000c  lwi  r3, r19, 12
3cc:  3063000c  addik r3, r3, 12
3d0:  e8930010  lwi r4, r19, 16
3d4:  f8830000  swi  r4, r3, 0
   return;
3d8:  80000000  or   r0, r0, r0
}
3dc:  10330000  addk  r1, r19, r0
3e0:  ea610004  lwi  r19, r1, 4
3e4:  30210008  addik r1, r1, 8
3e8:  b60f0008  rtsd  r15, 8
3ec:  80000000  or   r0, r0, r0
```
```
   000003f0 <level2>:
void level2(int *array, int s1) {
3f0:   3021ffe0  addik r1, r1, 32
3f4:   f9e10000  swi  r15, r1, 0
3f8:   fa61001c  swi  r19, r1, 28
3fc:   12610000  addk  r19, r1, r0
400:   f8b30024  swi  r5, r19, 36
404:   f8d30028  swi  r6, r19, 40
   s1 += s1;
408:   e8930028  lwi  r4, r19, 40
40c:   e8730028  lwi  r3, r19, 40
410:   10641800  addk  r3, r4, r3
414:   f8730028  swi  r3, r19, 40
   array[2] = s1;
418:   e8730028  lwi  r3, r19, 36
41c:   30630008  addik r3, r3, 8
420:   e8930028  lwi  r4, r19, 40
424:   f8830000  swi  r4, r3, 0
   level3(array, s1);//call level 3
428:   e8d30028  lwi  r6, r19, 40
42c:   e8b30024  lwi  r5, r19, 36
430:   b9f4ff78  brlid r15, -136
// 3a8 <level3>
434:   80000000  or   r0, r0, r0
   return;
438:   80000000  or   r0, r0, r0
}
43c:   e9e10000  lwi  r15, r1, 0
440:   10330000  addk  r1, r19, r0
444:   ea61001c  lwi  r19, r1, 28
448:   30210020  addik r1, r1, 32
44c:   b60f0008  rtsd  r15, 8
450:   80000000  or   r0, r0, r0
```

(a)

```
   00000454 <level1>:
   void level1(int *array, int s1) {
454:   3021ffe0  addik  r1, r1, -32
458:   f9e10000  swi  r15, r1, 0
45c:   fa61001c  swi  r19, r1, 28
460:   12610000  addk  r19, r1, r0
464:   f8b30024  swi  r5, r19, 36
468:   f8d30028  swi  r6, r19, 40
   s1 += 1;
46c:   e8730028  lwi  r3, r19, 40
470:   30630001  addik r3, r3, 1
474:   f8730028  swi  r3, r19, 40
   array[1] = s1;
478:   e8730024  lwi  r3, r19, 36
47c:   30630004  addik r3, r3, 4
480:   e8930028  lwi  r4, r19, 40
484:   f8830000  swi  r4, r3, 0
   level2(array, s1); // call level 2
488:   e8d30028  lwi  r6, r19, 40
48c:   e8b30024  lwi  r5, r19, 36
490:   b9f4ff60  brlid r15,-160
                  //3f0 <level2>
494:   80000000  or   r0, r0, r0
   return;
498:   80000000  or   r0, r0, r0
}
49c:   e9e10000  lwi  r15, r1, 0
4a0:   10330000  addk  r1, r19, r0
4a4:   ea61001c  lwi  r19, r1, 28
4a8:   30210020  addik r1, r1, 32
4ac:   b60f0008  rtsd  r15, 8
4b0:   80000000  or   r0, r0, r0
===================================
000004b4 <main>:
int main(void) {
4b4:   3021ffa8  addik  r1, r1, -88
4b8:   f9e10000  swi  r15, r1, 0
4bc:   fa610054  swi  r19, r1, 84
4c0:   12610000  addk  r19, r1, r0
   volatile int s1, s2, s3;
   int array[11];
   s1 = 0x1233; // Memory location sp+0
4c4:   30601233  addik  r3, r0, 4659
4c8:   f873001c  swi  r3, r19, 28
   while(1)  {
      level1(array, s1);
4cc:   e893001c  lwi  r4, r19, 28
4d0:   30730028  addik  r3, r19, 40
4d4:   10c40000  addk  r6, r4, r0
4d8:   10a30000  addk  r5, r3, r0
4dc:   b9f4ff78  brlid r15, -136  //
454 <level1>
4e0:   80000000  or   r0, r0, r0
      s1 = array[1];
4e4:   e873002c  lwi  r3, r19, 44
4e8:   f873001c  swi  r3, r19, 28
      s2 = array[2];
4ec:   e8730030  lwi  r3, r19, 48
4f0:   f8730020  swi  r3, r19, 32
      s3 = array[3];
4f4:   e8730034  lwi  r3, r19, 52
4f8:   f8730024  swi  r3, r19, 36
   }
4fc:   b800ffd0  bri  -48   // 4cc
```

(b)

Fig. 10.19 TRISC3MB nesting loop ASM test routine. (**a**) 3- and 2-level functions. (**b**) level 1 and main() program

Other than PICOBLAZE, the MICROBLAZE does not have a pc stack to host return addresses for loops. We only have one register r15=r1 to store a return address. If we just call one function, that would be sufficient (see level 3 in Fig. 10.19a). But if we call another function within our first called functions, we need to store the r15=r1 value into memory (see Fig. 10.19 lines 0x3f4 and 0x458) and restore r1=r15 at the end of the function (lines 0x43c and 0x49c). To index the memory location for the loop address, the stack register sp=r1 is used. Since sp is also used to index our local variables, a copy of sp is placed in r19 (line 0x4c0) to index our variables during function calls. At the end we will have in the memory region addressed starting at 0x1890 sp with increasing index the variable values and below sp address (0x1834 and 0x1854), the return addresses used for the function call. We run our main program loop one time and the monitor the memory content staring at 0x1830. After one loop iteration, we will observe the values as shown in Table 10.7.

At location 0x1890 we find s1=0x1234, and array[1] at 0x18A0. At address 0x1834, we see the return address 0x490 for the call to first level1 function and 0x4DC as the return address for the call to level2 within the level1 function at address 0x1854. This will be our test bench data for the HDL simulation later.

Now as a third example, let us re-code our PICOBLAZE LED flash example using MICROBLAZE. For GPIO we have several options how to handle I/O operations. So far, we have often used the three sequence of functions calls. For the leds we would for instance use:

Table 10.7 The memory contend after the nesting.c program run

0x1850 : 0x1850 <Hex Integer> ⊠ ✚ New Renderings...				
Address	0 – 3	4 – 7	8 – B	C – F
00001830	00001834	00000490	0000189C	00002469
00001840	00000000	00000000	00000000	00000000
00001850	00001854	000004DC	0000189C	00002468
00001860	00000000	00000000	00000000	00000000
00001870	00001874	00000374	0000189C	00001234
00001880	00000000	00000000	00000000	00000000
00001890	00001234	00002468	00002469	00000000
000018A0	00001234	00002468	00002469	00000000

Return addresses in red; scalars in green; array values in blue

```
XGpio sw, leds;    // from xgpio.h struct
...
XGpio_Initialize(&leds, XPAR_AXI_GPIO_LEDS_DEVICE_ID);
XGpio_SetDataDirection(&leds, 1, 0x0);
XGpio_DiscreteWrite(&leds, 1, count);
```

This will set the BaseAddress, IsReady, InterruptPresent, and IsDual values for the GPIO. We would need do to the same for switches sw. As an alternative we may also consider the function calls defined in xil_io.h, with the address valued listed in xparameters.h and can then code

```
volatile u32 LEDS = 0x40000000;
volatile u32 SWITCHES = 0x40010000;
...
SW_value =  Xil_In32(SWITCHES);
Xil_Out32(LEDS, SW_value);
```

The third alternative using pointer will be our preferred. It has the shortest code and does not need any function calls, and we have used for the Nios II also. Here is a code idea:

```
volatile int * LED_ptr = (int *) 0x40000000;//ZYBO green LED
address
  volatile int * SW_ptr = (int *) 0x40010000;//4 slider switch
address
  int SW_value;
...
SW_value = *(SW_ptr); // get SW value
*(LED_ptr) = SW_value; // Show SW on LEDs
```

Since we do not have UART or caches implemented, we will not need a call to init_platform(). The call to XGpio_Initialize() and XGpio_SetDataDirection() will also be simplified since we not really have a complete GPIO, such ports we need to read from and write to. Here is the simplified ANSI C code we will translate with GCC to build our assembler code.

ANSI C Program 10.9: LED Toggle Using a MicroBlaze Coding

```
1    #include <stdio.h>
2    #include "xparameters.h"
3    #include "xil_io.h"
4    #include "platform.h"
5    #include "xil_printf.h"
6    #include "xgpio.h"
7
8    #define LOOP_ITERATIONS 2000000
9
10   int main()
11   {
12     XGpio sw, leds;// from xgpio.h struct with: UINTPTR BaseAddress;
13                    // u32 IsReady; int InterruptPresent; int IsDual;
14
15   volatile unsigned int i = 0;//volatile compiler does not remove loop
16
17   volatile u32 LEDS = 0x40000000;
18   volatile u32 SWITCHES = 0x40010000;
19   volatile int * LED_ptr = (int *) 0x40000000;//ZYBO green LED address
20   volatile int * SW_ptr = (int *) 0x40010000; //4slider switch address
21   int SW_value;
22
23       init_platform(); // Enable UART and caches
24       //initialize SW/LEDs and set data direction to I/O
25       XGpio_Initialize(&leds, XPAR_AXI_GPIO_LEDS_DEVICE_ID);
26       XGpio_SetDataDirection(&leds, 1, 0x00000000);
27       XGpio_Initialize(&sw, XPAR_AXI_GPIO_SW_DEVICE_ID);
28       XGpio_SetDataDirection(&sw, 1, 0xFFFFFFFF);
29
30       SW_value = *(SW_ptr); // get SW value
31
32       while (1) {
33       *(LED_ptr) = SW_value; // Show SW on LEDs
34           for(i = 0; i < LOOP_ITERATIONS; i+=1){}
35           SW_value = SW_value ^ 0xFF; // Toggle LED
36       }
37
38       // Alternative Styles to handle GPIO I/O
39       SW_value =  Xil_In32(SWITCHES); // ok
40       Xil_Out32(LEDS, SW_value); // ok
41       SW_value = XGpio_DiscreteRead(&sw, 1); // ok
42       XGpio_DiscreteWrite(&leds, 1, SW_value); // ok
43
44       cleanup_platform();
45       return 0;
46   }
```

From the generate assembler code, we just select the portion that is necessary for I/O operations without the GPIO registers we not really have. This includes setting up our stack-point r1 or r19, the SW, and led address stored in DRAM for later use, followed by the forever while(1) loop. GCC select sp value to be 0x1db4 and the address for LEDs is therefore stored at sp + 32, i.e., location 1909_{10} and for slider switch address GCC picked location sp + 36, i.e., 1910_{10}. Since all jumps by GCC are relative, we can adjust the line numbers to match the pc. Starting label for MicroBlaze assembler is always the <start 1>: label.

With the assembler code rearranged into HDL code, we are ready to go through the assembler code (Fig. 10.20a) and at the same time monitor the progress in a HDL simulation below (Fig. 10.21). For the HDL simulation, we set the counter limit to one such we see the counter toggle within a reasonable time. The I/O address are listed in xparameters.h and we see that the slider switch at 0x40010000

```
     <_start1>:
          000:   b0000000     imm    0
          004:   30201db4     addik  r1, r0, 7604
          008:   12610000     addk   r19, r1, r0
     --... =========================================
     -- volatile int * LED_ptr =(int *) 0x40000000;//ZYBO green LED address
          00C:   b0004000     imm    16384
          010:   30600000     addik  r3, r0, 0
          014:   f8730020     swi    r3, r19, 32
     --volatile int * SW_ptr = (int *) 0x40010000;//4slider switch address
          018:   b0004001     imm    16385
          01C:   30600000     addik  r3, r0, 0
          020:   f8730024     swi    r3, r19, 36
     --... =========================================
          SW_value = *(SW_ptr); // get SW value
          024:   e8730024     lwi    r3, r19, 36
(a)       028:   e8630000     lwi    r3, r3, 0
          02C:   f873001c     swi    r3, r19, 28

          while (1) {
               *(LED_ptr) = SW_value; // Show SW on LEDs
          030:   e8730020     lwi    r3, r19, 32
          034:   e893001c     lwi    r4, r19, 28
          038:   f8830000     swi    r4, r3, 0
               for(i = 0; i < LOOP_ITERATIONS; i+=1){}
          03C:   f8130048     swi    r0, r19, 72
          040:   b8000010     bri    16             // 48c
          044:   e8730048     lwi    r3, r19, 72
          048:   30630001     addik  r3, r3, 1
          04C:   f8730048     swi    r3, r19, 72
          050:   e8930048     lwi    r4, r19, 72
          054:   b000001e     imm    30
          058:   3060847f     addik  r3, r0, -31617
          05C:   16441803     cmpu   r18, r4, r3
          060:   bcb2ffe4     bgei   r18, -28       // 480
               SW_value = SW_value ^ 0xFF; // Toggle LED
          064:   e873001c     lwi    r3, r19, 28
          068:   a86300ff     xori   r3, r3, 255
          06C:   f873001c     swi    r3, r19, 28
          }
          070:   b800ffc0     bri    -64            // 46c
```

(b)

0x1db4 : 0x1DB4 <Hex Integer> ⊠ ⊹ New Renderings...			
Address	0 – 3	4 – 7	8 – B
00001DD0	0000000C	40000000	40010000

Fig. 10.20 Trisc3mb flash ASM test routine. (a) main() flash assembler program. (b) Memory values for sw and leds

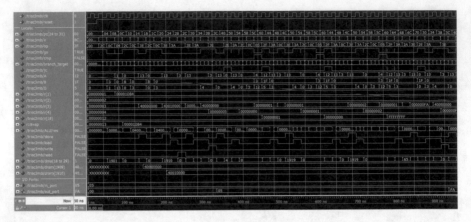

Fig. 10.21 First 1000 ns simulation steps for the toggle program showing reduced terminal count

and ZyBo LEDs at address $0x400000000$ share the same 16-bit LSBs. We load first the address value for slider switches and green LEDs and store these in memory (lines $00C$–020). Next the loop starts. We first write the SW value to the LEDs (lines 030–038). Now the loop counter is incremented and compared to our final values. If the comparison of $r3$–$r4$ becomes negative (line $5C$), the loop is completed, and the SW value is toggled using a XOR (lines 064–$06C$) and the loop is started from the begin using a branch.

Encoding and decoding of these instructions is left as an exercise at the end of the chapter (see Exercises 10.17–10.24).

Now let us have a look at the VHDL simulation in Fig. 10.21. The simulation shows the input signal for clock and reset, followed by local non-I/O signals and finally the I/O signals for switches and LEDs. After the release of the reset (active low), we see how the program counter (pc) continuously increases. The pc changes with the falling edge and instruction word is updated with the rising edge (within the ROM). The value 5 is read using the address in $r3$ from the input port in_port and put into register $r4$. The address register $r3$ value is loaded in the out_port register. The counter is incremented by 1 and is compared to the terminal count value. The result for the comparison can be found in register $r18$. Looping continues in general for many clock cycles until the 1-second simulation time is reached and the output toggles. We have reduced in the simulation the terminal count such that the output register out_port toggles already after two loops are completed. We need to restore the original terminal counter value before synthesis and downloading to the board. The counter becomes negative at 770 ns and at 930 ns the output toggles in the simulation.

HDL Implementation and Testing

The PICOBLAZE design `trisc2.vhd` or `trisc2.v` should be our starting point
for the tiny MICROBLAZE design we call TRISC3MB since MICROBLAZE is a three-
address machine. The HDL code is shown next.

VHDL Code 10.10: TRISC3MB (Final Design)

```
1    -- ========================================================================
2    -- Title: T-RISC 3 address machine
3    -- Description: This is the top control path/FSM of the
4    -- T-RISC, with a single 3 phase clock cycle design
5    -- It has a 3-address type instruction word
6    -- implementing a subset of the MicroBlaze architecture
7    -- ========================================================================
8    LIBRARY ieee; USE ieee.std_logic_1164.ALL;
9
10   PACKAGE n_bit_type IS                 -- User defined types
11     SUBTYPE U8 IS INTEGER RANGE 0 TO 255;
12     SUBTYPE U12 IS INTEGER RANGE 0 TO 4095;
13     SUBTYPE SLVA IS STD_LOGIC_VECTOR(0 TO 11); -- Address prog. mem.
14     SUBTYPE SLVD IS STD_LOGIC_VECTOR(0 TO 31); -- Width data
15     SUBTYPE SLVD1 IS STD_LOGIC_VECTOR(0 TO 32); -- Width data + 1
16     SUBTYPE SLVP IS STD_LOGIC_VECTOR(0 TO 31); -- Width instruction
17     SUBTYPE SLV6 IS STD_LOGIC_VECTOR(0 TO 5);  -- Full opcode size
18   END n_bit_type;
19
20   LIBRARY work;
21   USE work.n_bit_type.ALL;
22
23   LIBRARY ieee;
24   USE ieee.STD_LOGIC_1164.ALL;
25   USE ieee.STD_LOGIC_arith.ALL;
26   USE ieee.STD_LOGIC_signed.ALL;
27   -- ========================================================================
28   ENTITY trisc3mb IS
29    PORT(clk      : IN  STD_LOGIC; -- System clock
30         reset    : IN  STD_LOGIC; -- Active low asynchronous reset
31         in_port  : IN  STD_LOGIC_VECTOR(0 TO 7); -- Input port
32         out_port : OUT STD_LOGIC_VECTOR(0 TO 7) -- Output port
33    -- The following test ports are used for simulation only and should be
34    -- comments during synthesis to avoid outnumbering the board pins
35    --   r1_OUT    : OUT SLVD;  -- Register 1
36    --   r2_OUT    : OUT SLVD;  -- Register 2
37    --   r3_OUT    : OUT SLVD;  -- Register 3
38    --   r19_OUT   : OUT SLVD;  -- Register 19 aka 2. stack pointer
39    --   r15_OUT   : OUT SLVD;  -- Register 14 aka return address
40    --   jc_OUT    : OUT STD_LOGIC;   -- Jump condition flag
41    --   me_ena    : OUT STD_LOGIC;   -- Memory enable
42    --   i_OUT     : OUT STD_LOGIC;   -- constant flag
43    --   pc_OUT    : OUT STD_LOGIC_VECTOR(0 TO 11); -- Program counter
44    --   ir_imm16  : OUT STD_LOGIC_VECTOR(0 TO 15); -- Immediate value
45    --   imm32_out : OUT SLVD;                 -- Sign extend immediate value
46    --   op_code   : OUT STD_LOGIC_VECTOR(0 TO 5)   -- Operation code
47         );
48   END ENTITY;
```

```
49    -- ================================================================
50    ARCHITECTURE fpga OF trisc3mb IS
51    -- Define GENERIC to CONSTANT for _tb
52      CONSTANT WA : INTEGER := 11;  -- Address bit width -1
53      CONSTANT NR : INTEGER := 31;  -- Number of Registers -1; PC is extra
54      CONSTANT WD : INTEGER := 31;   -- Data bit width -1
55      CONSTANT DRAMAX : INTEGER := 4095; -- No. of DRAM words -1
56      CONSTANT DRAMAX4 : INTEGER := 16384; -- X"4000";
57                                           -- True DRAM bytes -1
58      COMPONENT rom4096x32 IS
59      PORT (clk   : IN STD_LOGIC;        -- System clock
60            reset : IN STD_LOGIC;        -- Asynchronous reset
61            pma   : IN STD_LOGIC_VECTOR(11 DOWNTO 0); -- Program mem. add.
62            pmd   : OUT STD_LOGIC_VECTOR(31 DOWNTO 0));--Program mem. data
63      END COMPONENT;
64
65      SIGNAL op  : SLV6;
66      SIGNAL dmd, pmd, dma : SLVD;
67      SIGNAL ir, pc, pc4, pc_d, branch_target, target_delay : SLVP;-- PCs
68      SIGNAL mem_ena, not_clk : STD_LOGIC;
69      SIGNAL jc, go, link, Dflag, Delay, cmp : boolean;-- controller flags
70      SIGNAL br, bra, bri, brai, condbr, condbri : boolean;-- branch flags
71      SIGNAL swi, lwi, rt : boolean; -- Special instr.
72      SIGNAL rAzero, rAnotzero, I, K, L, U, LI, D6, D11 : boolean;-- flags
73      SIGNAL aai, aac, ooi, xxi : boolean; -- Arith  instr.
74      SIGNAL imm, ld, st, load, store, read, write : boolean; -- I/O flags
75      SIGNAL D, A, B : INTEGER RANGE 0 TO 31; -- Register index
76      SIGNAL rA, rB, rD : SLVD := (OTHERS => '0');-- current Ops
77      SIGNAL rAsxt, rBsxt,   rDsxt : SLVD1; -- Sign extended Ops
78      SIGNAL rI, imm16 : STD_LOGIC_VECTOR(0 TO 15);   -- 16 LSBs
79      SIGNAL sxt16 : STD_LOGIC_VECTOR(0 TO 15); -- Total 32 bits
80      SIGNAL imm32 : SLVD; -- 32 bit branch/mem/ALU
81      SIGNAL imm33 : SLVD1; -- Sign extended ALU constant
82      SIGNAL c : STD_LOGIC;
83
84    -- Data RAM memory definition use one BRAM: DRAMAXx32
85      TYPE MTYPE IS ARRAY(0 TO DRAMAX) OF SLVD;
86      SIGNAL dram : MTYPE;
87
88    -- Register array definition 16x32
89      TYPE REG_ARRAY IS ARRAY(0 TO NR) OF SLVD;
90      SIGNAL r : REG_ARRAY;
91
92    BEGIN
93
94      rAzero <= true WHEN (rA=0) ELSE false; -- rA=0
95      rAnotzero <= true WHEN (rA/=0) ELSE false; -- rA/=0
96      WITH ir(8 TO 10) SELECT -- Evaluation of signed condition
```

```
 97         go <= rAzero                    WHEN "000",   -- BEQ =0
 98             rAnotzero                   WHEN "001",   -- BNE /=0
 99             rA(0)='1'                   WHEN "010",   -- BLT < 0
100             rA(0)='1' OR rAzero         WHEN "011",   -- BLE <=0
101             rA(0)='0' AND rAnotzero     WHEN "100",   -- BGT: > 0
102             rA(0)='0' OR rAzero         WHEN "101",   -- BGE >=0
103             false   WHEN OTHERS;   -- if not true
104
105       FSM: PROCESS (reset, clk) -- FSM of processor
106       BEGIN -- update the PC
107         IF reset = '0' THEN
108           pc <= (OTHERS => '0');
109         ELSIF falling_edge(clk) THEN
110           IF jc THEN
111             pc <= branch_target ; -- any current jumps
112           ELSIF Delay THEN
113             pc <= target_delay ; -- any jumps with delay
114           ELSE
115             pc <= pc4;  -- Usual increment by 4 bytes
116           END IF;
117           pc_d <= pc;
118           IF Dflag THEN  Delay <= true;
119           ELSE           Delay <= false;
120           END IF;
121           target_delay <= branch_target; -- store target address
122         END IF;
123       END PROCESS FSM;
124       pc4 <= pc + X"00000004"; -- Default PC increment is 4 bytes
125       jc <= NOT Dflag AND ((go AND (condbr OR condbri)) OR br
126                                      OR bri or rt); -- New PC; no delay?
127       branch_target <= rB WHEN bra -- Order is important !
128                         ELSE imm32 WHEN brai
129                         ELSE pc + rB WHEN condbr OR br
130                         ELSE rA + imm32 WHEN rt
131                         ELSE pc + imm32; -- bri, condbri etc.
132
133       rt <= true WHEN op= "101101" ELSE false; -- return from
134       br  <= true WHEN op= "100110" ELSE false; -- always jump
135       bra <= true WHEN br AND ir(12)='1' ELSE false;
136       bri <= true WHEN op= "101110" ELSE false;--always jump w imm
137       brai <= true WHEN bri AND ir(12)='1' ELSE false;
138       -- link = bit 13 for br and bri
139       link <= true WHEN (br OR bri) AND L ELSE false; -- save PC
140       condbr <= true WHEN op= "100111" ELSE false;-- cond. branch
141       condbri <= true WHEN op= "101111" ELSE false;--cond. b/w imm
142       cmp <= true WHEN op= "000101" ELSE false; -- cmp and cmpu
143
144       -- Mapping of the instruction, i.e., decode instruction
```

```
145        op    <= ir(0 TO 5);    -- Data processing OP code
146        imm16 <= ir(16 TO 31);          -- Immediate ALU operand
147
148        -- Delay (D), Absolute (A) Decoder flags not used
149        I <= true WHEN ir(2)='1' ELSE false;   -- 2. op is imm
150        K <= true WHEN ir(3)='1' ELSE false; -- K=1 keep carry
151        L <= true WHEN ir(13)='1' ELSE false; -- Link for br and bri
152        U <= true WHEN ir(30)='1' ELSE false; -- Unsigned flag
153        D6 <= true WHEN ir(6)='1' ELSE false; -- Delay flag condbr/i;rt;
154        D11 <= true WHEN ir(11)='1' ELSE false; -- Delay flag br/i
155        Dflag <= (D6 AND go AND (condbr OR condbri)) OR (rt AND D6) OR
156                 (D11 AND (br OR bri)); -- All Delay ops summary
157
158        -- I = bit 2; K = bit; 3 add/addc/or/xor with(out) imm
159        aai <= true WHEN ir(0 TO 1)= "00" AND ir(4 TO 5)= "00" ELSE false;
160        aac <= true WHEN ir(0 TO 1)= "00" AND ir(4 TO 5)= "10" ELSE false;
161        ooi <= true WHEN ir(0 TO 1)= "10" AND ir(3 TO 5)= "000" ELSE false;
162        xxi <= true WHEN ir(0 TO 1)= "10" AND ir(3 TO 5)= "010" ELSE false;
163        -- load and store:
164        ld  <= true WHEN ir(0 TO 1)= "11" AND ir(3 TO 5)= "010" ELSE false;
165        st <= true WHEN ir(0 TO 1)= "11" AND ir(3 TO 5)= "110" ELSE false;
166
167        imm <= true WHEN ir(0 TO 5)= "101100" ELSE false;-- always store imm
168        sxt16 <= (OTHERS => imm16(0)); -- Sign extend the constant
169        imm32 <=  rI & imm16 WHEN LI   -- Immediate extend to 32
170                   ELSE sxt16 & imm16; -- MSBs from last imm
171
172        A <= CONV_INTEGER('0' & ir(11 TO 15)); -- Index 1. source reg.
173        B <= CONV_INTEGER('0' & ir(16 TO 20));-- Index 2. source register
174        D <= CONV_INTEGER('0' & ir(6 TO 10));-- Index destination reg.
175        rA <= r(A); -- First operand ALU
176        rAsxt <= rA(0) & rA; -- Sign extend 1. operand
177        rB <= imm32 WHEN I -- 2. ALU operand maybe constant or register
178             ELSE r(B);   -- Second operand ALU
179        rBsxt <= rB(0) & rB; -- Sign extend 2. operand
180        rD <= r(D);  -- Old destination register value
181          rDsxt <= rD(0) & rD; -- Sign extend old value
182
183        prog_rom: rom4096x32        -- Instantiate a Block ROM
184        PORT MAP (clk  => clk,   -- System clock
185                  reset => reset,  -- Asynchronous reset
186                  pma  => pc(18 TO 29),-- Program memory address 12 bits
187                  pmd  => pmd);   -- Program memory data
188        ir <= pmd;
189
190        dma <= rA + imm32 WHEN I
191             ELSE rA + rB;
192        store <= st AND (dma <= DRAMAX4); -- DRAM store
```

```
145        op    <= ir(0 TO 5);    -- Data processing OP code
146        imm16 <= ir(16 TO 31);         -- Immediate ALU operand
147
148        -- Delay (D), Absolute (A) Decoder flags not used
149        I <= true WHEN ir(2)='1' ELSE false;  -- 2. op is imm
150        K <= true WHEN ir(3)='1' ELSE false; -- K=1 keep carry
151        L <= true WHEN ir(13)='1' ELSE false; -- Link for br and bri
152        U <= true WHEN ir(30)='1' ELSE false; -- Unsigned flag
153        D6 <= true WHEN ir(6)='1' ELSE false;  -- Delay flag condbr/i;rt;
154        D11 <= true WHEN ir(11)='1' ELSE false; -- Delay flag br/i
155        Dflag <= (D6 AND go AND (condbr OR condbri)) OR (rt AND D6) OR
156                 (D11 AND (br OR bri)); -- All Delay ops summary
157
158        -- I = bit 2; K = bit; 3 add/addc/or/xor with(out) imm
159        aai <= true WHEN ir(0 TO 1)= "00" AND ir(4 TO 5)= "00" ELSE false;
160        aac <= true WHEN ir(0 TO 1)= "00" AND ir(4 TO 5)= "10" ELSE false;
161        ooi <= true WHEN ir(0 TO 1)= "10" AND ir(3 TO 5)= "000" ELSE false;
162        xxi <= true WHEN ir(0 TO 1)= "10" AND ir(3 TO 5)= "010" ELSE false;
163        -- load and store:
164        ld  <= true WHEN ir(0 TO 1)= "11" AND ir(3 TO 5)= "010" ELSE false;
165        st <= true WHEN ir(0 TO 1)= "11" AND ir(3 TO 5)= "110" ELSE false;
166
167        imm <= true WHEN ir(0 TO 5)= "101100" ELSE false;-- always store imm
168        sxt16 <= (OTHERS => imm16(0)); -- Sign extend the constant
169        imm32 <=  rI & imm16 WHEN LI   -- Immediate extend to 32
170                  ELSE sxt16 & imm16; -- MSBs from last imm
171
172        A <= CONV_INTEGER('0' & ir(11 TO 15)); -- Index 1. source reg.
173        B <= CONV_INTEGER('0' & ir(16 TO 20));-- Index 2. source register
174        D <= CONV_INTEGER('0' & ir(6 TO 10));-- Index destination reg.
175        rA <= r(A); -- First operand ALU
176        rAsxt <= rA(0) & rA; -- Sign extend 1. operand
177        rB <= imm32 WHEN I --  2. ALU operand maybe constant or register
178             ELSE r(B);   -- Second operand ALU
179        rBsxt <= rB(0) & rB; -- Sign extend 2. operand
180        rD <= r(D);  -- Old destination register value
181          rDsxt <= rD(0) & rD; -- Sign extend old value
182
183        prog_rom: rom4096x32        -- Instantiate a Block ROM
184        PORT MAP (clk   => clk,    -- System clock
185                  reset => reset,  -- Asynchronous reset
186                  pma   => pc(18 TO 29),-- Program memory address 12 bits
187                  pmd   => pmd);   -- Program memory data
188        ir <= pmd;
189
190        dma <= rA + imm32 WHEN I
191             ELSE rA + rB;
192        store <= st AND (dma <= DRAMAX4); -- DRAM store
```

```
241              ELSE C <= '0';
242             END IF;
243           END IF;
244           -- Compute and store new register values
245           IF imm THEN -- Set flag: last was imm instruction
246             rI <= imm16; LI <= true;
247           ELSE
248             rI <= (OTHERS => '0'); LI <= false;
249           END IF;
250           IF D>0 THEN -- Do not write r(0)
251           IF link THEN -- Store LR for operation branch with link aka call
252             r(D) <= pc_d; -- Old pc + 1 op after return
253           ELSE
254             r(D) <= res(1 TO 32); -- Store ALU result
255           END IF;
256           END IF;
257           IF write THEN out_port <= rD(24 TO 31); END IF;--LSBs are right
258         END IF;
259       END PROCESS ALU;
260
261   --  -- Extra test pins:
262   --  pc_OUT <= pc(20 TO 31);
263   --  ir_imm16 <= imm16;
264   --  op_code <= op; -- Data processing ops
265   --  jc_OUT <= '1' WHEN jc ELSE '0';   -- Xilinx modified
266   --  i_OUT <= '1' WHEN I ELSE '0';   -- Xilinx modified
267   --  me_ena <= mem_ena; -- Control signals
268   --  r1_OUT <= r(1);        -- First two user registers
269   --  r2_OUT <= r(2); r3_OUT <= r(3);      -- Next two user registers
270   --  r15_OUT <= r(15); r19_OUT <= r(19);     -- Compiler registers
271
272   END fpga;
```

We can use a similar I/O interface as for the PICOBLAZE but would not monitor
the first register r0 since it is always zero. For the memory, we will monitor
locations around the sp=r1 value. A few additional flags such as the bit 2 constant
flag we call I will also be monitored. Operation codes are now all 6 bits, and pc,
data, and address width need to be adjusted to 32 bits. Note that vectors are enumerated from left to right using TO instead of the usual DOWNTO index, which does not
allow a standard mapping for weighted number with LSB on the right, but the
VHDL conversion function work, nevertheless. Since VHDL-1993 (ok in
VDHL-2008) does not allow coding 6-bit hex values, we always use a binary coding
(the default base specification) when decoding instructions. We often use Boolean
flags for the instructions to simplify instruction decoding later. We start the architecture coding with the evaluation of the conditions for conditional branches (lines
94–103). Without the link register stack, the processor controller FSM will be a little
shorter. pc increment counting is byte wise, so each standard instruction increases
the program counter pc by 4. For the branch target, we now have more choices: it
can come from the IMM32 directly for instructions such as brai, relative to pc
(bri, bnei, beqi, etc.) with IMM32 offset, relative to pc (br, bne, beq,
etc.) with rB offset, from rB register for the bra instruction, or from IMM32+rA
register for rtsd. An extra feature is the fact that all (conditional) branches may

have a so-called branch delay slot, coded as brd vs. br. The branch delay version of the instruction executes also the one instruction immediately after the branch before continuing at the new pc location. For the delay instructions, we therefore need to issue a delay flag (line 118–120) and save the branch_target value for one clock cycle (line 121). The evaluations of the conditional instructions are placed immediately after the processor FSM (lines 133–142).

Decoding the instructions is based on the two instruction formats with a few small exceptions. The first 6 bits are dedicated to the operations, and these are followed by the 15 bits for register index of destination and the two sources D, A, and B. Flags located at bit 2 (immediate), bit 3 (carry keep), bit 13 (link), and bit 30 (unsigned) are used by some but not all instructions. The delay slot code can be found in bit 6 (conditional branches and return) or in bit 11 for unconditional branch (lines 149–155). We combine the 3 registers and 2 registers with immediate instructions for add, addc, or, xor, load, and store (lines 159–165) to simplify ALU design. For constants, we have the special instruction imm allowing us to access the 16 MSB; otherwise the 16-bit constant is always sign extended to 32 bits (lines 167–170). The first ALU source will always be rA=r(A); the second ALU source may be a register or the (sign extended or zero appended) IMM16 value. The I flag identifies all instructions that use IMM16, i.e., located in the 16 bits to the right. Register values r(A), r(B), and old r(D) are made available (requiring each a 32 bit 32:1 mux) as special signals, rA, rB, and rD, and sign extension by one bit for arithmetic operations (line 172–181). Next comes the program memory (lines 183–188) that instantiates a 4 K word program ROM with 32-bit for each word. It follows the data RAM with the control signal and the I/O components. Since we only have one input and one output device, the decoding if the memory read or write involve the DRAM or I/O component is done with a simple threshold at the (maximum possible) size of the DRAM (lines 190–196). The DRAM (lines 197–210) is a synchronous negative edge triggered memory. The number of words depends on the global constant DRAMAX. For an easy transition for ANSI C programs, you may use the same size as the program ROM, i.e., 4095. The maximum DRAM size is limited by the first address used for the I/O component. For MicroBlaze, the first I/O address in use is at 0x4000, so last DRAM address is at 0x3FFF as specified in the byte constant DRAMAX4 used for decoding (line 56).

Now the ALU coding follows (lines 212–259). The R-type and I-type ALU instructions are combined to allow for resource sharing since only the second operand is different but the operation or logic gates are the same. The carry flag is not updated when the keep flag K is active. All logic, arithmetic, as well as data transfer are coded that write to a register. R-type instruction writes to r(D), while the imm operation writes to temporary register rI. pc_d is stored for operation with link flag L active. Register writes are controlled by the rising edge clock. Register zero is not written. All registers are set to their register number when reset is active. Finally output assignments are made to some additional test ports; these should be disabled at the end for final synthesis to avoid an overloading of the I/O pins of the FPGA. This concludes the HDL description of TRISC3MB, now let us run our example program in MODELSIM.

The simulation for the memory access example discussed in program Listings 10.7 is shown in Fig. 10.22. The memory values shown in the simulation match the memory content of the SDK Debugger from Fig. 10.18. The variables have been mapped by the compiler using the `sp` pointers and a global symbol table for DRAM addresses as follows:

Variable	s	g	sarray[0]	garray[0]
DRAM address (decimal)	1536	776	1537	777

The major events of the DRAM test simulation are s=1 @ 180 ns; g=2 @ 240 ns; `sarray[0]`=s @ 280 ns; `garray[0]`=   g @ 360 ns; `sarray[1]`=0x1357 @ 520 ns; `garray[2]`=2468 @ 680 ns.

The HDL simulation for the nesting loop code from Program 10.8 (ANSI C) and Fig. 10.19 (assembler) is shown in Fig. 10.23. The address values in the range of `sp` match the assembler debug display in Table 10.7. Special attention should be given to the return address register `r(15)`=`r1` for the loops. The major events of the loop behavior simulation can be summarized as follows: s1=0x1233 @ 200 ns; r15=0x4DC @ 310 ns; r15=4DC stored at DRAM(1557) @ 360 ns; r15=490 @ 650 ns; r15=0x490 stored at DRAM(1549) @ 700 ns; r15=430 @ 1010 ns; loading pc with 0x430+8 @ 1380 ns; r15=490 @ 1430 ns; loading pc with 490 + 8 @ 1520 ns; r15=4DC @ 1570 ns; loading pc with r15=04DC+8=4E4 @ 1660 ns. The subroutines are entered as follows: level1 @ 320 ns, level2 @ 660 ns, and level3 @ 1020 ns. The final DRAM values for s1, s2, and s3 are located at DRAM(1572), DRAM(1573), and DRAM(1574), respectively. The first while loop is completed after 1800 ns.

Finally Fig. 10.24 shows the overall TRISC3MB architecture together with the implemented instructions. Implementation of additional instructions are left to the reader as exercise at the end of the chapter (see Exercise 10.57–10.58).

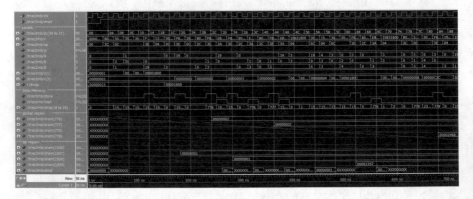

Fig. 10.22 Simulation shows the TRISC3MB memory access

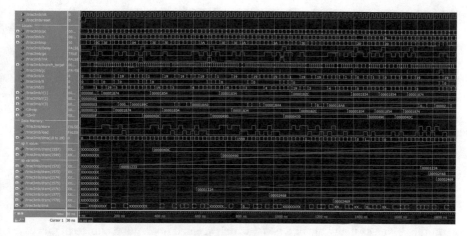

Fig. 10.23 Simulation shows the TRISC3MB nesting loop behavior

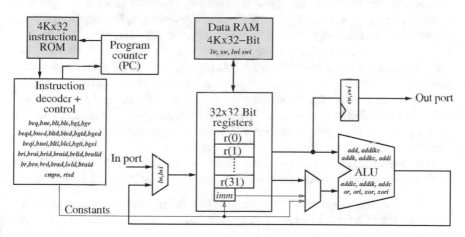

Fig. 10.24 The final implemented tiny MICROBLAZE aka TRISC3MB core. Instruction (in italic) mainly associated with major hardware units. Substantial memory blocks (excluding registers) in gray

Synthesis Results and ISA Lesson Learned

The synthesis results for the TRISC3MB are shown in Table 10.8. The columns 2 and 3 show Xilinx VIVADO synthesis results for VHDL and Verilog, respectively. The columns 4 and 5 show Altera QUARTUS synthesis results for VHDL and Verilog, respectively. The major differences in the four synthesis results come from the method how the Program ROM is implemented. Since our programs are small, the synthesis tool has in three out of four cases mapped the ROM to logic cells and not BRAM. Only Verilog QUARTUS version uses block RAM for RAM and ROM

Table 10.8 TRISC3MB synthesis results for Altera and Xilinx tools and devices

	VIVADO 2015.1		QUARTUS 15.1	
Target device	Zynq 7 K xc7z010t-1clg400		Cyclone V 5CSEMA5F31C6	
HDL used	VHDL	Verilog	VHDL	Verilog
LUT/ALM	957	965	766	1458
used as Dist. RAM	0	0	0	0
Block RAM RAMB18	4	4	–	–
or M10K	–	–	16	32
DSP blocks	0	0	0	0
HPS	0	0	0	0
Fmax/ MHz	54.05	57.14	63.04	40.17
Compile time	4:36	6:16	4:50	7:24

requiring twice as many block RAMs, which seemed not a good choice in this case since performance is also reduced. The Altera needs four times the number of Zynq block RAMs due to the smaller size of the embedded memory blocks. Compile time is reasonable, and speed is above 50 MHz. The VIVADO on ZyBo may need a clock divider by 2, but the design worked fine even with the 125 MHz system clock.

We finally have a working microprocessor with 53 out of 148 MICROBLAZE operations implemented. Major units such as register array, logical, arithmetic operation, input and output, data memory, jump, and call are included. Missing from the complete instruction set are those group we did not need for our tasks so far such as shift and rotate, Boolean AND and AND immediate, multiply and divide, interrupt handling, and floating-point operations. Conditional branches are needed for ANSI C control language elements such as if or switch.

The "lesson learned" from looking under the hood is that the MICROBLAZE architecture has (compared to the Nios II) some significant ISA additions, namely,

- The majority of branches (br vs. brd) have a delay slot simplifying pipelining and let the (pipelined) MICROBLAZE run at higher speed. All return operations have a delay slot too.
- Floating-point operations such as fadd, fmul, or fdiv are part of the instructions set and do not requires custom instructions handling and therefore improving the floating-point performance substantial (see Fig. 10.7).
- Some special instruction such as byte swap, direct cache writes, or 8 and 16 bit sign extension have been added.

However, these special instructions are not used in DMIPS routines and the DMIPS performance similar to Nios II at the same clock speed. The ARMv7 processor in the next chapter is known for high DMIPS performance, so it will be interesting to see how the ARMv7 achieves the higher DMIPS rating.

Review Questions and Exercises

Short Answer

10.1. What are the main differences between PicoBlaze and MicroBlaze?

10.2. Name three features of the MicroBlaze that are not part of the Nios II.

10.3. For which applications would you use the MicroBlaze and not PicoBlaze?

10.4. What are the five instruction groups in MicroBlaze assembler?

10.5. Name four addressing modes used by MicroBlaze branch operations. For each mode, specify an instruction example.

10.6. How are nested loops implemented by GCC for MicroBlaze?

10.7. Name five advantages of the Trisc3MB HDL design approach compared to MicroBlaze IP core design.

10.8. MicroBlaze has no sub, subi, nop, or clear register instructions. Define "pseudo-instructions" using the instructions from Trisc3MB that accomplish these tasks.

10.9. Give MicroBlaze architecture example for the RISC design principles: (a) Simplicity favors regularity; (b) Smaller is faster, and (c) good designs demands good compromise.

10.10. What are the major differences in the video standard for (a) VGA vs. HDMI? (b) VGA vs. DVI? (c) DVI vs. HDMI?

10.11. What are benefits to build a CIP vs. use the HDL directly.

Fill in the Blank

10.12. The first _____ bits of the MicroBlaze instruction opcode are used to encode the instruction.

10.13. The MicroBlaze is a ____ address machine.

10.14. The Trisc3MB uses _____ instructions of the MicroBlaze ISA.

10.15. The MicroBlaze first SDK manual was published _____.

10.16. The immediate constants in the Trisc3MB instructions are of length _____.

10.17. The MicroBlaze operation brlid r15, 0x94 will be encoded in hex as _____.

10.18. The MicroBlaze operation addik r1, r1, -4 will be encoded in hex as _____.

10.19. The MicroBlaze operation xor r5, r4, r3 will be encoded in hex as _____.

10.20. The MicroBlaze operation swi r5, r19, 0x1C will be encoded in hex as _____.

10.21. The MicroBlaze operation _____ will be encoded in hex as 0xeb160088.

10.22. The MicroBlaze operation _____ will be encoded in hex as
`0x12610000`.

10.23. The MicroBlaze operation _____ will be encoded in hex
as `0xa883ffff`.

10.24. The MicroBlaze operation _____ will be encoded in hex as
`0xbc030028`.

True or False

10.25. _____ The MicroBlaze is available as 3- and 5-pipeline versions.

10.26. _____ The bottom-up design approach is simplified using the Run
Connection Automation feature in Vivado.

10.27. _____ The top-down will benefit from the Run Block Automation feature
in Vivado.

10.28. _____ The ZyBo board can be programmed via USB, Wi-Fi, or Bluetooth.

10.29. _____ The SDK debugger can be used to monitor register values and memory content in a step-by-step mode.

10.30. _____ The MicroBlaze core will require more embedded LUTs if we add
hardware barrel shifter or divider.

10.31. _____ The SDK has templates for Hello World code, I/O peripheral tests,
and DMIPS tests.

10.32. _____ The MicroBlaze FPH improves speed single and double precision
FP operations.

10.33. _____ The MicroBlaze FPH uses predefined instructions in the instruction set.

10.34. _____ The `xparameters.h` file of SDK contains the I/O address of the
component.

10.35. _____ The MicroBlaze Vivado SDK allows programs in Java, Python,
and C/C++.

10.36. _____ To download the MicroBlaze to the ZyBo we need Vivado, the
SDK program cannot be used for that.

10.37. _____ Pmod connector on ZyBo can be used to add custom hardware.

10.38. Specify standard features (**T**rue/**F**alse) for the MicroBlaze and Trisc3mb
core in the table below

Feature	MicroBlaze	Trisc3mb
RISC architecture		
32 × 32-bit register file		
gp and sp registers		
Branch delay slot		
One clock cycle instructions		
Boolean AND instruction		
Add with keep carry instruction		

Projects and Challenges

10.39. Run an example from Xilinx SDK templates and write a short essay (1/2 page) what the examples is about. What I/O components are used?

10.40. Develop an assembler or ANSI C program to implement a left, right car turn signal and emergency light similar to the Ford Mustang: 00X, 0XX, XXX for a left turn where 0 is off and X is LED on. Use X00, XX0, XXX for right turn signal. Use the slider switch for the left/right selection. For emergency light both switches are on. The turn signal sequence should repeat once per second. See mustangQT.MOV or mustangWMP.MOV for a demo.

10.41. Write an ANSI C program that implements a running light. The speed of the running light should be determined by the SW values, and any change should be displayed at the SDK Terminal. See led_coutQT.MOV or led_countWMP.MOV for a demo.

10.42. Develop an ANSI C programs that lights up one LED only and moves to the left or right by pushing button 4 or 3, respectively.

10.43. Repeat the previous Exercise with a three-color bar on the VGA or HDMI monitor.

10.44. Develop an assembler or ANSI C program that has a random number generator, using ADD and XOR instructions. Display the random number on the LEDs. What is the period of your random sequence?

10.45. Build a stopwatch, which counts up in second steps. Use three buttons: start clock, stop or pause clock, and reset clock. Using the LEDs, LCD, OLED, or the seven-segment display (see Table 10.1) for display. See stop_watch_ledQT.MOV or stop_watch_ledWMP.MOV for a demo.

10.46. Use the pushbuttons to implement a reaction speed timer. Turn on one of 4 LEDs and measured time it takes until the associate button is pressed. Display the measurement after 10 tries on LED, LCD, OLED, VGA, or HDMI monitor (see Table 10.1).

10.47. Write ANSI C program to display all character *A-Z* on a 7-segment display. For ZyBo, you will need to buy 7 Euro/USD the extra 2xSSD Pmod. Take special attention for character X, M, and W. You may use a bar to indicated double length, e.g., $m = \bar{n}; w = \bar{v}$. Produce a short video that shows all characters on the 7-segment display. See ascii4sevensegment.MOV for a demo.

10.48. Use the character set from the last Exercise to generate a running text on the seven-segment display using the current month and year and/or our name or your initials if you name is too long. See idQT.MOV and idWMP.MOV from the CD for a demo.

10.49. Write ANSI C program to reproduce the ASCII from Fig. 5.2 for all printable character. (Decimal 32...127) on the HDMI display and the Terminal window.

10.50. Write an ANSI C program for the MICROBLAZE Basic Computer with HDMI text monitor CIP that draws a circle on the monitor using Bresenham circle algorithm: Start with $y = 0$; $x = R$. Move in y-direction one pixel in each iteration. Decrement x if error $|x^2 + y^2 - R^2|$ is smaller for $x' = x - 1$ otherwise keep x. Use * character for drawing. Use symmetry to complete all four segments of the circle.

10.51. Use the Bresenham algorithm from the last exercise to draw a filled circle with the same character.

10.52. Write an ANSI C program for the MICROBLAZE Basic Computer with HDMI text monitor CIP that draws a ellipse on HDMI text monitor using standard floating-point arithmetic: $x = a*\cos([0:0.1:\pi/2]); y = b*\sin([0:0.1:\pi/2])$. Use symmetry to fill the ellipse with horizontal lines. Is the ellipse complete filled? What is the Exercise with this type of algorithm?

10.53. Write an ANSI C program for the MICROBLAZE Basic Computer with HDMI text monitor CIP that draws an ellipse on VGA using Bresenham algorithm: Start with $y = 0$; $x = R$. Move in y-direction one pixel in each iteration. Decrement x if error $|x^2/a^2 + y^2/b^2 - R^2|$ is smaller for $x' = x - 1$ otherwise keep x. Use symmetry to fill the ellipse with horizontal lines. Is the ellipse complete filled? What is the Exercise with this type of algorithm?

10.54. Write an ANSI C program for the MICROBLAZE Basic Computer with HDMI text monitor CIP that implements a display of the four fractals (use range $x, y = -1.5...1.5$). The complex iteration equations are as follows:

(a) Mandelbrot: $c = x + i*y$; $z = \mathbf{0}$
(b) Douday rabbit: $z = x + iy$; $c = -0.123 + j0.745$
(c) Siegel disc: $z = x + i*y$; $c = -0.391 - j0.587$
(d) San marco: $z = x + i*y$; $c = -0.75$

Find the number of iteration (and code in unique color or character) $z = z.*z + c$; such that $|z|^2 > 5$, see Fig. 10.25. The Name or the fractal should be displayed in the center of the screen. See 4fractals.MOV for a demo.

10.55. Write an ANSI C program for the MICROBLAZE Basic Computer with VGA or HDMI text monitor CIP that implements a game such as:

(a) Tic-Tac-Toe (by Christopher Ritchie, Rohit Sonavane, Shiva Indranti, Xuanchen Xiang, Qinggele Yu, Huanyu Zang 2016 game 3/6/8; 2017 game 9/10/11). Use two players: Enter selecting with slider switches and the push button to submit your selection for player 1 (button 3) and player 2 (button 0). Alternative toggle by default and display which player turn it is with display on LED or seven segment.

(b) Reaction time (by Mike Pelletier 2016 game 4): Light up one of three squares on the screen or LEDs and measure the time until the correct button was pressed.

(c) Minesweeper (Sharath Satya 2016 game 7): In a 3 × 3 field 3 bombs are hidden. You need to find the six boxes that do not contain a bomb. If you find all six you win otherwise player loses.

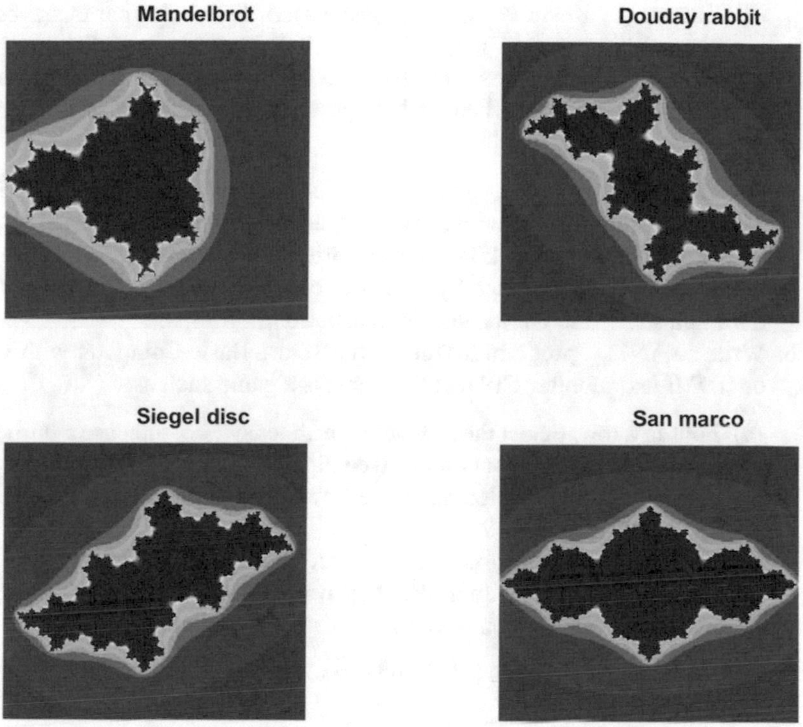

Mandelbrot

Douday rabbit

Siegel disc

San marco

Fig. 10.25 Four fractal examples from Project 10.54

(d) Pong (by Samuel Reich and Chris Raices 2017 game 4/5): Use one player only. Use one direction to move the ball and the pushbutton for moving the paddle on the other axis. Set a timer to count how long the games runs until the player misses the ball. Ball may be a small square.

(e) Simon says (by Melanie Gonzalez and Kiernan Farmer 2016 game 1/2): Repeat the three color sequence of VGA square with the push buttons 1, 2, and 3.

(f) Box Chaser (by Ryan Stepp 2017 game 7): A big box moves randomly over the display and you need to avoid that your small box is hit by this big box moving it to left or right with the push buttons.

(g) Space Invades (by Zlatko Sokolikj 2017 game 6). You have flying object that needs to be hit with your firing gun shouting in y direction. Your gun moves on the x axis, so a total of three buttons are in use. Counting the targets hit within a time frame gives the score.

(h) Flappy Bird (by Javier Matos 2017 game 3). Pressing a button move your bird in y-direction, while your bird continuously drops. Obstacle such as wall with a hole move from right to left. Game is over when the bird hits the wall or ground.

(i) Bounce (by Kevin Powell 2016 game 5). A three-color bar is moved by two push buttons to left or right. A ball bounces in y direction and changes color whenever the ball reaches the top. The ball color must match the bar color when the ball hits the ground, otherwise the game is over.

For some of these games, you find more detail description in Chap. 11, Sect. 11.2. All games have been designed by students within a semester project competition in 2016 and 2017. Write a brief description of your implementation including Computer System, I/O component used, and game rule. Submit a short video of a game demonstration.

10.56. Write an ANSI C program for the MicroBlaze Basic Computer with VGA or HDMI text monitor CIP that implements a game such as:

(a) Four in a row: Select the row with the slider switch, then use a push button for submitting your choice (see Fig. 10.26a for an example).
(b) Tetris: Let different element move slowly in y direction (see Fig. 10.26b for an example).
(c) Sudoku: Place easy, normal, or hard starting configurations on the board. Then allow with three, entry row, column, and number (1–9), to make your submission (see Fig. 10.26c for an example).

10.57. Run short sequence of ANSI C code to identify the missing instruction in Trisc3mb set for

(a) `if` and `switch` statements
(b) The factorial example

10.58. Add additional MicroBlaze features to the Trisc3mb architecture, i.e., instruction such as

(a) Sign extension sxt8 or 16
(b) Multiply

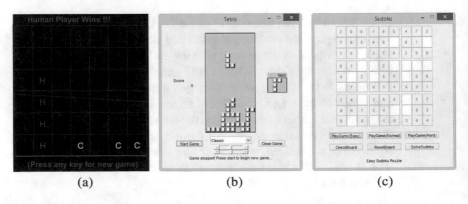

(a)   (b)   (c)

Fig. 10.26 Additional example Games. (**a**) Four in a row. (**b**) Tetris. (**c**) Sudoku

(c) Divide

(d) 32-bit barrel shift

(e) Matching byte pattern

Add the HDL code and test with a short assembler test program. Report the difference in synthesis (area and speed) compared to the original design.

10.59. Consider a floating-point representation with a sign bit, $E = $ 7-bit exponent width, and $M = 10$ bits for the mantissa (not counting the hidden one).

(a) Compute the bias using Eq. (9.2).

(b) Determine the (absolute) largest number that can be represented.

(c) Determine the (absolutely measured) smallest number (not including denormals) that can be represented.

10.60. Use the result from the previous Exercise

(a) Determine the representation of $f_1 = 9.25_{10}$ in this (1,7,10) floating-point format.

(b) Determine the representation of $f_2 = 10.5_{10}$ in this (1,7,10) floating-point format.

(c) Compute $f_1 + f_2$ using floating-point arithmetic.

(d) Compute $f_1 * f_2$ using floating-point arithmetic.

(e) Compute f_1/f_2 using floating-point arithmetic.

10.61. For the IEEE single-precision [I85, I08] format determine the 32-bit representation of:

(a) $f_1 = -0$.

(b) $f_2 = \infty$.

(c) $f_3 = -0.6875_{10}$.

(d) $f_4 = 10.5_{10}$.

(e) $f_5 = 0.1_{10}$.

(f) $f_6 = \pi = 3.141593_{10}$.

(g) $f_7 = \sqrt{3/2} = 0.8660254_{10}$.

10.62. Convert the following 32 bit floating-point number [I85, I08] into a decimal number:

(a) $f_1 = $ 0×44FC6000.

(b) $f_2 = $ 0×49580000.

(c) $f_3 = $ 0×40C40000.

(d) $f_4 = $ 0×C1AA0000.

(e) $f_5 = $ 0×3FB504F3.

(f) $f_6 = $ 0×3EAAAAAB.

10.63. Design a 32 bit floating-point ALU with a select sign sel to choose on the following operations: (0) fix2fp (1) fp2fix (2) add (3) sub (4) mul (5) div (6) reciprocal (7) power-of-2 scale. Test your design using the simulation shown in Fig. 10.27.

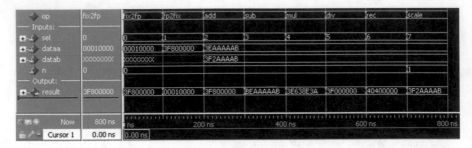

Fig. 10.27 MODELSIM RTL simulation of eight functions of the floating-point arithmetic unit fpu. 0001.0000 in sfixed becomes 3F800000 in hex code for FLOAT32. The arithmetic test data are 1/3 = 0x3EAAAAAB and 2/3 = 0x3F2AAAAA

10.64. Design a TMDS decoder according to Fig. 10.8b. Use the simulation similar to Figs. 10.9 and 10.10 to test your decoder.

10.65. Develop a 7-segment display using the flash program Listing 10.9 to count up in seconds. For ZyBo, you will need to buy for 7 Euro/USD the extra 2xSSD Pmod. This will require to modify the pin assignment that not only LEDs, but also the seven segments are driven by the TRISC3MB output port.

10.66. Write a short MatLab or C/C++ program to generate a new memory initialization file for the HDMI encoder. Check the file eupsd_hex.mif for required format. Use as new text:

(a) Your home country constitution
(b) Your favorite poem
(c) Your favorite joke
(d) Your favorite song text

10.67. Use the HDMI CIP files to build a standalone system not requiring a microprocessor. For the ports use the following assignment: reset=KEY[0]; WE=KEY[3]; DATA = { 4'h3, SW};COUNTER_X=X<<3;COUNTER_Y=Y<<3; LEDR[0]=CLOCK_25; LEDR[1] = CLOCK_250; LEDR[2] = DLY_RST; LEDR[3] = LOCKED;(keep SW[2]=REG; SW[3]=GRN;   SW[1:0]=font;CLOCK_125=sys   clock 125 MHz). Enable the ports in the XDC file. Keep HDMI output names. Use the text from the previous Exercise for display. Add your name in the banner.

10.68. Modify the HDMI CIP to be used as text terminal for the VGA port. If you have DE1 SoC or ZyBo, use the available VGA port; for gen 2 ZyBo buy a VGA Pmod (ca. 9 USD/Euro) (see Fig. 10.6a). You will also need a VGA cable. The VGA uses the same timing for row and column control but analog output signals. You will also need to forward the internal H/V sync signals to the output. The VGA port contains three DAC for this purpose. Study the ZyBo master XDC file for required ZyBo signals. Look for the ##VGA Connector section. You will use the 5 MSBs for red and blue and 6 MSBs for green.

10.69. Use the VGA code from the previous Exercise 10.68 to build a VGA text terminal CIP. Use you MICROBLAZE to test your CIP.

10.70. Write an ANSI C program for the MICROBLAZE Basic Computer that test a component:

    (a) AUDIO-IN: Use an amplitude tracking to determine sags or swells in a power signal

    (b) AUDIO-OUT: Implement an "echo" that changes with SW settings

    (c) Custom Instruction: Implement S-Boxes used in the DES algorithm

    (d) VGA: Implement text terminal using the HDMI CIP

    (e) Wi-Fi: Add WIFI Pmod and try to communicate with other devices

    (f) ACL: Add 3 axis accelerometer Pmod and build a tremor sensor

    (g) BT2: Add Bluetooth Pmod and try to communicate with other devices

    (h) JTEG-UART: Measure file transfer rate from/to the MICROBLAZE

    (i) ACL: Build a tremor sensor using a 3 axis accelerometer

The Project e-i may require to purchase small addition modules that are attached through the ZyBo Pmods connectors (see Table 10.1).

10.71. Develop a `srec2prom` utility that reads a Motorola `*.srec` file (SREC file format, see for instance [A01]) as generated by GCC for SDK. Generate a complete VHDL file that includes to code as constant values, or for Verilog a `*.mif` table that can be loaded with a `$readmemh()`. Test your program with the MICROBLAZE `mb_flash.c` example.

10.72. Identify missing instructions from TRISC3MB set that are needed if you use the `srec2prom` from the previous exercise. Implement the missing instruction in HDL and run the `srec2prom` generated program.

10.73. Develop a TRISC3MB program to implement a multiplication of two 32-bit numbers via repeated additions. Test you program by reading the SW values and computing $q = x * x$.

10.74. Add the missing instruction from the ISA to run the factorial program on the hardware. Test the single new instructions then run the factorial program.

References

[A01]   Altera, Converting .srec Files to .flash Files for Nios Embedded Processor Applications, White Paper (2001)

[D14]   Digilent, *ZYBO Reference Manual* (Pullman, 2014)

[D99]   DDWG, Digital Visual Interface DVI, Initial Specification Release 1.0 (1999) 76 pages

[F05]   B. Fletcher, FPGA Embedded Processors, in Embedded Systems Conference San Francisco, (2005), p. 18

[H19]   HDMI online: https://www.fpga4fun.com/HDMI.html (URL last accessed 8/2019)

[I85]   IEEE, Standard for binary floating-point arithmetic. IEEE Std. **754–1985**, 1–14 (1985)

[I08]   IEEE, Standard for binary floating-point arithmetic. IEEE Std. **754–2008**, 1–70 (2008)

[M14]   U. Meyer-Baese, *Digital Signal Processing with Field Programmable Gate Arrays*, 4rd edn. (Springer, Berlin, 2014), 930 pages

[P10]   V. Pedroni, *Circuit Design and Simulation with VHDL* (The MIT Press, Cambridge, MA, 2010)
[P10]   R. Prinz, PixelFontEdit, Version 2.7.0 at https://www.min.at/prinz/o/software/pixelfont/ 2010 (URL last accessed 8/2019)
[TI07]   Texas Instruments, *HDMI Design Guide*, (2007)
[X05]   Xilinx, MICROBLAZE – The Low-Cost and Flexible Processing Solution, www.xilinx. com (2005)
[X08]   Xilinx, *MICROBLAZE Processor Reference Guide UG081*, San Jose (2008)
[X18]   Xilinx, *MICROBLAZE Processor Reference Guide UG984*, San Jose (2018)
[X460]   Xilinx application note, Video Connectivity Using TMDS I/O in Spartan-3A FPGAs, by Bob Feng and Eric Crabill (2011)
[X495]   Xilinx application note, Implementing a TMDS Video Interface in the Spartan-6 FPGA, by Bob Feng (2010)

Chapter 11
ARM Cortex-A9 Embedded Microprocessor

Abstract This chapter gives an overview of the ARM Cortex-A9 hardcore microprocessor system design, used by both FPGA vendors, Altera and Xilinx, its architecture, and instruction set. It starts with a brief overview followed by ISA architecture and interesting design features.

Keywords ARM Cortex-A9 · ARM v7 · ZyBo · Xilinx · Vivado · TRISC3A · Mandelbrot fractal · ZYNQ · Game design · Custom Intellectual Property (CIP) · Multiplexed input/output (MIO) · GPIO · Top-down-design · Bottom-up-design · Nesting functions · Floating-point performance · Fast Fourier transform FFT · GCC · Bitreverse · ModelSim

11.1 Introduction

The family of ARM embedded microprocessor has become one of the dominant architectures in all embedded systems, providing high performance at low power for many devices ranging from your iPhone to game console. Chip sales with embedded ARM core IP in 2015 exceeded 14B [PH17]! ARM's newest core architectures are the 32-bit ARMv7 and the 64-bit ARMv8. It is interesting to notice that both Xilinx and Altera have introduced new device families that incorporate the same ARMv7 Cortex-A9 dual processor core. Altera's Arria V and Cyclone V devices and Xilinx Zynq-7000 devices include the new A9 dual core [A11a, X18]. The Cortex-A9 versions used by both vendors are almost the same such that most likely the additional features and hard IPs included on the devices may be the key point which vendor we will choose. Xilinx devices have the dual 12-bit 1 MSPS ADC and larger on-chip memory that may allow us to include the boot operating system on chip. Altera has a faster transmitter (up to 100 GBps) and more logic resources, i.e., LE and multipliers with their families.

Now let us have a closer look at the 3-address (Rd <= Rn □ Rm) ARM core that is shown in Fig. 11.1. It has many of the standard features [A11b, A14, X18, CEE14] we expect today from a modern 32-bit µP, such as:

- Dual issue superscalar pipeline with 2.5 DMIPS per MHz
- 800 MHz dual core processor

© Springer Nature Switzerland AG 2021

U. Meyer-Baese, *Embedded Microprocessor System Design using FPGAs*,
https://doi.org/10.1007/978-3-030-50533-2_11

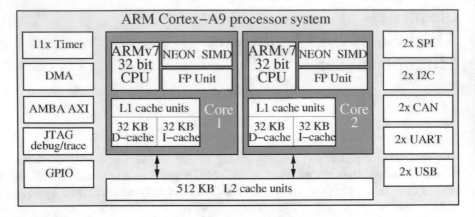

Fig. 11.1 The ARM Cortex-A9 overall architecture [A14]

- 32 KB instruction and 32 KB data L1 4-way set-associative cache
- Shared 512 KB, 8-way associate L2 cache for both processors
- 32-bit timer and watchdog

  And also some additional advanced features such as:

- Dynamic branch prediction
- Out-of-order multi-issue instruction queue with speculation
- Register renaming of 16 architectural to 56 physical registers
- NEON media processing accelerator for 128-bit SIMD processing
- Single and double precision floating-point operation support including square-root
- Thumb-2 technology for code compression
- Configurable 32-, 64-, or 128-bit AMBA AXI interface
- Hardware support for many I/O standards such as CAN, I2C, USB, Ethernet, SPI, and JTAG
- MMU that works with L1 and L2 to ensure coherent data

Floating-point performance is substantial higher than any other processors we have discussed so far; see Fig. 11.2. The operating system support for the ARM A9 is provided by multiple sources. There are open-source tools such as Linux, Android 2.3, and FreeR-TOS. In addition, commercial OS support is available such as WindRiver Linux or VxWorks, iVeia Android, or Xilinx PetaLinux.

11.2   Top-Down ARM System Design

The majority of FPGA boards come with a starting-point design for quick evaluation purpose. Since vendor want to demonstrate all the great features a board has, this starting-point design are often not minimum systems, instead usually a substantial

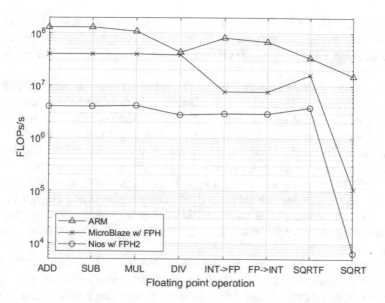

Fig. 11.2 The ARM Cortex-A9 vs. Nios and MICROBLAZE (both with maximum FPH hardware support) floating-point performance for basic operations

number of peripheral components are accessed in the start-up design to make the system more appealing and demonstrate the large number of functions of the board. Since these systems are configured such that all peripheral components are functionally correct, the mistakes of a wrong configuration of a peripheral driver are less likely in the top-down design approach. We simply remove all design components that are not needed in the system target application and/or modify the existing component as needed. The ZYBO board comes with a couple of demo starting-point designs that target a specific task. The three ZYBO demos that use the ARM SDK are:

- The HDMI-OUT demo uses four features: the USB-UART bridge, the HDMI Sink/Source port, the 16-bit VGA, and the DDR. Artificial images are produced and displayed over HDMI and VGA on a monitor. Three frame buffers are used. A menu over the UART allows to choose between eight display options.
- The HDMI-IN demo is similar to the HDMI-OUT and uses four features: the USB-UART bridge, the HDMI Sink/Source port, the 16-bit VGA, and the DDR. The HDMI video input is displayed over HDMI and VGA on a monitor. Again, a menu over the UART allows to choose between eight display options.
- The DMA-Audio uses the following four features: the USB-UART bridge, five push buttons, the audio CODEC, and the DDR. The demo records 5 seconds from a microphone signal and plays it back through a headphone.

As we notice the ZYBO demos do not use a full-featured multimedia computer; in particular, the following features are not in use: four switches, four user LEDs, MicroSD, User EEPROM, 10/100/1000 Ethernet, serial flash, five PMOD, or USB HID host [D14].

The TerASIC/Altera DE1 SoC board [T14] has a more complete starting-point design: A "DE1 SoC Computer" is a full-featured "multimedia computer" design that includes the ARM and two Nios II processors, the majority of I/Os (SDRAM, LEDs, 7-segment, switches, buttons, PS2, 4 × JTAG, IrDA, 4 × timer, ADC, audio, VGA, video). Three AXI bridges are used such that ARM and Nios II both can access the majority of I/O components: a 64-bit FPGA→HPS interface, a 128-bit HPS → FPGA interface, and a lightweight 32-bit HPS → FPGA interface are used as AXI bridges. Missing are only the I^{2}C FPGA peripheral and Gigabit Ethernet, MicroSD, and two-port USB on the HPS/ARM side. Nevertheless, the system is large and the possible applications huge; see also Fig. 9.2 in Chap. 9 for the included IPs [A15a].

Today's students differ from older generations in several respects [VTT09]. The top-down design approach is an excellent match for today's student. By *tinkering* and playing with the board, they gain a better understanding how software and hardware interact with each other. Since todays student are *impatient*, the fast development in one afternoon of a game or video is a good fit. Lastly students think *software is everything*. As part of the smartphone generation, students relate to computer screens extremely well, but this oftentimes results in the illusion that everything can be done by pressing a few keys and that somewhere else somebody else will take care of designing and building the hardware. This thinking is particular dangerous in a resource-limited embedded system design where a single C++ catch can add more than 2500 instructions to the code size.

To demonstrate the use of the system, two typical student projects seemed most beneficial: a video project would involve storing a sequence of precomputed images in memory and then to display these images in a rapid sequence. Computation time of such images may be substantially reduced if we take advantage of the high floating-point performance of the ARM when compared to the Nios II or MICROBLAZE; see Fig. 11.2. A switch between the softcore and hardcore can be easily achieved when using ANSI C code (no HAL functions) on a DE board: we only need to substitute the header-file "address_map_nios2.h" from the Nios II system with the header file "address_map_arm.h" for the ARM and recompile the ANSI C code, and the system should work fine without the need to make any other code changes. This is demonstrated with the example projects CLOCK, FLASH, FRACTAL_COLOR, FRACTAL, GREY, MOVIE_COLOR, and MOVIE_GREY located in the ARM/DE1_SoC_TopDown/app_C folder on the CD that was originally developed for the Nios II processor but work with the header file substitution also for the ARM processor. In the past student projects that included fractal videos or illusions (e.g., Lilac Chaser) were preferred. Let us briefly describe how we may get started – the fully developed examples are included in the book CD ROM files.

The DE default image has 16-bit color 320 × 240 QVGA size display. The pixel buffer DMA controller uses the X/Y addressing mode, i.e., each pixel can easily address with

```
pixel_ptr = FPGA_ONCHIP_BASE + (row << 10) + (col << 1);
*(short *) pixel_ptr =  pixel_color;
```

with the color coded as 5,6,5-bit RGB short 16-bit integer. In the fractal movie we will select the color based on the number of iterations it takes for $z = z.*z + c$ to reach $\sqrt{5}$. The core of the fractal movie generation is a zoom version of the original fractal stored in a reasonable number of frames, e.g., 25 in our example implementation; see Fig. 11.3. After the initial computation we can reuse the frame one after the other to give the impression of a zoom-in or zoom-out video that shows the self-similarity of the fractal which is the most stunning property of all these beautiful fractals. Here is the core of the ANSI C code; see also video: `MandelbrotZoom.MOV` on book CD.

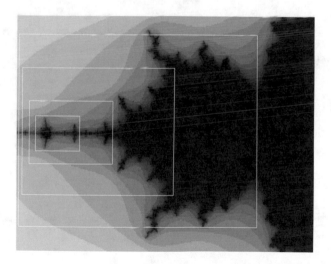

Fig. 11.3 The zoom scale for a fractal Mandelbrot movie

ANSI C Code 11.1: The Core of the Mandelbrot Video

```
1    …
2    for (r=0;r<240;r++) {
3      for (s=0;s<320;s++) {
4      //********** Mandelbrot ************
5      cr=(float) d*(s/320.0*2.5-2.2-x1)+x1; // x has range -1.5 .. 1
6      ci=(float) d*(1.5-r/240.0*3-y1)+y1; // y has range 1.5 .. -1.5
7      nr=0; ni=0;
8      k=0;sq=0;
9      while ((k<iterations) && (sq < 5)) {
10       k=k+2;
11       //z=z.*z + c;
12       tr = nr; ti=ni;
13       nr = tr*tr - ti*ti + cr;
14       ni = 2*tr*ti + ci;
15       // a=find(abs(z)>sqrt(5));
16       sq = nr*nr + ni*ni;
17     }
18     //map(r,s)=k*4;
19     g = 30-k; // Use dark color for small iteration k
20     pixel_color = g; // blue
21     if (g<=0) pixel_color = 0xF800; // red
22     if ((g>=2) & (g<=8)) pixel_color = 0x07FF; // light blue
23     pixel_ptr = FPGA_ONCHIP_BASE + (r << 10) + (s << 1);
24     *(short *) pixel_ptr =  pixel_color;
25     img[iframe][r][s] = pixel_color;
26     }
27   }
28   …
29     iframe=0;direction=1;frame=0;
30     while(1) {
31     /* output text message in the middle of the VGA monitor */
32     VGA_text (35, 30, " Fractal Video  \0");
33     sprintf(text_top_row0, "%2d: Zoom            \0",iframe);
34     VGA_text (35, 29, text_top_row0);
35     for (r=0;r<240;r++)  // copy the image to image buffer
36       for (s=0;s<320;s++) {
37         pixel_ptr = FPGA_ONCHIP_BASE + (r << 10) + (s << 1);
38         *(short *) pixel_ptr = img[iframe][r][s];
39         }
40     iframe += direction;
41     if (iframe>FMAX) {direction=-1;iframe=FMAX;}
42     if (iframe<0) {direction=1; iframe=0;}
43     sw = *(sw_ptr); // use switch value to change display speed
44     wait(sw); // Wait statement
45     }
```

Another great path to become familiar with the boards using the top-down approach and highly appreciated by many students is the design of a game with the embedded system. Most games can already be designed without any additional hardware. For some games additional component such as a PS2 mouse, joystick, or keyboard maybe useful. Let us have in the following a brief look at games students have designed during an in-class competition in 2016 and 2017. Winner was selected by double-blind voting. Here is a brief description of the games that all used the DE2 Media Computer to implement the game:

- Tic-Tac-Toe (by Christopher Ritchie, Rohit Sonavane, Shiva Indranti, Xuanchen Xiang, Qinggele Yu, Huanyu Zang 2016 game 3/6/8; 2017 game 9/10/11): Use two players. Enter selecting with slider switches and the push button to submit your selection for player 1 (button 3) and player 2 (button 0). For a DE2 you can use slider switches 0–8 for player 1 and 9–17 slider switches for player 2.
- Reaction time (by Mike Pelletier 2016 game 4): Light up one of three squares on the screen and measure the time until the correct button was pressed.
- Minesweeper (Sharath Satya 2016 game 7): In a 3×3 field, three bombs are hidden. You need to find the six boxes that do not contain a bomb. If you find all six, you win; otherwise, player loses.
- Pong (by Samuel Reich and Chris Raices 2017 game 4/5): Use one player only. Use one direction to move the ball and the pushbutton for moving the paddle on the other axis. Set a timer to count how long the game runs until the player misses the ball, which may be a small square.
- Simon Says (by Melanie Gonzalez and Kiernan Farmer 2016 game 1/2): Repeat the three-color sequence of HDMI/VGA square with the push buttons 1, 2, and 3.
- Box Chaser (by Ryan Stepp 2017 game 7): A big box moves randomly over the display, and you need to avoid getting your small box hit by this big box by moving it to the left or right with the push buttons.
- Space Invades (by Zlatko Sokolikj 2017 game 6): You have a flying object that needs to be hit with your firing gun shooting in y direction. Your gun moves on the x axis, so a total of three buttons are in use. Counting the target hit within a time frame gives the score.
- Flappy Bird (by Javier Matos 2017 game 3): Pressing a button moves your bird in y direction, while it continuously drops. Obstacles such as walls with a hole move from right to left. Game is over when the bird hits the wall or ground.
- Bounce (by Kevin Powell) 2016 game 5): A three-color bar is moved by two push buttons to the left or right. A ball bounces in y direction and changes color whenever the ball reaches the top. The ball color must match the bar color when the ball hits the ground; otherwise, the game is over.

Winner in 2017 was Tic-Tac-Toe and 2016 was the Bounce game by double-blind voting. Let us briefly review the Tic-Tac-Toe game since many other games can be developed in a similar way.

For the Tic-Tac-Toe, we need to decide if we like to have two players or one player against a computer. For two players, we would need 18 switches as on the DE2–115, or for the DE1 SoC, use a keyboard as input device. We may also use the switches and two additional push buttons to "submit" our choice alternating. Then we may also use the ZyBo to implement the game if the nine fields are binarily enumerated. For one player, we would need nine switches as provided by the DE1 SoC. Let us discuss the one-player version that we can use the DE1 SoC without any additional components. As for the games, we would run in an infinity loop with the following algorithm:

- Draw nine alternating color boxes.
- Read the SW values, and display the human player choice using O.
- For the computer choices, select the same number of fields by random choice, and place X such that we do not have twice the same field selected.
- For a simple strategy, we would use a random location and increment until we find an empty field. For a more sophisticated strategy, you may prefer corner place first, and/or see if the human has two in a row, we need to intervene such that not three can be selected in the next move.
- We would then draw the selection by the human and the computer on the monitor.
- We will update the text message if we have a winner, i.e., three in a row; otherwise, continue from top.

Figure 11.4 shows a possible outcome. We can code the game in ANSI C, or we may take advantage of the HAL functions to clear display and text buffer and draw a box or text.

11.3   Bottom-Up ARM System Design

Putting together an ARM-based FPGA system is not a trivial task and require a substantial amount of experience both with the design tools as well as the ARM microprocessor itself. Let us have a look at the major steps putting together a Basic Zynq ARM Computer system. A bottom-up approach for the DE1 SoC ARM-based system seemed to be substantially more complicated due to missing board package

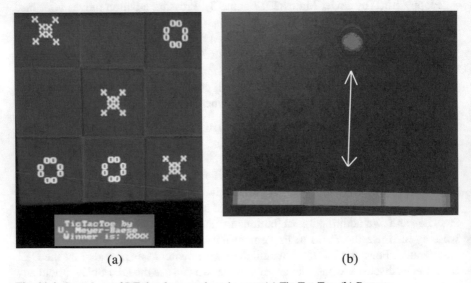

(a)   (b)

Fig. 11.4 Snapshots of DE development board game. (**a**) Tic-Tac-Toe. (**b**) Bounce

files available in VIVADO. There are two features in the VIVADO design that seemed to make the design a little easier than using QSYS.[1] One is the "block automation" of system components and "autocompletion" of the wiring in the graphic design process, and the second is the automatic generation of test programs after adding individual components. However, it still remains a substantial challenge to get a first ARM/DDRAM system to work. The reason is that the ARM requires a substantial number of information how the processor is embedded in the FPGA board. As you may recall from Chap. 2, the ARM uses a multiplexed I/O (MIO) to select between the many communication formats/IP blocks available on-chip; see Fig. 11.1. Therefore, it is essential to know how the I/O of the ARM were indeed wired in order to make the correct connections to the I/O components on the FPGA board. You may follow online tutorial such as "Getting Started with Zynq" from the Digilent Inc. web page to get you started: start *VIVADO*, and then click *Create New Project*. Choose your desired project location, and then click *Next*. Use *RTL Project type*, and click *Next* a few times until you can select *Parts and Boards*. Select the ZYBO board to get the correct I/O locations, MIO setting, and dynamic DDR timing. Make sure you download from `GitHub` or `digilentinc.com` the `master.zip` file and installed the board files in the VIVADO path under `...data/ boards/board_files`.

Now start *IP Integrator* → *Create Block Design*, and click the ⬚ button such that you can place a new library component of the *ZYNQ7 Processing System* in the block design. Tip: Use the search for Z..., then click on the library component with your left mouse button one time, and then use the *Enter* on your keyboard. On the top of your Diagram you should see *Designer Assistance available*; click the blue Run Block Automation link. VIVADO will then add the DDR and FIXED_IO ports; see Fig. 11.5b. If you have a predesigned system, you can use *IP Integrator* → *Open Block Design* of an example project, and then double click the ZYNQ7 processing systems; you will first see an overview of the processor. The left *Page Navigator* panel lets us select a configuration panel. You will notice that the just created (even with board specification) and the example design may not match. A substantial work will be needed to actually get to a working ZYNQ7 system with correct DDR interface if you have not previously installed the board files. A correct ZYBO configuration of you ZYNQ7 will have all the peripherals enable as shown in Fig. 11.6 indicated with a checkmark $\sqrt{}$.

Your default setting for MIO will only match the board if you have installed and selected the correct board files when you had set up the RTL project. Otherwise there will be no checkmarks. You need to follow the ZYBO board description carefully to make the right assignments in the *Peripheral I/O Pins panel*. You may configure all available components; however, if you do not use the Ethernet IP, you may want to turn this off since in the automatic generated test routines, you will need to implement a hardware Ethernet loop to get the test program working correctly. You need the correct setting for the MIO 500 port group for SPI Flash and LED4. For

---

[1] QSYS has recently been renamed by Intel to PLATFORM DESIGNER.

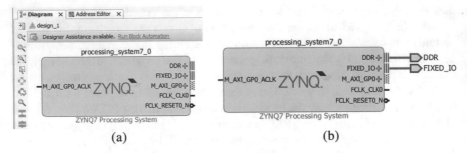

Fig. 11.5 The ZYNQ7 initial Vivado Diagram instantiation. (**a**) ZYNQ7 instantiation. (**b**) ZYNQ7 after block Automation run

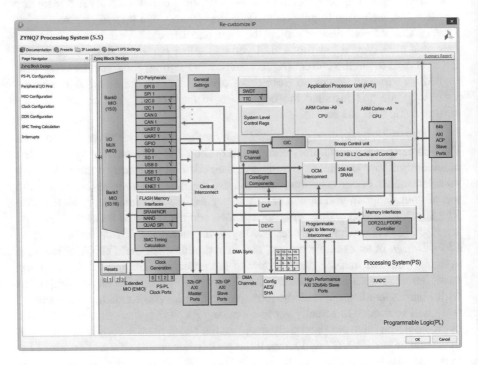

Fig. 11.6 The ARM Cortex-A9 configuration panel overview

MIO 501 port group, you need to configure Ethernet 0, USB 0, SDIO, UART1, and GPIO for button 4 and 5. Figure 11.7 shows the I/O pin assignments for all components of the ZYBO. If the configuration of the ARM and the DDR seemed too much for your first bottom-up design, you may also consider using an ARM example project (such as the DMA or HDMI we discussed in Sect. 11.2) you have tested with your board and removing all blocks and pins except the ZYNQ7 Processing System and the DDR connector.

The second nontrivial component we need to add is the DDR RAM. Here also a substantial number of parameter (panel 6: *DDR Configuration*) need to be provided.

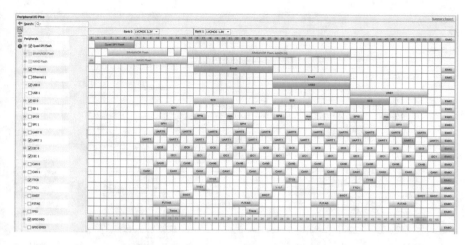

Fig. 11.7 The ARM Cortex-A9 pin connection for MIO components

Default setting for the RAM will most likely not match your ZYBO board requirements. Selecting the correct part, MT41K128N16 JT-125, should be no problem using the ZYBO user guide. Setting the correct delays such as *DQS to Clock Delay* will require detailed knowledge not even given in the ZYBO Reference Manual [D14]. You may want to have a demo project open as a reference for this data or use the board files.

After you have successfully configured your ARM and DDR memory interface, you may want to add to your Basic Computer some simple I/O such as switches, LEDs, buttons, and a timer. Right click on in the *Block Diagram* and select ⬚, or right click and select *Add IP...* look for *AXI GPIO* and click return. Repeat these two more times that you have three *AXI GPIO*. Double click the AXI GPIO, and set *Board Interface* to the three I/O we like to use: *leds_4bits*, *sws_4bits*, and *btns_4bits*. Then click *OK* to close the AXI GPIO window. Use one more time the *Add IP...* to look for the *AXI Timer*. Keep the default setting for the *AXI Timer*. You should see at the top of the block diagram that designer assistance is available. Click the <u>Run Connection Automation</u> underlined and highlighted in blue. The processor system reset, AXI switch, and I/O port will be added, and all connections are routed. Also, inside the IP an update of the *IP Configuration* within the *AXI GPIOs* takes place. You may then rename the *AXI GPIOs* to reflect the ports (e.g., led, sw, btn) and finally use *Validate Design* ⬚ to check for connection errors and the ⬚ button, or right click and select *Regenerate Layout* to have a nice aligned block diagram of your design. Figure 11.8 shows the final layout of the block diagram. The ARM and reset logic are to the left and the GPIO blocks and DDR port on the right.

The Basic Computer design did not need much resource since the hard-core ARM is used as microprocessor. For the GPIOs, a total of 928 LUTs were used with 62 as memory. No block RAM or embedded multiplier is needed. The ARM CPU runs at 650 MHz, DDR memory at 525 MHz, and GPIO at 125 MHz. Timing is not critical since a 1.712 slack was reported.

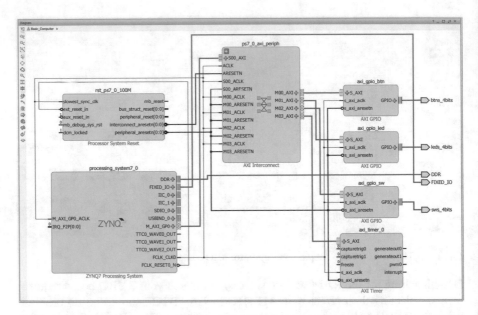

Fig. 11.8 The ARM Cortex-A9 Basic Computer design with added timer and GPIO

Now we are ready to translate our design into a bit-stream we can download to the FPGA. To compile the graphic design, we need to build a HDL wrapper first. In the *Block Design* panel, right click on `design_1.bd`, select *Create HDL Wrapper* ..., and choose the VIVADO *auto update* option. Now under the *Program and Debug* panel, select *Generate Bitstream*. This may take some time to complete.

It is now time to start with the program development. SDK is more or less a standalone tool, and we use *File → Export → Export Hardware...* to get started. Do not forget to select the *Include bitstream* option before export, and then click *OK*. Next select *File → Launch SDK* and click *OK*. Since we have new components, it seemed to be a good idea to take advantage of the SDK feature to generate a peripheral test program for us that tests all components, but only the components we have added to the system. Select *File → New → Application Project*, and choose a *Project name* such as `test_IO` click *OK* and then as *Templates* choose *Peripheral Tests*. You then run the test and you will see success message in the Terminal window for each block and a very short-running light on the four LEDs next to the slider switches of the ZYBO board. SDK will produce a project folder and a second with ..._*bsp* that includes system files and ANSI C service routines. Have a look at the main routine `test_IO/src/testperiph.c` and the associate test programs in the same folder. These contain many useful functions you may need in your program development.

For example, for each of the `XGpio` component, SDK provides an *Initialize* function and a *DataDirection* function. For the `leds`, we would, for instance, use

```
XGpio_Initialize(&leds, XPAR_AXI_GPIO_LED_DEVICE_ID);
XGpio_SetDataDirection(&leds, 1, OUTPUT_DIR);
```

The `DEVICE_IDs` and name we can look up in `xparameters.h` in the `bsp` include files. We can then display an integer `count` at the LEDs values using

```
XGpio_DiscreteWrite(&leds, 1, count);
```

As you see each GPIO block usually requires a minimum of three functional calls: `Initialize`, `direction`, and then `Write` or `Read` calls.

On the ARM we now have two timers working. The processor timer `XScuTimer` and the `XGpioTimer`. Here are some functions useful to set up the internal timer for measurements: `ScuTimer_LookupConfig()`, `XScuTimer_CfgInitialize()`, `XScuTimer_LoadTimer()`, and `XScuTimer_Start()`. Then each instance a timer interval needs to be measured, we would use `XScuTimer_GetCounterValue()` before and after the code segment to be measured and take the difference.

For our new AXI Timer, we would use initially `XTmrCtr_Initialize()`, `XTmrCtr_SelfTest()`, `XTmrCtr_SetOptions()`, `XTmrCtr_SetResetValue()`, `XTmrCtr_Reset`, and `XTmrCtr_Start()`. For our measurements we would make call before and after the code segment to `XTmrCtr_GetValue()`. The advantage of using the AXI Timer is that we can use 99% the same code for our measurement with ARM or MicroBlaze, eliminating any error that may occur when using the internal ARM vs. the AXI Timer. A typical measurement for the floating-point test (results were presented in Fig. 11.2) would now look as the listing below.

ANSI C Code 11.2: The Floating-Point Performance Measurement

```
1    …
2    //start timer
3    Value1 = XTmrCtr_GetValue(TmrCtrInstancePtr, TmrCtrNumber);
4    for(i = 0; i < 1000000; i++){
5      //conversion
6      result = (float) i;
7    }
8    //stop timer
9    Value2 = XTmrCtr_GetValue(TmrCtrInstancePtr, TmrCtrNumber);
10   //TODO get timer value
11   Value21 = abs(Value1 - Value2);
12   //Elapsed cycles for 1000.0000 single precision conversion
13   xil_printf("INT->FP 1000K tics: %d\r\n", (int) Value21);
14   xil_printf("INT->FP cycles: %d\r\n", (int) (Value21/1000000));
15   …
```

Results will be displayed in the terminal window (see below) of the ARM since the UART we used for programming can also be used in the SDK ARM configuration as UART terminal. No need for a Pmod UART and the additional USB cable as in the MicroBlaze design in Chap. 10, Fig. 10.6.

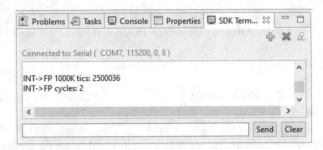

11.4   Custom Instruction ARM System Design

Modern highly sophisticated embedded processors such as the ARM processor have hardware support for several standards with custom hardware on-chip, and our ARM has two each of the SPI, I²C, CAN, UART, USB, and Ethernet, and we would not need to design custom logic for these I/O standards. But there are two instances when a FPGA-based integration of custom logic closely coupled with the micropro-cessor aka *custom intellectual property* (CIP) are beneficial. Think, for instance, of the large amount of data a video display such as VGA or HDMI continuously need to be provided. In the lowest resolution, a HDMI controller needs to deliverer 250 Mb/s. Even the fastest microprocessor will be overburden, and substantial per-formance decrease is expected when the microprocessor has to pick up these data from memory and forward it to the HDMI ports or VGA DACs. Such a HDMI CIP system was discussed in detail in Chap. 10 for the MICROBLAZE. A second, maybe less likely, case may happen if the algorithm requirements are not a good fit for our microprocessor. Such examples are switch boxes in cryptographic algorithms or the bit-reverse required in DCT or FFT algorithms [S04, MSC06]. Let us in the follow-ing have a closer look at the design of a bit-reverse CIP for the ARM and see if and how much performance improvement can be achieved.

Before we start with the details of the CIP design, let us briefly review the bit-reverse operations, typical software implementation, and HDL source code needed.

Bit-Reverse Application Example and Software Implementation

Digital signal processing (DSP) became popular in the 1980 mainly because of two applications: digital filtering and the Fast Fourier transform (FFT). The FFT is a fast version of the standard discrete time Fourier transform that allows a signal represen-tation by its (periodic) sin/cosine components. The standard equation of the DFT

$$X(k) = \sum_{n=0}^{N} x(n) e^{-j2\pi nk/N} \tag{11.1}$$

Can be rearranged in such a way that the number of complex multiply operations is reduced from N^2 to $N*\log_2(N)$ operations with N being the number of points in the transform. While that seemed not much, think of a simple example with $N = 1000$-point transform. Standard DFT would require $1000^2 = 1,000,000$ multiply, while the FFT only need $1000 \times \log_2(1000) = 1000 \times 10 = 10,000$, or a factor 100 less operations! Now the price to pay for the so-called radix-2 Cooley-Tukey FFT (name after their two (re-)inventors, which was actually Gauss 200 years earlier but did not publish it) is a permutation in the order of the output sequence. Luckily the output sequence is not completely random but appears in so-called bit-reverse order; see Fig. 11.9 on the right. If you write down the index in binary form, then all bit locations need to be switched. Let us assume we have an 8-point transform, and then the index range is 0...7, for bit-reverse we go, for instance, from binary $110 \rightarrow 011$. Memory location 6 and 3 need to swap places. Here is a complete list for all eight indices in binary:

Original	000	001	010	011	100	101	110	111
Reverse	000	100	010	110	001	101	011	111

Figure 11.9 shows the complete signal flow graph to compute the length-8 FFT. Note the reverse order of the output sequence $X[k]$ with input sequence $x[n]$ in natural order.

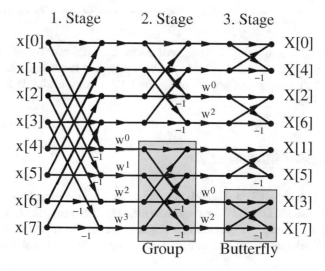

Fig. 11.9 The signal flow graph length-8 DIF FFT. The output data $X[k]$ appear in bit-reverse order [M14]. (The arrow indicates complex multiplication by $w = e^{j2\pi/N}$, and a dot with two inputs is a complex adder)

In a software test bench, we would use hex display for long strings. To simplify the verification, we will use multiple of length 4, i.e., 4, 8, 16, ..., 32 bit strings. The bit-reverse operations will then be applied to each half-byte separately, and the order of the hex digit will also switch, e.g., if the input string is 32-bit long, then we first build the bit-reverse of all half bytes and then reverse the order, e.g.,

Original 32-bit string in hex	1	2	3	4	5	6	7	8
Reverse	1	E	6	A	2	C	4	8

Since $1_{16} \rightarrow 0001_2 \rightarrow 1000_2 \rightarrow 8_{16}$, $2_{16} \rightarrow 0010_2 \rightarrow 0100_2 \rightarrow 4_{16}$, etc. Unfortunately, this bit-reverse operation is not very fast to run in software. We need first to separate each single bit, shift it to the new locations, and add it to the output word. Here is a possible solution in ANSI C.

ANSI C Code 11.4: The Bit-Reverse in Software

```
1   //=====================================================================
2   int SW_BITSWAP(int a)
3   { int lsb, k, r=0;
4     int t=a;
5     for (k=0; k<BITS; k++)
6     {
7         lsb = t & 1; // take LSB
8         r = r*2 + lsb; // add lsb and shift left
9         t >>= 1; // shift to right by one bit
10    }
11    return(r);
12  }
13  // =====================================================================
```

While the code is not very long, it will take substantial number of clock cycles since it is run for every index. Now let us have a look at hardware solution for the bit-reverse aka bit-swap and how to make the HDL design available to the microprocessor as CIP.

HDL Design of Bit-Swap CIP

The process to build a CIP for the bit-reverse operation is similar to the CIP procedure we have used in Chap. 10 for the MICROBLAZE HDMI interface. Another good tutorial on the procedure is provided by Digilent Inc. in the "Creating a Custom IP core suing IP Integrator" tutorial online. Here is the outline of the steps:

1. Create a New RTL project first. Alternatively you can continue using the Basic Computer system from the last section or use the HDMI-out or DMA as starting template. If you use HDMI-out or DMA, remove everything except the ZYNQ block and DDR port.
2. Go to *Tools → Create and Package IP*, click *Next*, and then select *Create a new AXI4 peripheral* and click *Next*. Under *Peripheral Detail* choose a name such as MY_SWAP. Start with MY... as such you can easily find your component later in the IP library; also check the box for *Overwrite existing*. As *AXI interface Type*, choose the default *Lite* and use Data Width 32. Click *Next*, and under the *Create Peripheral* panel, choose *Edit IP* and click *Finish*.
3. A completely new copy of VIVADO will pop up.
4. Expand Design Sources MY_SWAP_core_v1_0, and double click MY_SWAP_Core_v1_0_S00_AXI_inst, or right click it and select *Open*. There are three locations we typically need to modify the HDL file:

   - Add user parameter (top of file)
   - Add user ports (just after parameter)
   - Add user logic (cnd of file)

```
                   // Users to add parameters here
               parameter  integer BITS = 32,
                   // User parameters ends
       ...

                   // Users to add ports here
               // Nothing for SWAP here
                   // User ports ends
       ...

       // Add user logic here
           integer k;  // loop variable
           reg [31:0] result = 0;
           always @(posedge S_AXI_ACLK )
            begin
              for (k=0; k<BITS; k=k+1) begin
                result[k] = slv_reg0[BITS-k-1];
             end
            end
       // User logic ends
```

5. Now modify the wrapper file MY_SWAP_v1_0.v. There are again three locations we typically need to modify the HDL file:

   - Add user default parameter
   - Add mapping of user parameter
   - Add mapping for user ports (we do not have)

```
            // Users to add parameters here
            parameter integer BITS = 4,
            // User parameters ends
     ....

            .BITS(BITS)

     ...
```

6. Now the core design is complete, and we need to package the core that we can reuse the CIP in our designs. Click on the *Package IP – MY_SWAP* tap, and make sure that for selection *Compatibility*, the *zynq* is available. Now from the left panel, select *Customization Parameters*, and `BITS` with the default value 4 should be listed if you work on a revision. If not, then select *Customization GUI*. Right click **BITS** and select *Edit Parameter....* Check the box next to *Visible in Customization GUI*. Check the *Specify Range* box. Select *Range of Integers* from the *Type* drop-down menu. We have a max value of 32 and a min value of 0. Click *OK*. Finally drag the `BITS` into *Page 0* to get it into the list of *Customization Parameters*. Now click on *Review and Package* panel, and select the *Re-Package IP* button. Answer *YES* to allow closing the project. The second copy of Vivado will disappear.

7. Now we can integrate *MY_SWAP* into our Zynq system. It is highly recommended to use the Basic Computer from the last section or another working design such as HDMI-out tutor to avoid any MIO or DDR configuration problems. If you work on a revision and your IP block is looked, you need to run the *Upgrade IP* first. If it is still looked, then you may try to close Vivado and then start again. You should now be able to instantiate a copy of our new CIP from the IP library via 💬, parameterize using the GUI, and run <u>Connection Automation</u>. We do not need to create ports since we just use the local AXI bus (Fig. 11.10).

8. It is now time to start with the program development. In Vivado use *File → Export → Export Hardware...* to get started. Do not forget to select the *Include bitstream* option before export, and then click *OK*. Next select *File → Launch SDK* and click *OK*. Since we have new component, it seemed to be a good idea to take advantage of the SDK feature to generate a peripheral test program for us that tests all components. Select *File → New → Application Project*, and choose a *Project name* such as `Test_MY_SWAP`, click *OK*, and then as *Templates* choose *Peripheral Tests*. When you then run the test, you will see success message in the Terminal window for each peripheral and a very short-running light on the four LEDs of the ZYBO board. We then modify the main routine `Test_MY_SWAP/src/testperiph.c` to measure the swap performance.

Now with both hardware and software bit-swap working, let us do some comparisons in performance.

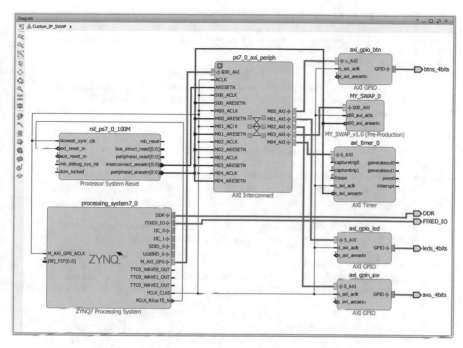

Fig. 11.10 CIP overall system in VIVADO. The new CIP is in the upper right corner

HW/SW Solution Performance Comparisons

Measuring the time for SWAP will require to measure a good number of swaps. We use 1 M test and report the time for the timer before and after the loop. The quotient of the time will give the speed-up factor of our CIP. Communication with the CIP is done via the `Xil_In32()` and `Xil_Out32()` function calls. The CIP bus naming and address definitions can be found in the `xparameters.h` file in the ..._bsp/ps7_cotex9_0/include folder.

ANSI C Code 11.5: Measurement of SWAP Performance with XScuTimer

```
1   #define LOOPS 1000000
2   #define BITS 32
3   ...
4   xil_printf("*** Measure the Software BITSWAP operation ...\n");
5   start_time = XScuTimer_GetCounterValue(TimerInstancePtr);
6   for (k=1;k<LOOPS;k++){
7       sv = SW_BITSWAP(sv);
8   }
9   finish_time = XScuTimer_GetCounterValue(TimerInstancePtr);
10  total_time1 = start_time - finish_time; // count down to zero
11  printf("SW_BITSWAP %d cycle %d ticks\n",LOOPS,total_time1);
12  printf("SW_BITSWAP %d time   %d/1000 ms\n",LOOPS, (int)
13  total_time1/325);
14
15  xil_printf("*** Measure the Custom IP BITSWAP operation ...\n");
16  start_time = XScuTimer_GetCounterValue(TimerInstancePtr);
17  swap = sv;
18  for (k=1;k<LOOPS;k++){
19      Xil_Out32(XPAR_MY_SWAP_0_S00_AXI_BASEADDR, swap); // write to CIP
20      swap = Xil_In32(XPAR_MY_SWAP_0_S00_AXI_BASEADDR + 4); // read CIP
21  }
22  finish_time = XScuTimer_GetCounterValue(TimerInstancePtr);
23  total_time2 = start_time - finish_time; // count down to zero
24  printf("CIP_BITSWAP %d cycle %d ticks\n",LOOPS,total_time2);
25  printf("CIP_BITSWAP %d time   %d/1000 ms\n",LOOPS, (int)
                                                  total_time2/325);
26  printf("CIP_BITSWAP speedup = %d (int)\n",(int) (total_time1*1.0
                                                  /total_time2));
27  printf("CIP_BITSWAP speedup = %d percent\n",(int) (100.0*total_time1
    /total_time2))
```

The measurement results are graphically represented in Fig. 11.11. For 32-bit we get a factor of 2 improvement, i.e., 100% faster. For 8-bit the software solution was faster, since the ARM runs with 650 MHz, while our AXI component only had a

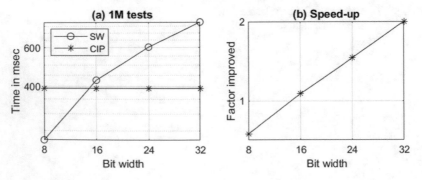

Fig. 11.11 CIP performance. (**a**) Time for 1 M swap operations. (**b**) Quotient time software/CIP time

125 MHz bus speed. We may try to synthesize a little higher speed since VIVADO reported 1.712 ns slack, but in general we should not except a substantial high speed for the CIP.

We may also want to know what the CIP will cost us in terms for additional logic resources. That would be mainly the AXI interface and the one extra port at the AXI mux, since the bit reverse in hardware is just a wiring not requiring much logic. Overall the number of LUT increases from 928 for the Basic Computer to 1021 for the Basic Computer with CIP.

11.5   Looking Under the Hood: ARMv7 Instruction Set Architecture

The DMIPS/MHz performance of the ARM processor is substantially higher than the MICROBLAZE or Nios II. Part of the advantages is obvious based on superior architecture features such as register renaming, dynamic branch predictions, virtual to physical register pool, or level-1 and level-2 cache with snoop control to solve memory conflicts. So, the question now is if the instructions set also have special features not seen in other microprocessors. To this end we should take a look at the generated assembler code to learn why the performance of the ARM is so much higher. To make this investigation a bit more fun, let us design our own tiny ARM that implements a reduced set of the 32-bit ARMv7 Instruction Set Architecture (ISA) and design a microprocessor (we will call TRISC3A) that allows us to run our basic flash program, global and local memory access, and a nested loop example. As we saw at the begin of this chapter, the overall ARM architecture is substantially more complex than anything we have seen before, with many I/O standards built-in hardware, L1 and L2 cache, dynamic DDR memory controller, etc. This high complexity also continues in the ISA. The ARMv7 ISA used by our Cortex-A9 processor is outlined in the 2736-page manual [A14]! Let us have a brief look at an example for basic add operation. While typical three-address machine may support instructions like

```
add r1, r2, r3
```

or using immediate

```
add r1, r2, #1
```

The ARM ISA also allows shift of the second operand before adding

```
add r1, r2, r3, lsl #3
```

or arithmetic shift or left/right shift, or rotate of the second operand as specified in a fourth register

```
add r1, r2, r3, ror r4
```

Another interesting instruction is the load/store multiple instructions: PUSH aka STM and POP aka LDM of a whole selection of registers. These instructions are very useful in subroutines and by servicing interrupts. Using a single instruction, any of the 16 registers can be stored into memory specified by the stack pointer or reloaded from memory.

Besides more sophisticated instructions, the ARM has over 9 operation modes and over 100 levels of interrupt control for the software and any I/O units and supports not one but actually 4 complete different instruction sets: the standard 32-bit ARM, the 16-bit Thumb, a 32-bit Thumb, and a 8-bit Java code. Switching between these sets is accomplished by a single BX instruction. While 16-bit Thumb and 8-bit Java code may reduce program size, high performance is usually achieved using the standard 32-bit ISA, and we will focus on the 32-bit ISA. Again, we will see that the ISA is much more complex than anything we have seen before. Some instructions use 12 flags to precisely specify the flavor of the instruction. Besides some exception, we can roughly put the ISA in four major groups specified by bits 27 and 26:

- 00 specifies data processing and miscellaneous instruction
- 01 contains load/store instructions
- 10 branch (with link) type instructions and block register transfer
- 11 coprocessor instructions

We will concentrate on the first three since these are most often used by GCC. Compared with other ISA, we can see two major differences in the ARMv7 format, one in registers and the other in condition codes. The programmer sees 16 registers and not 32 as in most other 32-bit processors. Three of these have special responsibilities: r13 is the stack pointer (sp), r14 is link register (lr), and r15 is program counter (pc), which enables pc relative addressing but also allows accidental overwriting of the pc. The 16 registers are coded in 4 bit which makes reading the hex assembler code also a little easier, but the reduced number of registers makes it harder for the ANSI C compiler to generate good code. Another difference is that r0 is a general-purpose register and does NOT contain a zero as in Nios II or MicroBlaze. The other major difference is that all instructions and not just branch instructions have a condition attached in bits 31–28. The condition code is shown in Table 11.1 together with the flag involved. These flag values are set by the previous instruction, indicated by an appended "s" in the assembler code, e.g., ADD vs. ADDS. The flag values are stored in the 32-bit current program status register (CPSR) in the following order:

31	30	29	28	27...8	7	6	5	4...0
N	Z	C	V	–	I	F	T	M

Table 11.1 Condition code summary

Code	Suffix	Flags	Meaning
0000	EQ	$Z = 1$	Equal
0001	NE	$Z = 0$	Not equal
0010	CS	$C = 1$	Unsigned higher or same
0011	CC	$C = 0$	Unsigned lower
0100	MI	$N = 1$	Negative
0101	PL	$N = 0$	Positive or zero
0110	VS	$V = 1$	Overflow
0111	VC	$V = 0$	No overflow
1000	HI	C AND NOT $Z = 1$	Unsigned higher
1001	LS	NOT C AND $Z = 1$	Unsigned lower or same
1010	GE	N EXOR $V = 0$	Greater or equal
1011	LT	N EXOR $V = 1$	Less than
1100	GT	Z OR (N EXOR V) = 0	Greater than
1101	LE	Z OR (N EXOR V) =1	Less than or equal
1110	AL	Ignored	Always

Besides the four arithmetic flags, we also see interrupt disable flags (I/F), Thumb flag (T), and 4-bit processor operation mode flag (M) are stored in CPSR. Since the majority of instructions are independent of a condition, we will see the always AL code $E_{16} = 1110_2$ often in the 4 MSBs of our assembler code.

Let us start with the easiest format how branch and branch with link aka call are encoded (bits 27...26 = 10). Here is the format:

31...28	**27...26**	**23**	**24**	**23...0**
cond	10	X	L	imm24

For X=1 and L=0, we have the usual branch instruction with a 24-bit pc relative offset. For X=1 and L=1, we have a call aka branch with link (bl), i.e., the current pc is stored in the link register (r14) before the CPU continue processing at the new program memory location specified relative to pc by imm24. For condition code "F," this branch format becomes a BLX instruction that switches to another instruction set (we do not support). If X=0, this code is used for single or multiple store/load aka push/pop. The push/pop instructions come in two different flavors. In the single-register format (A2), the source/destination register is defined in the usual target Rd location, i.e., bits 12 to 15. If more than one register is moved, then the other format A1 must be used. Now bits 0 to 15 specify all registers (1 = move; 0 = don't move) that should be moved. An instruction like PUSH {r1, r3-r5, r13} would move registers 1, 3, 4, 5, and 13 (aka link register) onto the stack (i.e., a location close to the end in our data RAM) using a multicycle operation coded with a single assembler instruction. The sp. register will be used for indexing incrementing by 4 for each memory access. The sp auto-update is implemented as decrement before for `push` and an increment after of `pop`. The four different push/pop operations each come with a 12-bit operation code. Here are the four formats.

Format	31...28	27...16	15...12	11...0
pop A1	cond	100010111101	register list	
pop A2	cond	010010011101	Rd	0x004
push A1	cond	100100101101	register list	
push A2	cond	010100101101	Rd	0x004

We will only implement single push/pop since multiple would require multicycle instructions which would complicate substantially the design. For convenience, we allow multi-register assembler code, but we will only move the register with the largest index, i.e., PUSH {r3, lr} will be implemented as PUSH {lr} only since 13 > 3.

The second format group for *data processing instructions* (bits 27...26 = 00) is substantially more complex and looks as follows:

31...28	27..26	25	24...21	20	19...16	15...12	11...0
cond	00	I	OP code	S	Rn	Rd	Operand 2

For operand 2 we have depending on the I-flag further differentiations. For I=1 we use a 12-bit immediate operand 2, i.e.,

11...0
imm12

and the operation becomes Rd <= Rn □ imm12, and for I=0 operand 2 is a register (with optional shift) coded as

11...7	6...5	4	3...0
Operand 3	type	R	Rm

and the operation becomes Rd <= Rn □ Rm, where `type` specifies an optional shift operation (ASR = 10, LSL = 00, LSR = 01, or ROR = 11). For R=0 operand 3 in Bits 11...7 may specify imm5, a 5-bit constant and for R=1 a forth register {Rs,0}. If the S bit is set, then N, Z, C, and V flags of the CPSR register are updated. The 4-bit operation code in bits 24...21 specifies 1 of the 16 operations shown in Table 11.2. It starts with two Boolean operations (AND and EXOR) followed by six arithmetic operations. The next four operations will not modify the Rd register, only update the flags. Next follows the Boolean OR operation. The move operation is often used to reset or copy register values but can include shift too. Bitwise clear and bitwise NOT complete the 16 data processing instructions.

Table 11.2 Data processing register instructions with flag update for type ...S instructions

Code	Ass. code		Meaning	Flag updates			
				N	Z	C	V
0000	AND	Rd = Op1 AND Op2	Bitwise AND	Y	Y	Y	N
0001	EOR	Rd = Op1 EOR Op2	Bitwise Exclusive OR	Y	Y	Y	N
0010	SUB	Rd = Op1 – Op2	Subtract	Y	Y	Y	Y
0011	RSB	Rd = Op2 – Op1	Reverse Subtract	Y	Y	Y	Y
0100	ADD	Rd = Op1 + Op2	Add	Y	Y	Y	Y
0101	ADC	Rd = Op1 + Op2 + C	Add with carry	Y	Y	Y	Y
0110	SBC	Rd = Op1 – Op2 + C -1	Subtract with carry	Y	Y	Y	Y
0111	RSC	Rd = Op2 – Op1 + C -1	Reverse subtract with carry	Y	Y	Y	Y
1000	TST	Check Op1 AND Op2	Test AND	Y	Y	Y	N
1001	TEQ	Check Op1 EOR Op2	Test equivalence	Y	Y	Y	N
1010	CMP	Check Op1 – Op2	Compare	Y	Y	Y	Y
1011	CMN	Check Op2 + Op1	Compare Negative	Y	Y	Y	Y
1100	ORR	Rd = Op1 OR Op2	Bitwise OR	Y	Y	Y	N
1101	MOV	Rd = Op1	Move (with optional shift)	Y	Y	N	N
1110	BIC	Rd = Op1 AND NOT Op2	Bitwise bit clear	Y	Y	Y	N
1111	MVN	Rd = NOT Op2	Bitwise NOT	Y	Y	Y	N

So far, we only have 12-bit immediate operations and do not have 16-bit immediate constant operations. While that is not possible for all arithmetic operations, we can add this for the move instructions. This are the move top MOVT and the MOVW instructions placing the 16-bit constant in the MSBs and zero extended 16-bit constant in the LSBs, respectively. The 4 bits at the Rn location are used for the additional imm4 constant. The bit 22 aka M flag is used to distinguish these two operations. The format looks as follows:

31...28	27..26	25	24...21	20	19...16	15...12	11...0
cond	00	1	10M0	0	imm4	Rd	imm12

Comparing with Table 11.2, these two instructions look like "borrowed" from the TST and CMP instructions. This is possible by noticing that S=0, i.e., not updating the flags is not useful in TST and CMP type instructions.

The third instruction format we discuss is the single data transfer aka load/store (bits 27...26 = 01). There are bytes, and 16-bit data instructions too, but we will only use and discuss the 32-bit data instructions. Since we have a load/store architecture, only moves between memory and registers are allowed. The ARMv7 ISA uses six flags to distinguish the many different addressing modes supported by our Cortex-A9 processor. A typical load operation LDR has the function Rd = Mem(Rn+offset), and a typical store instruction STR works as Mem(Rn+offset)=Rd. Here is the format:

31...28	27..26	25	24	23	22	21	20	19...16	15...12	11...0
cond	01	I	P	U	B	W	L	Rn	Rd	Operand 2

The functions of the six flags in bits 25...20 are as follows:

- The I bit 25 is used to specify if the offset specified in operand 2 is a 12-bit constant imm12 (I=0) or a register (I=1) with optional shift similar to the data processing instructions. (Note that the use of the I-flag is a little counterintuitive here since for I=1 imm12 is *not* used.)
- The P flag indicates if the offset is added before or after transfer. P=0 gives post-addition and P=1 gives pre-addition.
- For U=1 the offset is added and U=0 subtracted to address register Rn.
- For B=1 the byte size is used and B=0 the word size.
- The Write-back flag is W. For W=1 the index register is updated, and for W=0 we have no write-back.
- For L=1 we have a load operation and for L=0 a store operation.

For our tiny RISC, we will only use flags I, U, and L, and the other flags will be ignored. Rn, Rd, and Operand 2 have the same location and functionality as in the data processing instructions.

A full description of all instructions can be found in the processor reference manual [A14]. Make sure you download the correct ARMv7-A manual one matching your processor.

Now let us re-code our PICOBLAZE LED flash example in ARMv7 assembler language. Here is our first ARMv7 assembler program listing (using address values for the DE1 SoC board):

ASM Program 11.6: LED Toggle Using a ARMv7 Assembler Coding

```
1            .text                    /* ARM executable code follows */
2            .global     _start
3    _start:
4            mov    r1, #0             //=LEDR_BASE address of LEDR lights
5            movt   r1, #65312         //=SW_BASE address + 0x40=64
6            ldr    r2, [r1,#64]       // load SW switches
7    flash:  str    r2, [r1]          // write to red LEDs
8            movw   r3, #30784        //=25_000_000
9            movt   r3, #381
10   loop:   subs   r3, r3, #1         // delay counter
11           bne    loop
12           mvn    r2, r2             // toggle/bit inverse
13           b      flash
```

Starting label for ARM assembler is always the _start label. The I/O addresses are listed in address_map_arm.h, and we see that the slider switch at 0xFF200040 and red LEDs at address 0xFF200000 share the same 16-bit MSBs. We load first the slider switch value (line 6) and write the data to the red

LED outputs. The GCC translates the 32-bit loading instruction for the delay counter value 25,000,000 for register r3 (line 8,9) in a series of movw of LSBs and movt immediate instructions for the MSBs. If we use the pseudo-instruction LDR r3,=0x17D7840, then the assembler would place the constant in a literal pool behind the program, and the constant can be loaded via a *single* pc relative load from memory such as ldr r4, [pc, #20] as recommended by the Altera ARM Tutorial [A15b]. The instructions from lines 10 and 11 implement the delay loop: counting down r3 until it reaches zero. Then all bits in register 2 are inverted. This could have been done with an eor instruction also. In line 13, we branch back to line 7 (flash) to start another loop. We see that the data processing instructions are used most often. We have only used two times condition, so all other assembler code instruction should start with letter E. Encoding and decoding of these instructions is left as an exercise at the end of the chapter; see Exercise 11.17–11.24.

Now let us have a look at the VHDL simulation in Fig. 11.12. The simulation shows the input signal for clock and reset, followed by local non-I/O signals and finally the I/O signals for switches and LEDs. After the release of the reset (active low), we see how the program counter (pc) continuously increases until the end of the loop. The pc changes with the falling edge, and instruction word is updated with the rising edge (within the ROM). The port address is placed in register r1. Then the value 5 is read from the input port in_port and put into register r2. The register r2 value is loaded in the out_port register. Next the counter start value 0x17D7840 is loaded into register r3 in two steps: first the 16-bit LSBs via movw and then the 16-bit MSBs via a movt instruction. The counter is decremented by 1, and since the 32-bit value is not zero, the loop is repeated. This continues for many clock cycles until the 1-second simulation time is reached and the output toggles.

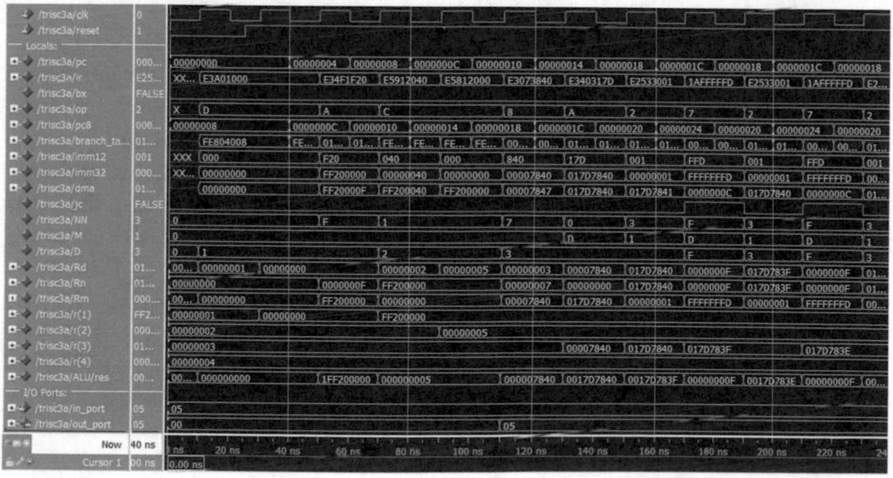

Fig. 11.12 First 200 ns simulation steps for the toggle program showing 32-bit counter in r(4) load and counting down ...40,...3F,...3E,...

Now with basic I/O of ARMv7 understood, let us have a look how GCC organizes our data for the ARM processor in memory. In ANSI C code, we can place our variable/data definitions in the initial section of our code, making the data "global" available in the main section as well as any of the functions we may have. The other option is that we can place the variable definition within the main code aka "local" variables. Let us have a look how GCC is handling these two cases since the ARM only uses a single stack pointer and not global pointer as Nios II. Here is a little ANSI C test program for this.

DRAM Program 11.7: DRAM Write Followed by Read

```
1    volatile int g;// Volatile allows us to see the full assembler code
2    volatile int garray[15];
3    int main(void) {
4      volatile int s;    // Declare volatile tells compiler do
5      volatile int sarray[14]; // not try optimization
6      s = 1; // Memory location sp+15*4
7      g = 2; // Memory location gp+0
8      sarray[0] = s; // Memory location sarray[0] is sp+4
9      garray[0] = g; // Memory location garray[0] is gp+4
10     sarray[s] = 0x1357; // use variable; requires *4 for byte address
11     garray[g] = 0x2468; // use variable; requires *4 for byte address
12   }
```

The assembler output code of interest in the Altera monitor is shown in Fig. 11.13. The local data defined within our main() sections are indexed via the stack pointer sp. All data (and program) access is done in increments of bytes and that is why our first integer variable has address 0(sp), the second 4(sp), third 8(sp), etc. Other than Nios II, the ARM does not have a global pointer, and GCC needs to track and find appropriate locations in the main memory that are located somewhere behind the program section and below sp section. The ARM assumes a von Neumann architecture and not separate data and program busses aka Harvard architecture, such that a whole 32-bit constant can be loaded with one pc-relative read from memory. Starting address of the global memory section is somewhere after the code sections, while sp use a starting address that is at the end of our physical memory decreased by the number of bytes needed to host our data. The value for sp is set by GCC at the beginning of the program; see Fig. 11.13a. We skip the other initialization done by GCC for ARM to keep the discussion brief. GCC starts the actual main() program at program memory location 25C; see Fig. 11.13b.

We ran the testdem.c program until it reaches garray[g]=0x2468 and then look up the memory content. At the end of our physical memory (see Fig. 11.13c), we find the local values indexed by sp. We see that sarray[1] can be found at address 0x3FFFFFB4, i.e., close to the DDR end address, and garray[2] at address 0xBB0 somewhere after the program code. This should also be the memory location we should monitor in a HDL simulation.

```
                                        _start:
        0x00000128    E51FD000          ldr    sp, [pc, #-0]    ; 130 <_start+0x8>
        0x0000012C    EAFFFFC4          b      44 <_cs3_start_c>
        0x00000130    3FFFFFFC          .word  0x3ffffffc
(a)
```

```
                                    int main(void) {
                                    main:
        0x0000025C    E24DD040          sub    sp, sp, #64      ; 0x40
                                    volatile int s;   // Declare volatile tells compiler do
                                    volatile int sarray[14]; // not try optimization
                                    s = 1; // Memory location sp+15*4
        0x00000260    E3A03001          mov    r3, #1
        0x00000264    E58D303C          str    r3, [sp, #60]    ; 0x3c
                                    g = 2; // Memory location gp+0
        0x00000268    E3003BA4          movw   r3, #2980        ; 0xba4
        0x0000026C    E3403000          movt   r3, #0
        0x00000270    E3A02002          mov    r2, #2
        0x00000274    E5832000          str    r2, [r3]
                                    sarray[0] = s; // Memory location sarray[0] is sp+4
        0x00000278    E59D203C          ldr    r2, [sp, #60]    ; 0x3c
        0x0000027C    E58D2004          str    r2, [sp, #4]
                                    garray[0] = g; // Memory location garray[0] is gp+4
        0x00000280    E5932000          ldr    r2, [r3]
(b)     0x00000284    E5832004          str    r2, [r3, #4]
                                    sarray[s] = 0x1357; // use variable; requires *4 for byte address
        0x00000288    E59D203C          ldr    r2, [sp, #60]    ; 0x3c
        0x0000028C    E28D1040          add    r1, sp, #64      ; 0x40
        0x00000290    E0812102          add    r2, r1, r2, lsl #2
        0x00000294    E3011357          movw   r1, #4951        ; 0x1357
        0x00000298    E502103C          str    r1, [r2, #-60]   ; 0xffffffc4
                                    garray[g] = 0x2468; // use variable; requires *4 for byte address
        0x0000029C    E5932000          ldr    r2, [r3]
        0x000002A0    E0833102          add    r3, r3, r2, lsl #2
        0x000002A4    E3022468          movw   r2, #9320        ; 0x2468
        0x000002A8    E5832004          str    r2, [r3, #4]
                                    }
```

(c)	+0x0	+0x4	+0x8	+0xc
0x3FFFFFB0	00000001	00001357	0000070C	00000002
0x3FFFFFC0	FFFFF014	FFFF5F08	000002CC	00000714
0x3FFFFFD0	0000058C	00000208	000006C8	00000B88
0x3FFFFFE0	00000758	FFFFFFFF	00000001	00000BE8
0x3FFFFFF0	000000D8	FFFFF014	FFFF1351	4B097224

(d)	+0x0	+0x4	+0x8	+0xc
0x00000BA0	00000000	00000002	00000002	00000000
0x00000BB0	00002468	00000000	00000000	00000000

Fig. 11.13 TRISC3A data memory ASM test routine. Local and global memory write are followed by reading the same locations. (**a**) Starting sequence. (**b**) main() program. (**c**) Memory at sp. (**d**) global memory section

Comparing the generated ARMv7 assembler code with the Nios II, we see two concepts producing shorter aka faster code: Since the array index computation is byte-wise, a "multiply" by four is required. In Nios II this is done with two add operation (recall $s = a + a$; $i = s + s$ gives $i = 4a$). The ARMv7 allows shift while adding, so the add r3, r3, r2, lsl #2 reduces index computation by one instruction. The second code size reduction can be achieved with the pc-relative read. A 32-bit constant can be loaded in a single operation which usually takes two instructions (load MSB and LSBs) due to the code word length of 32-bit microprocessors. The price to pay for pc-relative read (without doubling the memory requirement) is that we need to give up the concept of a Harvard architecture with separate program and data memory since now a constant is loaded from *program* memory into *data* register making a separation of program and data memory very difficult. For a Harvard design, we would need to duplicate the memory: leaving the upper part of the PROM unused and the lower part of the DRAM unused.

Challenging is to find a HDL of a true dual-port RAM memory with separate clock and initial data. There are multiple ways to define a dual-port RAM. It is very important to find a HDL description that synthesizes to Block RAM and not logic cells since our RAM has 4 K × 32 = 131 K bits, and the DE1 SoC FPGA only has 70 K flip-flops, so using BRAM is mandatory. We should study the vendor language template carefully that our HDL code works for all three: MODELSIM, QUARTUS, and VIVADO. We may use dual-port memory with no-change mode, read-first mode, or write-first mode. These modes handle the different cases when read/write is performed at the same time to the same address. This is not really an issue for our microprocessor since in one instruction we do only read or write but not both. MODELSIM compile fine for all three, and VIVADO synthesizes all three templates using four 36KBits BRAMs. QUARTUS worked fine for write-first mode and the no-change mode. The read-first mode template did not synthesize with QUARTUS. QUARTUS need 16 block RAM of size M10K and 1 LUT for the dual-port RAM. The HDL below shows the write-first mode style with a DMEM test program initialization.

VHDL Program 11.8: DMEM test program for TRISC3A translated to VHDL

```
1    -- =====================================================================
2    -- File is True Dual Port RAM with dual clock for DRAM and ROM
3    -- DMEM test program for the tiny ARMv7 processors
4    -- Copyright (C) 2019  Dr. Uwe Meyer-Baese.
5    -- =====================================================================
6    LIBRARY ieee; USE ieee.STD_LOGIC_1164.ALL;
7    USE ieee.STD_LOGIC_arith.ALL; USE ieee.STD_LOGIC_unsigned.ALL;
8    -- =====================================================================
9    ENTITY dpram4Kx32 IS
10   PORT (clk_a  : IN STD_LOGIC; -- System clock DRAM
11         clk_b  : IN STD_LOGIC; -- System clock PROM
12         addr_a : IN STD_LOGIC_VECTOR(11 DOWNTO 0); -- Data mem. address
13         addr_b : IN STD_LOGIC_VECTOR(11 DOWNTO 0); -- Prog. mem. address
14         data_a : IN STD_LOGIC_VECTOR(31 DOWNTO 0); -- Data in for DRAM
15         we_a   : IN STD_LOGIC := '0'; -- Write only DRAM
16         q_a    : OUT STD_LOGIC_VECTOR(31 DOWNTO 0); -- DRAM output
17         q_b    : OUT STD_LOGIC_VECTOR(31 DOWNTO 0)); -- ROM output
18   END ENTITY dpram4Kx32;
19   -- =====================================================================
20   ARCHITECTURE fpga OF dpram4Kx32 IS
21
22   -- Build a 2-D array type for the RAM
23   TYPE MEM IS ARRAY(0 TO 4095) of STD_LOGIC_VECTOR(31 DOWNTO 0);
24
25   -- Define RAM and initial values
26   SHARED VARIABLE dram : MEM  := (
27   -- This style would require a PC relative addressing
28   X"E51F_D000", --          ldr    sp, [pc, #0] /*initial stack pointer*/
29   X"EA00_0000", --          b main          /* initial global pointer */
30   X"3FFF_FFEC", --          .word 0x3fffffec    /* constant */
31   X"E24D_D040", -- main: sub sp, sp, #64      /* make space on stack */
32   --------------- s = 1 -------------
33   X"E3A0_3001", --          mov r3, #1       /* s= 1 */
34   X"E58D_303C", --          str  r3, [sp, #60] /* store s */
35   --------------- g = 2 -------------
36   X"E300_3BA4", --          movw    r3, #2980      /* pointer g */
37   X"E340_3000", --          movt    r3, #0         /* MSBs g */
38   X"E3A0_2002", --          mov     r2, #2         /* LSBS */
39   X"E583_2000", --          str     r2, [r3]       /* store g */
40   --------------- sarray[0] = s -------------
41   X"E59D_203C", --          ldr     r2, [sp, #60]  /* address sarray[0] */
42   X"E58D_2004", --          str     r2, [sp, #4]   /* sarray[0]=s */
43   --------------- garray[0] = q -------------
44   X"E593_2000", --          ldr     r2, [r3]       /* address garray[0] */
45   X"E583_2004", --          str     r2, [r3, #4]   /* store garray[0]=g */
46   --------------- sarray[s] = 0x1357 -------------
47   X"E59D_203C", --          ldr     r2, [sp, #60]  /* s */
48   X"E28D_1040", --          add     r1, sp, #64    /* mv sp to r1*/
49   X"E081_2102", --          add     r2, r1, r2, lsl #2 /* s*4+sp =
50   &sarray[s]*/
```

```
51   X"E301_1357", --        movw    r1, #4951          /* =0x1357 */
52   X"E502_103C", --        str     r1, [r2, #-60]  /* store sarray[s] */
53   --------------- garray[g] = 0x2468 -------------
54   X"E5932000", --         ldr     r2, [r3]           /* load g */
55   X"E0833102", --         add     r3, r3, r2, lsl #2 /* g*4+&garray[0] */
56   X"E3022468", --         movw    r2, #9320          /* =0x2468 */
57   X"E5832004", --         str     r2, [r3, #4]     /* store garray[g] */
58   X"EAFFFFEB", --         b       main               /* start loop again */
59   OTHERS => "UUUUUUUUUUUUUUUUUUUUUUUUUUUUUUUU"); -- Default unknown mem
60
61   BEGIN
62
63     -- Port A aka DRAM
64     PROCESS(clk_a)
65     BEGIN
66       IF rising_edge(clk_a) THEN
67         IF we_a = '1' THEN
68           dram(CONV_INTEGER('0'& addr_a)) := data_a;
69           q_a <= data_a;
70         ELSE
71           q_a <= dram(CONV_INTEGER('0'& addr_a));
72         END IF;
73       END IF;
74     END PROCESS;
75
76     -- Port B aka ROM
77     PROCESS(clk_b)
78     BEGIN
79       IF rising_edge(clk_b) THEN
80         q_b <= dram(CONV_INTEGER('0' & addr_b));
81       END IF;
82     END PROCESS;
83
84   END fpga;
```

Both memories use the same "shared" variable for the memory data definition. Note that the program definition (lines 28–59) data constants (line 30) become part of the program section. The DRAM is mapped to port A (lines 63–74) and uses a rising edge clock. The program ROM is mapped to port B (lines 77–82) and uses a falling clock edge. Since we do not write data to the ROM, we remove the write-enable portion of the template.

As a third example, let us check what instructions are needed for (nested) function calls. Again, we start with a short ANSI C example and then see what GCC requires as instructions. Here is our C code.

3 Level Nesting Program 11.9: Nesting Function Test Code

```
 1  void level3(int *array, int s1) {
 2      s1 += 1;
 3      array[3] = s1;
 4      return;
 5  }
 6  void level2(int *array, int s1) {
 7      s1 += s1;
 8      array[2] = s1;
 9      level3(array, s1); // call level 3
10      return;
11  }
12  void level1(int *array, int s1) {
13      s1 += 1;
14      array[1] = s1;
15      level2(array, s1); // call level 2
16      return;
17  }
18  int main(void) {
19    volatile int s1, s2, s3;
20    int array[11];
21    s1 = 0x1233; // Memory location sp+0
22    while(1)   {
23        level1(array, s1);
24        s1 = array[1];
25        s2 = array[2];
26        s3 = array[3];
27    }
28  }
```

As in our PicoBlaze studies, we use here also three levels of loops and only local, i.e., sp indexed variables. The starting sequence is setting up sp, and jumping to main() is similar to the previous example so we skip this section. In the function call, we include one array and one scalar. During the call we increment the scalar and assign the new value to our array. We use an initial value of 0x1233 such that it can be easily identified in memory. Figure 11.14a shows the assembler code for the three functions, while Fig. 11.14b shows assembler code for the main() code.

Other than PicoBlaze, the ARM Cortex-A9 does not have a pc stack to host return addresses for loops. We only have one register r13=lr to store a return address. If we just call one function having one return address register would be sufficient (see level 3 in Fig. 11.14a). The bx lr instruction (line 0x264) will restore the pc with the original value we stored in lr before entering the subroutine. But if we call another function within our first called functions, we need to store the lr value into memory and restore lr at the end of the subroutine. The save and restore of lr in ARMv7 can conveniently be done with a push and pop operations. These two operations use sp and perform an auto-update (decrement before for push; increment after of pop). Since r3 is not used in our short subroutine, saving and restoring of r3 as shown in the coding example is not really necessary (and our HDL will use one register in push/pop). To index the memory location for the loop address, the stack register sp is used. Since sp is also used to index our local variables, GCC needs to track our array indices and the return address carefully. At the end we will have in the memory region addressed by sp with increasing index the variable values and lower values the return address used for the function calls. We run our main program loop one time and then monitor the memory content starting at sp=0x3FFFFFA0. After one loop iteration, we will observe the values shown in Table 11.3.

At location 0x3FFFFFE0, we find s1=0x1234 and array[1] at 0x3FFFFFB0. At address 0x3FFFFFA8, we see the return address 0x2AC for the call to first level1 function, and at address 0x3FFFFFA0, we find 0x28C as the return address for the call to level2 within the level1 function. This will be our test bench data for the HDL simulation that follows next.

```
                    void level3(int *array, int s1) {                int main(void) {
                        s1 += 1;                                      main:
                        level3:                      0x00000290  E52DE004    push    {lr}        ; (str lr, [sp, #-4]!)
0x0000025C  E2811001        add     r1, r1, #1       0x00000294  E24DD03C    sub     sp, sp, #60  ; 0x3c
                            array[3] = s1;                               volatile int s1, s2, s3;
0x00000260  E580100C        str     r1, [r0, #12]                        int array[11];
0x00000264  E12FFF1E        bx      lr                                   s1 = 0x1233; // Memory location sp+0
                                                     0x00000298  E3013233    movw    r3, #4659    ; 0x1233
                        return;                      0x0000029C  E58D3034    str     r3, [sp, #52] ; 0x34
                    }                                    while(1) {
                    void level2(int *array, int s1) {                        level1(array, s1);
                        level2:                      0x000002A0  E59D1034    ldr     r1, [sp, #52] ; 0x34
0x00000268  E92D4008        push    {r3, lr}         0x000002A4  E1A0000D    mov     r0, sp
                        s1 += s1;                    0x000002A8  EBFFFFF3    bl      27c <level1>
0x0000026C  E1A01081        lsl     r1, r1, #1           s1 = array[1];
                            array[2] = s1;           0x000002AC  E59D3004    ldr     r3, [sp, #4]
0x00000270  E5801008        str     r1, [r0, #8]     0x000002B0  E58D3034    str     r3, [sp, #52] ; 0x34
                            level3(array, s1); // call level 3       s2 = array[2];
0x00000274  EBFFFFF8        bl      25c <level3>     0x000002B4  E59D3008    ldr     r3, [sp, #8]
0x00000278  E8BD8008        pop     {r3, pc}         0x000002B8  E58D3030    str     r3, [sp, #48] ; 0x30
                                                         s3 = array[3];
                        return;                      0x000002BC  E59D300C    ldr     r3, [sp, #12]
                    }                                0x000002C0  E58D302C    str     r3, [sp, #44] ; 0x2c
                    void level1(int *array, int s1) {0x000002C4  EAFFFFF5    b       2a0 <main+0x10>
                        level1:
0x0000027C  E92D4008        push    {r3, lr}
                        s1 += 1;
0x00000280  E2811001        add     r1, r1, #1
                            array[1] = s1;
0x00000284  E5801004        str     r1, [r0, #4]
                            level2(array, s1); // call level 2
0x00000288  EBFFFFF6        bl      268 <level2>
0x0000028C  E8BD8008        pop     {r3, pc}

                        return;
                    (a)                                              (b)
```

Fig. 11.14 TRISC3A nesting loop ASM test routine. Local and global memory write are followed by reading the same locations. (**a**) Three level functions. (**b**) main() program

Table 11.3 The ARM memory contend after running nesting.c

	+0x0	+0x4	+0x8	+0xc
0x3FFFFFA0	0000028C	00001233	000002AC	00000BB8
0x3FFFFFB0	00001234	00002468	00002469	00000002
0x3FFFFFC0	FFFFF014	FFFF5F08	000002E0	00000728
0x3FFFFFD0	000005A0	00000208	00002469	00002468
0x3FFFFFE0	00001234	FFFFFFFF	00000624	00000BB8
0x3FFFFFF0	000000D8	FFFFF014	FFFF1351	4B097264

HDL Implementation and Testing

The PICOBLAZE design `trisc2.vhd` or `trisc2.v` should be our starting point for the tiny ARM design we call TRISC3A, since ARM Cortex-A9 is a three-address machine. This is a single three-phase clock cycle design – a pipelined version would require substantially more HDL coding [L06]. The HDL code is shown next.

VHDL Code 11.10: TRISC3A (Final Design)

```
 1    -- ==============================================================
 2    -- Title: T-RISC 3 address machine
 3    -- Description: This is the top control path/FSM of the
 4    -- T-RISC, with a single 3 phase clock cycle design
 5    -- It has a 3-address type instruction word
 6    -- implementing a subset of the ARMv7 Cortex A9 architecture
 7    -- ==============================================================
 8    LIBRARY ieee; USE ieee.std_logic_1164.ALL;
 9
10    PACKAGE n_bit_type IS                   -- User defined types
11      SUBTYPE U8 IS INTEGER RANGE 0 TO 255;
12      SUBTYPE U12 IS INTEGER RANGE 0 TO 4095;
13      SUBTYPE SLVA IS STD_LOGIC_VECTOR(11 DOWNTO 0); -- Address prog. mem.
14      SUBTYPE SLVD IS STD_LOGIC_VECTOR(31 DOWNTO 0); -- Width data
15      SUBTYPE SLVD1 IS STD_LOGIC_VECTOR(32 DOWNTO 0); -- Width data + 1
16      SUBTYPE SLVP IS STD_LOGIC_VECTOR(31 DOWNTO 0);    -- Width instruction
17      SUBTYPE SLV4 IS STD_LOGIC_VECTOR(3 DOWNTO 0);   -- Full opcode size
18    END n_bit_type;
19
20    LIBRARY work;
21    USE work.n_bit_type.ALL;
22
23    LIBRARY ieee;
24    USE ieee.STD_LOGIC_1164.ALL;
25    USE ieee.STD_LOGIC_arith.ALL;
26    USE ieee.STD_LOGIC_unsigned.ALL;
27    -- ==============================================================
28    ENTITY trisc3a IS
29     PORT(clk      : IN  STD_LOGIC; -- System clock
30          reset    : IN  STD_LOGIC; -- Active low asynchronous reset
31          in_port  : IN  STD_LOGIC_VECTOR(7 DOWNTO 0); -- Input port
32          out_port : OUT STD_LOGIC_VECTOR(7 DOWNTO 0) -- Output port
33    -- The following test ports are used for simulation only and should be
34    -- comments during synthesis to avoid outnumbering the board pins
35    --   r0_OUT   : OUT SLVD; -- Register 0
36    --   r1_OUT   : OUT SLVD; -- Register 1
37    --   r2_OUT   : OUT SLVD; -- Register 2
38    --   r3_OUT   : OUT SLVD; -- Register 3
39    --   sp_OUT   : OUT SLVD; -- Register 13 aka stack pointer
40    --   lr_OUT   : OUT SLVD; -- Register 14 aka return address
41    --   jc_OUT   : OUT STD_LOGIC;   -- Jump condition flag
42    --   me_ena   : OUT STD_LOGIC;   -- Memory enable
43    --   i_OUT    : OUT STD_LOGIC;   -- constant flag
44    --   pc_OUT   : OUT STD_LOGIC_VECTOR(11 DOWNTO 0); -- Program counter
45    --   ir_imm12 : OUT STD_LOGIC_VECTOR(11 DOWNTO 0); -- Immediate value
46    --   imm32_out : OUT SLVD;              -- Sign extend immediate value
47    --   op_code  : OUT STD_LOGIC_VECTOR(3 DOWNTO 0)   -- Operation code
48             );
49    END;
```

```
50    -- ========================================================================
51    ARCHITECTURE fpga OF trisc3a IS
52    -- Define GENERIC to CONSTANT for _tb
53      CONSTANT WA : INTEGER := 11;  -- Address bit width -1
54      CONSTANT NR : INTEGER := 15;  -- Number of Registers -1; PC is extra
55      CONSTANT WD : INTEGER := 31;   -- Data bit width -1
56      CONSTANT DRAMAX : INTEGER := 4095; -- No. of DRAM words -1
57      CONSTANT DRAMAX4 : INTEGER := 1073741823; -- X"3FFFFFFF";
58                                            -- True DDR RAM bytes -1
59    COMPONENT dpram4Kx32 IS
60    PORT (clk_a  : IN STD_LOGIC; -- System clock DRAM
61          clk_b  : IN STD_LOGIC; -- System clock PROM
62          addr_a : IN STD_LOGIC_VECTOR(11 DOWNTO 0); -- Data mem. address
63          addr_b : IN STD_LOGIC_VECTOR(11 DOWNTO 0);-- Prog. mem. address
64          data_a : IN STD_LOGIC_VECTOR(31 DOWNTO 0); -- Data in for DRAM
65          we_a   : IN STD_LOGIC := '0'; -- Write only DRAM
66          q_a    : OUT STD_LOGIC_VECTOR(31 DOWNTO 0); -- DRAM output
67          q_b    : OUT STD_LOGIC_VECTOR(31 DOWNTO 0)); -- ROM output
68    END COMPONENT;
69
70      SIGNAL op  : SLV4;
71      SIGNAL dmd, pmd, dma : SLVD;
72      SIGNAL cond : STD_LOGIC_VECTOR(3 DOWNTO 0);
73      SIGNAL ir, tpc, pc, pc4_d, pc4, pc8, branch_target : SLVP;-- PCs
74      SIGNAL mem_ena, not_clk : STD_LOGIC;
75      SIGNAL jc, go, dp, rlsl  : boolean; -- jump and decoder flags
76      SIGNAL I, set, P, U, bx, W, L : boolean;-- Decoder flags
77      SIGNAL movt, movw, str, ldr, branch, bl : boolean; -- Special instr.
78      SIGNAL load, store, read, write, pop, push : boolean; -- I/O flags
79      SIGNAL popPC, popA1, pushA1, popA2, pushA2: boolean;--LDR/STM instr.
80      SIGNAL ind, ind_d : INTEGER RANGE 0 TO NR; --push/pop index
81      SIGNAL N, Z, C, V :  boolean; -- CPSR flags
82      SIGNAL D, NN, M : INTEGER RANGE 0 TO 15; -- Register index
83      SIGNAL Rd, Rdd, Rn, Rm, r_M : SLVD := (OTHERS => '0');-- current Ops
84      SIGNAL Rd1, Rn1, Rm1 : SLVD1; -- Sign extended Ops
85      SIGNAL imm4 : STD_LOGIC_VECTOR(3 DOWNTO 0); -- imm12 extended
86      SIGNAL imm5 : STD_LOGIC_VECTOR(4 DOWNTO 0); -- Within Op2
87      SIGNAL imm12 : STD_LOGIC_VECTOR(11 DOWNTO 0);   -- 12 LSBs
88      SIGNAL sxt12 : STD_LOGIC_VECTOR(19 DOWNTO 0); -- Total 32 bits
89      SIGNAL imm24  : STD_LOGIC_VECTOR(23 DOWNTO 0); -- 24 LSBs
90      SIGNAL sxt24  : STD_LOGIC_VECTOR(5 DOWNTO 0); -- Total 30 bits
91      SIGNAL bimm32, imm32, mimm32 : SLVD; -- 32 bit branch/mem/ALU
92      SIGNAL imm33 : SLVD1; -- Sign extended ALU constant
93
94    -- OP Code of instructions:
95    -- The 4 bit for all data processing instructions
96      CONSTANT opand : SLV4 := "0000"; -- X0
97      CONSTANT eor   : SLV4 := "0001"; -- X1
98      CONSTANT sub   : SLV4 := "0010"; -- X2
99      CONSTANT rsb   : SLV4 := "0011"; -- X3
100     CONSTANT add   : SLV4 := "0100"; -- X4
101     CONSTANT adc   : SLV4 := "0101"; -- X5
```

```
102    CONSTANT sbc    : SLV4 := "0110"; -- X6
103    CONSTANT rsc    : SLV4 := "0111"; -- X7
104    CONSTANT tst    : SLV4 := "1000"; -- X8
105    CONSTANT teq    : SLV4 := "1001"; -- X9
106    CONSTANT cmp    : SLV4 := "1010"; -- XA
107    CONSTANT cmn    : SLV4 := "1011"; -- XB
108    CONSTANT orr    : SLV4 := "1100"; -- XC
109    CONSTANT mov    : SLV4 := "1101"; -- XD
110    CONSTANT bic    : SLV4 := "1110"; -- XE
111    CONSTANT mvn    : SLV4 := "1111"; -- XF
112
113  -- Register array definition 16x32
114    TYPE REG_ARRAY IS ARRAY(0 TO NR) OF SLVD;
115    SIGNAL r : REG_ARRAY;
116
117  BEGIN
118
119    WITH ir(31 DOWNTO 28) SELECT -- Evaluation of condition bits
120    go <= Z WHEN "0000", NOT Z WHEN "0001",   -- Zero: EQ or NE
121         C WHEN "0010", NOT C WHEN "0011",    -- Carry: CS or CC
122         N WHEN "0100", NOT N WHEN "0101",    -- Negative: MI or PL
123         V WHEN "0110", NOT V WHEN "0111",    -- Overflow: Vo or VC
124         C AND NOT Z WHEN "1000",             -- HI
125         NOT C AND Z WHEN "1001",             -- LS
126         N=V WHEN "1010", N/=V WHEN "1011",   -- GE or LT
127         NOT Z AND N=V WHEN "1100",           -- GT
128         Z AND N/=V WHEN "1101",              -- LE
129         true WHEN OTHERS;                    -- Always
130
131    P1: PROCESS(ir) -- find last '1' for PUSH/POP format A1
132    BEGIN
133      ind <= 0;
134      FOR i IN 0 TO NR LOOP
135        IF ir(i)='1' THEN ind <= i; END IF;
136      END LOOP;
137    END PROCESS;
138
139    P2: PROCESS (reset, clk) -- FSM of processor
140    BEGIN -- update the PC
141      IF reset = '0' THEN
142        tpc <= (OTHERS => '0');pc4_d <= (OTHERS => '0');
143        popPC <= false;
144      ELSIF falling_edge(clk) THEN
145        IF jc THEN
146          tpc <= branch_target ; -- any jumps that use immediate
147        ELSE
148          tpc <= pc4;  -- Usual increment by 4 bytes
149        END IF;
150        pc4_d <= pc4;
151        popPC <= false;
152        IF (popA1 AND ind=15) OR (popA2 AND D=15) THEN
153          popPC <= true; -- Last op= pop PC ?
```

```
154          END IF;
155        END IF;
156      END PROCESS P2;
157      -- true PC in dmd register if last op is pop AND ind=15
158      pc  <= dmd WHEN popPC ELSE tpc;
159      pc4 <= pc + X"00000004"; -- Default PC increment is 4 bytes
160      pc8 <= pc + X"00000008"; -- 2 OP PC increment is 8 bytes
161      jc <= go AND (branch OR bl OR bx OR (pop AND ind=15)); -- New PC?
162      sxt24 <= (OTHERS => imm24(23)); -- Sign extend the constant
163      bimm32 <= sxt24 & imm24 & "00"; -- Immediate for branch
164      branch_target <= r_m WHEN bx ELSE
165                       bimm32 + pc8; -- Jump are PC relative
166
167      -- Mapping of the instruction, i.e., decode instruction
168      op    <= ir(24 DOWNTO 21);   -- Data processing OP code
169      imm4  <= ir(19 DOWNTO 16);   -- imm12 extended
170      imm5  <= ir(11 DOWNTO 7);    -- The shift values of Op2
171      imm12 <= ir(11 DOWNTO 0);    -- Immediate ALU operand
172      imm24 <= ir(23 DOWNTO 0);    -- Jump address
173      -- P, B, W Decoder flags  not used
174      set <= true WHEN ir(20)='1' ELSE false; -- update flags for S=1
175      I <= true WHEN ir(25)='1' ELSE false;
176      L <= true WHEN ir(20)='1' ELSE false; -- L=1 load L=0 store
177      U <= true WHEN ir(23)='1' ELSE false; -- U=1 add offset
178      movt <= true WHEN ir(27 DOWNTO 20)= "00110100" ELSE false;
179      movw <= true WHEN ir(27 DOWNTO 20)= "00110000" ELSE false;
180      branch  <= true WHEN ir(27 DOWNTO 24)= "1010" ELSE false;
181      bl <= true WHEN ir(27 DOWNTO 24)= "1011" ELSE false;
182      bx <= true WHEN ir(27 DOWNTO 20)= "00010010" ELSE false;
183      ldr <= true WHEN ir(27 DOWNTO 26)= "01" AND L ELSE false; -- load
184      str <= true WHEN ir(27 DOWNTO 26)= "01" AND NOT L ELSE false;--store
185      popA1 <= true WHEN ir(27 DOWNTO 16)= "100010111101" ELSE false;
186      popA2 <= true WHEN ir(27 DOWNTO 16)= "010010011101" ELSE false;
187      pop <= popA1 OR popA2;
188    -- load multiple (A1) or one (A2) update sp-4 after memory access
189      pushA1 <= true WHEN ir(27 DOWNTO 16)= "100100101101" ELSE false;
190      pushA2 <= true WHEN ir(27 DOWNTO 16)= "010100101101" ELSE false;
191      push <= pushA1 OR pushA2;
192    -- store multiple (A1) or one (A2) update sp+4 before memory access
193      dp <= true WHEN ir(27 DOWNTO 26)= "00" ELSE false;-- data processing
194
195      NN <= CONV_INTEGER('0' & ir(19 DOWNTO 16)); -- Index 1. source reg.
196      M  <= CONV_INTEGER('0' & ir(3 DOWNTO 0));-- Index 2. source register
197      D  <= CONV_INTEGER('0' & ir(15 DOWNTO 12));-- Index destination reg.
198      Rn <= r(NN); -- First operand ALU
199      Rn1 <= Rn(31) & Rn; -- Sign extend 1. operand by one bit
200      r_M <= r(M);
201      rlsl <= true WHEN ir(6 DOWNTO 4)= "000" ELSE false;--Shift left reg.
202      Rm <= imm32 WHEN I -- 2. ALU operand maybe constant or register
203            ELSE r_M(30 DOWNTO 0)& "0" WHEN imm5="00001" AND  rlsl --LSL=1
204            ELSE r_M(29 DOWNTO 0)& "00" WHEN imm5="00010" AND rlsl --LSL=2
205            ELSE r_M; -- Second operand ALU
```

```
206        Rm1 <= Rm(31) & Rm; -- Sign extend 2. operand by one bit
207        Rd <= r(D);   -- Old destination register value
208        Rd1 <= Rd(31) & Rd; -- Sign extend old value by one bit
209
210        mimm32 <= sxt12 & imm12; -- memory immediate
211        dma <= Rn + Rm WHEN I
212               ELSE r(13) - 4 WHEN push -- same as STMDB sp!, {Rx}
213               ELSE r(13) WHEN pop   -- same as LDMIA sp!, {Rx}
214               ELSE Rn + mimm32 WHEN U AND NN/-15
215               ELSE Rn - mimm32 WHEN NOT U AND NN/=15
216               ELSE pc8 + mimm32 WHEN U and NN=15 -- PC-relative is special
217               ELSE pc8 - mimm32;
218        store <= (str OR push) AND (dma <= DRAMAX4); -- DRAM store
219        load  <= (ldr OR pop) AND (dma <= DRAMAX4); -- DRAM load
220        write <= str AND (dma  > DRAMAX4); -- I/O write
221        read  <= ldr AND (dma  > DRAMAX4); -- I/O read
222        mem_ena <= '1' WHEN store ELSE '0';  -- Active for store only
223        Rdd <= r(ind) WHEN pushA1 ELSE Rd;
224        not_clk <= NOT clk;
225
226     -- ARM PC-relative ops require True Dual Port RAM with dual clock
227        mem: dpram4Kx32 -- Instantiate a Block DRAM and ROM
228        PORT MAP(clk_a  => not_clk, -- System clock DRAM
229                 clk_b  => clk, -- System clock PROM
230                 addr_a => dma(13 DOWNTO 2),-- Data memory address 12 bits
231                 addr_b => pc(13 DOWNTO 2),-- Program memory address 12 bits
232                 data_a => Rdd, -- Data in for DRAM
233                 we_a   => mem_ena, -- Write only DRAM
234                 q_a    => dmd, -- Data RAM output
235                 q_b    => pmd); -- Program memory data
236        ir <= pmd;
237
238        -- ALU imm computations:
239        sxt12 <= (OTHERS => imm12(11)); -- Sign extend the constant
240        imm32 <= imm4 & imm12 & Rd(15 DOWNTO 0) WHEN movt ELSE
241                 X"0000" & imm4 & imm12 WHEN movw ELSE
242                 sxt12 & imm12; -- Place imm16 in MSBs for movt
243        imm33 <= imm32(31) & imm32; -- sign extend constant
244
245        ALU: PROCESS (op,Rm1,Rn1,in_port,dmd,reset,clk,load,read,
246                                   C,Rd1,dp,movw,imm33,movt,pop)
247        VARIABLE res: STD_LOGIC_VECTOR(32 DOWNTO 0);
248        VARIABLE Cin: STD_LOGIC;
249        BEGIN
250          IF C THEN Cin := '1'; ELSE Cin := '0'; END IF;
251          res := Rd1; -- Keep old/default
252          IF DP THEN
253            CASE op IS
254            WHEN opand => res := Rn1 AND Rm1;
255            WHEN eor | teq => res := Rn1 XOR Rm1;
256            WHEN sub  => res := Rn1 - Rm1;
257            WHEN rsb  => res := Rm1 - Rn1;
```

```
258             WHEN add | cmn => res := Rn1 + Rm1;
259             WHEN adc  => res := Rn1 + Rm1 + Cin;
260             WHEN sbc  => res := Rn1 - Rm1 + Cin -1;
261             WHEN rsc  => res := Rm1 - Rn1 + Cin -1;
262             WHEN tst  => IF movw THEN res := imm33; ELSE
263                          res := Rn1 AND Rm1; END IF;
264             WHEN cmp  => IF movt THEN res := imm33; ELSE
265                          res := Rn1 - Rm1; END IF;
266             WHEN orr  => res := Rn1 OR Rm1;
267             WHEN mov  => res := Rm1;
268             WHEN bic  => res := Rn1 AND NOT Rm1;
269             WHEN mvn  => res := NOT Rm1;
270             WHEN OTHERS => res := Rd1;
271             END CASE;
272           END IF;
273           IF load OR pop THEN res := '0' & dmd; END IF;
274           IF read THEN res := "0" & X"000000" & in_port; END IF;
275     -- Update flags and registers -------------------------------
276           IF reset = '0' THEN          -- Asynchronous clear
277             Z <= false; C <= false; N <= false; V <= false;
278             out_port <= (OTHERS => '0');
279             FOR k IN 0 TO NR LOOP -- reset to zero
280               r(k)  <= conv_std_logic_vector(k,32); --X"00000000";
281             END LOOP;
282           ELSIF rising_edge(clk) THEN -- ARMv7 has 4 flags
283             IF dp AND set THEN -- set flags N and Z for all 16 OPs
284               IF res(31) = '1' THEN N <= true; ELSE N <= false; END IF;
285               IF res(31 DOWNTO 0) = X"00000000" THEN
286                             Z <= true; ELSE Z <=false; END IF;
287               IF res(32) = '1' AND op /= mov THEN
288                             C <= true; ELSE C <=false; END IF;
289                         -- Compute new C flag except of MOV
290               IF res(32) /= res(31) AND (op = sub OR op = rsb OR op = add
291               OR op = adc OR op = sbc OR op = rsc OR op = cmp OR op = cmn)
292               THEN -- Compute new overflow flag for arith. ops
293                             V <= true; ELSE V <=false; END IF;
294             END IF;
295             IF bl THEN -- Store LR for operation branch with link aka call
296               r(14) <= pc4_d; -- Old pc + 1 op after return
297             ELSIF push THEN
298               r(13) <= r(13) - 4;
299             ELSIF read OR load OR movw OR movt OR (dp AND
300             op /= tst AND op /= teq AND op /= cmp AND op /= cmn) THEN
301               r(D) <= res(31 DOWNTO 0);--Store ALU result (not for test ops)
302             IF popA1 AND ind /= 13 THEN
303                 r(13) <= r(13) + 4;
304                 r(ind) <= res(31 DOWNTO 0);
305             END IF;
306             IF popA2 AND D /= 13 THEN
307                 r(D) <= res(31 DOWNTO 0);
308                 r(13) <= r(13) + 4;
309             END IF;
```

```
310         END IF;
311            IF write THEN out_port <= Rd(7 DOWNTO 0); END IF;
312         END IF;
313      END PROCESS ALU;
314
315   --  -- Extra test pins:
316   --  pc_OUT <= pc(11 DOWNTO 0);
317   --  ir_imm12 <= imm12;
318   --  imm32_out <= imm32;
319   --  op_code <= op; -- Data processing ops
320   --  jc_OUT <= '1' WHEN jc ELSE '0';  -- Xilinx modified
321   --  i_OUT <= '1' WHEN I ELSE '0';   -- Xilinx modified
322   --  me_ena <= mem_ena; -- Control signals
323   --  r0_OUT <= r(0); r1_OUT <= r(1);      -- First two user registers
324   --  r2_OUT <= r(2); r3_OUT <= r(3);      -- Next two user registers
325   --  sp_OUT <= r(13); lr_OUT <= r(14);    -- Compiler registers
326
327   END fpga;
```

We can use a similar I/O interface as for PICOBLAZE. For the memory, we will monitor location around the sp values. Besides the first four registers, pc, lr, and sp, we monitor also a few additional flags such as the I, jc, and me_ena. Data processing operation codes are now all 4 bits, and pc, data, and address width need to be adjusted to 32 bits. After the ENTITY, all constants, signals, flags, and the dual-port RAM component are specified (lines 53–115). Then the architecture body starts with concurrent decode for the 4-bit conditions based on the four ALU flags: Z, N, V, and C (lines 119–129). In PROCESS P1, the MSB of the register list is detected and put into variable ind for the push and pop instructions. Without the link register stack, the FSM processor controller (PROCESS P2) will be a little shorter (lines 139–156). pc increment counting is byte-wise, so each standard instruction increases the program counter by 4. For the branch target, we now have more choices: it can come from the IMM24 relative to pc for instructions such as bl or b or from r_m register for the bx instruction. Since DRAM and pc update with the falling edge, we use directly the dmd register value if the last instruction was a pop pc.

Decoding of the instruction and register identification is performed in the next code section (lines 167–208). Decoding the instruction isn't straightforward since we have many exceptions in the coding formats. We have four constants of size 4, 5, 12, and 24 to identify (line 169–172). To simplify processing several of the instructions, get an additional flag for easy processing. N, M, and D get each 4 bits. The first ALU source will always be Rn=r (NN); (using NN here since N is the ALU flag for negative numbers) the second ALU source may be a (shifted) register or the IMM32 value. Currently we only support 0, 1, or 2 shifts left. The I flag identifies all instructions that use IMM32. Next comes the input signal for the dual-port RAM that hosts the program ROM and the data memory (lines 210–224). Instantiation of a 4 K word data and program RAM with 32-bit for each word is next (lines 226–236). Since our memory only includes 4 K words, the DPRAM works similar as a direct mapped cache. Only the last 12 bits of the pc and dma are used. However, for easy transition for ANSI C programs, the address processing for the I/O port uses the full DDR RAM size as specified in DRAMAX4. The ALU processing follows in lines 238–313. The ALU output for data processing instructions is computed based on the 4-bit op code. Besides the 16 operations, it also takes care for the 16-bit move instructions described earlier. All registers are set to their register number when reset is active (lines 279–281). The flags are then updated (lines 283–294) if and only if the S bit 20 was set.

Register write are controlled by the rising edge clock. For the test-only operations and some read/write, the destination register r(D) is not written. Register sp is also updated for push and pop. Finally output assignments are made to some additional test ports; these should be disabled at the end to avoid an overloading of the I/O pins of the FPGA.

The simulation for the memory access example discussed in ANSI C program listings 11.7 is shown in Fig. 11.15. The memory values shown in the simulation match the memory content of the Monitor program from Fig. 11.13. The variables have been mapped by the compiler using the sp pointer to DRAM addresses as follows:

Variable	s	g	sarray[0]	garrray[0]
DRAM address (decimal)	4090	745	4076	746

The major events of the DRAM test simulation are s=1 @ 120 ns; g=2 @ 200 ns; sarray[0]=s @ 240 ns; garray[0]=g @ 280 ns; sarray[1]=0x1357 @ 380 ns; garray[2]=2468 @ 460 ns.

The HDL simulation for the nesting loop code from Program 11.9 is shown in Fig. 11.16. The address values in the range of sp are very similar to the assembler debug display in Table 11.3. However, we do only push lr and not r3, so location of lr values in memory differs. Special attention should be given to the return link register r(14)=lr for the loops. The major events of the loop behavior simulation can be summarized as follows: s1=0x1233 @ 140 ns; lr=0x2AC @210 ns;

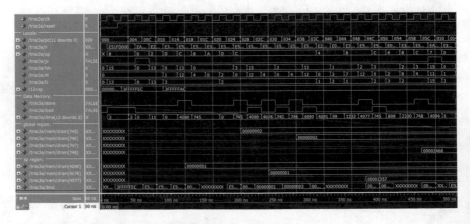

Fig. 11.15 Simulation shows the TRISC3A memory access

$lr=0x2AC$ stored at $DRAM(4074)$ @ 220 ns; $lr=0x28C$ @ 290 ns; $lr=0x28C$ stored at $DRAM(4073)$ @ 300 ns; loading pc with $lr=0x278$ @ 420 ns; loading pc with $0x28C$ @ 440 ns; and loading pc with $0x2AC$ @ 460 ns. The subroutines are entered as follows: level1 @ 200 ns, level2 @ 280 ns, and level3 @ 360 ns. The final DRAM values for s1, s2, and s3 are located at $DRAM(4088)$, $DRAM(4087)$, and $DRAM(4086)$, respectively.

Finally, Fig. 11.17 shows the overall TRISC3A architecture together with the implemented instructions. Implementation of additional instructions is left to the reader as exercise at the end of the chapter; see Exercise 11.58–11.61.

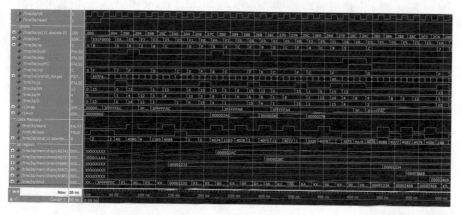

Fig. 11.16 Simulation shows the TRISC3A nesting loop behavior

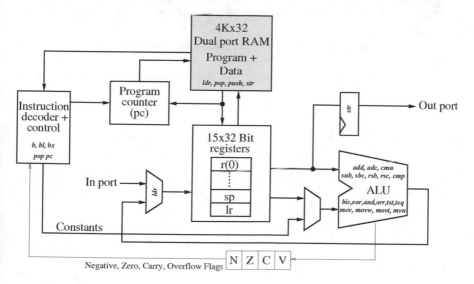

Fig. 11.17 The final implemented tiny ARMv7 Cortex-A9 aka TRISC3A core. Instruction (in italics) mainly associated with major hardware units. Substantial memory blocks (excluding registers) in gray

Synthesis Results and Lesson Learned

The synthesis results for the TRISC3A are shown in Table 11.4. The second and third columns show Xilinx VIVADO synthesis results for VHDL and Verilog, respectively. The fourth and fifth columns show Altera QUARTUS synthesis results for VHDL and Verilog, respectively. In Xilinx VIVADO the Dual-Port 4K×32 RAM can be implemented with four RAMB36 or eight RAMB18 blocks in the Xilinx device. The Altera BRAM is only 10 Kbits, and we need 16 to build the 4K×32 dual-port memory. The overall speed for the DE1 SoC board Altera is close to the default speed of 50 MHz. The Xilinx system clock with 125 MHz is most likely too high. We may need to add a PLL or use a four divider (i.e., 125/4 = 31.25 MHz) to meet the maximum clock rates. The Xilinx synthesizer used 31 addition LUTs to synthesize the divide by four.

We finally have a working microprocessor with 25 ARMv7 operations implemented. Major units such as register array, logical, arithmetic operation, input and output, data memory, branch, and stack management are included. Missing from the complete instruction set are (besides the three complete other ISA in 8 and 16 bits) those group we did not need for our tasks so far such as shift and rotate, multiply and divide, interrupt handling, and coprocessor control. Some of these are left to the reader as exercises at the end of the chapter.

The "lesson learned" from looking under the hood is that the ARMv7 architecture has four major instruction set improvements over the other processor discussed so far, namely:

- pc-relative addressing to load 32-bit constants using one instruction.
- PUSH operation to save the link register and at the same time update the stack pointer.
- POP of the program counter from data memory without first restoring the link register.
- Optional shift of the second operand (to implement multiply by 4) allows a faster address computation.

Table 11.4 TRISC3N synthesis results for Altera and Xilinx tools and devices

	VIVADO 2016.4		QUARTUS 15.1	
Target device	Zynq 7 K xc7z010t-1clg400		Cyclone V 5CSEMA5F31C6	
HDL used	VHDL	Verilog	VHDL	Verilog
LUT/ALM	1637	1615	1300	1301
used as Dist. RAM	0	0	0	
Block RAM RAMB36	4	4	–	–
or M10K	–	–	16	16
DSP Blocks	0	0	0	0
HPS	0	0	0	0
Fmax/ MHz	36.36	41.67	43.76	42.81
Compile time	4:20	4:14	5:25	4:17

The push/pop alone had reduced the level1 subroutine call from Nios II eight instructions to five instructions for the ARMv7, a 60% improvement in execution time. So, besides the architecture features, the ISA improvements are a major contributor to the higher DMIPS rating for the ARMv7 processor.

Review Questions and Exercises

Short Answer

11.1. What are the main differences between ARMv7 and ARMv8?
11.2. Describe two architectural features of the ARMv7 architecture different from MicroBlaze and Nios II.
11.3. What is MIO, and why do we need it?
11.4. What are the four instruction groups in ARMv7 assembler?
11.5. Name six addressing modes used by ARMv7. For each mode specify an instruction example.
11.6. How are nested loops implemented by GCC for ARMv7?
11.7. Explain the push and pop instructions in ARMv7? When would you use these? What is the price to pay for having these instructions?
11.8. Name five advantages of the ARMv7 HDL design approach compared to IP core.
11.9. ARMv7 has no nop or clear register instructions. Define "pseudo-instructions" using the instructions from ARMv7 that accomplish these tasks.
11.10. Give ARMv7 architecture examples for the RISC design principles: (a) simplicity favors regularity, (b) smaller is faster, and (c) good designs demand good compromise.

Fill in the Blank

11.11. The bits _____ of the ARMv7 instruction are used to encode the major instruction groups.
11.12. The bits _____ of the ARMv7 instruction are used to encode conditional code.
11.13. The ARMv7 is a ____ address machine.
11.14. The TRISC3A uses _____ instructions of the ARMv7 ISA.
11.15. The ARM IP core was used _____ times in 2015.
11.16. The four immediate constants in the TRISC3A instructions are of length _____.

11.17. The ARMv7 (pre add) operation `str r2, [r1, 0x40]` will be encoded in hex as ____

11.18. The ARMv7 operation `movw r5, 0x1234` will be encoded in hex as _____

11.19. The ARMv7 operation `add r3,r4,r2 #2` will be encoded in hex as _____

11.20. The ARMv7 operation `bne #-4` will be encoded in hex as _____

11.21. The ARMv7 operation _____ will be encoded in hex as 0xE3451432.

11.22. The ARMv7 operation _____ will be encoded in hex as 0x0AFFFFF8.

11.23. The ARMv7 operation _____ will be encoded in hex as 0xE25670FF.

11.24. The ARMv7 operation _____ will be encoded in hex as 0xE5912008.

True or False

11.25. _____ The ARMv7 has a lower DMIPS rating than MICROBLAZE.

11.26. _____ The ARMv7 can load a 12-bit, 16-bit, or 32-bit constant with one instruction.

11.27. _____ The ARMv7 has a usual and reverse subtract operation.

11.28. _____ The Altera uses ARMv7 and Xilinx ARMv8 architectures for their FPGA core.

11.29. _____ The Altera monitor program compiler messages show up in the Info & Errors window.

11.30. _____ The VIVADO "connection automation" will take care also of all I/O ports. XDC file is never needed.

11.31. _____ The Altera monitor program default start shows five windows: Disassembly, Memory, Registers, Terminal, and Info & Errors.

11.32. _____ The ARMv7 performance cannot be improved with a CIP.

11.33. _____ Floating-point CIP will speed up ARMv7 basic FP operations.

11.34. _____ The ARMv7 bottom-up design approach requires detailed knowledge of the board configuration to specify the correct MIO.

11.35. _____ The ARMv7 microprocessor can be programmed in assembler and C/C++.

11.36. _____ To download the ARM configuration to the DE1 SoC board, we need QUARTUS; the Altera monitor program cannot be used for that.

11.37. _____ The VIVADO block automation for ZYNQ will add additional hardware IP to the design.

11.38. Specify standard features (**T**rue/**F**alse) for the ARMv7 and TRISC3A cores in the table below.

Feature	ARMv7	Trisc3a
Softcore architecture		
I-Cache		
D-Cache		
Pipelined		
Used with different FPGA and vendors		
Harvard architecture		
Has a very small LE count		

Projects and Challenges

11.39. Run an example from Xilinx SDK templates, and write a short essay (1/2 page) what the examples are about. What I/O components are used?

11.40. Run an example from the Altera *app_software_nios2_C* folder from the *University_program* → *Computer_Systems*, and write a short essay (1/2 page) what the examples are about, what I/O components are used.

11.41. Develop an assembler or ANSI C program to implement a left, right car turn signal and emergency light similar to the Ford Mustang: 00X, 0XX, XXX for a left turn where 0 is off and X is LED on. Use X00, XX0, and XXX for right turn signal. Use the slider switch for the left/right selection. For emergency light both switches are on. The turn signal sequence should repeat once per second. See mustangQT.MOV or mustangWMP.MOV for a demo.

11.42. Write an ANSI C program that implements a running light. The speed of the running light should be determined by the SW values, and any change should be displayed at the Altera Monitor Terminal. See led_coutQT. MOV or led_countWMP.MOV for a demo.

11.43. Develop an ANSI C programs that lights up one LED only and moves to the left or right by pushing button 4 or 3, respectively.

11.44. Repeat the previous Exercise 11.43 with a three-color bar on the VGA or HDMI monitor.

11.45. Develop an assembler or ANSI C program that has a random number generator, using ADD and XOR instructions. Display the random number on the LEDs. What is the period of your random sequence?

11.46. Build a stopwatch that counts up in second steps. Use three buttons: start clock, stop or pause clock, and reset clock. Using the LEDs, LCD, OLED, or the seven-segment display (see Table 10.1) for display. See stop_watch_ledQT.MOV or stop_watch_ledWMP.MOV for a demo.

11.47. Use the push buttons to implement a reaction speed timer. Turn on one of the four LEDs, and measure the time it takes until the associate button is pressed. Display the measurement after ten tries on LED, LCD, OLED, seven-segment display, or VGA/HDMI.

11.48. Write ANSI C program to display all characters *A–Z* on a seven-segment display. Take special attention for characters *X, M,* and *W.* You may use a bar to indicate double length, e.g., $m = \bar{n}; w = \bar{v}$. Produce a short video that shows all characters on the seven-segment display. See `ascii4sevensegment.MOV` for a demo. For the ZyBo you will need the seven-segment expansion Pmod; see Sect. 10.2 and Table 10.1.

11.49. Use the character set from the last Exercise 11.48 to generate a running text on the seven-segment display using the current month and year and/or our name or your initials if your name is too long. See `idQT.MOV` and `idWMP.MOV` from the CD for a demo.

11.50. Write ANSI C program to reproduce the ASCII from Fig. 5.2 for all printable characters (Decimal 32…127) on the VGA/HDMI display and the Terminal window.

11.51. Write an ANSI C program for the Altera DE1 SoC Computer or Xilinx HDMI-out demo that draws a circle on VGA/HDMI monitor using Bresenham circle algorithm: start with $y = 0; x = R$. Move in *y*-direction one pixel in each iteration. Decrement *x* if error $|x^2 + y^2 - R^2|$ is smaller for $x' = x - 1$; otherwise, keep *x*. Use symmetry to complete all four segments of the circle.

11.52. Use the Bresenham algorithm from the last exercise to draw a filled circle in a specified color.

11.53. Write an ANSI C program for the Altera DE1 SoC ARM Computer or Xilinx HDMI-out demo that draws an ellipse on the VGA/HDMI monitor using standard floating-point arithmetic: $x = a*\cos([0:0.1:\pi/2]); y = b*\sin([0:0.1:\pi/2])$. Use symmetry to fill the ellipse with horizontal lines. Is the ellipse completely filled? What is the problem with this type of algorithm?

11.54. Write an ANSI C program for the Altera DE1 SoC Computer or Xilinx HDMI-out demo that draws an ellipse on VGA using Bresenham algorithm: start with $y = 0; x = R$. Move in *y* direction one pixel in each iteration. Decrement *x* if error $|x^2/a^2 + y^2/b^2 - R^2|$ is smaller for $x' = x - 1$; otherwise, keep *x*. Use symmetry to fill the ellipse with horizontal lines. Is the ellipse completely filled? What is the problem with this type of algorithm?

11.55. Write an ANSI C program for the Altera DE1 SoC ARM Computer or Xilinx ZyBo HDMI-out demo that implements a display of the four fractals (use range $x,y = -1.5..1.5$). The complex iteration equations are as follows:

(a) Mandelbrot: $c = x + i*y; z = 0$
(b) Douday rabbit: $z = x + iy; c = -0.123 + j0.745$
(c) Siegel disc: $z = x + i*y; c = -0.391 - j0.587$
(d) San marco: $z = x + i*y; c = -0.75$

Find the number of iteration (and code in unique color) $z = z.*z + c$, such that $|z|^2 > 5$; see Fig. 11.18. The name or the fractal should be displayed in the center of the VGA screen. See `4fractals.MOV` for a demo.

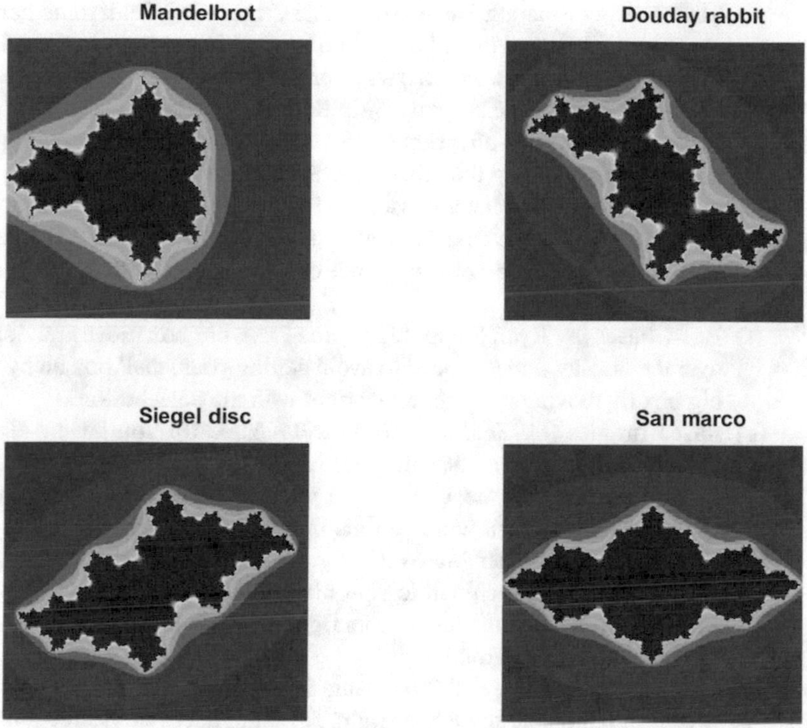

Fig. 11.18 Four fractal examples from Project 11.54

11.56. Write an ANSI C program for the Altera DE1 SoC ARM Computer or Xilinx HDMI-out demo that implements a zoom of a Mandelbrot fractals. Compute the fractal at 25 different zoom levels, and then display the 25 images in a series. The number of the frames should be displayed in the center of the VGA screen. See `MandelbrotZoomColor.MOV` and `MandelbrotZoomGrey.MOV` for a demo.

11.57. Write an ANSI C program for the Altera DE1 SoC ARM Computer or ZyBo Xilinx HDMI-out demo that implements a game such as:

(a) Tic-Tac-Toe (by Christopher Ritchie, Rohit Sonavane, Shiva Indranti, Xuanchen Xiang, Qinggele Yu, Huanyu Zang 2016 game 3/6/8; 2017 game 9/10/11): Use two players. Enter selecting with slider switches and the push button to submit your selection for player 1 (button 3) and player 2 (button 0). Alternatively toggle by default, and display which player turn it is on LED or seven segment. For a DE115 you can use slider switches 0–8 for player 1 and 9–17 slider switches for player 2.

(b) Reaction Time (by Mike Pelletier 2016 game 4): Light up one of the three squares on the screen, and measure the time until the correct button was pressed.

(c) Minesweeper (Sharath Satya 2016 game 7): In a 3×3 field, three bombs are hidden. You need to find the six boxes that do not contain a bomb. If you find all six, you win; otherwise, player loses.

(d) Pong (by Samuel Reich and Chris Raices 2017 game 4/5): Use one player only. Use one direction to move the ball and the push button for moving the paddle on the other axis. Set a timer to count how long the game runs until the player misses the ball. Ball may be a small square.

(e) Simon Says (by Melanie Gonzalez and Kiernan Farmer 2016 game 1/2): Repeat the three-color sequence of VGA square with the push buttons 1, 2, and 3.

(f) Box Chaser (by Ryan Stepp 2017 game 7): A big box moves randomly over the display, and you need to avoid getting your small box hit by this big box by moving it to the left or right with the push buttons.

(g) Space Invades (by Zlatko Sokolikj 2017 game 6): You have a flying object that needs to be hit with your firing gun shooting in y direction. Your gun moves on the x axis, so a total of three buttons are in use. Counting the target hit within a time frame gives the score.

(h) Flappy Bird (by Javier Matos 2017 game 3): Pressing a button moves your bird in y direction, while your bird continuously drops. Obstacles such as wall with a hole move from right to left. Game is over when the bird hits the wall or ground.

(i) Bounce (by Kevin Powell 2016 game 5). A three-color bar is moved by two push buttons to the left or right. A ball bounces in y direction and changes color whenever the ball reaches the top. The ball color must match the bar color when the ball hits the ground; otherwise, the game is over.

For some of these games, you will find more detailed description in Sect. 11.2. All games have been designed by students within a semester project competition in 2016 and 2017. Write a brief description of your implementation including Computer System, I/O component used, and game rule. Submit a short video of a game demonstration.

11.57. Write an ANSI C program for the Altera DE1 SoC ARM Computer or Xilinx ZyBo HDMI-out demo that implements a game such as:

(a) Four in a Row: Select the row with the slider switch, and then use a push button for submitting your choice; see Fig. 11.19a for an example.

(b) Tetris: Let different elements move slowly in y direction; see Fig. 11.19b for an example.

(c) Sudoku: Place easy, normal, or hard starting configurations on the board. Then allow with three entries: row, column, and number (1–9) to make your submission. See Fig. 11.19c for an example.

11.58. Run short sequence of ANSI C code to identify the missing instruction in TRISC3A set for:

(a) `if` and `switch` statements

(b) The factorial example

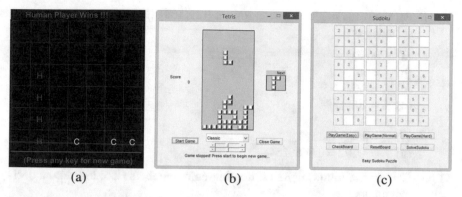

Fig. 11.19 Additional example games. (**a**) Four in a Row. (**b**) Tetris. (**c**) Sudoku

11.59. Add additional ARMv7 features to the TRISC3A architecture, i.e., instruction such as:

    (a) Multiply
    (b) Divide
    (c) 32-bit shift
    (d) Rotate

    Add the HDL code, and test with a short assembler test program. Report the difference in synthesis (area and speed) compared to the original design.

11.60. Develop a TRISC3A program to implement a multiplication of two 32-bit numbers via repeated additions. Test your program by reading the SW values and computing $q = x * x$ output to LEDs.

11.61. Add the missing instruction from the ISA to run the factorial program on the hardware. Test the single new instructions, and then run the factorial program.

11.62. Consider a floating-point representation with a sign bit, $E = 7$-bit exponent width, and $M = 10$ bits for the mantissa (not counting the hidden one).

    (a) Compute the bias using Eq. (9.2).
    (b) Determine the (absolute) largest number that can be represented.
    (c) Determine the (absolutely measured) smallest number (not including denormals) that can be represented.

11.63. Using the result from the previous exercise:

    (a) Determine the representation of $f_1 = 9.25_{10}$ in this $(1,7,10)$ floating-point format.
    (b) Determine the representation of $f_2 = 10.5_{10}$ in this $(1,7,10)$ floating-point format.
    (c) Compute $f_1 + f_2$ using floating-point arithmetic.
    (d) Compute $f_1 * f_2$ using floating-point arithmetic.
    (e) Compute f_1/f_2 using floating-point arithmetic.

11.64. For the IEEE single-precision format [I85, I08], determine the 32-bit representation of:

   (a) $f_1 = -0$
   (b) $f_2 = \infty$
   (c) $f_3 = -0.6875_{10}$
   (d) $f_4 = 10.5_{10}$
   (e) $f_5 = 0.1_{10}$
   (f) $f_6 = \pi = 3.141593_{10}$
   (g) $f_7 = \sqrt{3}/2 = 0.8660254_{10}$

11.65. Convert the following 32-bit floating-point numbers [I85, I08] into a decimal number:

   (a) $f_1 = 0\times44FC6000$
   (b) $f_2 = 0\times49580000$
   (c) $f_3 = 0\times40C40000$
   (d) $f_4 = 0\times C1AA0000$
   (e) $f_5 = 0\times3FB504F3$
   (f) $f_6 = 0\times3EAAAAAB$

11.66. Design a 32-bit floating point ALU with a select sign sel to choose one of the following operations: (0) fix2fp (1) fp2fix (2) add (3) sub (4) mul (5) div (6) reciprocal (7) power-of-2 scale. Test your design using the simulation shown in Fig. 11.20.

11.67. Develop a seven-segment display using the flash program listing 11.38 to count up in seconds. This will require to modify the pin assignment that not only LEDs but also the seven-segments are driven by the Trisc3A output port. For the ZyBo you will need the seven-segment expansion Pmod; see Sect. 10.2 and Table 10.1.

11.68. Write a short MATLAB or C/C++ program to generate a new memory initialization file for the HDMI encoder from Chap. 10 and add the CIP to your ARM µP system. Check the file eupsd_hex.mif for required format. Use as new text:

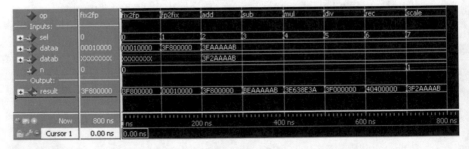

Fig. 11.20 MODELSIM RTL simulation of eight functions of the floating-point arithmetic unit fpu. 0001.0000 in sfixed becomes 3F800000 in hex code for FLOAT32. The arithmetic test data are 1/3 = 0×3EAAAAAB and 2/3 = 0×3F2AAAAA

      (a) Your home country constitution

      (b) Your favorite poem

      (c) Your favorite joke

      (d) Your favorite song text

11.69. Use the HDMI CIP files from Chap. 10 to build a standalone ZyBo system not requiring a microprocessor. For the ports use the following assignment: `reset=KEY[0]; WE=KEY[3]; DATA = { 4'h3, SW};COUNTER_X=X<<3;COUNTER_Y=Y<<3; LEDR[0]=CLOCK_25; LEDR[1] = CLOCK_250; LEDR[2] = DLY_RST; LEDR[3] = LOCKED;` (keep `SW[2]=REG; SW[3]=GRN; SW[1:0]=font;CLOCK_125=sys clock 125 MHz`). Enable the ports in the XDC file. Keep HDMI output names. Use the text from the previous problem for display. Add your name in the banner.

11.70. Modify the HDMI CIP from Chap. 10 to be used as text terminal for the VGA port. If you have DE1 SoC or ZYBO, use the available VGA port; for gen 2 ZYBO, buy a VGA Pmod (ca. 9 USD/Euro); see Fig. 10.6a. You will also need a VGA cable. The VGA uses the same timing for row and column control but analog output signals. You will also need to forward the internal H/V sync signals to the output. The VGA port contains three DAC for this purpose. Study the ZYBO master XDC file for required ZYBO signals. Look for the `##VGA Connector` section. You will use the five MSBs for red and blue and six MSBs for green.

11.71. Use the VGA code from the previous problem to build a VGA text terminal CIP. Use your ARM Basic Computer to test your CIP.

11.72. Write an ANSI C program for the ZyBo that test a component:

      (a) USB OTG: Check a device such USB memory stick for R/W errors, and measure maximum transfer rate.

      (b) SD Card: Check memory R/W and measure maximum transfer rate.

      (c) Custom Instruction: Implement S-Boxes used in the DES algorithm.

      (d) Ethernet: Try to communicate via Ethernet cable to other device.

      (e) JTEG-UART: Measure file transfer rate from/to the ARMv7.

      (f) XADC: Read data from XADC, and put into file on host system.

      (g) VGA: Implement text terminal using the HDMI CIP.

      (h) WiFi: Add WIFI Pmod (see Table 10.1), and try to communicate with other devices.

      (i) ACL: Add three-axis accelerometer Pmod (see Table 10.1), and build a tremor sensor.

      (j) BT2: Add Bluetooth Pmod (see Table 10.1), and try to communicate with other devices.

The Project g-j may require to purchase small addition modules that are attached through the ZyBo Pmod connectors (see Table 10.1).

11.73. Write an ANSI C program for the DE1 SoC Computer that test a component:

(a) AUDIO-IN: Use an amplitude tracking to determine sags or swells in a power signal.
(b) AUDIO-OUT: Implement an "echo" that changes with SW settings.
(c) USB: Check a device such as USB memory stick for R/W errors, and measure maximum transfer rate.
(d) InDA: Allow an optical device to transmit data.
(e) PS2: Input data from keyboard, and display on VGA or LED.
(f) VGA: Let "EuPSD" fly over the VGA monitor.
(g) SD Card: Check memory R/W, and measure maximum transfer rate.
(h) Custom Instruction: Implement S-Boxes used in the DES algorithm.
(i) Ethernet: Try to communicate via Ethernet cable with another device.
(j) JTEG-UART: Measure file transfer rate from/to the Nios II.
(k) G-sensor: Build a tremor sensor.
(l) Video: Edge detection or resampler.

11.74. Develop a `srec2prom` utility that reads a Motorola `*.srec` file (SREC file format, see, for instance, [A01]) as generated by GCC in the Altera Monitor program. Generate a complete VHDL file that includes the code as constant values or for Verilog a `*.mif` table that can be loaded with a `$readmemh()`. Test your program with the TRISC3A `flash.c` example.

11.75. Identify missing instructions from TRISC3A set that are needed if you use the `srec2prom` from the previous exercise. Implement the missing instruction in HDL and run the `srec2prom` generated program.

References

[A01]   Altera, Converting .srec Files to .flash Files for Nios Embedded Processor Applications, White Paper (2001)
[A11a]   Altera, *Cyclone V Device Handbook*, vol. 1–3, San Jose (2015)
[A11b]   ARM, *Cortex-A Series Programmer's Guide* (Cambridge, England, 2011), 455 pages
[A14]   ARM, *ARM Architecture Reference Manual*, ARMv7-A and ARMv7-R edn. (Cambridge, England, 2014)
[A15a]   Altera, *DE1- SoC Computer System with ARM Cortex-A9*, San Jose (2015)
[A15b]   Altera, *Introduction to the ARM ® Processor Using Altera Toolchain 15.0*, San Jose (2015)
[CEE14]   L. Crockett, R. Elliot, M. Enderwitz, R. Stewart, *The Zynq Book* (Strathclyde Academic Media, Glasgow, 2014)
[D14]   Digilent, *ZYBO Reference Manual* (Pullman, 2014)
[I85]   IEEE, Standard for binary floating-point arithmetic. IEEE Std. **754-1985**, 1–14 (1985)
[I08]   IEEE, Standard for binary floating-point arithmetic. IEEE Std. **754-2008**, 1–70 (2008)
[L06]   S. Lee, *Advanced Digital Logic Design: Using VHDL, State Machines, and Synthesis for FPGAs*, 1st edn. (Thomson, Toronto, 2006)

[MSC06] U. Meyer-Baese, D. Sunkara, E. Castillo, E.A. Garcia, Custom Instruction Set NIOS-Based OFDM Processor for FPGAs, in *Proceedings of SPIE International Society for Optical Engineering Orlando* (2006), pp. 6248o01–15

[M14]   U. Meyer-Baese, *Digital Signal Processing with Field Programmable Gate Arrays*, 4th edn. (Springer, Berlin, 2014) 930 pages

[PH17]   D. Patterson, J. Hennessy, *Computer Organization & Design: The Hardware/Software Interface, ARM Edition* (Morgan Kaufman Publishers, Inc., San Mateo, 2017)

[S04]   D. Sunkara, Design of Custom Instruction Set for FFT using FPGA-Based Nios processors, Master's thesis, Florida State University (2004)

[T14]   TerASIC, *DE1-SoC User Manual*, Altera University Program (2014)

[VTT09] J. Vandewalle, L. Trajkovic, S. Theodoridis, Introduction and outline of the special issue on circuits and systems education: experiences, challenges, and views. IEEE Circuits Syst. Mag. **Q1**, 27–33 (2009)

[X18]   Xilinx, Zynq-7000 SoC Data Sheet: Overview, Product Specification DS190 (2018)

Appendix A: Verilog Source Code and Xilinx Vivado Simulation

Abstract This appendix has the Verilog source code for the example in the book. It also shows the VIVADO XSIM behavior simulation results for the examples. It starts with an overview how the design files of the project have been named that are on the book CD or GitHub.

Keywords Verilog·Xilinx·Vivado·Quartus·Verilog source code·Simulation·Test bench·Tiny RISC·Custom Intellectual Property (CIP)·URISC·TRISC2·TRISC3N·T RISC3MB·TRISC3A·Scratchpad RAM·ROM program·Data RAM·Dual port RAM·ModelSim·LED toggle

Project Files Explained

The files on the CD include a minimum of eight VHDL and eight Verilog projects files for all synthesizable microprocessor examples: URISC, TRISC2, TRISC3NIOS, TRISC3MB, and TRISC3ARM for QUARTUS and VIVADO. The CD includes also project directories for Nios II, MicroBlaze, ARM Basic Computer System, as well as the three Custom IP designs for floating-point, FFT bit-reverse, and video HDMI decoder. A few additional "utility" directories for compiler design source code are also included. Each of the project's has at least the following VHDL files in directory `Quartus` for Altera/Intel and in directory `Vivado` for Xilinx:

`project.vhd`	Original VHDL top level design
`project.do`	MODELSIM VHDL stimuli files for design
`project_msim.gif`	The snapshot of the VHDL MODELSIM simulation (no TB)
`project_tb.vhd`	The 7 section VHDL testbench (TB) file including data stimuli
`project_tb.do`	MODELSIM testbench script with GSR delay and no data stimuli

© Springer Nature Switzerland AG 2021
U. Meyer-Baese, *Embedded Microprocessor System Design using FPGAs*,
https://doi.org/10.1007/978-3-030-50533-2

`project_tb_msim.gif`	The snapshot of the VHDL MODELSIM testbench simulation
`project_tb_behav.wcfg`	The waveform file of the VHDL behavior VIVADO TB simulation
`project_tb_behav.gif`	The snapshot of the VHDL behavior VIVADO TB simulation

All designs have been simulated with MODELSIM-Altera and XSIM VIVADO. In general, for each project, you will find over 8 files in the Verilog folders. The Verilog files in directory `vQuartus` for Altera/Intel and in directory `vVivado` for Xilinx are:

`project.v`	Original Verilog top level design
`project.do`	MODELSIM `Verilog stimuli files for` MODELSIM–Altera (no GSR delay)
`project_msim_v.gif`	The snapshot of the Verilog MODELSIM simulation (no TB)
`project_tb.v`	The 5 section Verilog testbench file including data stimuli
`project_tb.do`	MODELSIM TB script file with GSR delay and no data stimuli
`project_tb_v_msim.gif`	The snapshot of the Verilog MODELSIM TB simulation
`project_tb_v_behav.wcfg`	The waveform file of the Verilog behavior VIVADO TB simulation
`project_tb_v_behav.gif`	The snapshot of the Verilog behavior VIVADO TB simulation

In total you will find a minimum of eight files in the Verilog folders. A few designs also needed additional files such as memory initializations and components. The results for the simulations can be found in the directory `SimulationFigures`.

The following section has the (Tiny) microprocessor source code and simulation for URISC, TRISC2, TRISC3NIOS, TRISC3MB, and TRISC3ARM and required components. Not shown are testbench files for TRISC and TCL/DO file scripts.

Verilog Project Files URISC Processor

The first example from Chap. 2 describes the URISC processor. The Verilog code for the program ROM is shown first below followed by the URISC Verilog description. Finally, the simulation from XSIM shows reading from the in_port, storing the value to the out_port, and continues this in a loop forever.

Verilog Program ROM A.1: LED Test Program for URISC Translated to Verilog

```verilog
1   // ============================================================
2   // IEEE STD 1364-2001 Verilog file: rom128x16.v
3   // Author-EMAIL: Uwe.Meyer-Baese@ieee.org
4   // ============================================================
5   // Initialize the ROM with $readmemh. Put the memory
6   // contents in the file io.hex.  Without this file,
7   // this design will not compile. See Verilog
8   // LRM 1364-2001 Section 17.2.8 for details on the
9   // format of this file.
10  // ============================================================
11  module rom128x16
12    (input   clk,              // System clock
13     input [6:0] address,      // Address input
14     output[15:0] data);       // Data output
15  // ============================================================
16  // Declare the ROM variable
17    reg [15:0] rom[127:0];
18
19    initial
20
21    begin
22      $readmemh("io.mif", rom);
23    end
24
25    assign data = rom[address];
26
27  endmodule
```

The Verilog program ROM read the HEX memory initialization file (*.hex extension not working) data that are plain 16 bits, i.e., 4-digit HEX code where each line represents a single instruction for the URISC processor shown next.

Verilog Code A.2: The URISC Processor

```
1    // ============================================================
2    // IEEE STD 1364-2001 Verilog file: urisc.v
3    // Author-EMAIL: Uwe.Meyer-Baese@ieee.org
4    // ============================================================
5    // Title: URISC microprocessor
6    // Description: This is the top control path/FSM of the
7    // URISC, with a 3 state machine design
8    // ============================================================
9    module urisc
10    (input  clk,                 // System clock
11     input  reset,               // Active low asynchronous reset
12     input [7:0] in_port,        // Input port
13     output reg [7:0] out_port);// Output port
14    // ============================================================
15    // FSM States:
16      parameter FE=0, DC=1, EX=2;
17      reg [1:0] state;
18
19    // Register array definition
20      reg [7:0] r [15:0];
21
22    // Local signals
23      wire [15:0] data;
24      reg [15:0] ir;
25      reg mode;
26      wire jump;
27      reg [3:0] rd, rs;
28      reg [6:0] address, pc;
29      wire [7:0] dif;
30      reg [8:0] result;
31
32      rom128x16 prog_rom      // Instantiate the LUT
33      ( .clk(clk),            // System clock
34        .address(pc),         // Program memory address
35        .data(data));         // Program memory data
36
37      always @(posedge clk or negedge reset) //PSM w/ ROM behavioral style
38      begin : States                    // URISC in behavioral style
39          integer k;             // Temporary counter
40          if (~reset) begin     // all set register to -1
41            state <= FE;
42            for (k=1; k<=15; k=k+1) r[k] = -1;
43            pc <= 0;
44          end else begin      // use rising edge
45            case (state)
46            FE: begin          // Fetch instruction
47                  ir <= data; // Get the 16-bit instruction
48                  state <= DC;
49                end
```

```
50              DC: begin        // Decode instruction; split ir
51                  rd <= ir[15:12]; // MSB has destination
52                  rs <= ir[11:8];// second source operand
53                  mode <= ir[7];  // flag for address mode
54                  address <= ir[6:0]; // next PC value
55                  state <= EX;
56                end
57              EX: begin       // Process URISC instruction
58                  result = {r[rd][7], r[rd] } - {r[rs][7], r[rs] };
59                  if (rd>0) r[rd] <= result[7:0];// do not write input port
60                  if (~result[8]) begin pc <= pc + 1;//test is false inc PC
61                    end else begin   // result was negative
62                        if (mode) pc <= address; //absolute addressing mode
63                        else pc <= pc + address; //relative addressing mode
64                    end
65                  r[0] <= in_port;
66                  out_port <= r[15];
67                  state <= FE;
68                end
69            default : state <= FE;
70            endcase
71          end
72     end
73
74     // Extra test pins:
75     assign jump = result[8];
76     assign dif = result[7:0];
77
78   endmodule
79
```

The testbench for the URISC processor shown next as example for the other testbenches on the CD.

Test Bench A.3: URISC TB for Verilog

```
1    // Set reference time to 1 ns and precision to 1 ps
2    `timescale 1 ns / 1 ps
3    // ==================================================
4    module urisc_tb;
5    // ==================================================
6
7    // ========= Signal definition no initial values: =========
8      reg  clk;                // System clock
9      reg  reset;              // Asynchronous reset
10     reg [7:0] in_port;       // Input port
11     wire [7:0] out_port;     // Output port
12
13     parameter T = 20; // Simulation tick in ns
14
15   // ========= Unit under test instantiation: =========
16       urisc UUT(
17           .clk(clk),
18           .reset(reset),
19           .in_port(in_port),
20           .out_port(out_port));
21
22   // ========= Periodic signals: =============================
23     initial begin
24       clk = 0;
25       forever #(T/2) clk = ~clk;
26     end
27
28   // =================== Add stimuli: ========================
29     initial begin
30       reset = 1;
31       in_port = 5;
32       #105; // Time=105 ns
33       reset = 0;
34       #645; // Time=750 ns
35     end
36
37   endmodule
```

The Verilog VIVADO behavior simulation for the io.mif test program is shown next (Fig. A.1).

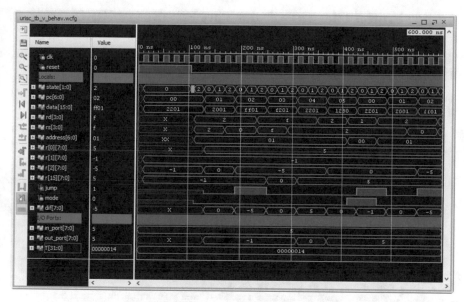

Fig. A.1 Verilog Vivado behavior simulation of URISC processor running the I/O program

Verilog Project Files TRISC2 aka Tiny PICOBLAZE RISC Processor

The final Verilog design for the TRISC2 is shown in the following listing.

Verilog Code A.4: TRISC2 (Final Design)

```
1    //==============================================================
2    // IEEE STD 1364-2001 Verilog file: trisc2.v
3    // Author-EMAIL: Uwe.Meyer-Baese@ieee.org
4    //==============================================================
5    // Title: T-RISC 2 address machine
6    // Description: This is the top control path/FSM of the
7    // T-RISC, with a single 3 phase clock cycle design
8    // It has a stack 2-address type instruction word
9    // ==============================================================
10   module trisc2
11     (input  clk,                // System clock
12      input  reset,              // Asynchronous reset
13      input [7:0] in_port,       // Input port
14      output reg [7:0] out_port  // Output port
     // The following test ports are used for simulation only and should be
15   // comments during synthesis to avoid outnumbering the board pins
16   //   output signed [7:0] s0_out,     // Register 0
17   //   output signed [7:0] s1_out,     // Register 1
18   //   output signed [7:0] s2_out,     // Register 2
19   //   output signed [7:0] s3_out,     // Register 3
20   //   output jc_out,                  // Jump condition flag
21   //   output me_ena,                  // Memory enable
22   //   output z_out,                   // Zero flag
23   //   output c_out,                   // Carry flag
24   //   output [11:0] pc_out,           // Program counter
25   //   output [11:0]  ir_imm12,        // Immediate value
26   //   output [5:0]  op_code           // Operation code
27     );
28   // ==============================================================
29
30     wire [4:0] op5;
31     wire [5:0] op6;
32     wire [7:0] imm8, dmd, x, y;
33     wire [11:0] imm12;
34     wire [7+1:0] x0, y0;
35     reg [7:0] s [0:15];
36     reg [11:0] lreg [0:30];
37     reg [11:0] pc;
38     reg [4:0] lcount;
39     wire [11:0] pc1;
40     wire [11:0] dma;
41     wire [17:0] pmd, ir;
42     wire  mem_ena, jc, not_clk, kflag;
43     reg z_new, c_new;
44     reg z, c;
45     reg [8:0] res;
46     wire [3:0] rd, rs;
47
```

```verilog
48      // OP Code of instructions:
49        parameter
50        add    = 5'b01000,    addcy  = 5'b01001,     sub  = 5'b01100,
51        subcy = 5'b01101,    opand  = 5'b00001,     opxor = 5'b00011,
52        opor  = 5'b00010,    opinput = 5'b00100, opoutput = 5'b10110,
53        store = 5'b10111,    fetch  = 5'b00101,     load = 5'b00000,
54        jump  = 6'b100010,   jumpz  = 6'b110010,   jumpnz = 6'b110110,
55        call  = 6'b100000, opreturn = 6'b100101;
56
57        always @(negedge clk or negedge reset) begin : P1
58            if (~reset) begin // update the program counter
59              pc <= 0;
60              lcount <= 0;
61            end else begin      // use falling edge
62              if (op6 == call) begin
63                lreg[lcount] <= pc1; // Use next address after call/return
64                lcount <= lcount + 1;
65              end
66              if (op6 == opreturn) begin
67                pc <= lreg[lcount-1]; // Use next address after call/return
68                lcount <= lcount -1;
69              end
70              else if (jc)
71                      pc <= imm12;
72                  else
73                      pc <= pc1;
74            end
75        end
76        assign pc1 = pc + 1;
77        assign jc = (op6==jumpz && z) || (op6==jumpnz && ~z)
78                              || (op6==jump) || (op6==call);
79
80        // Mapping of the instruction, i.e., decode instruction
81        assign op6  = ir[17:12]; // Full Operation code
82        assign op5  = ir[17:13]; // Reduced Op code for ALU ops
83        assign kflag = ir[12];   // Immediate flag 0=use register 1=use kk;
84        assign imm8 = ir[7:0];   // 8 bit immediate operand
85        assign imm12 = ir[11:0]; // 12 bit immediate operand
86        assign rd = ir[11:8];    // Index destination/1. source register
87        assign rs = ir[7:4];     // Index 2. source register
88        assign x = s[rd];        // first source ALU
89        assign x0 = {1'b0 , x};  // zero extend 1. source
90        assign y = (kflag) ? imm8 : s[rs]; // MPX second source ALU
91        assign y0 = {1'b0, y};
92
93        rom4096x18 brom
94        ( .clk(clk), .reset(reset), .address(pc), .q(pmd));
95        assign ir = pmd;
96
97        assign not_clk = ~clk;
98        assign mem_ena = (op5 == store) ? 1 : 0;  // Active for store only
99        data_ram bram
100       ( .clk(not_clk),.address(y), .q(dmd),
101         .data(x), .we(mem_ena));
```

```
102
103     always @(*)
104     begin : P2
105       case (op5)
106         add     :    res  = x0 + y0;
107         addcy   :    res  = x0 + y0 + c;
108         sub     :    res  = x0 - y0;
109         subcy   :    res  = x0 - y0 - c;
110         opand   :    res  = x0 & y0;
111         opor    :    res  = x0 | y0;
112         opxor   :    res  = x0 ^ y0;

113         load    :    res  = y0;
114         opinput :    res = {1'b0 , in_port};
115         fetch   :    res = {1'b0 , dmd};
116         default :    res  = x0;
117       endcase
118       z_new = (res == 0) ? 1 : 0;
119       c_new = res[8];
120     end
121
122     always @(posedge clk or negedge reset)
123     begin : P3
124       if (~reset) begin        // Asynchronous clear
125         z <= 0; c <= 0; out_port <= 0;
126       end else begin
127         case (op5) // Specify the stack operations
128           addcy, subcy : begin z <= z & z_new;
129                         c <= c_new; end    // carry from previous operation
130           add , sub : begin z <= z_new; c <= c_new; end
131                                           // No carry
132           opor , opand, opxor : begin z <= z_new; c <= 0; end
133                                           // No carry; c=0
134           default :  begin z <= z; c <= c; end    // keep old
135         endcase
136         s[rd] <= res[7:0];
137         out_port = (op5 == opoutput) ? x : out_port;
138       end
139     end
140
141     // // Extra test pins:
142     //   assign pc_out = pc;  assign ir_imm12 = imm12;
143     //   assign op_code = op6;  // Program control
144     // // Control signals:
145     //   assign jc_out = jc; assign me_ena = mem_ena;
146     //   assign z_out = z; assign c_out = c; // ALU flags
147     // // Two top stack elements:
148     //   assign s0_out = s[0]; assign s1_out = s[1];
149     //   assign s2_out = s[2]; assign s3_out = s[3];
150
151     endmodule
152
```

The Verilog data memory component for the TRISC2 is shown in the following listing.

Verilog Code A.5: Data RAM aka Scratch Pad RAM for TRISC2

```verilog
1    //===========================================================
2    // IEEE STD 1364-2001 Verilog file: data_ram.v
3    // Author-EMAIL: Uwe.Meyer-Baese@ieee.org
4    //===========================================================
5    module data_ram
6    #(parameter DATA_WIDTH=8, parameter ADDR_WIDTH=8)
7      (input  clk,                      // System clock
8       input we,                        // Write enable
9       input [DATA_WIDTH-1:0] data,     // Data input
10      input [ADDR_WIDTH-1:0] address,  // Read/write address
11      output [DATA_WIDTH-1:0] q);      // Data output
12   //===========================================================
13   // Declare the RAM variable
14      reg [DATA_WIDTH-1:0] ram[2**ADDR_WIDTH-1:0];
15
16   // Variable to hold the registered read address
17      reg [ADDR_WIDTH-1:0] addr_reg;
18
19      always @ (posedge clk)
20      begin
21        if (we)     // Write
22          ram[address] <= data;
23
24        addr_reg <= address; // Synchronous memory, i.e. store
25      end                    // address in register
26
27      assign q = ram[addr_reg];
28
29   endmodule
```

The TRISC2 program ROM file shown next is similar to the URISC ROM design. The program Hex file for TRISC2 can be directly generated from KCPSM6 or with psm2hex developed in Chap. 6. For VIVADO it is recommended to use MIF file extension instead of the HEX extension generated by KCPSM6.

Verilog Program ROM A.6: Verilog File `rom4096x18.v`

```
1    //============================================================
2    // IEEE STD 1364-2001 Verilog file: prog_rom.v
3    // Author-EMAIL: Uwe.Meyer-Baese@ieee.org
4    //============================================================
5    // Initialize the ROM with $readmemh. Put the memory
6    // contents in the file trisc0fac.txt.  Without this file,
7    // this design will not compile. See Verilog
8    // LRM 1364-2001 Section 17.2.8 for details on the
9    // format of this file.
10   module rom4096x18
11   #(parameter DATA_WIDTH=18, parameter ADDR_WIDTH=12)
12     (input  clk,                      // System clock
13      input  reset,                    // Asynchronous reset
14      input [(ADDR_WIDTH-1):0] address, // Address input
15      output reg [(DATA_WIDTH-1):0] q); // Data output
16   //============================================================
17   // Declare the ROM variable
18     reg [DATA_WIDTH-1:0] rom[2**ADDR_WIDTH-1:0];
19
20     initial
21     begin
22       $readmemh("flash.mif", rom);
23     end
24
25     always @ (posedge clk or negedge reset)
26       if (reset)
27         q <= 0;
28       else
29         q <= rom[address];
30
31   endmodule
```

The project uses three test programs files: led toggle (`flash.psm`), data memory access (`testdmem.psm`), and loop hierarchy (`testdnest.psm`). The program `psm2hex` will generate the `*.mif` files for Verilog and VHDL ROM files.

Simulation scripts are provided for the MODELSIM VHDL simulation (`trisc2.do`) and another script for the Verilog simulation using the testbench (`trisc2_tb.do`). A VIVADO VHDL simulation for the nesting subroutines is shown in Chap. 7, Fig. 7.8. The Verilog VIVADO behavior simulation for the `flash.mif` test program is shown next (Fig. A.2).

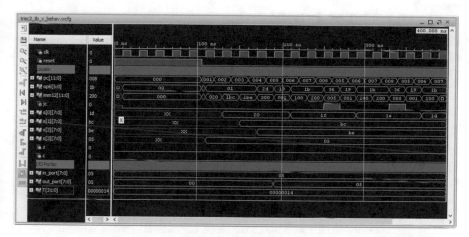

Fig. A.2 First Verilog VIVADO behavior simulation steps for the toggle program showing 24-bit counter load and counting one loop iteration

Verilog Project Files TRISC3N aka Tiny Nios II RISC Processor

The final Verilog design for the TRISC3N aka Tiny Nios II RISC is shown in the following listing.

Verilog Code A.7: TRISC3N (Final Design)

```
1     // ======================================================================
2     // IEEE STD 1364-2001 Verilog file: trisc3n.v
3     // Author-EMAIL: Uwe.Meyer-Baese@ieee.org
4     // ======================================================================
5     // Title: T-RISC 3 address machine
6     // Description: This is the top control path/FSM of the
7     // T-RISC, with a single 3 phase clock cycle design
8     // It has a stack 3-address type instruction word
9     // ======================================================================
10    module trisc3n
11     (input  clk,              // System clock
12      input  reset,            // Asynchronous reset
13      input [7:0] in_port,     // Input port
14      output reg [7:0] out_port // Output port
15    // The following test ports are used for simulation only and should be
16    // comments during synthesis to avoid outnumbering the board pins
17    //   output signed [31:0] r1_out,    // Register 1
18    //   output signed [31:0] r2_out,    // Register 2
19    //   output signed [31:0] r3_out,    // Register 3
20    //   output signed [31:0] r4_out,    // Register 4
21    //   output signed [31:0] sp_out,    // Register 27 aka stack pointer
22    //   output signed [31:0] ra_out,    // Register 31 aka return address
23    //   output jc_out,                  // Jump condition flag
24    //   output me_ena,                  // Memory enable
25    //   output k_out,                   // constant flag
26    //   output [11:0] pc_out,           // Program counter
27    //   output [15:0] ir_imm16,         // Immediate value
28    //   output [11:0] imm32_out,        // Sign extend immediate value
29    //   output [5:0]  op_code           // Operation code
30     );
31    // ======================================================================
32    // Define GENERIC to CONSTANT for _tb
33      parameter  WA = 4'd11;   // Address bit width -1
34      parameter  NR = 5'd31;   // Number of Registers -1
35      parameter  WD = 5'd31;   // Data bit width -1
36      parameter  DRAMAX = 12'd4095; // No. of DRAM words -1
37      parameter  DRAMAX4 = 14'd16383; // No. of DRAM bytes -1
38
39      wire [5:0] op, opx;
40      wire [WD:0] dmd, pmd, dma;
41      wire [4:0] imm5;
42      wire [15:0] sxti, imm16;
43      wire [25:0] imm26;
44      wire [WD:0] imm32;
45      wire [4:0] A, B, C;
46      wire [WD:0] rA, rB, rC;
47      reg  [WD:0] branch_target, pc, pc8;// PCs
48      wire [WD:0] ir, pc4,  pcimm26;// PCs
49      wire eq, ne, mem_ena, not_clk;
50      wire jc, kflag; // jump and imm flags
51      wire load, store, read, write; // I/O flags
52    //  Register array definition 32x32
53      reg [WD:0] r [0:NR];
54      reg [WD:0] res;
55    //Data RAM memory definition within component
```

```
56
57      // OP Code of instructions:
58      // The 6 LSBs IW for all implemented operations sorted by op code
59        parameter
60        call = 6'h00, jmpi = 6'h01, addi = 6'h04, br = 6'h06, andi = 6'h0C,
61        ori = 6'h14, stw = 6'h15, ldw = 6'h17, xori = 6'h1C, bne = 6'h1E,
62        beq = 6'h26, orhi = 6'h34, stwio = 6'h35, ldwio = 6'h37,
63        R_type = 6'h3A;
64
65      // 6 bits for OP eXtented instruction with OP=3A=111010
66        parameter
67        ret = 6'h05, jmp = 6'h0D, opand = 6'h0E, opor = 6'h16,
68        opxor = 6'h1E, add = 6'h31, sub = 6'h39;
69
70        always @(negedge clk or negedge reset)
71            if (~reset) begin // update the program counter
72              pc <= 0; pc8 <= 0;
73            end else begin    // use falling edge
74              if (jc)
75                pc <= branch_target;
76              else begin
77                pc <= pc4;
78                pc8 <= pc + 32'h00000008;
79              end
80            end
81        assign pc4 = pc + 4; // Default PC increment is 4 bytes
82        assign pcimm26 = {pc[31:28], imm26, 2'b00};
83        assign jc = (op==beq && rA==rB) || (op==jmpi) || (op==br)
84                    || (op==bne && rA!=rB) || (op==call)
85                    || (op==R_type && (opx==ret || opx==jmp));
86
87        always @* begin
88            if (op==jmpi || op==call) branch_target = pcimm26; else
89            if (op==R_type && opx==ret) branch_target = r[31]; else
90            if (op==R_type && opx==jmp) branch_target = rA; else
91            branch_target = imm32+pc4; // WHEN (op=beq OR op=bne OR op=br)
92        end
93
94      // Mapping of the instruction, i.e., decode instruction
95        assign op  = ir[5:0];      // Operation code
96        assign opx = ir[16:11];    // OPX code for ALU ops
97        assign imm5  = ir[10:6];   // OPX constant
98        assign imm16 = ir[21:6];   // Immediate ALU operand
99        assign imm26 = ir[31:6];   // Jump address
100       assign A  = ir[31:27]; // Index 1. source reg.
101       assign B  = ir[26:22]; // Index 2. source/des. register
102       assign C  = ir[21:17]; // Index destination reg.
103       assign rA = r[A]; // First source ALU
104       assign rB = (kflag) ? imm32 : r[B]; // Second source ALU
105       assign rC = r[C];  // Old destination register value
106     // Immediate flag 0= use register 1= use HI/LO extended imm16;
107       assign kflag = (op==addi) || (op==andi) || (op==ori) || (op==xori)
108                                  || (op==orhi) || (op==ldw) ||
```

```
109   (op==ldwio);
110     assign sxti = {16{imm16[15]}}; // Sign extend the constant
111     assign imm32 = (op==orhi)? {imm16, 16'h0000} :
112               {sxti, imm16}; // Place imm16 in MSbs for ..hi
113
114     rom4096x32 brom
115     ( .clk(clk), .reset(reset), .address(pc[13:2]), .q(pmd));
116     assign ir = pmd;
117
118     assign dma = rA + imm32;
119     assign store = ((op==stw) || (op==stwio)) && (dma <= DRAMAX4);//
120   DRAM store
121     assign load = ((op==ldw) || (op==ldwio)) && (dma <= DRAMAX4); //
122   DRAM load
123     assign write = ((op==stw) || (op==stwio)) && (dma > DRAMAX4); // I/O
124   write
125     assign read = ((op==ldw) || (op==ldwio)) && (dma > DRAMAX4);  // I/O
126   read
127     assign not_clk = ~clk;
128     assign mem_ena = (store) ? 1 : 0;  // Active for store only
129     data_ram bram
130     ( .clk(not_clk),.address(dma[13:2]), .q(dmd),
131       .data(rB), .we(mem_ena));
132
133     always @(*)
134     begin : P3
135       res = rC; // keep old/default
136       if ((op==R_type && opx==add) || (op==addi))  res = rA + rB;
137       if (op==R_type && opx==sub) res = rA - rB;
138       if ((op==R_type && opx==opand) || (op==andi)) res = rA & rB;
139       if ((op==R_type && opx==opor) || (op==ori) || (op==orhi))
140                                           res = rA | rB;
141       if ((op==R_type && opx==opxor) || (op==xori))  res = rA ^ rB;
142       if (load) res = dmd;
143       if (read) res = {24'h000000,  in_port};
144     end
145
146     always @(posedge clk or negedge reset)
147     begin : P4
148       integer k;        // Temporary counter
149       if (~reset)       // Asynchronous clear
150         begin
151           for (k=0; k<32; k=k+1) r[k] <= 0;
152           out_port <= 0;
153         end
154       else
155         begin
156           if (op==call) // Store ra for operation call
157             r[31] <= pc8; // Old pc + 1 op after return
158           else
159             begin if (kflag && B>0) // All I-type
160                     begin
161                       r[B] <= res;
162                     end else begin
```

```
163                           if (C > 0) r[C] <= res;
164                         end
165             end
166           out_port = (write) ? rB[7:0] : out_port;

167        end
168      end
169                                  .
170     // Extra test pins:
171     //  assign pc_out = pc[11:0];   assign ir_imm16 = imm16;
172     //  assign imm32_out = imm32;
173     //  assign op_code = op;  // Program control
174     //  // Control signals:
175     //  assign jc_out = jc; assign me_ena = mem_ena;
176     //  assign k_out = kflag;
177     //  // Two top stack elements:
178     //  assign r1_out = r[1]; assign r2_out = r[2]; // First two user
179     registers
180     //  assign r3_out = r[3]; assign r4_out = r[4]; // Next two user
181     registers
182     //  assign sp_out = r[27]; assign ra_out = r[31]; // Compiler
183     registers
184
185     endmodule
```

The Verilog data memory component for the TRISC3N is shown in the following listing.

Verilog Code A.8: Data RAM for TRISC3N

```
1   // ============================================================
2   // IEEE STD 1364-2001 Verilog file: data_ram.v
3   // Author-EMAIL: Uwe.Meyer-Baese@ieee.org
4   // ============================================================
5   module data_ram
6   #(parameter DATA_WIDTH=32, parameter ADDR_WIDTH=12)
7     (input  clk,                    // System clock
8      input we,                      // Write enable
9      input [DATA_WIDTH-1:0] data,   // Data input
10     input [ADDR_WIDTH-1:0] address,  // Read/write address
11     output [DATA_WIDTH-1:0] q);     // Data output
12  // ============================================================
13  // Declare the RAM variable
14    reg [DATA_WIDTH-1:0] ram[2**ADDR_WIDTH-1:0];
15
16  // Variable to hold the registered read address
17    reg [ADDR_WIDTH-1:0] addr_reg;
18
19    always @ (posedge clk)
20    begin
21      if (we)     // Write
22        ram[address] <= data;
23
24      addr_reg <= address; // Synchronous memory, i.e. store
25    end                    // address in register
26
27    assign q = ram[addr_reg];
28
29  endmodule
```

The TRISC3N program ROM file shown next is similar to the URISC ROM design. The program HEX file for TRISC3N can be generated with gcc directly or using the Altera Monitor Program. For VIVADO it is recommended to use MIF extension for the ROM initialization file.

Verilog Program ROM A.9: File `rom4096x32.v` with Nios Toggle Program Data

```verilog
1    // ============================================================
2    // IEEE STD 1364-2001 Verilog file: rom4096x32.v
3    // Author-EMAIL: Uwe.Meyer-Baese@ieee.org
4    // ============================================================
5    // Initialize the ROM with $readmemh. For Vivado
6    // use the MIF files extension. Without this file,
7    // this design will not compile. See Verilog
8    // LRM 1364-2001 Section 17.2.8 for details on the
9    // format of this file.
10   // ============================================================
11   module rom4096x32
12   #(parameter DATA_WIDTH=32, parameter ADDR_WIDTH=12)
13     (input  clk,                    // System clock
14     input  reset,                   // Asynchronous reset
15     input  [(ADDR_WIDTH-1):0] address, // Address input
16     output reg [(DATA_WIDTH-1):0] q); // Data output
17   // ============================================================
18   // Declare the ROM variable
19     reg [DATA_WIDTH-1:0] rom[2**ADDR_WIDTH-1:0];
20
21     initial
22     begin
23       $readmemh("flash_nios.mif", rom);
24     end
25
26     always @ (posedge clk or negedge reset)
27       if (~reset)
28         q <= 0;
29       else
30         q <= rom[address];
31
32   endmodule
```

Short simulation scripts are provided for the MODELSIM simulation (`trisc3n.do`) and another script for the simulation using the testbench (`trisc3n_tb.do`). The project uses three test programs files: led toggle (`flash.vhd`), data memory access (`dmem.vhd`), and loop hierarchy (`nesting.vhd`). A VHDL simulation for memory access is shown in Chap. 9, Fig. 9.22, and the nesting subroutines is shown in Chap. 9, Fig. 9.23. The Verilog VIVADO behavior simulation for the `flash_nios.mif` test program is shown next (Fig. A.3).

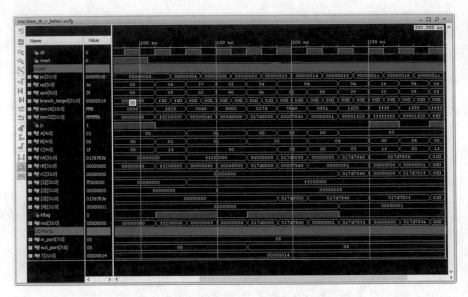

Fig. A.3 First Tiny Nios (trisc3n) Verilog Vivado behavior simulation steps for the toggle program showing counter load and 2 cycle counting loop iteration

Verilog Project Files Trisc3mb aka Tiny MicroBlaze RISC Processor

The final Verilog design for the Trisc3mb aka Tiny MicroBlaze RISC is shown in the following listing.

Verilog Code A.10: TRISC3MB (Final Design)

```
1    // =================================================================
2    // IEEE STD 1364-2001 Verilog file: trisc3mb.v
3    // Author-EMAIL: Uwe.Meyer-Baese@ieee.org
4    // =================================================================
5    // Title: T-RISC 3 address machine
6    // Description: This is the top control path/FSM of the
7    // T-RISC, with a single 3 phase clock cycle design
8    // It has a stack 3-address type instruction word
9    // implementing a subset of the MicroBlaze architecture
10   // =================================================================
11   module trisc3mb
12    (input   clk,              // System clock
13     input   reset,            // Asynchronous reset
14     input [0:7] in_port,      // Input port
15     output reg [0:7] out_port // Output port
16   // The following test ports are used for simulation only and should be
17   // comments during synthesis to avoid outnumbering the board pins
18   //   output signed [0:31] r1_out,  // Register 1
19   //   output signed [0:31] r2_out,  // Register 2
20   //   output signed [0:31] r3_out,  // Register 3
21   //   output signed [0:31] r19_out, // Register 19 aka 2. stack pointer
22   //   output signed [0:31] r14_out, // Register 14 aka return address
23   //   output jc_out,             // Jump condition flag
24   //   output me_ena,             // Memory enable
25   //   output i_out,              // constant flag
26   //   output [0:11] pc_out,      // Program counter
27   //   output [0:15] ir_imm16,    // Immediate value
28   //   output [0:11] imm32_out,   // Sign extend immediate value
29   //   output [0:5]  op_code      // Operation code
30    );
31   // =================================================================
32   // Define GENERIC to CONSTANT for _tb
33     parameter  WA = 4'd11;  // Address bit width -1
34     parameter  NR = 5'd31;   // Number of Registers -1
35     parameter  WD = 5'd31;   // Data bit width -1
36     parameter  DRAMAX = 12'd4095; // No. of DRAM words -1
37     parameter  DRAMAX4 = 14'd16383; // No. of DRAM bytes -1
38
39
40     wire [0:5]  op;
41     wire [0:WD] ir, dmd, pmd, dma, pc4, branch_target;
42     reg  [0:WD] pc, pc_d, target_delay; // PCs
43     wire mem_ena, not_clk;
44     wire jc, link, Dflag, cmp; // controller flags
45     reg go, Delay;
46     wire br, bra, bri, brai, condbr, condbri;// branch flags
47     wire swi, lwi, rt; // Special instr.
48     wire rAzero, rAnotzero, I, K, L, U, D6, D11;// flags
49     reg LI;
```

```verilog
50    wire aai, aac, ooi, xxi; // Arith.  instr.
51    wire imm, ld, st, load, store, read, write; // I/O flags
52    wire [0:4] D, A, B; // Register index
53    reg [0:4] T;
54    wire [0:WD] rA, rB, rD;// current Ops
55    wire [0:32] rAsxt, rBsxt, rDsxt; // Sign extended Ops
56    reg  [0:15] rI;   // 16 LSBs
57    wire [0:15] imm16, sxt16; // Total 32 bits
58    wire [0:WD] imm32; // 32 bit branch/mem/ALU
59    wire [0:32] imm33; // Sign extended ALU constant
60    reg C;
61
62    // Data RAM memory definition use one BRAM: DRAMAXx32
63
64    // Register array definition 16x32
65    reg [0:WD] r [0:NR];
66    reg [0:32] res;
67
68    assign rAzero = (rA==0)? 1'b1 : 1'b0; // rA=0
69    assign rAnotzero = (rA!=0)? 1'b1 : 1'b0; // rA/=0
70    always @* begin : P1 // Evaluation of signed condition
71      case (ir[8:10])
72        3'b000 :   go <= rAzero;                      // BEQ =0
73        3'b001 :   go <= rAnotzero;                   // BNE /=0
74        3'b010 :   go <= (rA[0]==1'b1);               // BLT < 0
75        3'b011 :   go <= (rA[0]==1'b1) || rAzero;     // BLE <=0
76        3'b100 :   go <= (rA[0]==1'b0) && rAnotzero; // BGT: > 0
77        3'b101 :   go <= (rA[0]==1'b0) || rAzero;     // BGE >=0
78        default:   go <= 1'b0;                         // if not true
79      endcase
80    end
81
82    always @(negedge clk or negedge reset) // FSM of processor
83    begin : FSM // update the PC
84      if (~reset) begin // update the program counter
85        pc = 32'h00000000;
86      end else begin     // use falling edge
87        if (jc) begin
88          pc <= branch_target; // any current jumps
89        end else if (Delay) begin
90          pc <= target_delay; // any jumps with delay
91        end else begin
92          pc <= pc4;  // Usual increment by 4 bytes
93        end
94        pc_d <= pc;
95        if (Dflag)  Delay <= 1'b1;
96        else        Delay <= 1'b0;
97        target_delay <= branch_target; // store target address
98      end
```

```
 99        end
100        assign pc4 = pc + 32'h00000004; // Default PC increment is 4 bytes
101        assign jc = ~Dflag && ((go && (condbr || condbri)) || br
102                                       || bri || rt); // New PC; no delay?
103        assign branch_target = (bra) ? rB:  // Order is important !
104                               (brai)? imm32 :
105                               (condbr || br)? pc + rB :
106                               (rt)? rA + imm32 :
107                                 pc + imm32;   // bri, condbri etc.
108
109        assign rt = (op==6'b101101)? 1'b1 : 1'b0; // return from
110        assign br  = (op==6'b100110 )? 1'b1 : 1'b0; // always jump
111        assign bra = (br && ir[12]==1'b1)? 1'b1 : 1'b0;
112        assign bri = (op==6'b101110)? 1'b1 : 1'b0; //always jump w imm
113        assign brai = (bri && ir[12]==1'b1)? 1'b1 : 1'b0;
114        // link = bit 13 for br and bri
115        assign link = ((br || bri) && L )? 1'b1 : 1'b0; // save PC
116        assign condbr = (op==6'b100111)? 1'b1 : 1'b0; // cond. branch
117        assign condbri = (op==6'b101111)? 1'b1 : 1'b0; //cond. b/w imm
118        assign cmp = (op==6'b000101)? 1'b1 : 1'b0; // cmp and cmpu
119
120        // Mapping of the instruction, i.e., decode instruction
121        assign op     = ir[0:5];   // Data processing OP code
122        assign imm16 = ir[16:31];          // Immediate ALU operand
123
124        // Delay (D), Absolute (A) Decoder flags not used
125        assign I = ir[2];  // 2. op is imm
126        assign K = ir[3]; // K=1 keep carry
127        assign L = ir[13]; // Link for br and bri
128        assign U = ir[30]; // Unsigned flag
129        assign D6 = ir[6]; // Delay flag condbr/i;rt;
130        assign D11 = ir[11]; // Delay flag br/i
131        assign Dflag = (D6 && go && (condbr || condbri)) || (rt && D6) ||
132                 (D11 && (br || bri)); // All Delay ops summary
133
134        // I = bit 2; K = bit; 3 add/addc/or/xor with(out) imm
135        assign aai = (ir[0:1]==2'b00 && ir[4:5]==2'b00)? 1'b1 : 1'b0;
136        assign aac = (ir[0:1]==2'b00 && ir[4:5]==2'b10)? 1'b1 : 1'b0;
137        assign ooi = (ir[0:1]==2'b10 && ir[3:5]==3'b000)? 1'b1 : 1'b0;
138        assign xxi = (ir[0:1]==2'b10 && ir[3:5]==3'b010)? 1'b1 : 1'b0;
139        // load and store:
140        assign ld = (ir[0:1]==2'b11 && ir[3:5]==3'b010)? 1'b1 : 1'b0;
141        assign st  = (ir[0:1]==2'b11 && ir[3:5]==3'b110)? 1'b1 : 1'b0;
142
143        assign imm = (op==6'b101100)? 1'b1 : 1'b0;// always store imm
144        assign sxt16 = {16{imm16[0]}}; // Sign extend the constant
145        assign imm32 =  (LI)? {rI, imm16} :  // Immediate extend to 32
146                    {sxt16, imm16}; // MSBs from last imm
147
```

```
148       assign A = ir[11:15]; // Index 1. source reg.
149       assign B = ir[16:20]; // Index 2. source register
150       assign D = ir[6:10]; // Index destination reg.
151       assign rA = r[A]; // First operand ALU
152       assign rAsxt = {rA[0], rA}; // Sign extend 1. operand
153       assign rB = (I)? imm32 : // 2. ALU operand maybe constant or
154   register
155                 r[B];    // Second operand ALU
156       assign rBsxt = {rB[0], rB}; // Sign extend 2. operand
157       assign rD = r[D];   // Old destination register value
158       assign rDsxt = {rD[0],  rD}; // Zero extend old value
159
160
161       rom4096x32 brom( // Instantiate a Block ROM
162         .clk(clk),    // System clock
163         .reset(reset), // Asynchronous reset
164         .address(pc[18:29]), // Program memory address 12 bits
165         .q(pmd));    // Program memory data
166       assign ir = pmd;
167
168
169       assign dma = (I)? rA + imm32 : rA + rB;
170       assign store = st && (dma <= DRAMAX4); // DRAM store
171       assign load  = ld && (dma <= DRAMAX4); // DRAM load
172       assign write = st && (dma  > DRAMAX4); // I/O write
173       assign read  = ld && (dma  > DRAMAX4); // I/O read
174       assign mem_ena = store;        // Active for store only
175       assign not_clk = ~ clk;
176
177        data_ram bram ( // Use one BRAM: 4096x32
178         .clk(not_clk), // Write to RAM at falling clk edge
179         .address(dma[18:29]),
180         .q(dmd),
181         .data(rD),
182         .we(mem_ena)); // Read from RAM at falling clk edge
183
184       always @(*)
185       begin : P3
186         res = rDsxt; // keep old/default
187         if (aai) res = rAsxt + rBsxt;
188         if (aac) res = rAsxt + rBsxt + C;
189         if (ooi) res = rAsxt || rBsxt;
190         if (xxi) res = rAsxt ^ rBsxt;
191         if (cmp) begin res = rBsxt - rAsxt; // ok for signed
192                 if (U) begin // unsigned special case
193                     if ({1'b0, rA} > {1'b0,  rB}) res[1] = 1'b1;
194                     else                           res[1] = 1'b0;
195                 end
196         end
```

```
197            if (load) res = dmd;
198            if (read) res = {24'h000000,  in_port};
199        end
200
201     // Update flags and registers =============================
202        always @(posedge clk or negedge reset)
203        begin : P4
204          integer k;           // Temporary counter
205          if (~reset)          // Asynchronous clear
206            begin
207              LI <= 1'b0; C <= 1'b0; rI <= 32'h00000000;
208              for (k=0; k<32; k=k+1) r[k] <= k;
209              out_port <= 8'h00;
210            end
211          else begin
212            if (~K) begin // Compute new C flag for add if Keep=false
213              if ((res[0] == 1'b1) && (aai || aac))  C <= 1'b1;
214              else                                   C <= 1'b0;
215            end
216            // Compute and store new register values
217            if (imm) begin // Set flag: last was imm operation
218              rI <= imm16; LI <= 1'b1;
219            end else begin
220              rI <= 16'h0000; LI <= 1'b0;
221            end
222            if (D>0) begin // Do not write r(0)
223                if (link) begin // Store LR for operation branch with link aka
224     call
225                  r[D] <= pc_d; // Old pc + 1 op after return
226                end else begin
227                  r[D] <= res[1:32]; // Store ALU result
228                end
229            end
230            if (write) out_port <= rD[24:31]; // LSBs on the right
231          end
232        end
233
234     //  // Extra test pins:
235     //  assign pc_out = pc[20:31];
236     //  assign ir_imm16 = imm16;
237     //  assign op_code = op; // Data processing ops
238     //  assign jc_out = jc;
239     //  assign i_out = I;
240     //  assign me_ena = mem_ena; // Control wires
241     //  assign r1_out = r[1]; // First two user registers
242     //  assign r2_out = r[2]; assign r3_out = r[3]; // Next two user
243     registers
244     //  assign r15_out = r[15]; assign r19_out = r[19]; // Compiler
245     registers
246
247     endmodule
```

The Verilog data memory component for the TRISC3MB is shown in the following listing.

Verilog Code A.11: Data RAM for TRISC3MB

```
1   // ============================================================
2   // IEEE STD 1364-2001 Verilog file: data_ram.v
3   // Author-EMAIL: Uwe.Meyer-Baese@ieee.org
4   // ============================================================
5   module data_ram
6   #(parameter DATA_WIDTH=32, parameter ADDR_WIDTH=12)
7     (input  clk,                        // System clock
8      input we,                          // Write enable
9      input [0:DATA_WIDTH-1] data,       // Data input
10     input [0:ADDR_WIDTH-1] address,    // Read/write address
11     output [0:DATA_WIDTH-1] q);        // Data output
12  // ============================================================
13  // Declare the RAM variable
14    reg [DATA_WIDTH-1:0] ram[2**ADDR_WIDTH-1:0];
15
16  // Variable to hold the registered read address
17    reg [ADDR_WIDTH-1:0] addr_reg;
18
19    always @ (posedge clk)
20    begin
21      if (we)     // Write
22        ram[address] <= data;
23
24      addr_reg <= address; // Synchronous memory, i.e. store
25    end                    // address in register
26
27    assign q = ram[addr_reg];
28
29  endmodule
```

The TRISC3MB program ROM file shown next is similar to the URISC ROM design. The program HEX file for TRISC3MB can be directly generated from gcc or with the VIVADO SDK. For VIVADO, it is recommended to use MIF extension for the ROM initialization file.

Verilog Program ROM A.12: File `rom4096x32.v` with MicroBlaze Toggle Program Data

```
1    // ===========================================================
2    // IEEE STD 1364-2001 Verilog file: rom4096x32.v
3    // Author-EMAIL: Uwe.Meyer-Baese@ieee.org
4    // ===========================================================
5    // Initialize the ROM with $readmemh. For Vivado
6    // use the MIF files extension. Without this file,
7    // this design will not compile. See Verilog
8    // LRM 1364-2001 Section 17.2.8 for details on the
9    // format of this file.
10   // ===========================================================
11   module rom4096x32
12   #(parameter DATA_WIDTH=32, parameter ADDR_WIDTH=12)
13     (input  clk,                      // System clock
14      input  reset,                    // Asynchronous reset
15      input [0:(ADDR_WIDTH-1)] address, // Address input
16      output reg [0:(DATA_WIDTH-1)] q); // Data output
17   // ===========================================================
18   // Declare the ROM variable
19     reg [DATA_WIDTH-1:0] rom[2**ADDR_WIDTH-1:0];
20
21     initial
22     begin
23       $readmemh("flash_mb.mif", rom);
24     end
25
26     always @ (posedge clk or negedge reset)
27       if (~reset)
28         q <= 0;
29       else
30         q <= rom[address];
31
     endmodule
```

Short simulation scripts are provided for the MODELSIM simulation (`trisc3mb.do`) and another script for the simulation using the testbench (`trisc3mb_tb.do`). The project uses three test program files: led toggle (`flash.vhd`), data memory access (`dmem.vhd`), and loop hierarchy (`nesting.vhd`). A VHDL simulation for memory access is shown in Chap. 10, Fig. 10.22, and the nesting subroutine is shown in Chap. 10, Fig. 10.23. The Verilog VIVADO behavior simulation for the `flash_mb.mif` test program is shown next (Fig. A.4).

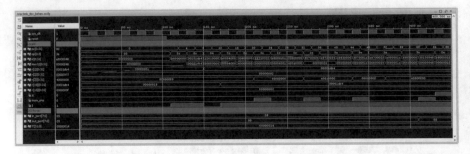

Fig. A.4 First Tiny MicroBlaze (trisc3mb) Verilog VIVADO behavior simulation steps for the toggle program showing counter load and 2 cycle counting loop iteration

Verilog Project Files TRISC3A aka Tiny ARM Cortex-A9 RISC Processor

The final Verilog design for the TRISC3A aka Tiny ARM Cortex-A9 RISC is shown in the following listing.

VHDL Code A.13: Trisc3a (Final Design)

```
1    // =================================================================
2    // IEEE STD 1364-2001 Verilog file: trisc3a.v
3    // Author-EMAIL: Uwe.Meyer-Baese@ieee.org
4    // =================================================================
5    // Title: T-RISC 3 address machine
6    // Description: This is the top control path/FSM of the
7    // T-RISC, with a single 3 phase clock cycle design
8    // It has a 3-address type instruction word
9    // implementing a subset of the ARMv7 Cortex A9 architecture
10   // =================================================================
11   module trisc3a
12     (input  clk,                // System clock
13      input  reset,              // Asynchronous reset
14      input [7:0] in_port,       // Input port
15      output reg [7:0] out_port // Output port
16   // The following test ports are used for simulation only and should be
17   // comments during synthesis to avoid outnumbering the board pins
18   //   output signed [31:0] r0_out,     // Register 0
19   //   output signed [31:0] r1_out,     // Register 1
20   //   output signed [31:0] r2_out,     // Register 2
21   //   output signed [31:0] r3_out,     // Register 3
22   //   output signed [31:0] sp_out,     // Register 13 aka stack pointer
23   //   output signed [31:0] lr_out,     // Register 14 aka link register
24   //   output jc_out,                   // Jump condition flag
25   //   output me_ena,                   // Memory enable
26   //   output i_out,                    // Constant flag
27   //   output [11:0] pc_out,            // Program counter
28   //   output [11:0] ir_imm12,          // Immediate value
29   //   output [31:0] imm32_out,         // Sign extend immediate value
30   //   output [3:0]  op_code            // Operation code
31     );
32   // =================================================================
33   // Define GENERIC to CONSTANT for _tb
34     parameter WA = 11;   // Address bit width -1
35     parameter NR = 15;   // Number of Registers -1; PC is extra
36     parameter WD = 31;   // Data bit width -1
37     parameter DRAMAX = 12'd4095; // No. of DRAM words -1
38     parameter DRAMAX4 = 30'd1073741823; // X"3FFFFFFF";
39                                         // True DDR RAM bytes -1
40
41     wire [3:0] op;
42     wire [WD:0] dmd, pmd, dma;
43     wire [3:0] cond;
44     wire [WD:0] ir,  pc,  pc_dd, pc4, pc8, branch_target;
45     reg [WD:0] tpc, pc4_d;
46     wire mem_ena, not_clk;
47     wire jc, dp, rlsl; // jump and decoder flags
48     wire I, set, P, U, bx, W, L; // Decoder flags
49     wire movt, movw, str, ldr, branch, bl; // Special instr.
50     wire load, store, read, write, pop, push; // I/O flags
51     wire popA1, pushA1, popA2, pushA2; // LDR/STM instr.
52     reg popPC, go;
```

```
53      reg [3:0] ind, ind_d; // push/pop index
54      reg N, Z, C, V; // reg flags
55      wire [3:0] D, NN, M; // Register index
56      wire [31:0] Rd, Rdd, Rn, Rm, r_m; // current Ops
57      reg  [31:0] r_s;
58      wire [32:0] Rd1, Rn1, Rm1; // Sign extended Ops
59      wire [3:0] imm4; // imm12 extended
60      wire [4:0] imm5; // Within Op2
61      wire [11:0] imm12; // 12 LSBs
62      wire [19:0] sxt12; // Total 32 bits
63      wire [23:0] imm24; // 24 LSBs
64      wire [5:0] sxt24; // Total 30 bits
65      wire [WD:0] bimm32, imm32, mimm32; // 32 bit branch/mem/ALU

66      wire [32:0] imm33; // Sign extended ALU constant
67
68  // OP Code of instructions:
69  // The 4 bit for all data processing instructions
70      parameter opand = 4'h0, eor = 4'h1, sub = 4'h2, rsb = 4'h3, add =
71  4'h4,
72      adc = 4'h5, sbc = 4'h6, rsc = 4'h7, tst = 4'h8, teq = 4'h9, cmp =
73  4'hA,
74      cmn = 4'hB, orr = 4'hC, mov = 4'hD, bic = 4'hE, mvn = 4'hF;
75
76   // Register array definition 16x32
77   reg [WD:0] r [0:NR];
78   reg [32:0] res;
79
80      always @* begin : P1 // Evaluation of condition bits
81        case (ir[31:28]) // Shift value 2. operand ALU
82          4'b0000 :  go <= Z;          // Zero: EQ or NE
83          4'b0001 :  go <= ~Z;
84          4'b0010 :  go <= C;          // Carry: CS or CC
85          4'b0011 :  go <= ~C;
86          4'b0100 :  go <= N;          // Negative: MI or PL
87          4'b0101 :  go <= ~N;
88          4'b0110 :  go <= V;                // Overflow: Vs or VC
89          4'b0111 :  go <= ~V;               // Overflow: Vs or VC
90          4'b1000 :  go <= C && ~Z;     // HI
91          4'b1001 :  go <= ~C && Z;        // LS
92          4'b1010 :  go <= (N==V);      // GE
93          4'b1011 :  go <= (N!=V);      // LT
94          4'b1100 :  go <= ~Z && (N==V); // GT
95          4'b1101 :  go <= Z && (N!=V);  // LE
96          default :  go <= 1'b1;        // Always
97        endcase
98      end
99
100     always @* begin : P2
101       integer i;       // Temporary counter
102       ind <= 0;
103       for (i=0;i<=NR;i=i+1)
104         if (ir[i] == 1'b1)  ind <= i;
105     end
```

```
106
107           always @(negedge clk or negedge reset) // FSM of processor
108           begin : P3
109               if (~reset) begin // update the program counter
110                   tpc <= 0; pc4_d <= 0; popPC <= 0;
111               end else begin      // use falling edge
112                   if (jc)
113                       tpc <= branch_target;
114                   else begin
115                       tpc <= pc4;
116                   end
117                   pc4_d <= pc4;
118                   popPC <= 0;
119                   if ((popA1 && ind==15) || (popA2 && D==15))
120                       popPC <= 1; // Last op= pop PC ?
121               end
122           end
123       // true PC in dmd register if last op is pop AND ind=15
124           assign pc  = (popPC)? dmd : tpc;
125           assign pc4 = pc + 32'h00000004; // Default PC increment is 4 bytes
126           assign pc8 = pc + 32'h00000008; // 2 OP PC increment is 8 bytes
127           assign jc = go && (branch||bl||bx|| (pop && ind==15)); // New PC?
128           assign sxt24 = {6{imm24[13]}}; // Sign extend the constant
129           assign bimm32 = {sxt24, imm24, 2'b00}; // Immediate for branch
130           assign branch_target = (bx)? r_m:bimm32 + pc8;//Jump are PC relative
131
132       // Mapping of the instruction, i.e., decode instruction
133           assign op   = ir[24:21]; // Data processing OP code
134           assign imm4 = ir[19:16]; // imm12 extended
135           assign imm5 = ir[11:7];  // The shift values of Op2
136           assign imm12 = ir[11:0]; // Immediate ALU operand
137           assign imm24 = ir[23:0]; // Jump address
138       // P, B, W Decoder flags  not used
139           assign set = (ir[20])? 1'b1 : 1'b0; // update flags for S=1
140           assign I = (ir[25])? 1'b1 : 1'b0;
141           assign L = (ir[20])? 1'b1 : 1'b0; // L=1 load L=0 store
142           assign U = (ir[23])? 1'b1 : 1'b0; // U=1 add offset
143           assign movt = (ir[27:20] == 8'b00110100)? 1'b1 : 1'b0;
144           assign movw = (ir[27:20] == 8'b00110000)? 1'b1 : 1'b0;
145           assign branch = (ir[27:24] == 4'b1010)? 1'b1 : 1'b0;
146           assign bl = (ir[27:24] == 4'b1011)? 1'b1 : 1'b0;
147           assign bx = (ir[27:20] == 8'b00010010)? 1'b1 : 1'b0;
148           assign ldr = (ir[27:26] == 2'b01 && L)? 1'b1 : 1'b0; // load
149           assign str = (ir[27:26] == 2'b01 && ~L)? 1'b1 : 1'b0; // store
150           assign popA1 = (ir[27:16] == 12'b100010111101)? 1 : 0;
151           assign popA2 = (ir[27:16] == 12'b010010011101)? 1 : 0;
152           assign pop = popA1 || popA2;
153       // load multiple (A1) or one (A2) update sp-4 after memory access
154           assign pushA1 = (ir[27:16] == 12'b100100101101)? 1'b1 : 1'b0;
155           assign pushA2 = (ir[27:16] == 12'b010100101101)? 1'b1 : 1'b0;
156           assign push = pushA1 || pushA2;
157       // store multiple (A1) or one (A2) update sp+4 before memory access
158           assign dp = (ir[27:26] == 2'b00)? 1'b1 : 1'b0; // data processing
159
160           assign NN = ir[19:16]; // Index 1. source reg.
161           assign M  = ir[3:0]; // Index 2. source register
162           assign D  = ir[15:12]; // Index destination reg.
163           assign Rn = r[NN]; // First operand ALU
```

```
164      assign Rn1 = {Rn[31], Rn}; // Sign extend 1. operand by 1 bit
165      assign r_m = r[M];
166      assign rlsl = (ir[6:4] == 3'b000)? 1'b1 : 1'b0; //Shift left reg.
167      // determine the 2 register operand
168      always @*
169        case (imm5) // Shift value 2. operand ALU
170          5'b00001 :  r_s <= {r_m[30:0], 1'b0}; // LSL=1
171          5'b00010 :  r_s <= {r_m[29:0], 2'b00}; // LSL=2
172          default  :  r_s <= r_m;
173        endcase
174      assign Rm = (I) ? imm32 : r_s; // 2. ALU operand maybe constant or
175    register
176      assign Rm1 = {Rm[31], Rm}; // Sign extend 2. operand by 1 bit
177      assign Rd = r[D]; // Old destination register value
178      assign Rd1 = {Rd[31], Rd}; // Sign extend old value by 1 bit
179
180      assign mimm32 = {sxt12, imm12}; // memory immediate
181      assign dma = (I)? Rn + Rm :
182              (push)? r[13] - 4 : // use sp
183              (pop)? r[13] :        // use sp
184              (U && NN!=15)? Rn + mimm32 :
185              (~U && NN!=15)? Rn - mimm32 :
186              (U && NN==15)? pc8 + mimm32 : // PC-relative is special
187                            pc8 - mimm32;
188      assign store = (str || push) && (dma <= DRAMAX4); // DRAM store
189      assign load  = (ldr || pop) && (dma <= DRAMAX4); // DRAM load
190      assign write = str && (dma > DRAMAX4); // I/O write
191      assign read  = ldr && (dma > DRAMAX4); // I/O read
192      assign mem_ena = (store) ? 1'b1 : 1'b0; // Active for store only
193      assign Rdd = (pushA1)? r[ind] : Rd;
194      assign not_clk = ~clk;
195
196      // ARM PC-relative ops require True Dual Port RAM with dual clock
197      dpram4Kx32 mem // Instantiate a Block DRAM and ROM
198      (   .clk_a(not_clk), // System clock DRAM
199          .clk_b(clk), // System clock PROM
200          .addr_a(dma[13:2]), // Data memory address 12 bits
201          .addr_b(pc[13:2]), // Program memory address 12 bits
202          .data_a(Rdd), // Data in for DRAM
203          .we_a(mem_ena), // Write only DRAM
204          .q_a(dma), // Data RAM output
205          .q_b(pmd)); // Program memory data
206      assign ir = pmd;
207
208      // ALU imm computations:
209      assign sxt12 = {20{imm12[11]}}; // Sign extend the constant
210      assign imm32 = (movt)? {imm4, imm12, Rd[15:0]} :
211                      (movw)? {16'h0000, imm4, imm12}  :
212                      {sxt12, imm12}; // Place imm16 in MSBs for movt
```

```
213        assign imm33 = {imm32[31], imm32}; // sign extend constant
214
215        always @(*)
216        begin : P4
217          res <= Rd1; // Default old value
218          if (dp)
219            case (op)
220              opand : res <= Rn1 & Rm1;
221              eor   : res <= Rn1 ^ Rm1;
222              sub   : res <= Rn1 - Rm1;
223              rsb   : res <= Rm1 - Rn1;
224              add   : res <= Rn1 + Rm1;
225              adc   : res <= Rn1 + Rm1 + C;
226              sbc   : res <= Rn1 - Rm1 + C -1;
227              rsc   : res <= Rm1 - Rn1 + C -1;
228              tst   : if (movw) res <= imm33; else
229                                res <= Rn1 & Rm1;
230              teq   : res <= Rn1 ^ Rm1;
231              cmp   : if (movt) res <= imm33; else
232                                res <= Rn1 - Rm1;
233              cmn   : res <= Rn1 + Rm1;
234              orr   : res <= Rn1 | Rm1;
235              mov   : res <= Rm1;
236              bic   : res <= Rn1 & ~Rm1;
237              mvn   : res <= ~Rm1;
238              default : res <= Rd1;
239            endcase
240          if (load || pop) res <= { 1'b0, dmd};
241          if (read) res <= {25'h0000000, in_port};
242        end
243
244      //=========== Update flags and registers ============================
245        always @(posedge clk or negedge reset)
246        begin : P5
247          integer k;        // Temporary counter
248          if (~reset) begin     // Asynchronous clear
249            Z <= 1'b0; C <= 1'b0; N <= 1'b0; V <= 1'b0;
250            out_port <= 8'h00;
251            for (k=0; k<16; k=k+1) r[k] <= k;
252          end
253          else begin // ARMv7 has 4 flags
254            if (dp && set) begin // set flags N and Z for all 16 OPs
255              if (res[31] == 1'b1) N <= 1'b1; else N <= 1'b0;
256              if (res[31:0] == 32'h00000000) Z <= 1'b1; else Z <=1'b0;
257              if ((res[32] == 1'b1) && (op != mov)) C <= 1'b1; else C
258  <=1'b0;
259      // ========== Compute new C flag except of MOV =====================
260              if ((res[32] != res[31]) && ((op == sub) || (op == rsb) ||
261              (op == add) || (op == adc) || (op == sbc) || (op == rsc) ||
```

```
262                  (op == cmp) || (op == cmn))) V <= 1'b1; else V <= 1'b0;
263                  // Compute new overflow flag for arith. ops
264             end
265             if (bl) // Store LR for operation branch with link aka call
266                r[14] <= pc4_d; // Old pc + 1 op after return
267             else if (push)
268                r[13] <= r[13] - 4;
269              else if (read || load || movw || movt || (dp &&
270              (op != tst) && (op != teq ) && ( op != cmp ) && (op != cmn)))
271             begin
272                r[D] <= res[31:0];  // Store ALU result (not for test ops)
273                if (popA1 && (ind != 13)) begin
274                   r[13] <= r[13] + 4;
275                   r[ind] <= res[31:0];
276                end
277                if (popA2 && (D != 13)) begin
278                   r[D] <= res[31:0];
279                   r[13] <= r[13] + 4;
280                end
281             end
282             out_port = (write) ? Rd[7:0] : out_port;
283          end
284       end
285
286    // -- Extra test pins:
287    // assign pc_out <= pc[11:0];
288    // assign ir_imm12 <= imm12;
289    // assign imm32_out <= imm32;
290    // assign op_code <= op; // Data processing ops
291    // assign jc_OUT <= jc;
292    // i_OUT <= I;
293    // me_ena <= mem_ena; // Control wires
294    // r0_out <= r[0]; r1_out <= r[1]; // First two user registers
295    // r2_out <= r[2]; r3_out <= r[3]; // Next two user registers
296    // sp_out <= r[13]; lr_out <= r[14]; // Compiler registers
297
298    endmodule
```

The TRISC3A program ROM file shown next is similar to the URISC ROM design. The program HEX file for TRISC3A can be directly generated from gcc, with Altera Monitor Program, or the VIVADO SDK. For VIVADO, it is recommended to use MIF extension for the ROM initialization file. The Verilog dual port data and program memory component for the TRISC3A is shown in the following listing.

Verilog Dual Port Memory A.14: File `dpram4Kx32.v` with ARM Toggle Program Data

```verilog
1   // ================================================================
2   // File is True Dual Port RAM with dual clock for DRAM and ROM
3   // Copyright (C) 2019  Dr. Uwe Meyer-Baese.
4   // ================================================================
5   module dpram4Kx32
6     (input  clk_a,         // System clock DRAM
7      input  clk_b,         // System clock PROM
8      input [11:0] addr_a,  // Data memory address
9      input [11:0] addr_b,  // Program memory address
10     input [31:0] data_a,  // Data in for DRAM
11     input we_a,           // Write only DRAM
12     output reg [31:0] q_a,  // DRAM output
13     output reg [31:0] q_b); // ROM output
14  // ================================================================
15
16  // Build a 2-D array type for the RAM
17    reg [31:0] dram[4095:0];
18
19    initial
20    begin
21      $readmemh("flash_arm.mif", dram);
22    end
23
24    // Port A aka DRAM
25    always @(posedge clk_a)
26      if (we_a) begin
27        dram[addr_a] <= data_a;
28        q_a <= data_a;
29      end else
30        q_a <= dram[addr_a];
31
32    // Port B aka ROM
33    always @(posedge clk_b)
34      q_b <= dram[addr_b];
35
36  endmodule
```

Short simulation scripts are provided for the MODELSIM simulation (trisc3a.do) and another script for the simulation using the testbench (trisc3a_tb.do). The project uses three test program files: led toggle (flash.vhd), data memory access (dmem.vhd), and loop hierarchy (nesting.vhd). A VHDL simulation for memory access is shown in Chap. 11, Fig. 11.15, and the nesting subroutine is shown in Chap. 11, Fig. 11.16. The Verilog VIVADO behavior simulation for the flash_arm.mif test program is shown next (Fig. A.5).

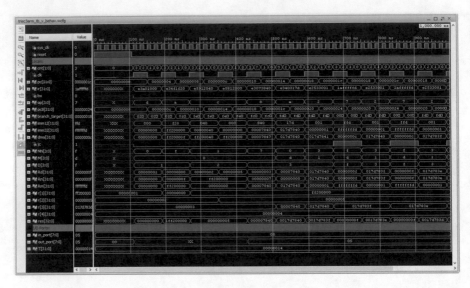

Fig. A.5 First Tiny ARM (`trisc3a`) Verilog Vivado behavior simulation steps for the toggle program showing counter load and 2 cycle counting loop iteration

Appendix B: Glossary

ABS	Anti-lock braking system
ACT	Actel FPGA family
ADC	Analog-to-digital converter
ADCL	All-digital CL
ADPLL	All-digital PLL
ADSP	Analog devices digital signal processor family
AES	Advanced encryption standard
AHDL	Altera HDL
ALM	Adaptive logic module
ALU	Arithmetic logic unit
AMPP	Altera megafunction partners program
AMBA	Advanced microprocessor bus architecture
AMD	Advanced Micro Devices, Inc.
ANSI	American National Standards Institute
ASCII	American Standard Code for Information Interchange
ASIC	Application-specific integrated circuit
BCD	Binary-coded decimal
BGA	Ball grid array
BIT	Binary digit
BMP	Bitmap
BNF	Backus-Naur form
BP	Bandpass
BRAM	Block RAM
CAD	Computer-aided design
CAE	Computer-aided engineering
CAM	Content addressable memory
CBIC	Cell-based IC
CCITT	Comíte consultatif international telephonique et telegraphique
CD	Compact disc

© Springer Nature Switzerland AG 2021
U. Meyer-Baese, *Embedded Microprocessor System Design using FPGAs*,
https://doi.org/10.1007/978-3-030-50533-2

CIF	Common intermediate format
CISC	Complex instruction set computer
CLB	Configurable logic block
CMOS	Complementary metal oxide semiconductor
CODEC	Coder/decoder
CORDIC	Coordinate rotation digital computer
COTS	Commercial off-the-shelf
CPLD	Complex PLD
CPU	Central processing unit
CRT	Cathode ray tube
CSD	Canonical signed digit
DA	Destination address
DAC	Digital-to-analog converter
DC	Direct current
DCO	Digital controlled oscillator
DDRAM	Double data rate RAM
DES	Data encryption standard
DFT	Discrete Fourier transform
DIP	Dual in-line package
DM	Delta modulation
DMA	Direct memory access
DMIPS	Dhrystone MIPS
DOD	Department of defense
DPLL	Digital PLL
DSP	Digital signal processing
EAB	Embedded array block
ECG	Electrocardiography
ECL	Emitter-coupled logic
EDA	Electronic design automation
EDIF	Electronic design interchange format
EPF	Altera FPGA family
EPROM	Electrically programmable ROM
ERA	Plessey FPGA family
ESB	Embedded system block
FCS	Frame check sequence
FC2	FPGA compiler II
FF	Flip-flop
FFT	Fast Fourier transform
FIFO	First-in first-out
FIR	Finite impulse response
FLEX	Altera FPGA family
FPGA	Field-programmable gate array
FPL	Field-programmable logic (combines CPLD and FPGA)
FPLD	FPL device
FPS	Frames per second

FSM	Finite state machine
GAL	Generic array logic
GB	Gigabyte
GCC	GNU C-compiler
GDB	GNU debugger
GHz	Giga Hertz (i.e., 10^9 Hz)
GIF	Graphic interchange format
GNU	GNU's not Unix
GPP	General purpose processor
GPR	General purpose register
GPS	Global positioning system
GUI	Graphical user interface
HDL	Hardware description language
HDMI	High-definition multimedia interface
HDTV	High-definition television
HP	Hewlett Packard
HPS	Hard processor system
HW	Hardware
IBM	International Business Machines (corporation)
IC	Integrated circuit
IEC	International Electrotechnical Commission
IEEE	Institute of Electrical and Electronics Engineers
I²C	Inter-Integrated Circuit
IMA	Interactive Multimedia Association
IP	Intellectual property
ISA	Instruction Set Architecture
ISDN	Integrated Services Digital Network
ISE	Integrated Synthesis Environment
ISO	International Organization for Standardization
ISS	Instruction set simulator
ITU	International Telecommunication Union
JPEG	Joint Photographic Experts' Group
JTAG	Joint test action group
KB	Kilobyte (i.e., 1024 bits)
KCPSM	Ken Chapman PSM
kHz	Kilohertz (1000 Hz)
LAB	Logic array block
LALR	Look-ahead left-to-right (derivation)
LAN	Local area network
LC	Logic cell
LCD	Liquid-crystal display
LE	Logic element
LED	Light-emitting diode
LIFO	Last-in first-out
LISA	Language for instruction set architecture

LFSR	Linear feedback shift register
LPM	Library of parameterized modules
LRM	Language reference manual
LSB	Least significant bit
LSI	Large-scale integration
LUT	Lookup table
MAC	Multiplication and accumulate
MACH	AMD/Vantis FPGA family
MAX	Altera CPLD family
MB	Megabyte
MHz	Megahertz (10^6 Hz)
MIF	Memory initialization file
MIMD	Multiple instruction, multiple data
MIO	Multiplexed input/output
MIPS	Microprocessor without interlocked pipeline
MIPS	Million instructions per second
MMU	Memory management unit
MMX	Multimedia extension
MSI	Medium-scale integration
MSPS	Mega samples per second
MOS	Metal-oxide semiconductor
μP	Microprocessor
MPEG	Moving Picture Experts' Group
MPX	Multiplexer
MSPS	Millions of sample per second
MSB	Most significant bit
NCO	Numeric controlled oscillators
NRE	Nonrecurring engineering costs
NRZI	Non-return-to-zero-inhibit
NTSC	National television system committee
OP-AMP	Operational amplifier
OVI	Open Verilog International
PAL	Phase alternating line
PC	Personal computer
PCA	Principle component analysis
PCI	Peripheral component interconnect
PD	Phase detector
PDF	Probability density function
PDSP	Programmable digital signal processor
PFD	Phase/frequency detector
PGA	Pin grid array
PLA	Programmable logic array
PLD	Programmable logic device
PLI	Programming language interface
PLL	Phase-locked loop

PNG	Portable network graphic
PPC	Power PC
PREP	Programmable Electronic Performance (cooperation)
PROM	Programmable ROM
PSM	Programmable state machine
QCIF	Quarter CIF
QDR	Quad Data Rate (SRAM)
QVGA	Quarter VGA
RAM	Random-access memory
RGB	Red, green, and blue
RISC	Reduced instruction set computer
RNS	Residue number system
ROL	Rotate left
ROM	Read-only memory
ROR	Rotate right
RSA	Rivest, Shamir, and Adelman
RTL	Register Transfer Language
SA	Source address
SAR	Successive approximation register
SD	Secure digital (card)
SDC	Synopsys design constraints
SDRAM	Synchronous dynamic RAM
SECAM	Sequential color with memory
SFD	Start-of-frame delimiter
S/H	Sample and Hold
SLA	Shift left arithmetic
SLL	Shift left logical
SIMD	Single instruction multiple data
SoC	System on a chip
SOPC	System-on-a-programmable-chip
SPEC	System performance evaluation cooperation
SPLD	Simple PLD
SRA	Shift right arithmetic
SRL	Shift right logical
SRAM	Static random-access memory
SSE	Streaming SIMD extension
SSI	Small-scale integration
SVGA	Super VGA
SW	Software
SW	(Slides) switches
SXGA	Super extended graphics array
TB	Test bench
T/H	Track and Hold
TLB	Translation look-aside buffer
TLU	Table look-up

TMS	Texas Instruments DSP family
TI	Texas Instruments (cooperation)
TOS	Top of stack
TQFP	Thin quad flat pack
TSMC	Taiwan semiconductor manufacturing company
TTL	Transistor-transistor logic
TVP	True vector processor
UART	Universal asynchronous receiver/transmitter
UMTS	Universal mobile telecommunications system
URISC	Ultimate reduced instruction set computer
USB	Universal serial bus
UUT	Unit under test
VCO	Voltage-control oscillator
VGA	Video graphics array
VHDL	VHSIC Hardware Description Language
VHSIC	Very High-Speed Integrated Circuit
VLIW	Very long instruction word
VLSI	Very large integrated ICs
WWW	World wide web
XC	Xilinx FPGA family
XNOR	Exclusive NOR gate
YACC	Yet another compiler-compiler
ZBT	Zero bus turnaround

Index

© Springer Nature Switzerland AG 2021
U. Meyer-Baese, *Embedded Microprocessor System Design using FPGAs*,
https://doi.org/10.1007/978-3-030-50533-2